# Processes and Materials of Manufacture

# Processes and Materials of Manufacture———2nd edition

**Roy A. Lindberg**
*University of Wisconsin*

**Allyn and Bacon, Inc.**
Boston, London, Sydney, Toronto

Second printing.....June, 1978

**Library of Congress Cataloging in Publication Data**

Lindberg, Roy A
    Processes and materials of manufacture.

    Includes bibliographies and index.
    1. Manufacturing processes.  2. Materials.  I. Title.
TS183.L56 1977       671      76-6488
ISBN 0-205-05414-5

# Contents

# *Preface*

The engineer is constantly confronted with the challenge of bringing ideas and designs into reality. To do this effectively, he requires a broad background of materials and manufacturing. He must be able to recommend methods of fabrication that will be the most feasible considering the wide selection of materials, machines, and processes available. He must be able to see the whole picture, both technically and economically, and modify his designs accordingly. I have written this text to give the manufacturing engineer and designer a basic understanding of modern materials and of up-to-date processes that will be of immediate practical value.

Traditionally, basic manufacturing processes have been isolated into separate chapters. I believe the similarities and tradeoffs can be more easily understood when interrelationships are shown. For example, all materials joining processes are placed in one chapter and all casting processes in another, including the use of molds for forming plastics and dies for forming powdered metals.

I have devoted a larger portion of this text to automation and numerical control than in previous texts of this type because of the rapid advancement of these factors and the emphasis now placed on them by industry.

Also I have made an effort to indicate that industry is very much aware of the problems technology has created and to show how steps are being taken for environmental improvement, safety, and the conservation of energy.

I have given considerable emphasis to problems at the end of each chapter. Not all of the problems have pat answers that can be found in the text. Rather, I hope that these problems will stimulate class discussions and further investigation.

I have placed very little emphasis on nomenclature, but rather on understanding processes and how they can best be used to develop engineering designs into production.

I wish to acknowledge the help of my colleagues for their review and constructive criticism, particularly Professors Norman R. Braton and Marvin F. DeVries. My sincere thanks also go to my wife, Priscilla, for her patience in deciphering the rough copy and typing it into an acceptable manuscript.

# Processes and Materials of Manufacture

# 1

# *Environment, Energy, and Safety—Challenges to Manufacturing*

In the peaceful era between World War I and II, visionaries extolled the coming "Age of Technocracy," when the work week would be three days long and all the dull, routine, monotonous tasks would be performed by robot-type machines. Production would be high enough to allow sufficient income for the average individual to truly enjoy leisure living.

Obviously this utopian dream that was to be a gift of science and technology has not materialized. On the contrary, rather than being hailed as the great benefactor of mankind, technology is being singled out by some critics for many of the present ills of society, particularly those concerned with our environment.

Since its inception, technology has been largely self-directed and profit-motivated. It has enjoyed an era of laissez-faire. Now an alert citizenry is demanding many changes.

## CHANGES AND SAFEGUARDS

Industry, the Corps of Engineers, pipeline builders, utilities, and others must now answer to an informed citizenry before making changes that have an environmental bearing. Technological and scientific programs that affect a great segment of society are now aired and debated. In retrospect, it is hard to believe that a huge program such as "National Goals in Space" was launched with little public debate. On the other hand, an informed constituency helped Congress debate and defeat the issue of a supersonic jet transport plane.

It is reasonable to expect that the procedures for safeguarding our environment have now been set up and that they will operate to the overall good of the country.

1

### Industry and Environment

The predominant ecological problems that relate directly to manufacture are water and air pollution.

### Water Pollution

Industrial water pollution is due largely to chemical wastes discharged into sewers or into adjoining lakes, rivers, and streams. Of major concern in the industrial sector have been plating, pickling (chemical cleaning of oxide scale) and coolant wastes.

***Plating and Pickling Wastes.*** Typical plating operations utilize salt solutions of cadmium, nickel, copper, tin, indium, and chromium. Operations allied with plating that pose similar waste disposal problems include chemical cleaning, chemical etching (chemical stock removal), and chemical stripping.

Various solutions have been worked out for the chemical treatment of effluent from plating and cleaning processes, as shown schematically in Fig. 1-1(*a*).

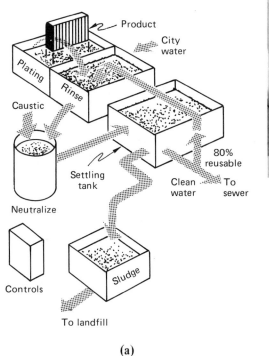

(a)

(b)

FIGURE 1-1. *Liquid wastes from plating operations are treated as shown in the diagram before discharging into the sewer system (a). Dry sludge used for landfill (b). Courtesy Chemfix Division of Environmental Sciences, Inc.*

2

Some appreciation for the magnitude of the waste disposal problem can be gained by citing two examples. The Kodak plant at Rochester, N.Y., processes about 28 million gallons of waste water per day and more than 1500 tons of solid waste per week. Ford Motor Company at Lorraine, Ohio, processes about 55,000 gallons of slurry per day. Now, by the use of additives and settling tanks or large lagoons, the processed slurry becomes a gelatin in about three days. This claylike solid, Fig. 1-1(*b*), can then be used for landfill.

***Coolant Wastes.*** Considerable heat is generated in normal metal cutting and forming operations. Cutting fluids, often referred to as coolants, are used to carry off this heat and prolong the life of the tool, as shown in Fig. 1-2. Water provides an excellent cooling quality but does not provide lubrication for the cutting tool or prevent rust. Water-soluble oils are the most frequently used coolants and *straight oils* or undiluted oils are used for some applications.

In recent years, coolants have been developed that are entirely biodegradable and thus more easily disposed of. However, in some metalworking industries, various oils and greases contaminate the coolant so that it is not easily disposed of. The process of incineration is used in this case. It operates on the principle that at a given temperature, in the presence of enough oxygen, chemical substances are converted to $CO_2$ and water. Hydrocarbons can be disposed of in this way. The incinerator itself is often fired from solid wastes collected from the plant. A schematic presentation of the incinerator principle is shown in Fig. 1-3.

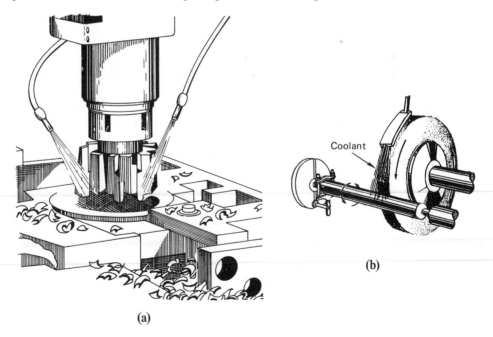

Coolant

(b)

(a)

FIGURE 1-2. *Coolants and cutting oils are used primarily to dissipate the heat in metal cutting (a), and grinding (b).*

3

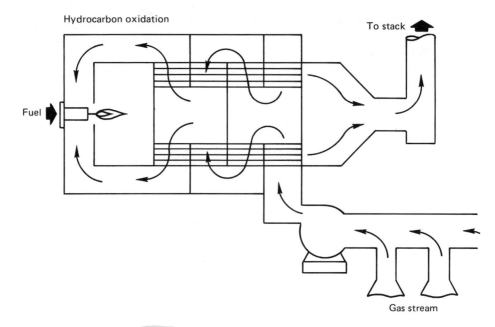

FIGURE 1-3. *A gas incinerator is used to raise the temperature of the hydrocarbons to 1205–1500°F (649–816°C) in the presence of oxygen. The hydrocarbons are oxidized to $CO_2$ and water.*

### Air Pollution

Although a great deal of attention is directed toward air pollution, industry contributes only about 17 % of it. The major source is transportation, which contributes 60 %. Power plants contribute 14 %, heating 6 %, and refuse disposal 3 %. The Clean Air Act of 1963, with further amendments in 1967 and 1970, has established emission standards of performance for hazardous pollutants from both moving and stationary sources.

The most reliable information on air-pollutant concentrations is obtained by continuous sampling with a wet chemical. This method can be automated using a continuous-flow technique. Reagents are added in their proper sequence, a reaction follows and is detected, and results are read out in a prescribed form.

In metal cutting, smoke is sometimes generated when straight oils are used. This occurs when metal is removed at relatively high volume rates. Even though the ventilation system is such that the concentration of smoke particles in the working environment is not excessive or does not cause discomfort, the exhaust of this smoke to the atmosphere is not permitted. The smoke is usually prevented from entering the atmosphere by *electrostatic precipitators*, or *scrubbers* (Fig. 1-4), or by some other means of removal.

An electrostatic precipitator is a means of putting an electric charge on the particles. The charge difference between the particle and the collecting plate causes attraction and collection.

Where the air becomes laden with large particles, gravity or simulated gravity provides a good means of separation. A gravity settling chamber, Fig. 1-5(a), is merely a "wide spot in the line" where the velocity of the smoke or gas is reduced to allow the particles to settle out. A cyclone separator subjects the particles to forces several times that of gravity as shown in part (b) of Fig. 1-5.

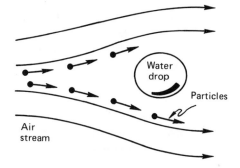

FIGURE 1-4. *The air stream may be subjected to a scrubber for cleaning. Particles in the air are impinged against the water droplets. The gas velocity or opposing water velocity must be high enough to prevent particles from following the slipstream around the water droplets.*

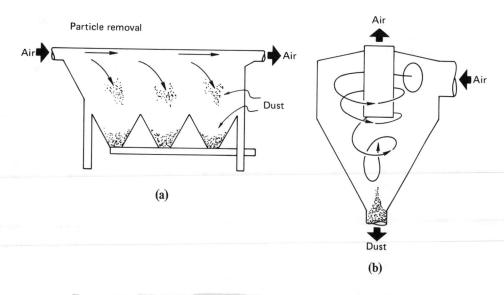

FIGURE 1-5. *Where air is contaminated by large, heavy particles, a gravity-type settling chamber (a) will do a good job of separating them out. A cyclone separator (b) forces the particles to external walls where they are collected in a dust chamber.*

### Industry and Energy

Until recent years the supply of energy was not seen as a crucial item in manufacturing. In a recent survey, the Battelle Columbus Laboratories found that industry consumes about 40 % of all the energy used in the United States, and that 22% of this consumption is returned to the atmosphere. A large portion of the energy used in industry is used in producing the basic materials of steel and aluminum. Next in order of energy consumption are the "hot-working" processes such as casting, forging and other hot-forming processes, welding, and heat treatment. Paint drying also uses considerable energy. Since energy, in whatever form, adds to the overall cost of the product, it has not been wasted. However, only recently has attention been seriously directed to effective energy conservation. The results of some of these studies will be presented to indicate what is being done in some manufacturing areas.

*Industrial Heat Treatment.* Heat treatment, used extensively in industry to change the properties of metals, ranks high on the energy consumption list. Furnaces have been found to be extremely inefficient. Typically, less than 15 % of the heat energy (natural gas or electric power) ends up in the form of heated parts. It is estimated that by careful redesign and better scheduling of parts, furnace efficiency can be improved 50 %.

Concerning energy, other responsibilities of the manufacturing engineer involve:

1. Investigation of the use of product components. Heat treatment may not be necessary. The specification may be a carry-over from a previous design.
2. Examination of the possibility of replacing the heat-treated part with materials that may have more desirable properties.
3. Traditionally, severely cold-worked parts had to be annealed to eliminate built-up stresses. Some steels are now available that allow a greater degree of cold working.
4. Parts are traditionally left in the furnace 1 hour per inch of cross section to ensure a uniform temperature throughout. More recently, metallurgists have observed the heating of metals with a hot-stage microscope and have found that these metals reach an equilibrium stage much faster than originally thought.

*Paint Finishing.* Industrial paint finishing, as presently done, is inefficient for several reasons. First, moist, solvent-based paints must be cured at elevated temperatures. However, standard convection ovens, where most of the curing is done, are only 30 to 35 % efficient. In addition, clean-air legislation requires that certain pollutants be removed from exhaust air. Afterburners convert harmful nonphotochemically reactive solvents into water and carbon dioxide, but increase typical fuel consumption by 75 to 100 % for a painting line. Finally, the solvent base in lacquers and enamels is derived from petroleum, which is better used as an energy source.

New kinds of paints and processes offer significant reductions in curing times and eliminate or reduce afterburning. Powder coatings, for example, contain no solvents, require no afterburning, and can be cured rapidly in infrared ovens that require 42 % less energy than standard hot-air systems. The powdered paint particles are electrostatically charged and attracted to the base part, so that only a small amount leaving the gun misses the target. Even this overspray is recoverable when one color is used at a station. There is also a trend to go from solvent-type to water-based paints. This reduces energy requirements significantly.

In-plant painting may even be eliminated on some items by using precoated sheet and coil stock. Energy is reduced since the coating process is done much more efficiently at the mill.

*Welding.* The energy consumed during welding is a function of welding time and volume of deposited filler metal. Joints should be designed to permit the fastest possible welding. Automated welding systems require at least 50 % less energy than manual welding. Manual weldors deposit metal only 15 to 30 % of the machine's running time; during the remaining time, the equipment draws wasted no-load power. Automated wire processes reduce no-load time to 40 % or less.

*Material Selection.* Material selected for a part contributes to the total energy consumed in two ways. First, energy is required to create the raw material. For some materials, such as aluminum, this is substantial. The material also dictates how much energy is consumed in manufacturing the product. Some metals are easily machined or formed, while others require several stages to obtain the final size.

Plastics not only require energy to create the raw material, but their product is directly derived from petroleum and natural gas. Plastics are selected over other material options because of their unique properties. The cost in terms of "energy in a pound of plastic" is not the only consideration for an energy-conscious engineer. Other factors such as lighter weight or ease of fabrication contribute to the overall energy consumption picture. The long-range outlook suggests that new sources for polymer materials should be explored, such as cellulose, which occurs naturally in plants and trees.

Aluminum requires a great deal of energy to be produced. According to Alcoa, 108,000 Btu are required to produce one pound of primary or "virgin" aluminum. However, recycled aluminum can be produced for 13.8 % of the thermal energy required for the primary metal.

A pound of steel, including a typical quantity of remelted scrap, can be produced for 16,500 Btu. Thus aluminum requires 6.54 times more energy by weight, or 1.57 times more energy by volume, to produce. But, as with other materials, comparisons on an energy-per-pound basis are misleading. Many steels, for example, require heat treatments to bring them up to maximum strength or toughness. Other steels may be difficult to work. In such cases, optional materials may compare more favorably with steels on an energy-per-product basis.

*Design and Manufacturing Engineering Responsibility.* Many designs overspecify, either by habit or for "good measure." Overweight parts waste material and are more difficult to produce. Energy is not only wasted in the manufacture of the part, but also in the operational life of the product.

The manufacturing engineer must be alert to the use of optimum cutting speeds, feeds, and depth of cut. As will be shown in this text, the use of the proper cutting-tool materials can result in substantial savings in the time required for machining. By working together, design and manufacturing engineers can plan the products to consume as little energy as possible. This will usually result in good practice and in overall savings to the manufacturer as well as in conservation of energy. Fig. 1-6 is a diagram showing how overdesign adds up to energy waste.

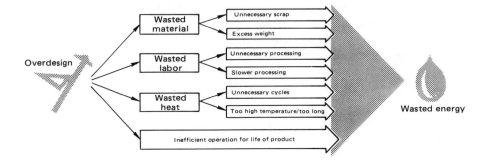

FIGURE 1-6. *Overdesign results in wasted energy in many ways.*

### Industry and Safety

In December of 1970 the United States government stepped forcefully into the arena of insuring safety and healthful conditions for workers throughout industry by passing the Occupational Safety and Health Act (OSHA). Since the signing of the Act, the Secretary of Labor (under whose jurisdiction the Act resides) has incorporated standards that had previously been adopted by the National Fire Protection Association (NFPA) and the American National Standards Institute (ANSI). Under the provisions of this Act, all employers must provide their employees with safe working conditions and protect them from their own carelessness.

The Act allows states to assume full responsibility for administering and enforcing their occupational safety and health laws. However, state standards must be equal to or better than the federal laws and must be accepted by the federal government. The Act may be divided into three parts: investigation, enforcement, and penalization.

*Investigation.* Investigation by OSHA may take place as an unannounced visit or it may come at the request of one of the employees. Any employee who believes a violation of health or safety standards exists that may cause him serious physical harm can request an investigation by a federal representative of OSHA.

The federal representative can take one of two courses of action: (1) he may proceed with the investigation accompanied by an authorized representative of the employee or, (2) he may feel the complaint does not merit investigation and will give a written explanation of his reasons to the employee. If an investigation is made, it may involve private audiences with the person or persons involved. The private examination of witnesses is provided to maintain the employee's anonymity from his employer, thus guaranteeing his protection under the Act.

*Enforcement.* The Secretary of Labor or any one of his representatives may issue a citation if the results of his investigation show noncompliance with the standards established under OSHA or if there are any proven "recognized hazards," even though there are as yet no standards covering them.

Citations will contain a written explanation of the violation or violations and set a realistic deadline for their elimination. The employer will have 15 days to respond to the Secretary of Labor's office; failure to do so eliminates any future review by the court.

Should the employer be unable to correct the violation by the date set because of extenuating circumstances, an extension may be granted. This is called a *variance*.

*Penalization.* In order to make the law effective, a system of penalties was set up as follows:

1. A serious violation, fine of up to $1000.00.
2. Failure to correct violations, fine of up to $1000.00 per day until corrected.
3. Willful or repeated violations, fine of up to $10,000.00 for each violation.
4. Willful violation resulting in death, fine up to $10,000.00 and/or imprisonment up to six months.
5. Failure to display requirements, fine up to $1000.00.
6. Falsification in any application or report, fine up to $10,000.00 and/or up to six months imprisonment.

In addition to the stated penalties, the U.S. District Court has the right to shut down any operation or place of business found to be under imminent danger until that danger has been eliminated.

*Additional Legal Powers.* Under the Act several clerical duties must be performed:

1. All work-related deaths, as well as injuries and illnesses requiring medical care, must be reported.
2. The employer must monitor employee exposure to health hazards.
3. The employer must provide for physical examinations at regular intervals if environmental conditions warrant it.
4. The employer must disclose to all employees all hazards encountered in his work and provide educational programs and safety equipment to help guard against them.

### OSHA and Industrial Noise

Tight standards have been set by OSHA in regard to industrial noise. Noise, which has been defined as unwanted excessive sound, is considered a major industrial health problem.

Noise-induced hearing losses are often insidious, usually occurring in the higher sound frequencies without discomfort or warning. Since these are not within the speech frequencies, the loss goes unnoticed. Although some people claim they become accustomed to noise, continued exposure to high levels spreads losses to speech frequencies, and the ability to understand and communicate is impaired. There is a growing concern that irritability, nervousness, headaches, stomach upsets or ulcers, high blood pressure, fatigue, poor morale, and other effects can be traced to noise pollution.

*Fundamentals of Sound.* Sound is made up of variations in pressure that radiate from a vibrating object. The variations alternately compress and rarify the atmosphere and are referred to as sound waves. The frequency of the sound waves is commonly measured in hertz (Hz), or cycles per second. The lower the frequency, the longer the wavelength. Sound can generally be detected by the human ear when frequencies are in the range of 20 to 20,000 Hz, but the ear is less responsive to very low or very high frequency sounds than to sounds of medium frequency.

*Decibels.* Sound is generally measured in decibels (db). The decibel is a dimensionless unit used in comparing the magnitude of sound pressures or powers. Shown in Table 1-1 are various sound pressure levels with the corresponding decibels. The sound of 80 db is shown as the sound pressure of only 0.00003 psi, or that of a busy office. The threshold of pain is generally considered to be near 140 db.

Decibels are logarithmic ratios and cannot be added or subtracted directly. To predict the noise from several different sources, the decibels can be converted to either sound pressures or sound powers, added or subtracted, and then converted back to decibels. Graphs are available to simplify this procedure.

*Octave-Band Analysis.* For engineering purposes, it is often helpful to know the frequency distributions or composition of noise. This is done by determining the sound-pressure levels of each set of frequency bands and plotting them as a function of the center of frequency of each band. Preferred octave bands, as seen in Fig. 1-7, are generally used. An octave band is a range of frequencies in which the highest is twice the lowest. Levels in each octave are measured in decibels and are referred to as octave band levels.

*Noise Measurement.* To control noise, it is often necessary to measure both sound and vibration near the source. Sound is picked up by microphones and vibration by accelerometers. These meters are generally provided with slow or fast response. The slow response facilitates reading the meter by reducing the fluctuating speed of the indicating needle. Shown in Fig. 1-8(*a*) is a sound-level meter being used for noise analysis at a punch press. When fixed-position noise measurement is impractical, lightweight docimeters (noise exposure monitors) may be worn by the individual (*b*). The docimeters accumulate noise exposure data for the workday.

TABLE 1-1. *Sound Pressure levels in Decibels and psi.* *Courtesy Sperry Vickers*

| Sound Pressure Level in Decibels (db) | Sound Pressure in Pounds per Square Inch (psi) | Common sounds (depending on distance) |
|---|---|---|
| | | Common sounds (depending on distance) |
| 160 | $3 \times 10^{-1}$ | Medium jet engine |
| 140 | $3 \times 10^{-2}$ | Large propeller aircraft<br>Air raid siren<br>Riveting and chipping |
| 120 | $3 \times 10^{-3}$ | Discotheque |
| 100 | $3 \times 10^{-4}$ | Punch press (impact)<br>Canning plant<br>Heavy city traffic; subway<br>Average machine shop |
| 80 | $3 \times 10^{-5}$ | Busy office |
| 60 | $3 \times 10^{-6}$ | Normal speech |
| | | Private office |
| 40 | $3 \times 10^{-7}$ | Quiet residential neighborhood |
| 20 | $3 \times 10^{-8}$ | Whisper |
| 0 | $3 \times 10^{-9}$ | Threshold of hearing |

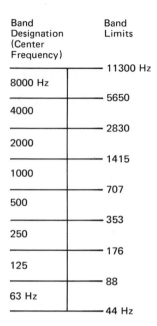

| Band Designation (Center Frequency) | Band Limits |
|---|---|
| | 11300 Hz |
| 8000 Hz | |
| | 5650 |
| 4000 | |
| | 2830 |
| 2000 | |
| | 1415 |
| 1000 | |
| | 707 |
| 500 | |
| | 353 |
| 250 | |
| | 176 |
| 125 | |
| | 88 |
| 63 Hz | |
| | 44 Hz |

FIGURE 1-7. *Preferred octave bands are generally used in determining the frequency distribution or composition of noise. Courtesy Sperry Vickers.*

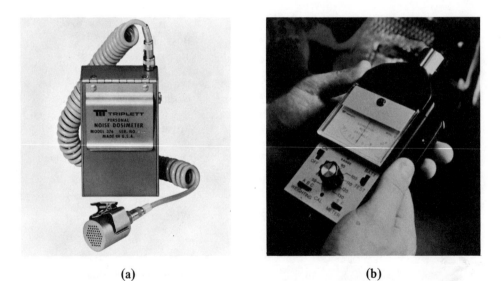

(a)  (b)

FIGURE 1-8. *A sound-level meter being used for noise analysis at a punch press (a). An audio dosimeter that can be worn in the shirt pocket with a microphone clipped to the shirt collar (b). A memory cell is used to store the data. Courtesy Triplett Corp.*

TABLE 1-2. *Permissible Noise Exposure*

| Duration Per day (hours) | Sound Level, db, Slow Exposure |
|:---:|:---:|
| 8 | 90 |
| 6 | 92 |
| 4 | 95 |
| 3 | 97 |
| 2 | 100 |
| 1-1/2 | 102 |
| 1 | 105 |
| 1/2 | 110 |
| 1/4 or less | 115 |

*Legal Requirements.* OSHA standards, which specify protection against the effects of noise exposure, are shown in Table 1-2. When employees are subjected to sound exceeding the limits shown, effort must be made to reduce the sound levels or to provide personal protective equipment.

*Controlling Industrial Noise.* Engineers, in compliance with the OSHA Act, have been seeking ways of controlling industrial noise. The control of noise is based on a knowledge of vibration isolation and damping. One method used to change the frequency of vibration is that of changing the mass or stiffness. Dampers consisting of a spring mass system, which can be tuned by altering the mass or the spring constant, are being used to improve the dynamic stiffness of machine tools.

Many materials are used for vibration damping, including sand, lead, asphalt, and impregnated felts. Materials that provide intermolecular friction to convert vibrational energy into heat are good for damping. Examples of this type of damping material are leaded vinyl sheets, sheet lead, polyurethane sandwiches, and thermoplastic materials sandwiched between steel sheets. A major plastics damping material is polyvinyl acetate that can be sprayed or troweled on surfaces.

Sound absorbing materials reduce only a portion of the noise reaching them. Part of the energy is converted to heat, some is reflected, and some filters through. In addition to their use as an acoustic treatment for walls and ceilings, sound absorbing materials are used for lining ducts and for enclosure panels, as shown in Fig. 1-9.

### Control Standards and Responsibility

There is little question that many changes will continue to be made in manufacturing technology that will help eliminate some of the pressing pollution, energy, and noise problems we now face. However, to expect industry to do it all is unrealistic. Everyone must be responsive, from the farmer who, in the interest of better crops, uses herbicides and fertilizer, to the consumer who strews cans and other debris along the countryside. As Pogo stated, "We have met the enemy and they is us."

Our environmental problems will require the cooperation and expertise of industry, government, and many individuals. Much of the work can be done by

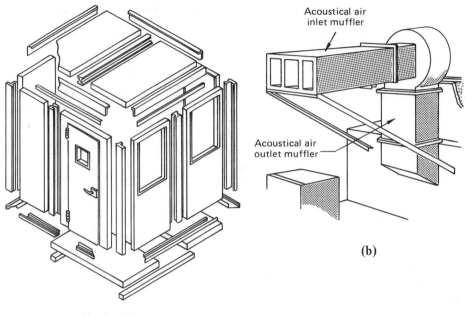

Acoustical air
inlet muffler

Acoustical air
outlet muffler

(b)

Total enclosure

(a)

FIGURE 1-9. *Sound absorbing materials are built in sheet form such as leaded vinyl sheets or polyurethane sandwich material for enclosures (a), or polyvinyl acetate that can be sprayed or troweled onto walls, ducts, etc., (b).* Courtesy Automation.

technically capable standards organizations, mentioned previously, with the Environmental Protection Agency (EPA) providing leadership in the perspective of costs for each increment of improvement. It must be pointed out that some of these agencies, such as EPA and OSHA, are still in their infancy, with major impacts yet to be worked out.

For further study of this topic, see the references at the end of the chapter.

## THE CHALLENGE OF MANUFACTURING

The need for qualified manufacturing engineers has never been greater. To meet today's challenges, it is necessary to gain a broad background in both material sciences, design, and manufacturing processes. Considerable emphasis is also being placed on computer sciences and automatic controls.

The manufacturing engineer must be able to visualize the whole problem of production in the correct perspective: from the basic component materials and tooling through the step-by-step operational planning, final assembly, inspection, and shipment.

The manufacturing engineer is responsible for the expenditure of large sums of money for tooling and machinery to keep production on a competitive basis. He not only should be familiar with the theoretical aspects of fabrication, but must be able to work out practical "bread and butter" problems of the production floor.

A few of the areas in which a manufacturing engineer can be expected to work are in planning process layouts, plant layouts, equipment specification, material specification, tool design, methods development, work standards, value analysis, and cost control.

As the manufacturing engineer gains experience, and if he has a high degree of creativity and innovative ability, he can expect to be involved in the generation of manufacturing systems concepts. He will also be involved in the development of novel and specialized equipment, research into the phenomena of fabricating technologies, and manufacturing feasibility studies of proposed new products.

As an example, shown in Fig. 1-10 is an assembly schematic of a rotary-combustion (RC) or Wankel engine. Although the engine is considered to be basically simple, it presented some extraordinary challenges in manufacture due to the unique trochoid combustion chamber or rotor housing. Also, the extremely close fit of the components and metal-to-metal sealing (no gaskets are used on the rotor housing) called for a degree of accuracy that had not been achieved on a production line basis. Tolerances are assumed to be in the area of "tenths of thousands" for the machined parts and in the low "millionths" on surface finish. New machines, with combination milling/grinding, were designed and built to achieve the shape and tolerance desired for the trochoidal configuration.

Since the trochoidal wall surface is a high-wear area, some companies are using a cast iron liner in the aluminum casting and others are using a sprayed-on coating of tungsten carbide put on with specially adapted flame spraying equipment. The process, when originally used, cost about $5.00 per square inch, but through manufacturing development was reduced to pennies per square inch.

Before assembly all the parts are thoroughly cleaned. After assembly, each engine is tested at 7000 to 8000 rpm for one minute.

The RC engine is considered to be one of the most dramatic mechanical revolutions of this century, with the full impact due between the years 1980 and 1985. Its uses, besides automobiles, include snowmobiles, aircraft, motorcycles, outboard motors, and a full complement of small engines for snow blowers, lawnmowers, garden tractors, and chain saws. As production requirements increase, it is expected that many manufacturing innovations will also develop that will make the process even more efficient.

Not all of the innovations that increase efficiency of manufacture are in the area of machine tools, inspection, or automated assembly per se. Increasing emphasis will be in the challenge of the individual worker. Men on production lines can become teams, collectively responsible for the end product rather than for just so many repetitious operations. Under this concept, it is the team's responsibility to help determine the distribution of work, the occupation of machines, and the speed and quality of work. It is assumed that the personnel will be given broader training

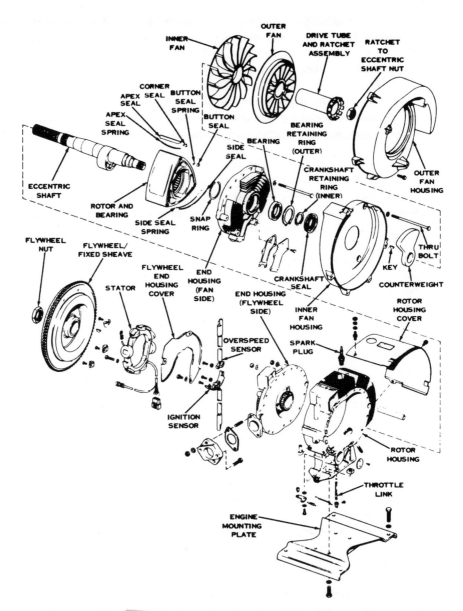

FIGURE 1-10. *An assembly schematic of a 35-horsepower rotary combustion (Wankel) engine. This new engine presented many challenges to the designer and manufacturing engineer. Courtesy Outboard Marine Corporation.*

so that they might have the flexibility and all-around skill to be able to step in where help is needed.

Modern manufacturing planning also includes the use of the computer to handle the vast amount of data needed to utilize raw information, such as sales forecasts and customer orders, in determining what is to be purchased, made, stored, etc. (Shown schematically in Fig. 1-11.)

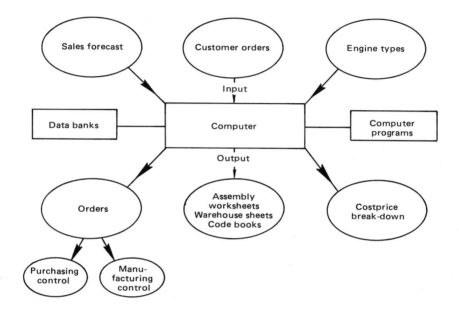

FIGURE 1-11. *The computer, with its data banks and programs, is an essential part of manufacturing planning. Courtesy* Engineering.

The computer, including the minicomputer and programmable controllers (as discussed later in the text), will have increased impact in the areas of product, tool, and process design for optimum production. Computer-aided design (CAD) and computer-aided manufacture (CAM) have the potential of being the largest single development in manufacturing since the Industrial Revolution. Whether the full potential will be realized is dependent on its acceptance by industry and by colleges and universities where future manufacturing engineers are trained. This is not to imply that considerable learning does not take place outside of the formal classroom. On the contrary, because of today's flood of new scientific and technical knowledge, and the astonishing pace of technological change, the engineer risks obsolescence unless he makes some plans for self study. However, even short courses and colloquiums may not be sufficient. For many areas of rapid technological change, a work cycle of four or five years followed by a year on campus may be a normal practice.

## PROBLEMS

**1-1.** Is industry the principal cause of air and water pollution? Explain.

**1-2.** (a) Does industry use most of the energy that is generated in the United States?
(b) Where is most of the energy used in industry?
(c) What are some ways that industry can conserve energy?

**1-3.** Does the OSHA Act supercede state safety laws? Explain.

**1-4.** Who is authorized to request an investigation by an OSHA inspector?

**1-5.** Does an OSHA inspector come out for every employee request? Explain.

**1-6.** Is there any protection from being fired for an employee if he turns in a request for inspection? Explain.

**1-7.** What happens if the employer refuses to recognize anything is unsafe in his plant?

**1-8.** Why may an individual not be aware of hearing losses?

**1-9.** According to OSHA standards, what is the daily limit of noise a person should be exposed to in the case of punch-press impact noise, or discotheque noise?

**1-10.** What is meant by a decibel?

**1-11.** What makes some materials good for vibration damping?

**1-12.** What may be a way of building employee morale in a production line situation?

## BIBLIOGRAPHY

**Books**

DE NEVERS, N., *Technology and Society*, Addison-Wesley Publishing Company, Reading, Massachusetts, 1972.

HOLDREN, J. P. and P. R. EHRLICH (editors), *Global Ecology*, Harcourt Brace Jovanovich, Inc., New York, 1971.

PARKMAN, R., *The Cybernetic Society*, Pergamon Press, Inc., New York, 1972.

WILSON, T. W., JR., *International Environmental Action*, Harvard University Press, Cambridge, Massachusetts, 1971.

**Periodicals**

BERANEK, L. L. and L. N. MILLER, "The Anatomy of Noise," *Machine Design*, Sept. 14, 1967.

CORLISS, O. S. and W. W. OLSEN, "OSHA Factors in Design," *Mechanical Engineering*, Dec. 1973.

GREEN, R. G., "OSHA—How Is Safety By Law Working?," *Automation*, Jan. 1974.

HANKEL, K. M., "Controlling In-Plant Noise," *Automation*, April 1974.

# 2

# *Materials, Structures, Properties, and Fabricating Characteristics*

## Introduction

Man discovered the art of smelting metal towards the end of the Stone Age. The first crude bronzes were probably the result of accidental mixtures of copper and tin ores. The Bronze Age was ushered in about 2500 B.C., when the art of extracting relatively pure tin had advanced to a point where intentional additions to copper were possible. Brass was not introduced until about 500 B.C., when copper and zinc ores were smelted. The Bible makes reference to Tubal Cain, a descendant of Cain, as an "instructor of every artificer in brass and iron" (Genesis 4:22).

Gold was used for jewelry and utensils as early as 3500 B.C. Silver was used about 2400 B.C. A wrought iron sickle blade was found beneath the base of a sphinx in Karnak near Thebes, Egypt, and a blade probably 5000 years old was found in one of the pyramids.

The origin of forced draft in the making of wrought iron is unknown, but prior to about 1500 B.C. the Egyptians had developed a bellows out of goat skin with a bamboo nozzle and an air inlet valve. Forced draft was the first major development in wrought iron manufacture. It was not until the end of the thirteenth century that water power was used to drive a bellows. Water power was also used at this time to drive a forging hammer.

Guilds sprang up in the early 1400s to perpetuate the skills and art of working metals. The iron and steel workers considered themselves very aristocratic and apprenticeships were reserved for legitimate sons of honorable parents. Children of night and gate watchmen, barbers, musicians, millers, tanners, weavers, shepherds, and tax collectors were not eligible.

## CRYSTAL STRUCTURE

The solidification of metal from the molten state requires that a sufficient number of atoms shall exist in the proper arrangement to form a crystal that will grow. The number of atoms that will form a nucleus depends on the decrease of energy when the liquid is replaced by a solid, and the increase of energy resulting from the formation of an interface between liquid and solid phases.

After nucleation, the structure grows until crystals, or grains as they are often called in metallurgy, are formed. The grains join in a dendritic or "pine tree" columnar structure (see Fig. 2-1). This dendritic freezing mode is the pattern of ferrous alloys.

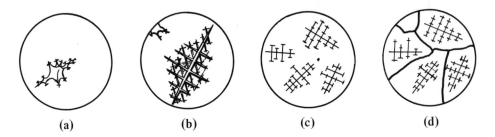

(a)          (b)          (c)          (d)

FIGURE 2-1. *Solidification of pure metal. Nucleation and the first dendritic structure is formed at (a). The dendritic structures join together at (b) and (c). Development is stopped by interference with adjacent structures and the container or the grain boundaries.*

### Space Lattices

Each grain consists of millions of tiny unit cells made up of atoms that are arranged in a definite geometric pattern of imaginary points. The atomic arrangements found in most metals are: (1) body-centered cubic (BCC), (2) face-centered cubic (FCC), and (3) hexagonal close-packed (HCP). (See Fig. 2-2.)

Upon cooling, the space lattice structures expand in all directions of the axes of the lattice until development is stopped by contact with the container or by contact with adjacent growing grains. Where the crystals meet or intersect, it is impossible for another space lattice to develop. Thus, at the grain boundaries the atomic dis-

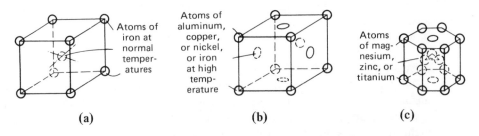

(a)          (b)          (c)

FIGURE 2-2. *The atomic structure of some metals.*

tances are not correct and the atoms are not in a stable position but are in a higher state of energy. The lower density of the atoms at the boundaries provides a place for impure atoms to reside. The size of the grain depends upon the temperature at which the metal is poured, the cooling rate, and the nature of the metal. If the metal cools slowly, large grains will result; if the metal cools fast, smaller grains are obtained.

### Crystal-Lattice Imperfections

Even if metal solidified very slowly, a large number of atoms would be laid down in the crystal structure each second. It is not surprising therefore that imperfections in the structure should occur. The most important of these are *vacancies*, *interstitials*, and *dislocations*.

Vacancies are, as the term implies, vacant sites in the crystal atomic structure [Fig. 2-3(*a*)]. Interstitials are extra atoms that fit into the interstices between the normal atom structure [Fig. 2-3(*b*)]. Both vacancies and interstitials produce local distortion and interrupt the regularity of the space lattice structure.

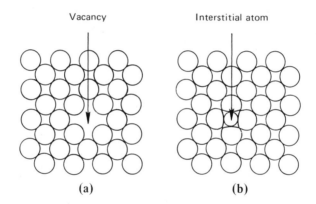

FIGURE 2-3. *Two types of crystal lattice imperfections: (a) vacancy, (b) interstitial.*

Vacancies help explain the phenomenon of *solid state diffusion*. This is why one metal, such as copper, will diffuse through a second metal, such as nickel, when placed in intimate contact and heated to a high temperature. It is assumed that vacancies move through the lattice producing random shifts of atoms from one lattice to another.

An example of an interstitial solid solution is the introduction of carbon atoms into an iron atomic lattice. The carbon atoms are much smaller than the iron atoms and some of them can squeeze into the structure. At room temperature, iron has a body-centered cubic (BCC) lattice and can contain only a few carbon atoms in "solution." However, at elevated temperatures, the structure changes to face-centered cubic (FCC) and there is a considerable increase in the number of carbon atoms that form the solid solution.

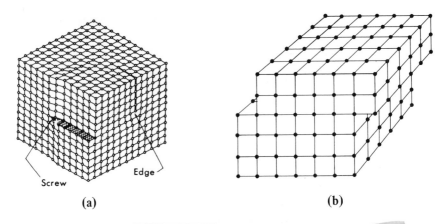

(a)                                        (b)

FIGURE 2-4. *Two types of line defects that occur in the crystal structure: screw and edge (a). Shown at (b) is an enlarged view of the screw dislocation that has passed through the crystal, causing unit slip.*

Dislocations, a third defect, may be thought of as imperfections caused by the mismatching of atomic planes within a crystal. These line-type defects are referred to as *edge* and *screw* dislocations, as shown in Fig. 2-4.

## The Crystal Structure Under Stress

One of the mysteries surrounding the enormous discrepancy between the theoretical and observed shear strength of metals was clarified with the aid of the electron microscope. Theoretical calculations show that a pure crystal of magnesium, for example, should have a yield strength of approximately 1,000,000 psi, whereas it may actually yield at 1000 psi. The advent of the electron microscope made it possible to see that it was not necessary for the whole plane of atoms to move at once, but that the movement could start at one place and only a few atoms shift progressively, as in the screw-type dislocation.

*Elastic and Plastic Deformation.* When a crystal lattice is subjected to a stress below its elastic limit, the crystal structure will temporarily yield a small amount but will recover when the load is released. If a sufficiently large load is applied, plastic deformation takes place. This means the atomic structure has to slip. Slip takes place along certain crystal planes called *slip planes*. When slip planes are readily available, the metal is considered very formable. With few slip planes and directions of slip available, the metal is thought of as being difficult to form. FCC metals, for example, have 12 systems of slip, BCC have 48, and HCP have only 3.

Twinning takes place in some metals when they are subjected to external stresses, as shown in Fig. 2-5. In the case of slip, the offset is a multiple of the interatomic spacing; but in twinning, the offset produced by sliding of one plane against its neighbor is a fraction-of-unit slip. This causes a difference in the orientation between the twinned and untwinned regions in the crystal.

COLD ROLLING

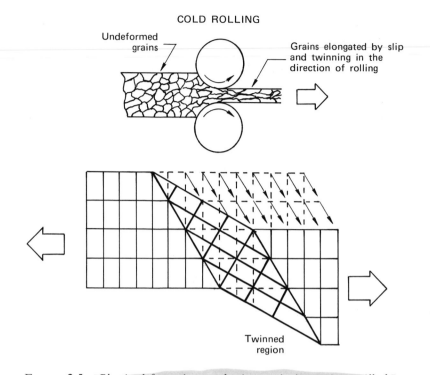

FIGURE 2-5. *Plastic deformation results in a twinning action. All the atoms in the twinned region move a given amount and change orientation as shown.*

Deformation by twinning is most common in HCP metals and its effect on others is to move parts of grains to a more favorable position for slip to occur. Thus, some metals deform by slip and others by twinning and some by both twinning and slip. Twinning always takes place in pairs of lattices, hence the name.

***Strain Hardening.*** As external forces are applied to the metal, slip occurs in the atomic structure. It begins at the points of imperfection or dislocation and twinning. As the force continues and the crystals deform, more dislocations are formed along the slip planes. Strong interactions arise as more and more dislocations are forced to intersect on the various slip planes. The net result is increasing resistance to further deformation, or strain hardening. In this condition, the metals are harder, stronger, less ductile, and have reduced electrical resistance.

***Fracture.*** Every solid material, from the most frail eggshell china to the toughest steel, has its breaking point. It resists stress up to a point, perhaps yielding somewhat and stretching, but then suddenly it fractures. Just what happens at the instant of fracture has been difficult to investigate due to the speed and finality of the reaction. For the engineer, fracture was simply a calamity to be avoided by careful design.

However, the fracture of entire ship hulls during World War II brought the subject under concerted investigation.

Brittle fracture is characterized by the small amount of work absorbed and by a crystalline appearance of the surfaces of the fracture.

Even though most metallic and some nonmetallic crystals have no cracks initially, they do have dislocation defects which make them vulnerable to crack formation. If slip takes place freely, the result is just a change in shape of the crystal. However, if sliding is blocked by a hard particle inside the crystal or at the boundary between crystals or grains, a high concentration of stress collects. The atomic bonds in this situation are under great stress. They stretch beyond their limit and rupture, forming a tiny crack. Once a crack is started, it takes comparatively little force to carry it through to a complete fracture. The force is of course dependent on the specific case. As an example, consider a plate of steel 6 in. wide and 0.25 in. thick. Assume it has a 2-in. crack running into one side. Then the force required to fracture

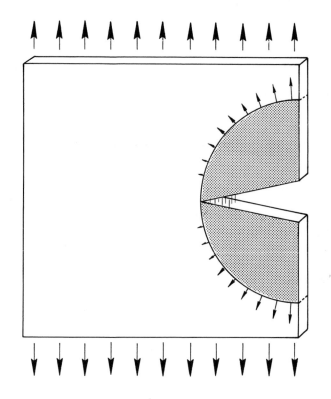

FIGURE 2-6. *The leverage effect of a crack helps explain the fracture mechanism. Forces applied to the top and bottom of the steel plate are transmitted to the vertex of the crack through schematically depicted lever arms. As the crack grows the lever arms have a multiplying effective force at the crack vertex.*

the remaining 4 in. would be only about 400 lb. Without the aid of the crack, it would require a force of about 450,000 lb to pull the plate apart if it were made of the best commercial steel available. It is a large leverage effect (Fig. 2-6) that makes it possible for a relatively small force to crack an entire ship in two.

Plastic flow or slip can be both detrimental and beneficial to the strength of a material. If slip is blocked, it may lead to crack formation. If it is not blocked, it may act to relieve the stress concentrations at the tip of a crack. As temperature decreases, the tendency to flow also decreases. With plastic flow inhibited, stresses build up at the vertex of the crack, stretching the atomic bonds to the breaking point. Some materials, such as ordinary steel, are normally resistant to cracking, but become highly susceptible to it at low temperatures. Referring again to the wartime ships: those used in southerly waters posed no problems, but those used in the northern waters were prone to cracking under shocks from heavy seas in the winter season.

Sudden and seemingly inexplicable fractures are often the culmination of a long series of discrete steps. A sheet of glass, for example, may support a load for a long time without apparent damage and then break without warning. The load does not exert sufficient stress to fracture the glass outright but exerts only enough force to open tiny invisible cracks in its surface. Atmospheric water vapor lays a thin film of moisture on the glass. The smooth surface resists penetration, but as the water reacts with silicon–oxygen bonds within the crack, it breaks those bonds and forms reaction products that have a larger volume than the glass. The reaction products act as a wedge, spreading the crack faces apart. This stretches the bonds at the vertex of the crack, encouraging the next reaction. Thus the process is both self-sustaining and self-accelerating. The crack which at one time was advancing only a few atomic diameters at a time now surges forward at a speed approaching that of sound, and complete fracture is the result. The same so-called static fatigue fractures take place in other materials.

Ductile fractures evidence a substantial plastic deformation prior to failure. The fractured surfaces are characterized by a cup and cone appearance (Fig. 2-7). Microcracks form in the necked-down region that is progressively strain-hardened. The cracks join together and fracture occurs. The size of the cup depends on the relative shear and cleavage strength values. High-strength metals produce a smaller cup, a topic discussed in connection with metal forming in Chapter 4.

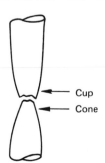

Cup

Cone

FIGURE 2-7. *Ductile fracture. The final shear slip action produces a cup and cone fracture.*

## PROPERTIES OF METALS

Metal properties are often spoken of as separate entities; however, they are closely interwoven. For example, if a specified hardness is desired, it will be accompanied by a corresponding strength.

Before the specific properties of metals can be defined, it would be advantageous to be familiar with three basic terms: stress, strain, and elasticity.

**Stress.** Stress may be defined as the load per unit area or $S = P/A_o$. Thus a rod whose original area is 0.5 sq in., subjected to a load of 2000 lb, would have the following stress:

$$S = \frac{2000\,\text{lb}}{0.5\,\text{sq in.}} = 4000\,\text{psi}$$

Stresses may be either tensile (tending to pull apart), compressive (tending to make shorter), or shear (tending to divide the material in layers). Bending stresses and torsional stresses (twisting) are combinations of the three main stresses.

**Strain.** Strain is the percent change in the unit length during elongation or contraction of the specimen and is expressed as a measure of deformation under load. As an example, a weight of 2000 lb is suspended by a 0.25-in. diameter wire. The wire changes in length from 25 in. to 25.05 in.

$$n = \frac{(l_f - l_o) \times 100}{l_o} = \frac{(25.05 - 25) \times 100}{25} = 0.2\,\%\ \text{strain}$$

*where:*   $n$ = strain
$l_f$ = final length
$l_o$ = original length

**Elasticity.** Most engineering materials are used under what is known as elastic conditions, which means that when a material is deformed, the deformation is not permanent. Elasticity was expressed as a theory by Robert Hooke and therefore bears his name. Hooke's Law states that the degree to which a body bends or stretches out of shape (strain) is in direct proportion to the force acting on it (stress). This law is applicable within a given range of stresses, this range being termed the *elastic limit.* Loads beyond the elastic limit cause permanent deformation.

## MECHANICAL PROPERTIES OF MATERIALS AND DESIGN

The basic modes of loading a component are tension, compression, bending, torsion, and shear. Mixed modes may also occur in service. Each of these five modes of loading is the key to determining certain mechanical properties of a component during its design stage.

### Tension

Several basic properties of a material can be determined from a tensile test. A standard-type specimen (Fig. 2-8) is shown before and after a tensile test. The specimen is gripped at both ends and stretched at a slow, controlled rate of extension until rupture. Usually, frequent or continuous measurements of load and extension are made, tensile stress and strain are calculated, and a *stress–strain diagram* is constructed. From the diagram (Fig. 2-9) may be calculated the modulus of elasticity, proportional limit, tensile strength, and yield strength.

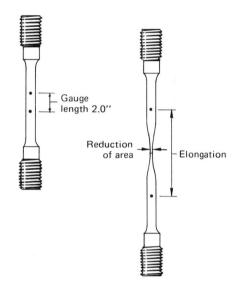

FIGURE 2-8. *A standard-type tensile specimen showing elongation and reduction of area.*

*Modulus of Elasticity (E).* The modulus of elasticity is a measure of stiffness of the material and is represented by the area under the stress–strain curve up to the proportional limit. Steels have an average modulus of elasticity of $30 \times 10^6$. $E = S/n$.

*Proportional Limit.* The proportional limit is the highest point on the stress–strain curve, where the strain is proportional to the stress.

*Tensile Strength.* Tensile strength is defined as the maximum load in tension which a material can withstand prior to fracture. This is the value most often used for listing the strength of a material and is given in pounds per square inch. In older metric terms, it is expressed as kilograms of force per millimeter squared ($kgf/mm^2$); but in the newer SI* units it is expressed as pascals or kilopascals. (See Table 9-3 for conversion factors.)

* SI is the abbreviation for " Le Systeme International d'unites," which is the internationally adopted metric system of weights and measures.

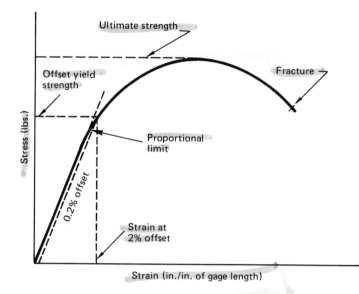

FIGURE 2-9.  *A simple stress–strain plot.*

***Yield Strength.***   Yield strength is the stress at which a material exhibits a specified limiting permanent deformation.  A practical approximation of it is usually determined by using an "offset."  Offset yield strength, as shown in Fig. 2-9, corresponds to the intersection of the stress curve to a straight line parallel to the curved line at a specified offset, usually 0.2 %.  In the absence of a diagram, the yield strength may be obtained at a specified strain under load.  The *yield point* is that point at which the material shows a marked increase in strain in relation to the stress.  Yield strength may be calculated from the equation:

$$S_y = \frac{L_y}{A_o}$$

*where:*   $L_y$ = yield load
$A_o$ = original area

***True Stress and Strain.***   The relationship of properties just described is based on the original cross-sectional area of the specimen and is referred to as *nominal*.  If each time the load was recorded it represented the instantaneous cross section of the specimen or the instantaneous length, it would be termed *true* stress and *true* strain.

$$\text{Stress, } \sigma = \frac{L}{A_i} \qquad \text{Strain, } \varepsilon = \int_{l_o}^{l_f} \frac{dl}{l} = \ln\frac{l_f}{l_o}$$

*where:*    $\sigma$ = true stress
$L$ = load
$A_i$ = instantaneous area
$\varepsilon$ = true strain
$l_f$ = final length; $l_o$ = original length
ln = natural log

The true-stress–true-strain curve exhibits no maximum; stress continues to rise until fracture (Fig. 2-10).

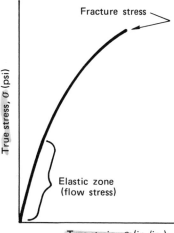

FIGURE 2-10. *True stress and true strain plot.*

***Percent Elongation.***    The percent elongation refers to the increase in length over the original gage length.

$$\% \text{ Elongation} = \frac{l_f - l_o}{l_o} \times 100$$

***Percent Reduction of Area.***    The percent reduction in area is the original cross-sectional area compared to the final cross-sectional area taken at fracture.

$$\% \text{ RA} = \frac{A_o - A_f}{A_o}$$

***Ductility.***    Ductility may be defined as the ability of a material to withstand plastic deformation without rupture. It is expressed in terms of percent elongation and reduction of area.

Ductility is also thought of in terms of bendability and drawability. The forming of cups as discussed in Chapter 4 or the drawing of wire to a smaller size is also a measure of ductility. Brittleness is the opposite of ductility. Usually, if two materials have approximately the same strength, the one that has the higher ductility is the more desirable.

29

## Compression

The compressive strength of a material is a measure of the extent it deforms under a compressive load prior to rupture. The total strain (%) of a specimen immediately before rupture is indicated by direct measurement on a universal testing machine or by the corresponding stress–strain diagram (Fig. 2-11).

Buckling of a compression specimen may take place if the length-to-diameter ratio is greater than 2.5, as shown in Fig. 2-12.

## Bending

Bending is characterized by the outside fibers being placed in tension and the inner fibers in compression. The stress goes to zero at the neutral axis, as shown in Fig. 2-13. The stress on the outer fibers depends on the section geometry, bend radius, and loading.

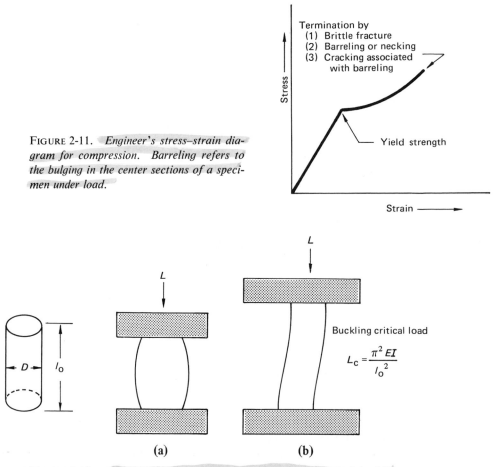

FIGURE 2-11. *Engineer's stress–strain diagram for compression. Barreling refers to the bulging in the center sections of a specimen under load.*

Termination by
(1) Brittle fracture
(2) Barreling or necking
(3) Cracking associated with barreling

Yield strength

Stress

Strain

$L$

$L$

Buckling critical load

$$L_c = \frac{\pi^2 EI}{l_o{}^2}$$

$D$   $l_o$

(a)           (b)

FIGURE 2-12. *Conditions for barreling (a) when $l_o = 2d_o$ and buckling (b) when $l_o = 2.5d_o$.*

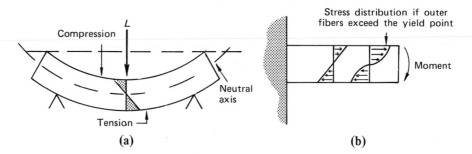

FIGURE 2-13. *Structural members subjected to transverse loads are classed according to the method of support. A simple beam supported at each end is shown at (a) and a cantilever beam is shown at (b). The moment at a section is resisted by the internal couple of compressive and tensile stresses.*

The deflection is dependent on the loading, section geometry, and modulus of elasticity of the material. To decrease the deflection at a given load, the section may be increased or a material with a higher modulus of elasticity may be selected.

The bending stress ($\sigma$) in a cantilevered shaft, as shown in Fig. 2-13(*b*), may be found by:

$$\sigma = \frac{My}{I}$$

*where:*    $M$ = the bending moment
$y$ = half the shaft diameter or the distance from the neutral axis to the outer edge
$I$ = area moment of inertia

The bending moment is equal to the force exerted on the shaft, multiplied by the distance from the supported end (PL). The formula can be further broken down as follows:

$$\sigma = \frac{My}{I} = \frac{(PL)d/2}{\pi d^4/64} = \frac{(PL)(32)}{\pi d^3}$$

One of the principal problems encountered in bending is *springback*. This topic is discussed in Chapter 4.

## Torsion

Torsion is the application of torque to a member to cause it to twist about its longitudinal axis, as shown in Fig. 2-14. It is usually referred to in terms of torsional moment or torque ($T$), which is basically the product of the externally applied force and the moment arm of the force. The moment arm is the distance of the centerline of rotation from the line of force and perpendicular to it. The principal deflection

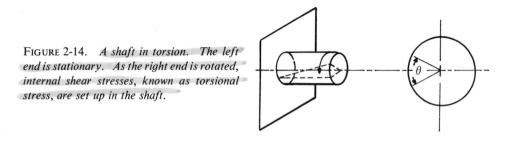

FIGURE 2-14. *A shaft in torsion. The left end is stationary. As the right end is rotated, internal shear stresses, known as torsional stress, are set up in the shaft.*

caused by torsion is measured by the angle of twist $(\theta)$ or by the vertical movement of one corner in framed sections.

For a solid circular shaft, torque is related to stress at the outer fibers as:

$$T = .196\tau d^3$$

*where:*   $T$ = torque (lb in.)
$\tau$ = shear stress at the outer fibers (psi)
$d$ = diameter of the shaft (in.)

For a hollow shaft:

$$T = .196\tau \times \frac{(d_o^4 - d_i^4)}{d_o}$$

*where:*   $d_o$ = outside diameter (in.)
$d_i$ = inside diameter (in.)

The angle of twist $(\theta)$ may be related to torque for a solid shaft as follows:

$$\theta = 584\frac{TL}{Gd^4}$$

*where:*   $L$ = length of shaft (in.)
$G$ = modulus of elasticity in shear (Note: for steel—$12 \times 10^6$).

For hollow circular shafts:

$$\theta = 584\,\frac{TL}{G(d_o^4 - d_i^4)}$$

*Example:*   What is the minimum diameter of a steel shaft that can be used if it is 3 ft long and subjected to a torque of 50,000 lb in. The maximum shearing stress should not exceed 8000 psi and the angle of twist will not be greater than 2°.

*Solution:*

$$T = .196\tau d^3$$
$$50,000 = .196 \times 8000 \times d^3$$
$$d = 3.17\,\text{in.}$$

and

$$\theta = 584 \left( \frac{TL}{Gd^4} \right)$$

$$L = 584 \left( \frac{50,000 \times 12 \times 3}{12,000,000 \times d^4} \right)$$

$$d = 3.17 \text{ in.}$$

Therefore, to satisfy both conditions given in the problem, the shaft should be at least 3.17 in. in diameter.

### Shear Strength

The ultimate shear strength of a material is the maximum load it can withstand without rupture when subjected to a shearing action. A method of determining the shear strength is by using a small sheet or disk of the material of known thickness which is clamped over a die. A corresponding punch is brought down on the material. A gradually increasing load is applied until the material is completely punched through. The shear stress ($S_s$) may be calculated as follows:

$$S_s = \frac{P}{\pi dt}$$

*where:* $P$ = the punch load in lb

$d$ = punch diameter in inches

$t$ = the specimen thickness in inches

The shear strength of mild steels ranges from about 60 to 80 % of the tensile strength.

### Testing

Standardized tests have been made to determine the properties of materials. Discussed here are those used for determining hardness, tensile strength, impact strength, and fatigue.

In selecting a material to withstand wear or erosion, the properties most often considered are hardness and toughness. Hardness is the property of a material that enables it to resist penetration and scratching.

Hardness testing can be done by several standard methods such as Brinell, Rockwell, Vickers, and sleroscope.

*Brinell.* Brinell tests are based on the area of indentation a steel or carbide ball 10 mm in diameter makes in the surface for a given load (Fig. 2-15). Loads used are 3000, 1500, or 500 kg. The load is applied for 15 sec on ferrous metals and at least 30 sec on nonferrous metals. When the load is released, the diameter of the spherical impression is measured with the aid of a Brinell microscope. From the diameter, a

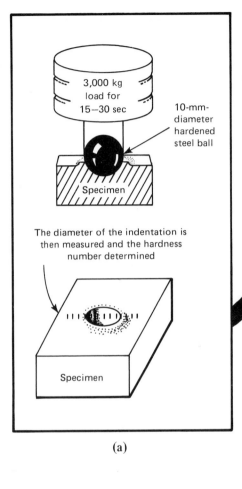

(a)

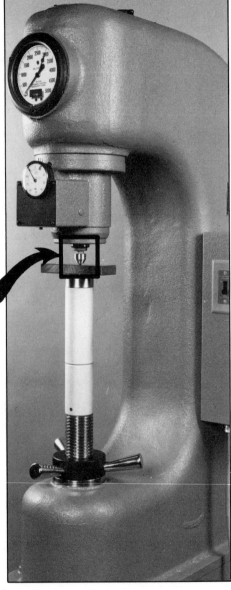

(b)

FIGURE 2-15. *Principles of Brinell hardness testing (a). A Brinell hardness-testing machine (b). Courtesy Acco, Wilson Instrument Division.*

Brinell hardness number is obtained by consulting standard tables such as that made by the Society of Automotive Engineers (SAE), Table 2-1.

Brinell hardness tests are especially good for materials that are coarse-grained or nonuniform in structure, since the relatively large indenter gives a better average reading over a greater area.

*Rockwell.* The Rockwell hardness tester uses two types of penetrators, steel balls and a diamond cone or Brale (Fig. 2-16). The ball indenter is normally $\frac{1}{16}$ in. diameter, but larger diameters such as $\frac{1}{8}$, $\frac{1}{4}$, or $\frac{1}{2}$ in. may be used for soft materials. The Brale is used for hard materials. The principle of this tester is based on measuring the difference in penetration between a minor and a major load. The minor load is 10 kg and the major load varies with the material being tested. If the material is known to by relatively soft, the ball penetrator is used with a 100-kg load in what is known as the "B scale," ($R_b$). If the material is relatively hard, the diamond Brale is used with a 150-kg load on what is known as the "C scale," ($R_c$). Other scales are available and are useful for checking extremely hard or soft surfaces. Very thin sections such as razor blades or parts that have just a thin, hard outer surface may be checked with a Rockwell Superficial hardness tester. The loads range from 15 to 45 kg on what is known as the "T scale."

*Microhardness Tests.* Microhardness tests usually refer to tests made with loads ranging from 1 to 1000 grams. The indenter is either a 136° diamond pyramid or a Knoop diamond indenter (Fig. 2-17). The Knoop indenter is a diamond, ground to a pyramidal form that makes an indentation having an approximate ratio between the long and short diagonals of 7:1.

Prior to the advent of the microhardness tester, it had been assumed that the 136° diamond indenter produced a hardness number which was independent of the indenting load. In general terms, this can be accepted for loads of 1 kg and up. However, microhardness tests (performed with loads of 500 g or lighter with the Knoop indenter, and 100 g or lighter with the diamond pyramid indenter) are a function of the test load.

The Knoop hardness number is the applied load divided by the unrecovered projected area of the indentation. Both hard and brittle materials may be tested with the Knoop indentor.

The diamond pyramid hardness number (DPH) is also the applied load divided by the surface area of the indentation. The 136° diamond pyramid is often referred to as a "Vickers" test.

*Shore Scleroscope.* The scleroscope presents a fast, portable means of checking hardness. The hardness number is based on the height of rebound of a diamond-tipped metallic hammer. The hammer falls free from a given height. The amount of rebound is observed on a scale in the background. The harder the material, the higher the rebound, and vice versa. Thin materials may be checked if a sufficient number of layers are packed together to prevent the hammer from penetrating the metal to the extent that the rebound is influenced by the steel anvil. This is known as *anvil effect.*

TABLE 2-1. *Hardness Conversion Table for Hardenable Carbon and Alloy Steels.* Reprinted with permission, Copyright © 1966, Society of Automotive Engineers. All rights reserved.

| Brinell, 10 mm Carbide Ball, 3000 kg Load | | Diamond Pyramid Hardness Number | Rockwell | | Shore | Tensile Strength, 1000 psi |
| | | | C Scale 150 kg Brale | B Scale 100 kg 1/16-in. Ball | | |
| Indentation Diam, mm | Hardness Number | | | | | |
|---|---|---|---|---|---|---|
| — | — | 940 | 68 | — | 97 | — |
| — | 767 | 880 | 66.5 | — | 93 | — |
| 2.25 | 745 | 840 | 65.5 | — | 91 | — |
| 2.30 | 712 | — | — | — | — | — |
| 2.35 | 682 | 737 | 61.5 | — | 84 | — |
| 2.40 | 653 | 697 | 60 | — | 81 | — |
| 2.45 | 627 | 667 | 58.5 | — | 79 | — |
| 2.50 | 601 | 640 | 57.5 | — | 77 | — |
| 2.55 | 578 | 615 | 56 | — | 75 | — |
| 2.60 | 555 | 591 | 54.5 | — | 73 | 298 |
| 2.65 | 534 | 569 | 53.5 | — | 71 | 288 |
| 2.70 | 514 | 547 | 52 | — | 70 | 274 |
| 2.75 | 495 | 528 | 51 | — | 68 | 264 |
| 2.80 | 477 | 508 | 49.5 | — | 66 | 252 |
| 2.85 | 461 | 491 | 48.5 | — | 65 | 242 |
| 2.90 | 444 | 472 | 47 | — | 63 | 230 |
| 2.95 | 429 | 455 | 45.5 | — | 61 | 219 |
| 3.00 | 415 | 440 | 44.5 | — | 59 | 212 |
| 3.05 | 401 | 425 | 43 | — | 58 | 202 |
| 3.10 | 388 | 410 | 42 | — | 56 | 193 |
| 3.15 | 375 | 396 | 40.5 | — | 54 | 184 |
| 3.20 | 363 | 383 | 39 | — | 52 | 177 |
| 3.25 | 352 | 372 | 38 | (110)* | 51 | 171 |
| 3.30 | 341 | 360 | 36.5 | (109) | 50 | 164 |
| 3.35 | 331 | 350 | 35.5 | (108.5) | 48 | 159 |
| 3.40 | 321 | 339 | 34.5 | (108) | 47 | 154 |
| 3.45 | 311 | 328 | 33 | (107.5) | 46 | 149 |
| 3.50 | 302 | 319 | 32 | (107) | 45 | 146 |
| 3.55 | 293 | 309 | 31 | (106) | 43 | 141 |
| 3.60 | 285 | 301 | 30 | (105.5) | — | 138 |
| 3.65 | 277 | 292 | 29 | (104.5) | 41 | 134 |
| 3.70 | 269 | 284 | 27.5 | (104) | 40 | 130 |
| 3.75 | 262 | 276 | 26.5 | (103) | 39 | 127 |
| 3.80 | 255 | 269 | 25.5 | (102) | 38 | 123 |
| 3.85 | 248 | 261 | 24 | (101) | 37 | 120 |
| 3.90 | 241 | 253 | 23 | 100 | 36 | 116 |
| 3.95 | 235 | 247 | 21.5 | 99 | 35 | 114 |
| 4.00 | 229 | 241 | 20.5 | 98 | 34 | 111 |
| 4.05 | 223 | 234 | (19) | 97.5 | — | — |
| 4.10 | 217 | 228 | (17.5) | 96.5 | 33 | 105 |
| 4.20 | 207 | 218 | (15) | 94.5 | 32 | 100 |

(Continued)

TABLE 2.1—*continued*

| Brinell, 10 mm Carbide Ball, 3000 kg Load | | Diamond Pyramid Hardness Number | Rockwell | | Shore | Tensile, Strength, 1000 psi |
|---|---|---|---|---|---|---|
| Indentation Diam, mm | Hardness Number | | C Scale 150 kg Brale | B Scale 100 kg 1/16-in. Ball | | |
| 4.30 | 197 | 207 | (12.5) | 93 | 30 | 95 |
| 4.40 | 187 | 196 | (10) | 90.5 | — | 90 |
| 4.50 | 179 | 188 | (8) | 89 | 27 | 87 |
| 4.60 | 170 | 178 | (5) | 87 | 26 | 83 |
| 4.70 | 163 | 171 | (3) | 85 | 25 | 79 |
| 4.80 | 156 | 163 | (1) | 83 | — | 76 |
| 5.00 | 143 | 150 | — | 78.5 | 22 | 71 |
| 5.20 | 131 | 137 | — | 74 | — | 65 |
| 5.40 | 121 | 127 | — | 70 | 19 | 60 |
| 5.60 | 111 | 117 | — | 65.5 | 15 | 56 |

* Values in parentheses are beyond normal range and are for information only.

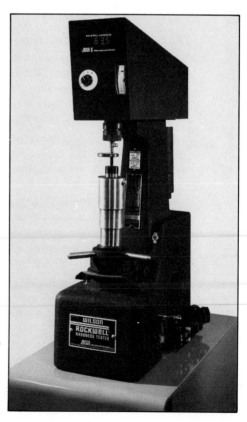

FIGURE 2-16. *Rockwell hardness tester with Brale penetration used in testing hard materials. Courtesy Acco, Wilson Instrument Division.*

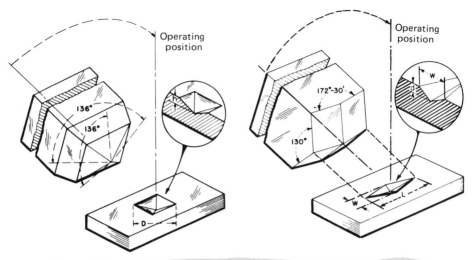

FIGURE 2-17. *The Knoop indenter with the long impression is shown at the left. At the right is the diamond-pyramid or Vickers indenter. Hardness in both systems are in units of kg/mm². Courtesy Acco, Wilson Instrument Division.*

***Tensile Testing.*** Tensile tests, when properly conducted, come closer to evaluating fundamental properties of a material for use in design than any other tests. This implies the use of standardized specimens with regard to shape and size, and a standardized testing procedure. Both tension and compression tests are made on a universal testing machine such as that shown in Fig. 2-18.

Although standard shapes have been designated for tensile or compression tests, this does not preclude making tests on full-size manufactured parts. In many cases, tests of this type are essential and are the best type to determine the properties of tubing, selected wires, reinforcement bars, fibers, fabrics, brick, tile, metal castings, etc.

Specimens are usually round; however, flat sheet stock is also used. The central portion is usually of smaller cross section than the ends in order to cause failure where it is not affected by the gripping device. The round specimen is often made 0.505 in. in diameter in order to have an even 0.200-sq in. cross section.

***Impact Tests.*** Many machines or parts of them must be designed to absorb impact loading. The aim of the designer is to provide for the absorption of as much of the energy as possible through *elastic* action and then to provide for some type of damping to dissipate it.

Standardized tests do not determine the energy-absorption characteristics of a whole mechanism, but they can provide information on a given material by subjecting it to impact loading. Two tests developed to provide a measure of rupture strength of the material or *toughness* are the Izod and Charpy tests as shown in Fig. 2-19.

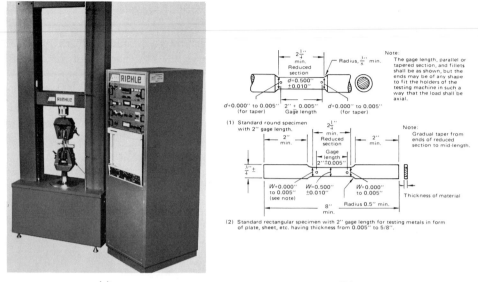

(a)                                                                                              (b)

FIGURE 2-18.  *A universal testing machine being used to pull a tensile specimen (a).  Standard ASTM round and flat specimens for ductile materials are shown at (b).  The machine may also be used for compression tests.  Courtesy Acco, Wilson Instrument Division.*

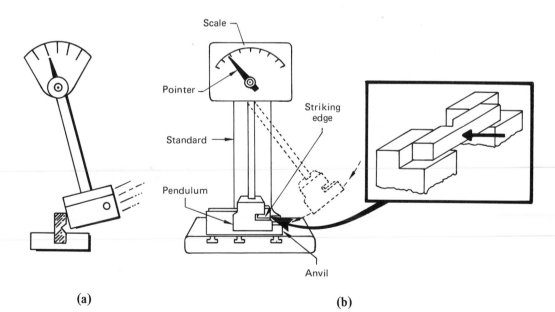

(a)                                                                                              (b)

FIGURE 2-19.  *The Izod (a) and Charpy (b) impact-toughness tests.*

The tests are conducted on a machined specimen that is usually notched and struck a single blow. The energy absorbed in breaking the specimen is measured and converted into engineering design calculations. If the material distributes the stress uniformly, the impact value in terms of foot-pounds of energy absorbed will be high. Three tests are normally performed and the average of the three is used as the impact value.

*Fatigue.* Fatigue is a phenomenon that begins with minute cracks in a material that develop into fractures under repeated or fluctuating stresses. The maximum stress value is less than the tensile strength of the material. Normally the fatigue failures start at the surface, since the stresses are higher at that point.

The most common type of fatigue test is the rotating beam test. In this test, the specimen is rotated while being subjected to a bending moment. The purpose of the rotation is to cause an alternate shift of the uniform bending stress from tension to compression every 180° of rotation (Fig. 2-20).

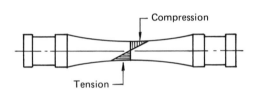

FIGURE 2-20. *A rotating-beam specimen. A uniform bending moment is applied so that the upper fibers are always in compression and the lower fibers in tension. Thus in one cycle all the fibers have been subjected to both compression and tensile stresses. The stress ratio R, which is the algebraic ratio of minimum to maximum stress, is −1.*

Great care must be taken in making the specimen, especially in machining the gage surface, since fatigue failure is very sensitive to surface influences.

To determine the endurance limit of a metal, it is necessary to prepare a number of similar specimens that are representative of the material. The first specimen is tested at a relatively high stress so that failure will occur with a relatively small number of stress applications. Succeeding specimens are then tested, each at a lower stress. Specimens stressed below the endurance limit will not rupture. The test results are usually plotted on S–N diagrams as shown in Fig. 2-21. For all ferrous metals tested and for most nonferrous metals, the S–N diagrams become horizontal, as nearly as can be determined for values of N ranging from 1,000,000 to 50,000,000 cycles, indicating a well-defined endurance limit.

As discussed later in this chapter, alloying elements and heat treatment have considerable effect on metals. When the static strength of a metal is increased by these methods, so also is the fatigue life.

A general rule in design is to avoid, as much as possible, stress concentrations or abrupt changes in geometrical continuity such as holes, notches, threads, and rough machine marks.

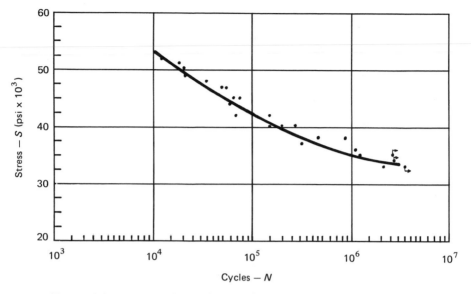

FIGURE 2-21. *A typical S–N diagram for determining the endurance limit (fatigue) of metals under reverse flexural stress. The metal used for this curve was 7075-T6 aluminum.*

## ALLOYING ELEMENTS AND THEIR EFFECTS ON METAL PROPERTIES

*General.* Every alloying element has a particular effect on a metal. Where a combination of two or more elements exist in the metal, the total result is usually an increase in each of the characteristics that is greater than the sum of their individual effects. Common alloying elements used in steel are carbon, nickel, chromium, molybdenum, vanadium, tungsten, manganese, copper, sulfur, boron, aluminum, and phosphorous. The effect of each of these elements on steel will be treated from a qualitative rather than a quantitative basis.

*Carbon.* Technically, carbon should not be considered an alloying element of steel, since without it steel would not exist, but would remain iron. Varying amounts of carbon in the steel have a profound effect on its properties. Therefore, plain carbon steels are usually referred to by their carbon content as being low-, medium-, or high-carbon steels. These classifications are: 0.05 to 0.30 % (low), 0.30 to 0.60 % (medium), and 0.60 to 1.2 % (high). Above this range of carbon content are the cast irons and below it are the wrought irons.

Among the properties influenced by the carbon content are hardness, tensile strength, yield strength, impact strength, reduction of area, and elongation, as shown in Fig. 2-22. It can be seen that tensile strength, yield strength, and hardness increase with carbon content. Impact strength, reduction of area, and elongation are reduced by carbon content.

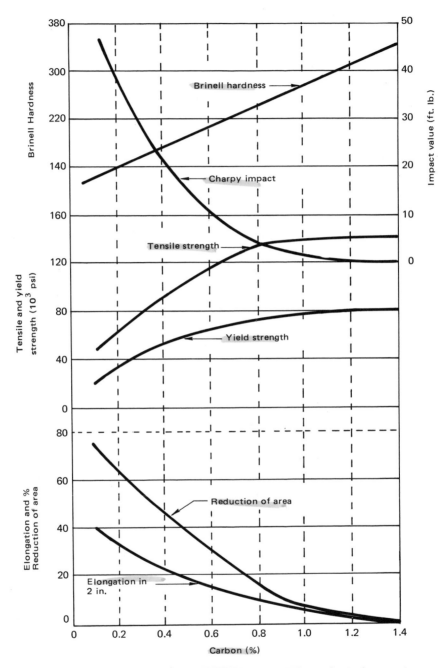

FIGURE 2-22. *The effect of the carbon content on the mechanical properties of steel. The data represented is from 1-in. diameter rolled-steel bars.*

**Nickel.** Nickel increases toughness and resistance to impact. It lessens distortion in quenching and improves corrosion resistance. With nickel, a given strength value often can be obtained with considerably lower carbon content. Nickel is often included as an alloying element in high-strength structural steels used in the as-rolled condition or in heavy forgings that harden by air cooling rather than by quenching in oil.

**Chromium.** Chromium, unlike nickel, joins with carbon to form chromium carbides, thus improving the depth to which a metal may be hardened (termed *hardenability*) and increasing its resistance to abrasion and wear. Chromium steels are stable at relatively high temperatures and are therefore used in steels where heat is a consideration. Chromium is also useful in resisting corrosion.

**Molybdenum.** Molybdenum, like chromium, joins with carbon and therefore promotes hardenability. It also has a tendency to hamper grain growth when the steel is at elevated temperatures, making the steel finer grained and tougher. Shown in Fig. 2-23 is the direct relationship of hardness and tensile strength of molybdenum steels.

Some familiar items that are made from molybdenum alloys are: high-speed cutting tools, forged gears and crankshafts, turbine rotors, high-pressure boiler plate, and high quality tubing.

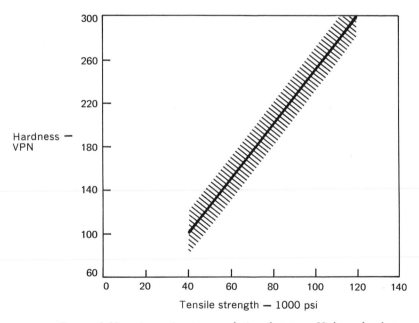

FIGURE 2-23. *Approximate correlation between Vickers hardness and tensile strength of molybdenum-based steels at room temperatures. Courtesy Climax Molybdenum Company.*

43

*Vanadium.* Vanadium is noted for providing a fine-grained structure over a broad range of temperatures. Parts made out of steels containing various amounts of vanadium are: certain types of spring steel, gears, high-temperature steels, forged axles, shafts, turbine rotors, and other items that require impact and fatigue resistance.

*Tungsten.* Tungsten increases hardness, promotes fine grain, and is excellent for resisting heat. Tungsten has a body-centered lattice and thus dissolves in steel at elevated temperatures, forming tungsten carbides. These carbides are very hard and stable. When used in higher percentages (18 %) and combined with lesser percentages of chromium and vanadium, the resulting steel is known as a high-speed steel that is used as a cutting tool material for drills, reamers, lathe tools, etc.

*Manganese.* Manganese is one of the basic alloying elements of steel. It is second only to carbon in frequency of use. Manganese contributes markedly to strength and hardness, but to a lesser extent than carbon. Actually the effectiveness of manganese is dependent on carbon. High manganese content, about 13 %, causes the steel to strain or "work harden" easily, that is, it possesses the property of increasing hardness as the metal is worked. Thus the wear life of such items as power shovel teeth, crushing machinery, and railroad switch frogs is greatly increased.

*Copper.* Copper is added to steel in varying amounts, generally from 0.2 to 0.5 %. It is used primarily to increase resistance to atmospheric corrosion, but it also acts as a strengthening agent.

*Sulfur.* Sulfur is considered an impurity that unites with steel to form iron sulphides, which contribute to cracking while being hot-worked. With manganese, this disadvantage is largely overcome due to the formation of manganese sulphide. Sulfur increases the free-cutting action of low-carbon steels because it breaks up the ferrite structure and helps the chips break free of the tool. Thus short chips are formed rather than long, hazardous types.

*Boron.* Boron is used to increase the hardenability of the steel. It is very effective when used with low-carbon alloy steels, but its effect is reduced as the amount of carbon increases.

*Aluminum.* Aluminum has an affinity for oxygen, therefore it is used in steel as an effective deoxidizer. It is also used to help form more and smaller grains.

*Phosphorous.* Phosphorous is also considered an impurity in steel. In large amounts, it increases the strength and hardness of steel but reduces ductility and impact resistance. Phosphorous in the low-carbon steels improves machinability.

# HEAT TREATMENT

Heat treatment is a term used to denote a process of heating and cooling materials in order to obtain certain desired properties, often to obtain the best properties the material can offer. As an example, a shear blade must have its structure controlled to produce just the right combination of hardness, wear resistance, and toughness in order for it to successfully cut other metals. The reasons for heat treating ferrous metals may be listed briefly as:

1. To relieve internal stresses.
2. To refine grain size or to produce uniform grains throughout the metal structure.
3. To alter the microstructure.
4. To change the surface chemistry by adding or deleting elements.

### Normalized Steel

The most common condition of steel is as a *normalized* structure. After the steel is produced in the furnace, it is poured into ingots for cooling and then rolled into sheets, plates, rods, etc. Thus the steel has proceeded from an elevated temperature to room temperature in a normal process. This process, along with other common heat treating processes such as annealing, hardening, tempering, stress relieving, and spheroidizing, can more easily be understood by following the iron–carbon equilibrium diagram shown in Fig. 2-24.

As steel cools from the elevated temperature, it passes through several stages, as shown on the diagram: from liquid, to liquid and austenite, and then to austenite. On the left, under the $A_3$ line, is austenite plus ferrite and on the right austenite plus cementite. Finally, below the $A_1$ line is ferrite plus pearlite, pearlite, and pearlite plus cementite. The formation of each of these structures will be discussed.

*Austenite.* Austenite is shown in the area above the line GSE and is a solid solution of carbon in a face-centered cubic iron. The $A_3$ line represents the initial precipitation of ferrite from the austenite. The line SE indicates the primary deposition of cementite ($Fe_3C$) from austenite.

*Ferrite.* Ferrite is an alpha iron with a body-centered cubic lattice that exists in the very narrow area at the extreme left of the diagram below 1674°F (911°C).

*Cementite.* Cementite is a very hard, brittle compound of iron and carbon, $Fe_3C$, containing 6.67 % carbon.

*Pearlite.* Pearlite is a two-phase structure consisting of thin, alternate layers of iron carbide (cementite) and ferrite, as shown in Fig. 2-25 and in the enlarged structural views in the diagram in Fig. 2-24.

Normally, as the metal cools slowly from austenite, there is an automatic separation of ferrite and the ferrite–cementite mixture (pearlite), as shown in the diagram. (The white areas represent ferrite and the lined areas pearlite.) As the

45

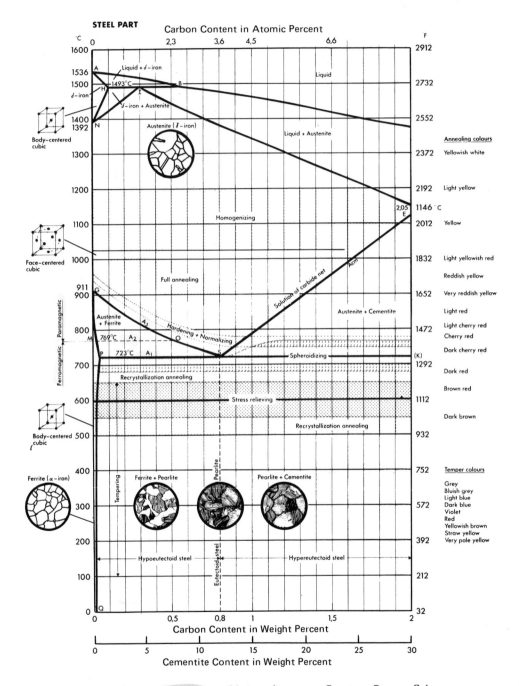

FIGURE 2-24. *Iron-carbon equilibrium diagram.* Courtesy Struers Scientific Instruments.

FIGURE 2-25. *Pearlite structure of steel.*
*Courtesy The International Nickel Co. Inc.*

carbon content increases, it unites with greater amounts of ferrite, thus increasing the pearlite. At a point where all of the ferrite is in combination with carbon, the structure is entirely pearlite, as shown by the centered microstructural view. Theoretically, this is at 0.83 % carbon but may range from 0.75 to 0.85 % in plain carbon steels. It is represented on this diagram at 0.80 % carbon. This combination of iron and carbon is known as a *eutectoid* steel. Eutectoid, taken from the Greek, means "most fusible." This combination occurs in binary alloys when a complete solid solubility does not exist. A eutectic transforms at a lower temperature than the melting point of either of the pure components. Steels with more than 0.80 % carbon are called *hypereutectoid* and those below 0.80 % carbon are *hypoeutectoid*. During slow cooling, the $Fe_3C$ precipitates out of the austenite until the temperature of 1333°F (723°C) is reached, at which time the remaining austenite changes to pearlite. The result is a pearlite structure surrounded by a network of cementite at the grain boundaries, as shown in the enlargement on the iron–carbon diagram.

When steel containing more than 2.11 % carbon is slowly cooled, the alloy freezes out into iron and graphite. The fine, soft graphite forms flakes in the iron matrix, producing a gray cast iron. If cooling is more rapid, a brittle structure known as white cast iron will result. Cast irons will be discussed in more detail later in this chapter.

If a specimen of hypoeutectoid steel is heated uniformly, at approximately 1333°F (723°C) the temperature of the steel will stop rising, even though the heat is still being applied. After a short time, the temperature will continue to rise again. Ordinarily, metals expand as they are heated, but it is found that at 1333°F a slight contraction takes place and then, after the pause, expansion again takes place. This indicates the lower critical temperature of the steel, shown in the iron–carbon diagram as line $A_1$. Actually, an atomic change takes place at this point, and the structure changes from a body-centered arrangement (alpha iron) to a face-centered arrangement (gamma iron).

As the heating continues, another pause will be noted at line $A_3$, at which time the transformation of ferrite to austenite is complete. This varies with the carbon content, as shown in the diagram, and is known as the upper critical temperature. At this point all of the carbon goes into solution with the iron, so that it is now more evenly distributed.

### Hardening Steel

Three requirements are necessary in order to successfully harden a piece of steel: first, the steel must contain enough carbon; second, it must be heated to the correct temperature; third, it must be cooled or quenched.

*Carbon Content.* In order to get extreme hardness in a piece of steel, it is necessary that the steel contain 80 points or more of carbon (one point of carbon = 0.01 % carbon). Low-carbon steels (less than 25 points carbon) will not be materially affected by heat treatment. Medium-carbon steels (30 to 60 points carbon) may be toughened considerably by heat treatment, but they will not be hardened to a very great extent. High-carbon steels (60 to 150 points carbon) may be successfully hardened by simple heat-treating methods.

*Heating.* The steel must be heated above the upper critical ($A_3$) temperature. The actual hardening range is about 100 to 200 degrees Fahrenheit above the upper critical temperature. Notice that this curve levels out beyond 0.80 % carbon.

After the steel has been heated at the hardening range for a sufficient period of time to equalize the heat (about 30 minutes per inch of cross section), it is taken out of the furnace and quenched in a cooling medium, which may be water, brine, oil, molten salts, or lead baths. The rate at which the material cools will largely determine its hardness. In order to avoid undue stress caused by uneven cooling during the quenching action, the part should be vigorously agitated to prevent steam pockets from forming on the surface.

Given the proper heat treatment, the hardness will be determined largely by the carbon content. However, the depth of hardness, known as *hardenability*, will be determined largely by the alloying elements. The effect of alloying elements on the hardenability is shown in Fig. 2-26.

When austenite is immersed in a water quench, it does not have time to separate out and form pearlite. Some of the austenite transforms almost instantaneously to martensite, an interlaced, needlelike structure as shown in Fig. 2-27 that is hard and brittle. How this change takes place can be shown by means of another diagram, termed a time–temperature–transformation diagram or TTT diagram (Fig. 2-28). The diagram shows the various structures that occur as the metal is cooled and the times at which they occur. As an example, if a high-carbon steel is quenched in water such that it reaches a temperature of 550°F (288°C) in about 5 sec or less, a martensite transformation would start ($M_s$). The martensite transformation would be finished ($M_f$) at about 325°F (160°C). The martensite that results is a super-saturated solid solution of carbon trapped in a body-centered tetragonal structure. It involves no chemical change from the austenite stage. The unit cells of the martensite structure are highly distorted because of the entrapped carbon, which is the principal hardening

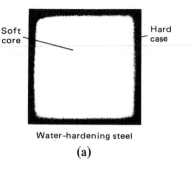

Soft core — Hard case

Water-hardening steel

**(a)**

FIGURE 2-26. *The effect of alloys on the hardenability of steel. A plain high-carbon steel gets an outer layer of hardness when heat treated (a). If manganese is added to plain carbon steel, it can be cooled at a slower rate, as in oil, and the hardness penetration will be deeper (b). As more alloys are added, the steel can be cooled more slowly, as in air, and the hardness will penetrate through the section (c).*

Type analysis
Carbon, plus
Manganese 1.60%

**(b)**

Type analysis
Carbon, plus
Manganese 2.00%
Chromium 1.00%
Molybdenum 1.35%

**(c)**

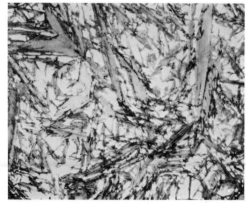

FIGURE 2-27. *Martensite structure (500X).*

mechanism. The body-centered cubic tetragonal structure is less densely packed than the FCC of austenite, hence there is about a 4% expansion causing still further localized stresses that distort the austenite matrix. Normally these internal stresses would have to be relieved as soon as possible to prevent cracking by a tempering process.

If the high-carbon steel had been cooled slightly slower, other structures may have formed in addition to martensite. This can be shown by a continuous-cooling-transformation diagram (C-T), Fig. 2-29. An end-quench hardenability bar is super-imposed on the top of the diagram. The letters *A, B, C,* and *D* represent various

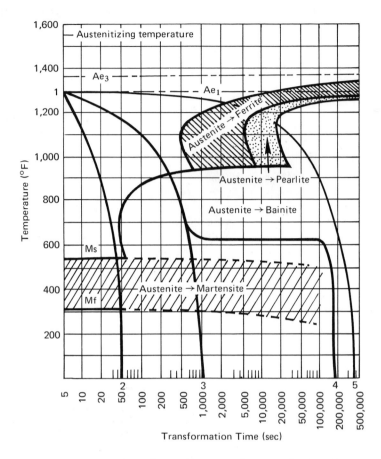

FIGURE 2-28. *An isothermal-transformation diagram showing the trans-formation temperatures in relation to time as austenite is cooled. Courtesy United States Steel Corp.*

distances from the quenched end and the corresponding cooling curves are shown on the transformation and C-T diagrams. Thus, the hardness of the bar at *A* will be the highest, as it has missed the "nose" of the transformation curve. At points *B*, *C*, *D*, and *E* the cooling will be progressively slower, so that it passes through the austenite–ferrite, austenite–bainite, and austenite–martensite transformation curves. The metal will consist of martensite, ferrite, and bainite, as shown by the microstructures below the diagram.

*Bainite.* Bainite is an intermediate-type structure, feathery in appearance, that forms between pearlite and martensite when the cooling rate is slowed to exceed the critical rate. The critical cooling rate is that rate which is just fast enough to miss the nose of the transformation curve.

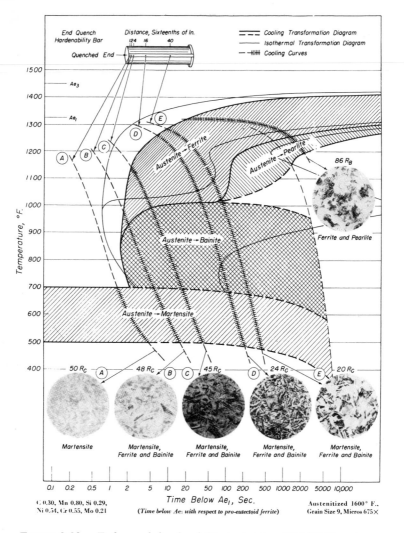

FIGURE 2-29. *End-quench hardenability tests of a AISI 8630 steel correlated with continuous-cooling-transformation and isothermal-transformation diagrams. Courtesy United States Steel Corp.*

## Tempering

As mentioned previously, after quenching the steel is in a highly stressed, unstable condition. To avoid, or minimize, problems of cracking and distortion, the metal is reheated or tempered. There are several different ways this may be done: conventional reheating after quenching, martempering, and austempering (Fig. 2-30).

The conventional method consists of reheating the metal immediately after quenching to a temperature less than critical. Since tempering softens the steel, the relationship of hardness to strength should be known. As an example, a steel that may have a hardness reading of $R_c$ 62 (file hard) after quenching may be reheated to

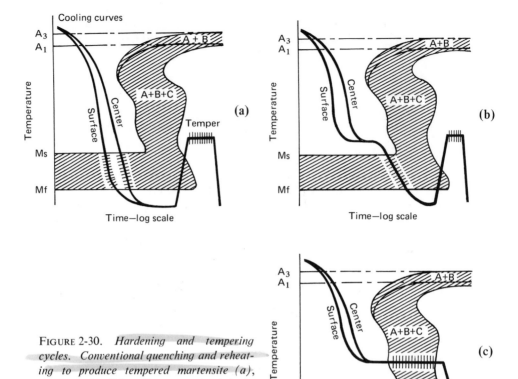

FIGURE 2-30. *Hardening and tempering cycles. Conventional quenching and reheating to produce tempered martensite (a), martempering (b), and austempering (c).*

a temperature of 450°F (232°C) and then have a hardness reading of $R_c$ 56. This would be hard enough to have high strength and good wear resistance, with the advantage of having a more stable structure.

In martempering, air-hardening steels may be quenched in a bath of molten salts at a temperature just above the $M_s$ line. This allows the inside of the steel to arrive at the same temperature as the outside. The steel is then removed from the bath and allowed to cool at room temperature. For straight-carbon and some alloy steels a quench after the hold time is necessary. This process has the advantage of reducing the stress on the steel as it changes to martensite, since the temperature has already been equalized throughout. Tempering follows (see Fig. 2-30).

Austempering also pauses in the cooling curve to equalize the outside temperature of the part with that of the inside. The part then is kept at that temperature to produce a bainite structure. This eliminates the additional step of reheating the metal. The hardness can be the same as that of tempered martensite, depending on the transformation temperature. If a temperature of 450°F (232°C) is used as in the example of conventional tempering, the same relative hardness of $R_c$ 56 would be expected.

### Annealing

Annealing in general refers to softening a metal by heating and cooling. It is generally of two types, *full annealing* and *stress-relief annealing*, as shown in the iron–carbon diagram, Fig. 2-24.

Full annealing consists of heating the steel about 100°F (38°C) above the hardening and normalizing range. It is held at that temperature for the desired length of time and then allowed to cool very slowly, usually in the furnace. This produces a soft, coarse, pearlitic structure that is stress-free.

In stress-relief annealing, the metal is heated to a temperature close to the lower critical temperature, followed by any desired rate of cooling. Stress-relief annealing is usually done to soften a metal that has become strain-hardened or work-hardened during a forming operation. As shown on the iron–carbon diagram, there is a recrystallization zone. At this temperature, the grains of a cold-worked steel will recrystallize and form fine grains. There are two zones shown for recrystallization. This does not mean that recrystallization occurs only at these temperature ranges, but rather that it can occur at different temperatures depending upon the amount of prior cold work. It takes energy to make the grains recrystallize, and if they have a considerable amount of energy already in the structure, it requires less to make them recrystallize. If there is no stress on the metal, the grains will not recrystallize until just below the lower critical temperature.

### Spheroidizing

When steel is tempered at a temperature just below the lower critical or $A_1$ line, as shown in the iron–carbon diagram, the cementite will consist of small spheroids surrounded by ferrite (Fig. 2-31). Prolonged heating (16 to 72 hours) at this temperature may be required. The process may be speeded up, particularly on smaller items, by alternately heating them to temperatures slightly below and then slightly above the lower critical. The length of time and the number of cycles will be dependent on the original structure of the steel. A fine pearlite is preferred.

Spheroidizing results in greater ductility, which improves the forming qualities as well as machinability.

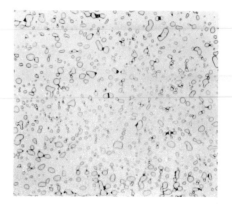

FIGURE 2-31. *Spheroidized iron carbides in a matrix of annealed steel, at a magnification of 750X. Courtesy The International Nickel Co. Inc.*

## Surface Hardening of Steels

The heat treating processes discussed were generally applicable to medium- and high-carbon steels. The depth of hardness was based on the alloy content and the quench rate. It is often desirable to have a hard surface accompanied by a softer, tougher core. This can be accomplished by flame hardening or induction hardening on steels of medium-carbon content or higher and cast irons of suitable composition. Carburizing and nitriding processes are used on low-carbon steels.

*Flame Hardening.* Flame hardening consists of moving an oxyacetylene flame over a part, followed by a quenching spray. The rate at which the flame is moved over the part will determine the depth to which the material is being heated to a critical temperature or higher. Quenching can be built into the burner as shown in Fig. 2-32.

In flame hardening, there is no sharp line of demarcation between the hardened surface zone and the adjacent layer, so there is little likelihood that it will chip out or break during the service. The flame is kept at a reasonable distance from sharp corners to prevent overheating and drilled or tapped holes are normally filled with wet asbestos for edge protection. Stress relieving at about 400°F (204°C) is recommended for all flame-hardened articles except those made from air-hardened steel.

Some advantages of flame hardening are:

1. Large machined surfaces can be surface hardened economically.
2. Surfaces can be selectively hardened with minimum warping and freedom from quench cracking. Examples of selective hardening are gear teeth, machine ways, cam surfaces, and engine push rod ends.
3. Scaling of the surface is only superficial because of the relatively short heating cycle.
4. The equipment may be controlled electronically to provide precise control of the hardness.
5. The depth of case may vary to suit the part, from $\frac{1}{8}$ to $\frac{1}{2}$ in.

Disadvantages include:

1. To obtain optimum results, a technique would have to be developed for each design.
2. Overheating can cause cracks, or excessive distortion, especially where thin sections are involved.

*Induction Hardening.* Induction hardening is done by heating the metal with a high-frequency alternating magnetic field. Heat is quickly generated by high-frequency eddy currents and hysteresis currents on the surface layer. The primary current is carried by a water-cooled copper tube, the workpiece serves as the secondary circuit. The depth of penetration decreases as the frequency of the current increases, for example, the approximate minimum hardness depth for 3000 Hz is 0.060 in. and for 500,000 Hz is 0.020 in.

Induction hardening is fast, even on comparatively large surfaces. As an example, a large truck crankshaft can be brought to the proper temperature and spray-quenched in 5 sec. There is very little distortion due to the short cycle time.

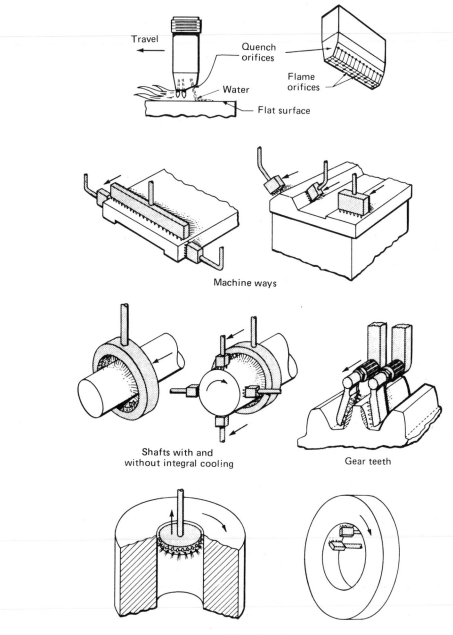

Travel

Quench
orifices

Flame
orifices

Water

Flat surface

Machine ways

Shafts with and
without integral cooling

Gear teeth

Internal surfaces

FIGURE 2-32. *Flame hardening with various-shaped burners.*

*Carburizing.* Carburizing, as the name implies, is a method of adding carbon to the surface of low-carbon steels. It may be done by pack carburizing, gas carburizing, or liquid carburizing (cyaniding).

*Pack carburizing* is done by placing the low-carbon steel in a heat-resistant metal box and surrounding it completely with a carburizing compound. The container is then heated to the austenitic temperature. The length of time is dependent on the depth of hardness desired. Three hours at 1650°F (899°C) will produce a case depth of about 0.060 in. Since this process is rather slow and dirty, it has been largely replaced by gas and liquid carburizing. After carburizing, the parts are quenched and tempered just as high-carbon steels.

*Gas carburizing* is done by placing the low-carbon steel in a heated retort in which carburizing gas (propane, natural gas or methane) is admitted. Continuous-type furnaces are available in which the parts are placed on a conveyor and are carburized, quenched, and tempered in a set sequential process.

*Liquid carburizing* is done by immersing the parts in a molten potassium cyanide salt bath which is kept at a temperature of 1550–1570°F (843–854°C). The process is often referred to as *cyaniding*. A case depth of about 0.025 in. can be obtained in 2 hours. Both carbon and nitrogen are added to the surface of the steel, producing a harder case than by gas carburizing. The parts are usually quenched directly from the cyanide bath and then tempered to the desired toughness.

*Nitriding.* Nitriding is a patented commercial process that is similar to gas carburizing. It has the advantage of requiring a relatively low temperature, 950°F (510°C). No scaling occurs, so the parts can be finished before hardening. The case formed is very thin, ranging from about 0.001 to 0.005 in. thick, but is very hard. Carburizing may range from $R_c$ 65 to 67, whereas nitriding may be in excess of $R_c$ 72. Since the case is very thin, it is measured on a Rockwell superficial hardness tester using either the 15N or 30N scale. The reading may be changed from the N scale to the C scale by the use of the hardness conversion table.

A newer method of nitriding has been developed, known as *glow-discharge* or *ion-nitriding*. Instead of placing the steel workpieces in an externally heated furnace, the pieces become the negative electrode of a low-pressure glow discharge in a mixture of nitrogen and hydrogen gases. Under the action of an applied voltage, positive ions bombard the surface of the steel, accelerating the nitrogen to form hard-alloy nitrides and delivering sufficient energy to heat the steel without requiring any external heating elements.

In addition to lower operating costs per unit of time, the time required for the process is shorter. Conventional nitriding requires about 30 hours in the furnace for a 0.010-in. case depth. A depth of 0.055 in. can be achieved in the same time by ion-nitriding. In addition, selective hardening is simplified. Conventionally, surfaces that are to be kept soft must be copper-plated and later ground. In ion-nitriding, masking is accomplished by placing a sheet of mild steel in front of the area that is to be masked.

TABLE 2-2. *Summary of Surface-Hardening Methods and Applications*

| Case Depth (in.) | Hardness, Rockwell C* | Remarks and Applications |
|---|---|---|
| **Carburizing** | | |
| To 0.020 | S 60+ C 18+ | Light case depths used for high wear resistance and low loads. Typical applications are push-rod balls and sockets, shifter forks, small gears, and water-pump shafts. |
| 0.020–0.040 | S 60+ C 18+ | Moderate case depths used for high wear resistance and moderate to heavy service loads. Applications include steering-arm bushings, valve rocker arms and shafts, gears, and brake-pedal shafts. |
| 0.040–0.060 | S 60+ C 18+ | Heavy case depths used for high wear resistance to sliding, rolling, or abrasive action, and for high resistance to crushing or bending loads. Applications include ring gears, transmission and slide gears, piston pins, gear shafts, roller bearings, and kingpins. |
| 0.060+ | S 60+ C 18+ | Extra heavy case for maximum wear and shock resistance. Typical applications are camshafts, armor plate, and cam surfaces. |
| **Carbonitriding** | | |
| 0.003–0.020 | S 62–65 C 32–35 | Produces a hard, wear-resistant, clean surface. Used on thin-wall tubing, ratchet wrenches, bolts, screws, small gears, and pneumatic cylinders. |
| **Nitriding** | | |
| To 0.030 | ... | Cycle of 50 hr produces 0.015-in. case, where initial 0.006 in. has Vickers hardness over 900. Cycle of 100 hr produces 0.030-in. case where initial 0.011 in. has 900+ Vickers hardness. Applications include aircraft exhaust valves, instrument shafts, pump shafts, and steam valves. |
| **Flame Hardening** | | |
| 0.030–0.125 | S 37–55 C 20+ | Produces high surface hardness with unaffected core. Surface is relatively free from scaling and pitting. Used on sprocket and gear teeth, track rails, and lathe beds and centers. |
| **Induction Hardening** | | |
| 0.030–0.125 | S 60+ | Produces high surface hardness with ductile core. Parts have good fatigue resistance. Applications include camshafts, sprocket and gear teeth, rocker-arm shafts, mower and shear blades, lathe beds, and bearing surfaces of axle shafts and crankshafts. |

* S = surface; C = core.

*Evaluation of Surface-Hardening Treatments.* If the prime consideration of surface hardening is wear resistance, then it is best to choose processes such as nitriding and cyaniding, where sufficient carbides are developed at the surface. If impact or torsion is involved, excess carbides will be detrimental, since cracks, chipping, and spalling will develop. As the case depth increases, the ability to withstand this type of loading also increases. Surface hardening procedures and applications are summarized in Table 2-2.

## FABRICATING CHARACTERISTICS OF FERROUS METALS

Fabricating characteristics of a metal refer to the ability or ease with which it can be formed, machined, cast, or welded.

Before considering the fabricating characteristics of ferrous metals, it is important to know how they are classified, so that references can be made to the qualities of a specific metal.

The common ferrous alloys that will be discussed are: plain carbon steels, low-alloy steels, tool steels, steel castings, cast iron (gray, malleable, and ductile), and stainless steel.

### Classification of Alloy Steels

The first coded numbering system for alloy steels was made by the Society of Automotive Engineers (SAE). It consists of a four- or five-digit system: the first two digits refer to the type and percent of alloy and the last two (or three) refer to the percent or points carbon. The American Iron and Steel Institute (AISI) developed a code similar to that of SAE, with some prefixes and suffixes as shown in Table 2-3. Identification of steels in this code system is relatively simple; for example, AISI 4340 steel is of a nickel–chromium–molybdenum type with 0.40 % carbon.

In addition, steels are often referred to as hot-rolled steels (HRS) or cold-rolled steels (CRS). Hot-rolled steels are easily identified because of an oxide scale that is formed during the heating process. Cold-rolled steels have the scale removed by acid baths or other methods and are then rolled to final size. The CRS are generally easier to machine, especially in the low-carbon range, and the dimensions are held within closer tolerances.

In addition to the standard code given, other organizations that have issued metal specifications are: American Society for Testing Materials (ASTM), Aerospace Materials Specifications (AMS), American Society of Mechanical Engineers (ASME), Index of Federal Specifications and Standards, and Department of Defense. An excellent Information Guide on Metals is given in the May 1970 issue of *Materials Research and Standards*. The one most comprehensive publication is the *Annual Book of ASTM—Standards*. It contains over 1200 standards relating to metals and metal products published in 10 of the 33 parts of the book. These standards are widely used by engineers and, for purchasing specifications, by the major metal industries.

TABLE 2-3. *AISI (American Iron and Steel Institute) Standard Steels*

| AISI Series | Nominal Composition or Range |
|---|---|
| **Carbon Steels** | |
| 10XX Series | Non-resulphurized carbon steels with 44 compositions ranging from 1008 to 1095. Manganese ranges from 0.30 to 1.65 %; if specified, silicon is 0.10 max. to 0.30 max., each depending on grade. Phosphorus is 0.040 max., sulphur is 0.050 max. |
| 11XX Series | Resulphurized carbon steels with 15 standard compositions. Sulphur may range up to 0.33 %, depending on grade. |
| B11XX Series | Acid Bessemer resulphurized carbon steels with 3 compositions. Phosphorus generally is higher than 11XX series. |
| 12XX Series | Rephosphorized and resulphurized carbon steels with 5 standard compositions. Phosphorus may range up to 0.12 % and sulphur up to 0.35 %, depending on grade. |
| **Alloy Steels** | |
| 13XX | Manganese, 1.75 %. Four compositions from 1330 to 1345. |
| 40XX | Molybdenum, 0.20 or 0.25 %. Seven compositions from 4012 to 4047. |
| 41XX | Chromium, to 0.95 %, molybdenum to 0.30 %. Nine compositions from 4118 to 4161. |
| 43XX | Nickel, 1.83 %, chromium to 0.80 %, molybdenum, 0.25 %. Three from 4320 to E 4340. |
| 44XX | Molybdenum, 0.53 %. One composition 4419. |
| 46XX | Nickel to 1.83 %, molybdenum to 0.25 %. Four compositions from 4615 to 4626. |
| 47XX | Nickel, 1.05 %, chromium, 0.45 %, molybdenum to 0.35 %. Two compositions, 4718 and 4720. |
| 48XX | Nickel, 3.50 %, molybdenum, 0.25 %. Three compositions from 4815 to 4820. |
| 50XX | Chromium, 0.40 %. One composition, 5015. |
| 51XX | Chromium to 1.00 %. Ten compositions from 5120 to 5160. |
| 5XXXX | Carbon, 1.04 %, chromium to 1.45 %. Two compositions, 51100 and 52100. |
| 61XX | Chromium to 0.95 %, vanadium to 0.15 % min. Two compositions, 6118 and 6150. |
| 86XX | Nickel, 0.55 %, chromium, 0.50 %, molybdenum, 0.20 %. Twelve compositions from 8615 to 8655. |
| 87XX | Nickel, 0.55 %, chromium, 0.50 %, molybdenum, 0.25 %. Two compositions, 8720 and 8740. |
| 88XX | Nickel, 0.55 %, chromium, 0.50 %, molybdenum, 0.35 %. One composition 8822. |
| 92XX | Silicon, 2.00 %. Two compositions, 9255 and 9260. |
| 50BXX | Chromium to 0.50 %, also containing boron. Four compositions from 50B44 to 50B60. |
| 51BXX | Chromium to 0.80 %, also containing boron. One composition, 51B60. |
| 81BXX | Nickel, 0.30 %, chromium, 0.45 %, molybdenum, 0.12 %, also containing boron. One composition, 81B45. |
| 94BXX | Nickel, 0.45 %, chromium, 0.40 %, molybdenum, 0.12 %, also containing boron. Two compositions, 94B17 and 94B30. |

Additional notes: When a carbon or alloy steel also contains the letter "L" in the code, it contains from 0.15 to 0.35 % lead as a free-machining additive, i.e., 12L14 or 41L40. The prefix "E" before an alloy steel, such as E4340, indicates the steel is made only by electric furnace. The suffix "H" indicates an alloy steel made to more restrictive chemical composition than that of standard steels and produced to a measured and known hardenability requirement (i.e. 8630 H or 94B30H). XXs indicate nominal carbon content within range.

## Fabricating Characteristics

The fabricating characteristics of each of the metals that follow will be discussed, where applicable, from the standpoint of machinability, formability, weldability, and castability.

### *Machinability*

Machinability is an involved term with many ramifications. Simply stated, however, it refers to the ease with which the metal may be sheared in such operations as turning, drilling, reaming, threading, sawing, etc.

Ease of metal removal implies, among other things, that the forces acting against the cutting tool will be relatively low, that the chips will be easily broken up, that a good finish will result, and that the tool will last a reasonable period of time before it has to be replaced or resharpened. Another way of expressing this is to give each material a *machinability rating*. This has been done for most ferrous metals, using AISI B1112 as the basis of 100 % machinability. Thus another metal may be said to have a machinability rating of 60 %, as in the case of one type of stainless steel. Factors that increase or decrease machinability of a metal are shown in Table 2-4.

### *Formability*

The ability of a metal to be formed is based on the ductility of the metal, which, in turn, is based on its crystal structure. Metals that have a face-centered cubic

TABLE 2-4. *Factors Affecting Machinability of Metals*

|  | Factors that increase machinability | Factors that decrease machinability |
|---|---|---|
| Structure | Uniform microstructure<br>Small, undistorted grains<br>Spheroidal structure in high-carbon steels<br>Lamellar structure in low- and medium-carbon steels | Nonuniformity<br>Presence of abrasive inclusion<br>Large, distorted grains<br><br>Spheroidal low- and medium-carbon steels<br>Lamellar high-carbon steels |
| Treatment | Hot working of alloys that are hard, such as medium- and high-carbon steels<br>Cold working of low-carbon steels<br>Annealing, normalizing, tempering | Hot working of low-carbon steels<br><br>Cold working of higher-carbon steels<br>Quenching |
| Composition | Small amounts of lead, manganese, sulphur, phosphorus<br>Absence of abrasive inclusions such as $Al_2O_2$ | Carbon content below 0.30 % or above 0.60 %<br>High alloy content in steels |

crystal have the greatest opportunity for slip—four distinct nonparallel planes and three directions of slip in each plane.

Other factors that govern, to a large extent, the flowability or ductility of the material are: grain size, hot and cold working, alloying elements, and softening heat treatments such as annealing and normalizing.

*Grain Size.* If all metals consisted of a single grain or crystal, the tendency to slip in any pure metal would depend solely upon the number of slip planes in the crystal and upon the directions of ready slip in each plane. However, every metal contains many separate grains, and the planes of slip or direction of slip rarely coincide with each other. The tendency to slip in any single grain, therefore, is obstructed, to a certain extent, by the resistance of opposing slip planes in adjacent grains. As will be seen later, small grain sizes are recommended for shallow drawing of copper, and relatively large grains for heavy drawing on the thicker gages.

*Hot and Cold Working.* The tremendous pressures encountered in hot working tend to reduce the size of the crystals, either by preventing the growth of the crystals at elevated temperatures or by breaking up the existing crystals. Generally, the grains are distorted in the process. The amount of distortion will be a determining factor in the ductility of the metal.

Cold working also results in varying degrees of distorted crystals. Generally, cold-worked crystals are more distorted than are the hot-worked, and therefore cold-worked metals are usually less ductile than the hot-worked.

*Alloying Elements.* Most alloying elements in a pure metal reduce its ductility. Whether the alloying element has replaced atoms of pure metal or has found room for itself in the spaces between the atoms of pure metal, the effect is to reduce the number of slip planes in which ready slippage can occur. For example, steel, which is an alloy of carbon and iron, is less ductile than iron. As steel solidifies, it can hold only an extremely small amount of carbon in solution. The excess carbon which is forced out of the individual cells of the atomic structure combines immediately with some iron to form iron carbides. Not only are the slip planes somewhat reduced by the presence of the alloy, but, as in the case of steel, the iron carbides offer increased resistance to slip. Therefore, the ductility of the steel decreases as the amount of carbon increases.

*Heat Treatment.* Heating cold-worked metal to the recrystallization temperature will restore most of the ductility. The distorted crystals are reformed so that slip can occur more easily.

*Quantitative Formability Tests.* Just how formability of a metal may be judged quantitatively is discussed in Chapter 4.

### Weldability

It may be said that all metals are weldable by one process or another. However, the real criterion for judging the weldability of a metal is *weld quality and the ease with which it can be obtained.*

The characteristics commonly considered in weldability are the heating and cooling effects on the metal, oxidation, and gas vaporization and solubility.

*Heat and Cooling.* The effect of heat in determining the weldability of a material is related to the change in microstructure that results. For example, steels are sometimes considered weldable or not weldable on the basis of the hardness of the weld. The deposited weld metal may pick up carbon or other alloys and impurities from the parent metal, making the weld hard and brittle so that cracks result upon cooling.

The opposite effect may also be considered. A metal may have a certain hardness temper that will be changed by the heat of the weld. Although both of these conditions can be corrected by added precautions and heat treatment, they add to the cost and hinder the simplicity of the weld.

Hot shortness, a characteristic which is indicated by lack of strength at high temperature, may result in weld failures during cooling of certain metals.

*Oxidation.* Oxidation of the base metal, particularly at elevated temperatures, is an important factor in rating the weldability of a metal. Metals that oxidize rapidly, such as aluminum, interfere with the welding process. The oxide has a higher melting point than the base metal, thus preventing the metal from flowing. It also may become entrapped in the weld metal, resulting in porosity, reduced strength, and brittleness. Again, this condition can be overcome by a choice of the proper welding process, but it is a consideration that must be recognized.

*Gas.* Large volumes of troublesome gas may form in the welding of some metals. These gases may become trapped in the weld because certain elements vaporize at temperatures below those needed for welding. Not only will this cause porosity, but some of the beneficial effects of these elements are lost.

### Castability

The castability of a metal is judged to a large extent on the following factors: solidification rate, shrinkage, segregation, gas porosity, and hot strength.

*Solidification Rate.* The ease at which a metal will continue to flow after it has been poured in the mold depends on its analysis and pouring temperature. Some metals, such as gray iron, are very fluid and can be poured into thin sections of complex castings.

*Shrinkage.* Shrinkage refers to the reduction in volume of a metal when it goes from a molten to a solid state. For steel, the amount of contraction amounts to about 6.9 to 7.4 % by volume, or 0.25 in. per ft; gray iron contracts half as much. This shrinkage factor has to be taken into account by the pattern maker and designer, not only to allow for the proper finished size, but also to see that undue strains will not be encountered during shrinkage due to the mold design. Various elements can be added to the alloy to control fluidity and shrinkage, as discussed later in this chapter.

*Segregation.* As ferrous metals start to solidify, tiny crystal structures resembling pine trees and referred to as dendrites start to form at the mold edges. As they form they tend to exclude alloying elements. Subsequent crystals that form are progressively richer in alloy content as the metal solidifies. Thus the surface of the casting is not the same quality as the center. This is overcome in part at least by subsequent heat treatment, or very slow cooling.

*Gas Porosity.* Some metals in the molten state have a high affinity for oxygen and nitrogen. These gases become trapped as the metal solidifies, creating voids or pinholes.

*Low Hot Strength.* Metals are very low in strength right after solidification. This is especially true of the nonferrous metals. Precautions must be taken at the time of casting to avoid stress concentrations that cause flaws and hot tears to develop as the metal solidifies.

## PLAIN CARBON STEELS

Plain carbon steels are steels whose principal alloy is carbon with no other special alloying elements.

Plain carbon steels are allowed to solidify under varying conditions which alter their characteristics. Under this classification the terms *killed*, *semikilled*, and *rimmed* are often used.

*Killed.* Killed steels are those that have been chemically deoxidized so that they lie quietly in the mold as they cool. They are characterized by relatively uniform chemical composition and properties. Strips and sheets made of killed steel have excellent forming and drawing qualities.

*Semikilled.* These steels are partly deoxidized before pouring into the mold. Thus, some gas evolution takes place during solidification. These steels are satisfactory for all but the most severe drawing operations.

*Rimmed.* Rimmed steels get their name from the fact that the outer rim of the ingot has considerable gas evolution during the initial solidification period from the liquid state. During solidification, a layer of iron about three inches thick forms around the outer edge of the ingot. The liquid in the center retains the carbon, phosphorous, and sulfur. This rimmed steel is often known as drawing quality (DQ) steel. However, where draws or forming are more severe, killed steel may be required.

*Temper.* When cold-rolled sheets are specified to a hardness range, they are designated by tempers as follows: quarter-hard ($R_b$ 60 to 70 range), half-hard ($R_b$ 70 to 80 range), and full-hard ($R_b$ 84).

*Machinability.* The machinability of plain carbon steels may be approximated by the curve shown in Fig. 2-33. Very low carbon steels do not machine well because the structure is mostly ferrite, which is too soft to produce a good shearing action. These steels must have manganese sulphides to help break up the ferrite structure, as shown in Fig. 2-34. Lead will also help break up the ferrite structure.

A number of grades of lead-bearing steels are available and are classified as free-machining steels. A small percentage of lead, 0.15 to 0.35 %, is added to the steel just shortly before it solidifies. It is immediately dispersed throughout the metal. Leaded steels can be used at 50 % higher cutting speeds than corresponding plain carbon steels. However they are more expensive and should not be used unless production conditions warrant the extra cost.

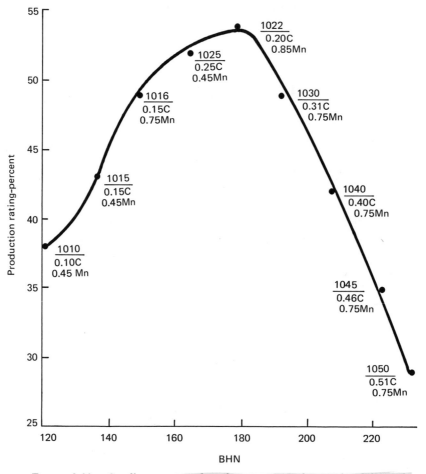

FIGURE 2-33. *Small amounts of carbon improve the machinability of plain carbon steels up to a point. Beyond this point, carbon additives increase hardness but are detrimental to machinability.* Courtesy American Machinist.

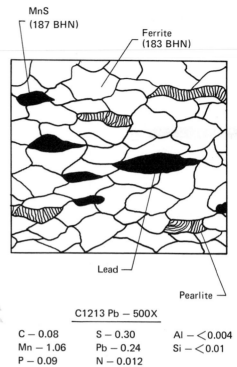

FIGURE 2-34. *Lead and sulphur additions form inclusions in the low-carbon steel microstructure that promote machinability. The sulphur combines with manganese and the lead appears either by itself or as a tail in MnS inclusions.*

C1213 Pb — 500X

| C — 0.08 | S — 0.30 | Al — <0.004 |
| Mn — 1.06 | Pb — 0.24 | Si — <0.01 |
| P — 0.09 | N — 0.012 | |

*Formability.*  Low-carbon steels have good forming qualities because there are less carbon and alloys to interfere with the slip planes.  Steels in this class can be given a class-2 bend, which is a 90° bend with a radius of $t$, or the thickness of the metal.  Classes of bends have been defined by the American Society for the Testing of Metals (ASTM).

Medium-carbon cold-rolled steels are too low in ductility for any practical degree of cold forming.  Hot-rolled, medium-carbon steels are somewhat more ductile and can be bent with a 1-$t$ radius ($t$ = material thickness) up to 0.009 in. thick.

*Weldability.*  Plain carbon steel is the most weldable of all metals.  It is only as the carbon percentages increase that there is a tendency for the metal to harden and crack.  Fortunately, 90 % of the welding is done on low-carbon (0.15 %) steels.  This amount of carbon presents no particular difficulties in welding.  Near the upper end of the low-carbon range (0.24 to 0.30 %), there may be some formation of martensite when extremely rapid cooling is used.

Medium- and high-carbon steels will harden when welded if allowed to cool at speeds in excess of the critical cooling rates.  Preheating to 500 or 600°F (260–316°C) and postheating between 100 and 1200°F (38–649°C) will remove any of the brittle microstructures.

**65**

The extra-high-carbon steels, or tool steels, having a carbon range of 1.00 to 1.70 %, are not recommended for high-temperature welding applications. These metals are usually joined by brazing with a low-temperature silver alloy. Because of the lower temperature of this process, it is possible to repair or fabricate tool-steel parts without affecting their heat-treated condition.

## ALLOY STEELS

### Classification

In addition to the SAE–AISI code classification discussed, alloy steels may be classified as through-hardenable and surface-hardenable. Each type includes a broad family of steels whose chemical, physical, and mechanical properties make them suitable for specific product applications.

Through-hardening types are used when maximum hardness and strength must extend deep into the part.

Surface-hardening grades are used where a tough core and relatively shallow hardness are needed. These steels, after nitriding or carburizing, are used in truck transmission gears and steering worm gears—parts that must withstand wear as well as high stresses. Shown in Table 2-5 is a classification of the various types and uses of through-hardenable and surface-hardenable steels.

### Fabricating Characteristics

*Machinability.* One reason for the widespread use of the alloy steels in the lower carbon range, such as 4024 and 4028, is that good machinability is obtained. The machinability drops off as the carbon content increases. For example, the machinability rating of 4023 is 70 % (based on B1112 as 100 %), 4032 is 65 %, and 4047 is 55 %. This holds true for the other alloy steels also. Shown in Table 2-6 is the machinability rating for carbon and alloy steels.

*Formability.* Alloy steels are not usually used where forming operations other than forging are required. Here they are employed extensively for gears, bearings, crankshafts, connecting rods, axle shafts, and any other uses where good strength and toughness are necessary. These steels are often induction-hardened to provide a surface that can take high compressive loads and still have a core with great toughness.

*Weldability.* Weldability for alloy steels is generally good as long as the carbon content is in the low range. As the carbon content increases, preheating and postheating are often used to reduce the stress. Heat treatment after welding will help produce a uniform structure in the weld and parent metal. For most of these steels, the best results are obtained by arc welding using a low-hydrogen electrode. Reducing the hydrogen content of the weld helps to eliminate brittleness. A general guide to weldability is shown in Table 2-7.

TABLE 2-5. *Classification and Use of Plain-carbon, Through, and Surface-hardening Steels*

| Type | Carbon (%) | Hardness Range | General Characteristics and Applications |
|---|---|---|---|
| **Through Hardenable** | | | |
| Medium-carbon | 0.3 to 0.5 | 250 to 400 BHN | Strength and toughness; for shafts, miscellaneous forgings, some gears (frequently cyanided), bolts, nuts, flanges, bearing cages. |
| High-carbon | 0.5 to 0.7 | 375 to 500 BHN | Strength and moderate wear resistance; for springs and collets. |
| Bearing | 1.0 | 60 to 64 $R_c$ | High strength and high resistance to wear and scuffing; for bearings, balls, rollers, spacers, pins, and bushings. |
| **Surface Hardenable** | | | |
| Carburized | 0.15 to 0.3 | Case, over 60 $R_c$ Core, to 45 $R_c$ (case, 0.15 to 0.125 in. thick) | Wear-resistant case, high endurance strength; used for gears, shafts, bearings, and inspection fixtures. |
| Flame- or induction-hardened | Over 0.4 | Case, over 55 $R_c$ Core, to 45 $R_c$ (case over 0.050 in thick) | Wear-resistant case; used for shafts, pins, some gears, cages, and bushings. |
| Nitrided | 0.3 to 0.5 and Nitralloys | Case, 500 to 1000 DPH Core, over 30 $R_c$ (case, to 0.030 in. thick) | Wear-resistant case, high endurance strength, some resistance to corrosion and elevated temperatures; used for shafts, gears, couplings, and bushings. |

TABLE 2-6.  *Steel Machinability (B1112 = 100)*

**Class 1**

| | | | | | |
|---|---|---|---|---|---|
| C-1109 | 85 | C-1115 | 85 | C-1117 | 85 |
| C-1118 | 80 | C-1120 | 80 | C-1132 | 75 |
| C-1137 | 70 | C-1022 | 70 | C-1016 | 70 |
| C-1018 | 70 | B-1111 | 95 | B-1112 | 100 |
| B-1113 | 135 | C-1213 | 130 | A-4023 | 70 |
| A-4027 | 70 | A-8620* | 90 | A-4140* | 75 |

**Class 2**

| | | | | | |
|---|---|---|---|---|---|
| C-1141 | 65 | C-1020 | 65 | C-1030 | 65 |
| C-1035 | 65 | C-1040 | 60 | C-1045 | 60 |
| A-2317 | 55 | A-3115 | 60 | A-3130 | 55 |
| A-3140 | 55 | A-3145 | 50 | A-4032 | 65 |
| A-4037 | 65 | A-4042 | 60 | A-4047 | 55 |
| A-4130 | 65 | A-4140 | 60 | A-4145 | 55 |
| A-4150 | 50 | A-4320 | 55 | A-4615 | 65 |
| A-4640 | 55 | A-4815 | 50 | A-5120 | 65 |
| A-5170 | 60 | A-5150 | 55 | NE-80B40 | 60 |
| NE-8140 | 60 | NE-81B45 | 55 | NE-8620 | 65 |
| NE-8640 | 60 | NE-8740 | 60 | NE-8745 | 55 |
| | | NE-9440 | 60 | | |

**Class 3**

| | | | | | |
|---|---|---|---|---|---|
| C-1008 | 50 | C-1010 | 50 | C-1015 | 50 |
| C-1050 | 50 | C-1095 | 45 | C-1320 | 50 |
| C-1330 | 45 | C-1340 | 45 | A-2330 | 50 |
| A-2340 | 45 | A-4340 | 45 | A-6120 | 50 |
| A-6145 | 45 | A-6150 | 45 | A-9255 | 45 |
| | | E-52100 | 40 | | |

* Leaded.

A — Basic open hearth alloy steel

B — Acid Bessemer carbon steel

C — Basic open hearth carbon steel

E — Electrical furnace steel

*Note: B before the carbon content denotes boron.*

TABLE 2-7.  *General Guide to Weldability*

| Steel Composition | General Weldability | Preheating | Post Heating |
|---|---|---|---|
| Carbon steel, below 0.30 % C, and Low-alloy steel, below 0.15 % C | Readily welded | None | None |
| Carbon steel, 0.30 % to 0.50 % C, and Low-alloy steel, 0.15 % to 0.30 % C | Weldable with care | Preferable | Preferable |
| Carbon steel, above 0.50 % C, Low-alloy steel, carbon above 0.30 %, Alloy steel, total alloy above 3 % | Difficult to weld | Necessary | Necessary |

*Castability.* An important characteristic of alloy steels is their ability to air-harden. Thus, complicated castings can be hardened when using these alloys, and tensile strengths from 70,000 to 100,000 psi can be obtained without quenching. Nickel and molybdenum with manganese increase the capacity to air-harden.

Combinations of chromium, nickel, manganese, and vanadium are used to produce wear resistance and high strength.

## TOOL STEELS

Tool steels are so named because they have the properties needed in making tools such as dies for cutting and forming metal, for jigs and fixtures, and for precision gages and molds. In general, they have a higher alloy content and thus have high wear resistance, stability at high temperatures, and toughness.

There are about 75 types of tool steels in the AISI list, which is divided into seven main types. Tool steels are identified by letter and number. The AISI classification is shown in Table 2-8. The letters are used to indicate the type of quench required for hardening or the main characteristic of the steel: O, for oil-hardening; A, for air-hardening; and S, for shock-resisting.

*High-Speed Steels.* High-speed steels are used extensively for metal cutting tools such as lathe tools, milling cutters, drills, reamers, taps, etc. These steels have excellent heat resistance, wear resistance, and compressive strength. They may be used in metal cutting operations up to a temperature of 1100°F (593°C) without softening below $R_c$ 60.

TABLE 2-8. *AISI Tool Steels*

| |
|---|
| High-speed |
|     Tungsten types: T1, T2, - - - |
|     Molybdenum types: M1, M2, - - - |
| Hot-work |
|     Chromium types: H10, H11, - - - |
|     Tungsten types: H21, H22, - - - |
|     Molybdenum types: H41, H42, H43 |
| Cold-work |
|     High carbon |
|       High-chromium types: D2, D3, - - - |
|     Medium alloy |
|       Air-hardening types: A2, A3, A4, - - - |
| Oil-hardening types: O1, O2, O6, - - - |
| Shock-resisting types: S1, S2, S5, - - - |
| Mold steels: P2, P3, P4 |
| Water-hardening: W1, W2, W5 |
| Special purpose |
|     Low-alloy types: L2, L3, L6 |
|     Carbon–tungsten types: F1, F2 |

Tool steels have a wide variety of applications other than for tools. For example, some applications of high-speed steels are for high-temperature bearings, springs, fasteners for aircraft, precision hydraulic pumps, and blades for commercial food-chopping equipment.

***Hot-Work Steels (H).*** Hot-work steels are those that are used for hot-blanking dies, hot-extrusion dies, hot-heading dies, hot-punching dies, and forging and die-casting dies.

Three main types of hot-work steels are available. Two of these are very similar to the high-speed steels described previously—the tungsten and molybdenum types. Both of these have high heat resistance and do not soften until the temperature reaches approximately 1100 to 1200°F (593–649°C).

The third steel in this class is a chrome–molybdenum type. It has 5 % chromium, 1.3 % molybdenum, and 0.30 % carbon. Although this steel does not withstand the high temperatures of the two previously mentioned, it has a wide range of applications.

***Cold-Work Steels [High Carbon–High Chromium] (D).*** These steels are second only to the high-speed steels in importance. They are used for all types of die work, including blanking, coining, drawing, forming, and thread rolling, as well as for reamers, gages, and rolls for beading and forming sheet metal. These steels are substantially lower in cost than the high-speed steels.

***Medium-Alloy Air-Hardening Steels (A).*** Some of the outstanding features of this steel are:

1. It can be air hardened in large sections.
2. It is safe to harden in intricate sections.
3. There is little distortion in hardening.
4. It is easier to machine than the high carbon–high chromium types.
5. It falls in the intermediate price range.

The air-hardening steels are particularly good for the larger sizes of tools and dies where the oil-hardening nondeforming steels will not harden, or where high carbon–high chromium steels are not necessary for the performance required.

***Oil-Hardening Nondeforming Die Steels (O).*** This group of steels has a substantially reduced alloy content compared to the previously mentioned types. Their characteristics include the following:

1. They have a wide range of uses for all types of medium-life tools and dies.
2. They are easy to machine and harden.
3. They harden uniformly where the mass of the section does not exceed the equivalent of 2.5-in. diameter by 6-in. length.
4. They are moderate in cost.

These steels are generally useful where the die sections are of limited mass, and long runs are not required. The oil quench provides a slower cooling action than does water, and therefore there is less strain on the workpiece.

***Shock-Resisting Steels (S).*** Shock-resisting steels are not only capable of taking a great deal of shock but also have high fatigue resistance and good wearing qualities. The most common type is probably S1 with 0.60 % carbon and tungsten, chromium, or vanadium. Where shock is severe but heating is intermittent or nominal, a 0.60 % carbon with 2 % or more tungsten, 1.2 % chromium, and 0.20 % vanadium is often used.

These steels are generally hardened to 58 to 60 $R_c$ by oil quenching. They give extended performance where higher-carbon types of die steels fail.

***Water-Hardening Steels (W).*** Water-hardening steels are often referred to as straight carbon. They have some manganese and silicon, but this is not in excess of 0.30 %. The carbon range is from 0.60 to 1.30 %; however, 80 % of all carbon tool steels lie within a carbon range of 0.90 to 1.10 %.

These steels are particularly useful in a large number of cold-punching and stamping operations where shock is severe. The steel hardens to about a $\frac{1}{8}$-in. depth, which provides a hard outer surface with a tough, shock-dampening core. Many die sections are hardened and tempered to 60 or 62 $R_c$.

Carbon tool steels are low in cost and easy to machine and harden. However, specifying the proper amount of carbon to harden just deep enough to give the desired depth and strength of case, and yet to retain a soft, tough, shock-dampening core, will require considerable experience or advice from the steel supplier.

## Tool-Steel Selection

As a means of summarizing some of the important facts about tool steels, the following points are given as a basis of selection.

1. Always start with the plain carbon steels, since they are the cheapest and easiest to machine. For example, if W8 does not satisfy the requirement in that steel, the hardness may not be deep enough. Then go to W9V, which is not only higher in carbon content but has vanadium for deeper hardness penetration and finer grain structure. This may still be short of some of the requirements needed, so it will be necessary to proceed to the oil- or air-hardening steels.
2. The oil- or air-hardening steels were developed for maximum safety in hardening and minimum dimensional change after heat treatment. They are preferred for dies with adjacent thin and thick sections, sharp corners, or numerous holes. Wear resistance is better than in the W steels, but toughness is not as good. They are generally not suited for elevated-temperature use. The oil-hardening grades are relatively low in cost, whereas the air-hardening ones are rather expensive.
3. The hot-work steels must combine red hardness, wear resistance, and shock resistance. They have relatively lower carbon and smaller quantities of alloying elements than do the high-speed steels.

4. The shock-resisting steels are used to withstand cold battering operations where abrasion resistance is of secondary importance.

5. The high-speed steels were developed primarily to provide red hardness and high abrasion resistance along with some shock resistance.

## CAST STEEL

Steel, though often associated only with the wrought form, as in plates, rods, bars, and tubes, is also extensively used in the cast form. Just as with other metals, casting steel provides a way of obtaining the desired product easily. The physical properties of the same steels, whether wrought or cast, are comparable. One advantage a cast steel has over the wrought is that the properties are the same in any direc-

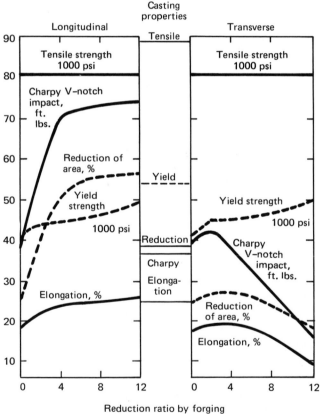

FIGURE 2-35. *The longitudinal and transverse properties of a 1035 wrought steel compared with the same steel in the cast condition (center section). Courtesy Steel Founders' Society of America.*

TABLE 2-9. *ASTM Steel Casting Specifications*

| ASTM Designation | Meaning |
|---|---|
| 27-55 | Medium-strength carbon-steel castings for general application |
| 148-55 | High-strength steel castings for structural purposes |
| 216-53T | Carbon-steel castings suitable for fusion welding for high-temperature service (tentative) |
| 217-55 | Alloy-steel castings for pressure-containing parts suitable for high-temperature service |
| 351-52T | Ferritic-steel castings for pressure-containing parts suitable for low-temperature service |
| 356-58 | Heavy-walled carbon and low-alloy-steel castings for steam turbines (tentative) |
| 389-59T | Alloy-steel castings specially heat-treated for pressure-containing parts suitable for high-temperature service |

tion, whether longitudinal or transverse. Shown in Fig. 2-35 is a comparison of the longitudinal and transverse properties of a 1035 steel. In between the two diagrams are the properties of 1035 cast steel.

## Classification

The specifications and standards predominantly used in the steel castings industry are those issued by ASTM, as shown in Table 2-9. The designation number may be followed by a second number, such as 27-55. The 55 indicates the year of adoption or latest revision. Class numbers may be used in addition to the regular identification number. Class numbers indicate the tensile strength and yield point. Thus an 80-50 cast steel would be one with a tensile strength of 80 ksi (1 ksi = 1000 psi) and a yield point of 50 ksi.

## Fabricating Characteristics

*Machinability.* Most often, steel castings are delivered to the customer in the normalized condition, but this does not necessarily coincide with optimum machining qualities. Heat treatment to obtain the optimum microstructure for machining is not always realistic due to the time required and other considerations. In general, steel castings have the same machinability as comparable wrought metals. The "skin" or outside surface of a casting has an oxide scale which is detrimental to the machinability. This skin should be removed by pressure blasting prior to machining. When this is done, skin cutting and base-metal cutting are about the same, as shown in Fig. 2-36. If the skin is not removed, the cutting tool generally should be set to a depth of $\frac{1}{4}$ to $\frac{3}{8}$ in. for the first cut.

*Weldability.* Cast steels are judged weldable by the same processes used in wrought steels. These are manual arc welding with coated electrodes, submerged-arc welding, and gas-metal–arc (GMA) welding. The shielding gases used for GMA welding are

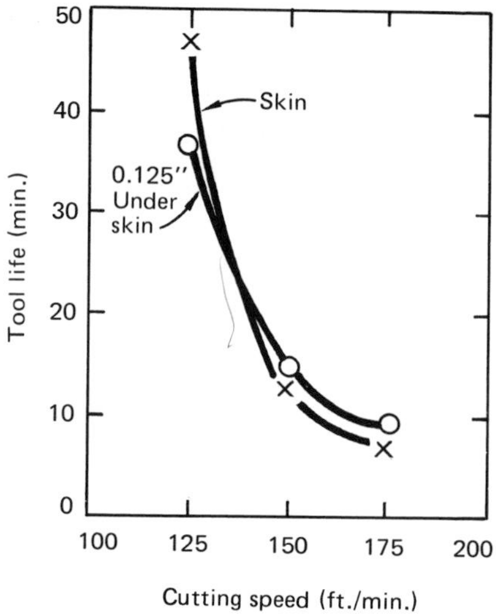

FIGURE 2-36. *Tool-life plots of high-speed, steel cutting tools machining 0.030 %-carbon cast steel. The skin has been cleaned by pressure blasting, therefore there is no marked difference in tool life whether the tool is used for a light skin cut or if it stays well below the surface. Courtesy Steel Founders' Society of America.*

argon, helium, or carbon dioxide. In addition, the electroslag welding process is well suited for welding vertical joints in heavy-sectioned castings. Although welding processes are discussed in detail in Chapter 6, the processes mentioned are shown in Fig. 2-37 to help differentiate them.

Standard welding procedures used on steel apply to steel castings. With a relatively few exceptions, all castings welded are subsequently stress-relieved, or given a full heat treatment (for example, normalized or quenched and tempered). This eliminates excessive hardness next to the weld as well as severe weld stresses. Another precaution to eliminate both hot and cold cracking of the weld along with permanent deformation of the part is the use of a suitable preheat temperature.

Preheating can be done by placing the casting in a furnace or by localized preheating with a torch. General preheat is preferred, since it minimizes localized stresses. The preheat temperature should be as low as is practical, since too much heat will make welding more difficult. Preheats of 200 to 400°F (93 to 204°C) for the medium-carbon steels are recommended. Preheats over 400°F are recommended for castings containing more carbon and other alloying elements.

Quenched and tempered castings are used when high strengths, resistance to abrasion, and sliding friction are needed. Heat-treated castings have not been involved in welding until recently. Shown in Fig. 2-38 is a heat-treated cast bull gear as is welded to a normalized car body, using the submerged-arc process. The first pass is put in by manual arc welding using the stick electrode. To evaluate the weld and the adjacent area, a Vickers microhardness test is made, as shown at (*b*) of the figure.

The wide range of metallurgical qualities of steel castings, with their related physical properties and extensive design flexibility, have made steel castings universally accepted for applications throughout industry. Shown in Fig. 2-39 are seven major steel castings used in constructing the chassis of an off-highway hauler with load capacities of 28 to 150 tons.

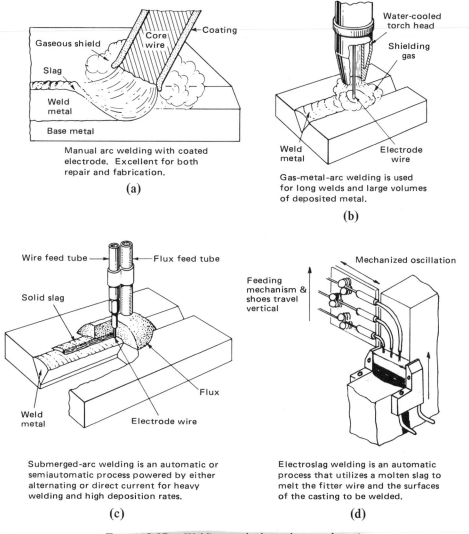

Manual arc welding with coated electrode. Excellent for both repair and fabrication.

(a)

Gas-metal-arc welding is used for long welds and large volumes of deposited metal.

(b)

Submerged-arc welding is an automatic or semiautomatic process powered by either alternating or direct current for heavy welding and high deposition rates.

(c)

Electroslag welding is an automatic process that utilizes a molten slag to melt the fitter wire and the surfaces of the casting to be welded.

(d)

FIGURE 2-37. *Welding methods used on steel castings.*

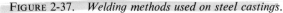

## CAST IRON

In order to understand the fabricating characteristics of cast iron, it is necessary to become familiar with the characteristics of the metal and the various types and classifications that are available.

One of the distinguishing features of all cast irons is that they have a relatively high carbon content. Steels range up to about 2 % carbon. Cast irons overlap with the steels somewhat and range from about 1.5 up to 5 % carbon. It is principally

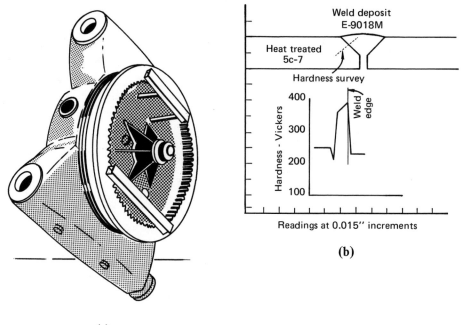

(a)

FIGURE 2-38. *A heat-treated, cast-steel bullgear as is welded to a normalized car body using the submerged-arc process (a). An evaluation of the weld procedure is made with the Vickers microhardness test as shown at (b). Courtesy Steel Founders' Society of America.*

the *form of the carbon*, which is governed by thermal conditions and alloying elements, that provides various structures that may be classified into the following main types:

Gray cast iron
White cast iron
Malleable iron
Ductile (nodular) iron

**Gray Cast Iron.** The terms *gray* and *white* cast iron refer to the appearance of the fractured area. The gray cast iron has a grayish appearance because of the large amount of flake graphite on the surface. Typical matrix structures are shown in Fig. 2-40. The dark sections show the graphite flakes. The pearlitic type may be made fine by faster cooling, or coarse by slow cooling. The size of the section will also determine the structure of the metal: the thinner the section the faster it cools. Thus a casting having large variations in section will also have large variations in hardness and strength unless special precautions are taken to ensure uniform cooling.

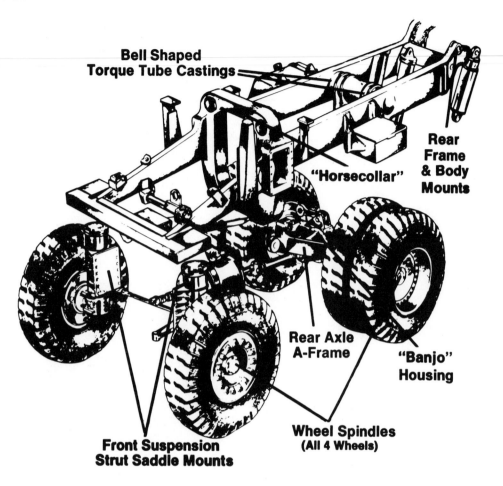

**Bell Shaped Torque Tube Castings**

**Rear Frame & Body Mounts**

**"Horsecollar"**

**Rear Axle A-Frame**

**"Banjo" Housing**

**Wheel Spindles** (All 4 Wheels)

**Front Suspension Strut Saddle Mounts**

FIGURE 2-39. *Seven steel castings are used in the construction of this chassis for an off-highway hauler made to handle load capacities ranging from 28 to 150 tons. Courtesy Steel Founders' Society of America.*

The basic composition of gray cast irons is often described in terms of carbon equivalent (CE). This factor gives the relationship of the percentage of carbon and silicon in the iron to its capacity to produce graphite. Thus

$$CE = C_t + \tfrac{1}{3}(\% \; Si + \% \; P)$$

*where:* $\quad C_t$ = total percentage carbon

*Example:* What is the CE of a gray iron containing 3.35 % carbon and 0.65 % silicon?

*Solution:*

$$CE = 3.35 + \tfrac{1}{3}(0.65)$$
$$= 3.57$$

**(a)**

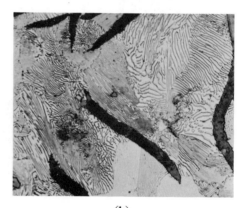

**(b)**

FIGURE 2-40. *The microstructure of three variations of gray iron. The matrix surrounding the graphite flakes is ferritic (a), pearlitic (b), and bainitic or acicular (c). The corresponding Brinell hardness is also shown. Etched, at a magnification of 500 X.*

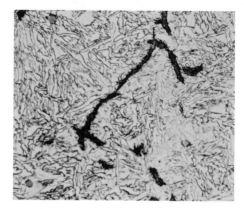

**(c)**

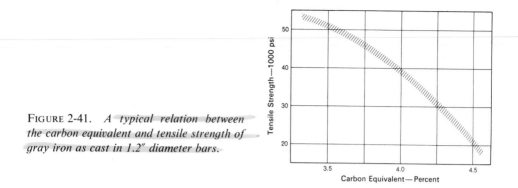

FIGURE 2-41. *A typical relation between the carbon equivalent and tensile strength of gray iron as cast in 1.2″ diameter bars.*

The CE value may then be related to the tensile strength of the metal, as shown in Fig. 2-41. Irons with a carbon equivalent of over 4.3 are called *hypereutectic* and are particularly good for thermal-shock resistance, such as for ingot molds. The higher strength gray irons, with less than 4.3 CE, are termed *hypoeutectic*.

The compressive strength of gray iron is one of its outstanding features. In general, it ranges from 3 to 5 times the tensile strength. As an example, a class-20 gray iron which has a tensile strength of 20,000 psi has a compressive strength of 83,000 psi.

**Gray-Iron Classification.** The most-used classification of gray irons is that of ASTM Specification A48-48. The last number indicates the year of original adoption as standard, or the year of last revision. This specification covers seven classes that are numbered in increments of 5, from 20 to 60. The class number refers to its tensile strength. For example, a class-40 cast iron refers to one having a minimum tensile strength of 40,000 psi. The carbon content becomes less as the strength requirement goes up. Alloys in increasing quantities are used to achieve strengths higher than 45,000 psi. Although high-strength cast irons have been developed, they are not widely used because they are still quite brittle.

Gray iron is the most widely used of all cast metals. Typical applications are engine blocks, pipes and fittings, agricultural implements, bathtubs, household appliances, electric motor housings, machine tools, etc.

**White Cast Iron.** White cast iron is produced by cooling the cast iron rapidly so that the carbon is in the combined form as cementite. It is often referred to as *chilled cast iron.* The structure is very hard and brittle and the fractured surface produces a shiny appearance. If the structure contains alloying elements such as nickel, molybdenum, or chromium, martensite will be formed. Irons of this type are extremely hard and wear-resistant, and are used on hammer mills, crusher jaws, crushing rollers, balls, wear plates. etc.

**Malleable Iron.** Malleable iron is made from white cast iron, which has the carbon in the combined form. The white cast iron is subjected to a prolonged annealing process. It may take two days to bring the castings to a temperature of about 1600°F

(871°C). The castings are held at this temperature for a period of 48 to 60 hours. They are then cooled at a rate of 8 to 10 degrees per hour until the temperature is about 1300°F. The temperature is held in the 1250 to 1300°F (677–704°C) range for a period of up to 24 hours. The furnace is then opened and allowed to cool to room temperature. During this time all the combined carbon has separated into graphite and ferrite. The graphite flake is shown in Fig. 2-42.

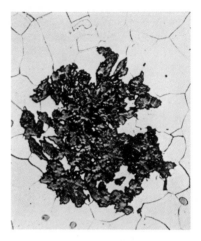

FIGURE 2-42. *Malleable iron showing the graphite flakes in a ferrite matrix.*

A pearlitic malleable iron can be made by allowing the cooling rate to be fast enough to retain the desired amount and type of combined carbon in the matrix. In most grades, the matrix is later modified by further treatment which tempers the as-quenched structure, raises its ductility, and improves its machinability.

As the name implies, malleable irons are softer than gray irons and are therefore used where greater toughness and shock resistance are required. Malleable irons are used extensively on farm implements, automobile parts, small tools, hardware, and pipe fittings.

***Classification of Malleable Irons.*** About 90 % of all the malleable iron made is classified as 32510, which means that it has a yield point of 32,500 psi and an elongation of 10 %. Grade 35018 has somewhat lower carbon content, but it has somewhat higher strength and better ductility. Grades 32510 and 35018 are known as standard types of malleable iron. There is an increasing demand for a newer type termed *pearlitic malleable*. Instead of having a ferrite matrix, it is pearlitic, which is stronger and harder than ferrite. These irons are obtained by interrupting the second-stage annealing process, by introducing larger quantities of manganese which prevents graphitization of the pearlite, or by air cooling followed by tempering.

Pearlitic malleables have yield strengths up to 80,000 psi, but low ductility. They are used where higher strengths and wear resistance are needed, along with less resistance to shock. Class number examples for pearlitic malleables are 45010, for the lowest yield strength, and 80002 for the highest yield strength.

***Ductile Iron.*** For generations, foundrymen and metallurgists have searched for some way to transform brittle cast iron into a tough, strong material. Finally, it was discovered that magnesium could greatly change the properties of cast iron. A small amount of magnesium, about one pound per ton of iron, causes the flake graphite to take on a spheroidal shape (Fig. 2-43). Graphite in a spheroidal shape presents the minimum surface for a given volume. Therefore there are less discontinuities in the surrounding metal, giving it far more strength and ductility. The processing advantages of steel, including high strength, toughness, ductility, and wear resistance, can be obtained in this readily cast material.

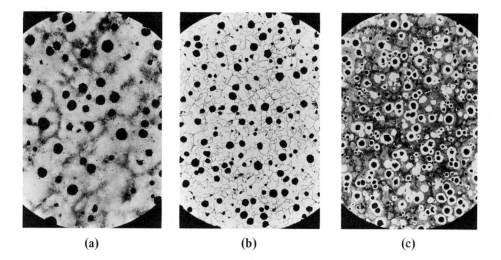

(a)                              (b)                              (c)

FIGURE 2-43. *The matrix structure surrounding the spheroidal graphite of ductile iron may be pearlitic (a), ferritic (b), or a ferritic band surrounding the spherulite, termed a* bulls-eye *structure (c).*

Ductile iron can be heat-treated in a manner similar to steel. Gray irons can too, but the risk of cracking is much greater.

There has been a rapid growth in the number of foundries licensed to produce this relatively new material, which goes by other names such as nodular iron and spheroidal iron. The combination of excellent casting qualities, excellent machinability, high strength, toughness, and wear resistance have led to wide acceptance of ductile irons for both small intricate castings, as used on a car door lock, to huge press rolls weighing 150 tons or more. Shown in Fig. 2-44 are a few representative parts made from ductile iron.

***Classification of Ductile Irons.*** The general classification numbers for ductile irons are similar to those for malleable cast iron, but a third figure is added which represents the minimum percent of elongation that can be expected. For example, an

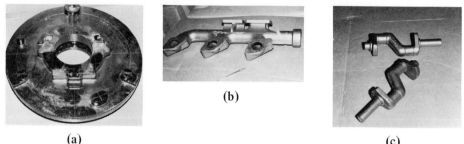

(b)

(a)                                                    (c)

FIGURE 2-44. *Examples of ductile iron castings. Shown at (a) is an 87-lb (39.115 kg) casting that has layout lines scribed on it as a means of checking the dimensional tolerances. Shown at (b) is a manifold and at (c) are crankshafts which are commonly made of ductile iron. The crankshafts are subjected to localized induction hardening on the journal bearings for maximum wear resistance. Courtesy of Hamilton Foundry.*

80-60-03 has 80,000 psi minimum tensile strength, 60,000 psi minimum yield strength, with 3 % minimum elongation. There are four regular types of ductile irons: 60-45-10, 80-60-03, 100-70-03, and 120-90-02. In addition, there are two special types: one is for heat resistance, the other (Ni-Resist*) has resistance to both corrosion and heat.

### Fabricating Characteristics

*Gray Iron.* The usual range of carbon in gray iron is between 2.50 and 3.75 %. The graphite flakes cause discontinuities in the ferrite. This helps the chips to break up easily. The graphite also furnishes lubricating qualities to the cutting action of the tool. Although gray cast iron is considered quite machinable, it varies considerably because of the microstructure. The ratings may range from 50 to 125 % of that of B1112.

One of the easiest ways to determine the machinability of a given piece of material is the Brinell test. The normal hardness range of 130 to 230 Brinell will present no machining difficulties. Beyond this range, however, even a few points will have a large effect on the cutting speed that should be used. Hardnesses above 230 Brinell indicate that the structure contains free carbides, which greatly reduce tool life.

Tungsten-carbide tools are usually used to machine cast irons because of the abrasive qualities of the material.

Machinability of gray cast irons can be improved through annealing. When this is done, its machinability exceeds that of most other ferrous materials.

*Malleable Iron.* Malleable iron has a machinability rating of 120 % and is considered one of the most readily machined ferrous metals. The reason for this good machinability is its uniform structure and the nodular form of the tempered carbon. The

* Registered trademark of the International Nickel Company, Inc.

TABLE 2-10. *Machinability Ratings For Cast Irons*

| Type of Iron | Machinability Index |
|---|---|
| Malleable (standard) | 110–120 |
| Gray iron (flake graphite) | 110 |
| Ductile iron | 90–110 |
| Gray iron (pearlitic) | 68 |

pearlitic malleables, which have a different annealing that leaves some of the carbon content in the form of combined carbides, have machinability ratings of 80 to 90 %. (The term pearlitic is just a convenient term and does not mean that the microstructure is necessarily in pearlite form.]

***Ductile Iron.*** The machinability of ductile iron is similar to that of gray cast iron for equivalent hardnesses and better than that of steel for equivalent strengths. Type 60-45-10 has both maximum machinability and toughness. Although the cutting action is good, the power factor is higher owing to the toughness of the material.

The machinability ratings for each of the main types of cast iron are given in Table 2-10.

### Formability

Cast irons are usually not considered for forming operations; however, the ductility of malleable irons makes it possible to use many production shortcuts which add up to time and material savings. Some typical examples are:

1. Sizing and straightening the casting in a press.
2. Coining—striking the part either hot or cold in a press. The operation can often bring the dimensions within ±0.010-in. tolerance when done cold and within ±0.004 in. when done hot.
3. Staking (heading over) and crimping can be used to eliminate the need for fasteners.

### Weldability

***Gray Iron.*** Gray cast iron can be welded with the oxyacetylene torch or with the electric arc. However, because of its low ductility, special precautions should be taken to avoid cracking in the weld area. Preheating is done, in the case of oxyacetylene welding, to make expansion and contraction more uniform. The opposite effect is used in arc welding, where short welds are made in order to keep the heat at a minimum. Slow cooling is essential to permit the carbon to separate out in the form of graphite flakes. Failure to do this will result in mottled or chilled areas.

Generally, the welding of gray cast iron is limited to repair work rather than fabrication. However, braze welding with bronze or nickel copper is frequently used in fabrication.

***Malleable Irons.*** Malleable irons are not considered weldable in the same sense that gray cast iron is weldable. The heat necessary to melt the edges of the break will completely destroy the malleable properties. Due to the long-term anneal required to produce these properties originally, a simple annealing process would not restore them. There are times, however, where stresses are low or are in compression only; in this case, successful arc welding can be accomplished. A commercially pure nickel rod or a 10 %-aluminum bronze rod is used.

Since brazing can be done at a temperature of 1700°F (927°C) or less, it is a preferred method of repair. Silver brazing is also used, since this procedure can be accomplished with even less heat.

**Ductile Irons.** Ductile iron, being a high-carbon-content material, should be given special consideration in welding, just as is gray cast iron. It can be welded by a carbon-arc process and most other fusion processes, either to itself or to other metals such as carbon steel, stainless steel, and nickel. The most easily welded types are 60-45-10 and the high-alloy variety. High-quality welds are made with flux-cored electrodes having 60 % nickel and 40 % iron. Noncritical applications may be welded with ordinary steel electrodes.

This material can also be brazed with silver or copper brazing alloys. Crack-free overlays can be made with commercial hard-surfacing rods. This will provide special abrasion and corrosion resistance.

### Castability

**Gray Iron.** From the iron–carbon diagram (Fig. 2-24) it can be seen that carbon markedly lowers the melting point of iron. Many benefits accrue from this fact. The pouring temperature, for example, can be several hundred degrees higher than the melting point, making for excellent fluidity. This makes it possible to produce thin-section castings over a large area. Shrinkage is considerably less than that of steel because the rejection of graphite causes expansion during solidification. The amount of solidification shrinkage differs with the various casting alloys. Class-20 and class-25 irons contain sufficient graphite so that there is virtually no shrinkage at all. The higher classes have one-half to two-thirds the solidification shrinkage of any ferrous alloys.

**Malleable Irons.** The molten white irons from which malleable irons are made have high fluidity, allowing complicated shapes to be cast. Good surface finish and close dimensional tolerances are possible. Since malleable irons require a prolonged heat treatment, internal stresses, which may have occurred as the casting solidified and cooled, are removed.

**Ductile Irons.** Ductile iron is similar to gray cast iron when it comes to casting qualities. It has a low melting point and good fluidity in the molten state. Therefore, it can be poured in intricate shapes and thin sectional parts. The metal will flow several inches, in sections as small as $\frac{1}{16}$ in., in greensand molds at normal foundry pouring temperatures. Many parts are now being cast from ductile iron that formerly could not be cast, because gray iron did not possess adequate properties and steel could not be cast in such intricate shapes.

## Stress-Strain Behavior Comparison

Typical stress–strain behavior of medium-carbon annealed steel, a ferritic malleable iron, an annealed ductile iron, and a class-40 gray iron are plotted in Fig. 2-45. Also included is a typical stress–strain curve for cast austenitic stainless steel (discussed in the next section).

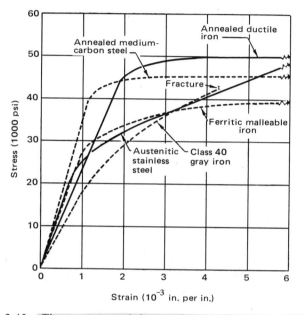

FIGURE 2-45. *The stress–strain behavior of cast ferrous materials.* Courtesy Machine Design.

# STAINLESS STEELS

The principle difference between stainless steels and mild steels is, of course, corrosion resistance. The mechanical properties of stainless steels are also superior to those of ordinary steels. Ultimate strengths of the annealed 300-series stainless steels vary from 80,000 to 100,000 psi, whereas mild steel is about 55,000 to 60,000 psi. In a full-hard condition, type 301 exceeds 300,000 psi tensile strength.

What is it that makes stainless steel stainless? Current theory holds that a thin, transparent, and very tough film forms on the surface. It is inert or passive and does not react with many corrosive materials that attack mild steel and other metals.

Unique to any other alloy system, the stainless steel family serves a temperature range from −454°F (−234°C) to above 1800°F (982°C). Within this range it exhibits strength, toughness, and corrosion resistance superior to most metals. Considerable processing and research is now being done at very low (cryogenic) temperatures. Stainless steel 18-8 (18 % chromium–8 % nickel) type 304 is a virtual "workhorse" for handling and storage of liquid helium, hydrogen, nitrogen, and oxygen that exist at cryogenic temperatures.

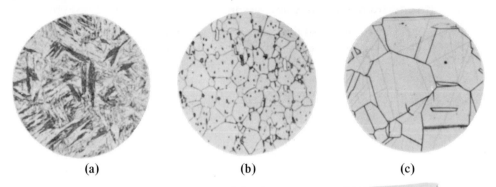

(a)           (b)           (c)

FIGURE 2-46. *Microstructures of the three main types of stainless steels. Courtesy Republic Steel.*

## Classification

The three main types of stainless steels are *martensitic, ferritic,* and *austenitic* (Fig. 2-46). In addition there is a special class of stainless steels referred to as "precipitation-hardening." A brief description of each of these types of steels follows.

*Class I: Martensitic Type (Hardenable).* The martensitic stainless steels are primarily straight chromium steels containing 11.5 to 18 % chromium together with enough carbon to render them hardenable by heat treatment. As the carbon content goes up, so does the range of mechanical properties which can be obtained by heat treatment. However, the increased carbon content ties up additional amounts of chromium in the form of chromium carbides, which, in turn, lowers the corrosion resistance. Therefore, as carbon content goes up, so must the chromium content.

AISI grades of Class I:

| | |
|---|---|
| 403 | 431 |
| 410 | 440-A |
| 414 | 440-B |
| 420 | 440-C |

Typical uses:

| | |
|---|---|
| Valves | Aircraft parts |
| Springs | Cutlery |
| Screws, nuts, bolts | Pump parts |

*Class II: Ferritic Type (Nonhardenable).* The ferritic stainless steels are straight chromium steels containing 12 to 27 % chromium with carbon controlled to the lowest practical limit to minimize its harmful effect on corrosion resistance. The ferritic steels can be forged and hot worked easier than the martensitic steels and after hot working may be air cooled without danger or cracking. The resistance to corrosion and scaling at elevated temperatures is better than that of Class I.

AISI grades of Class II:

| | |
|---|---|
| 405 | 430-F |
| 430 | 446 |

Typical uses:

| | |
|---|---|
| Heat exchangers | Screws, nuts, bolts |
| Boiler tubing | Furnace parts |
| Automotive trim | Chemicals and food |
| Household appliances | Processing industries |

*Class III: Austenitic Type (Nonhardenable).* The austenitic stainless steels are chromium-nickel steels containing from 16 to 26 % chromium and from 6 to 22 % nickel. As in the ferritic types, the carbon is there as a residual element and is controlled to the lowest practical limit. These steels can be work-hardened to high-strength levels. As a group, the austenitic steels have considerably better corrosion resistance than martensitic or ferritic steels and are characterized by excellent strength and oxidation resistance at elevated temperatures. However, these steel are susceptible to intergranular corrosion after prolonged heating within the temperature range of 750–1650°F (399–899°C). Chromium precipitates at the grain boundaries in the form of chromium carbides. Thus the corrosion resistance at the grain boundaries is greatly reduced. This problem can be avoided by using types 321 and 347 steels, which are stabilized with titanium and columbium respectively. These elements act to combine with carbon in the form of titanium or columbium carbides rather than chromium carbides.

AISI grades of Class III:

| | |
|---|---|
| 301 | 314 |
| 302 | 316 |
| 303 | 314 |
| 304 | 317-B |
| 305 | 321 |
| 308 | 347 |
| 309 | 348 |
| 310 | |

The use of Class III is practically unlimited. A few applications follow:

| | |
|---|---|
| Aircraft industry | Household items |
| Architectural trim | Pharmaceuticals |
| Chemical processing | Textile industry |
| Dairy industry | Transportation industry |
| Food processing | Paper industry |

*Precipitation–Hardening (PH) Stainless Steel.* Precipitation, as the name implies, is a strengthening process that takes place during aging. One type, called 17-7 PH, contains 17 % chromium, 7 % nickel, 1 % aluminum, and a small amount of titanium. It is martensitic at room temperature and is aged at 900°F (482°C).

During the aging process, aluminum–titanium compounds precipitate out of solution and distort the basic lattice structure. Thus, dislocations are produced, strengthening and hardening the alloy. Because the aging process is done while the steel is in a martensitic structure, the steels are often referred to as "maraging steels." Depending upon composition, tensile strengths of up to 310,000 psi, with 210,000 psi yield, are possible with 5 to 15 % elongation.

Precipitation-type stainless steels are incorporated into products ranging from race-car suspension members to spacecraft components.

### Formability

Stainless steels may be classified into two groups for forming purposes: straight chromium types (400 series) and chrome–nickel types (300–200 series). The latter type is the most formable and is capable of a 50 % reduction. Recall that: percent drawing reduction = (blank dia − cup dia) × 100 ÷ blank dia. Examples of percent reduction per draw are shown in Fig. 2-47. As several draws are made to decrease the diameter and increase the depth, the metalwork hardens. Chrome–nickel steels work-harden more than the straight chromium type. However, an outstanding characteristic of the chrome–nickel steels is their ability to deep-draw or cold-work beyond the drawing limit of both the straight-chromium stainless and plain carbon steels.

The drawing limit can be easily estimated by measuring the hardness of the part. Usually, for chrome–nickel steels, the hardness should not exceed about $R_c$ 40; and for chromium–stainless or carbon steel, the hardness will be about $R_c$ 20. If further drawing is to be done, the metal must be annealed.

Formability ratings of the various stainless steels can be made as described in Chapter 4. Also, forming information is obtained from a tensile test which ordinarily gives values for ultimate strength, yield strength, and elongation.

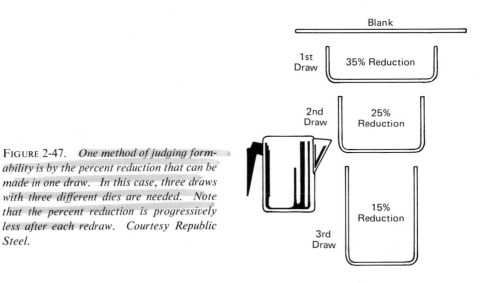

FIGURE 2-47. *One method of judging formability is by the percent reduction that can be made in one draw. In this case, three draws with three different dies are needed. Note that the percent reduction is progressively less after each redraw. Courtesy Republic Steel.*

TABLE 2-11. *Recommended Cutting Speeds for Free Machining Stainless Steels*

| AISI Type | Cutting Speed (sfpm) for HSS Tool |
|---|---|
| Austenitic, 303, 303 Se* | 85–120 |
| Martensitic, 416, 416 Se | 110–160 |
| Ferritic, 430F, 430 Se | 80–115 |

* The addition of a small amount of selenium to improve machinability.

## Machinability

Stainless steel is generally selected for its performance in service rather than for its machining qualities; however free-machining types have been developed in each of the three common types, as shown in Table 2-11.

Sulfur and selenium improve machinability of stainless steels, forming soft, nonmetallic sulfides and selenides in the steel. These appear in the microstructure as inclusions or stringers. They act to lubricate the cutting tool and produce non-clogging chips.

The recommended tool geometry for turning stainless steel is shown in Fig. 2-48. Tools should be kept sharp, particularly when turning austenitic grades, since they work-harden rapidly and make further cutting difficult.

## Weldability

Stainless steels are not particularly difficult to weld if the basic properties of the metal are considered. This can be done, in part, by comparing some of the important physical properties of stainless steel with those of the most weldable of metals, mild steel.

The expansion and contraction of metals must be considered in all fabricating processes, but especially so in welding. The coefficients of expansion are given as a means of comparing one metal with another. You will note, by studying Appendix 2A, that stainless steel has a considerably higher rate of expansion than that of mild steel.

Also the thermal diffusivity of the metal (ability of the metal to absorb heat) must be considered. It is used in welding to estimate the amount of energy required to bring the metal to the fusion point. Metals with low thermal conductivity, such as stainless steel, tend to localize the stress. That is, the hot metal wants to expand but is hemmed in by the adjacent cold metal. When the heat is removed, the metal cools and contracts. However, the cooler adjacent metal does not yield and results in cracks at the edge of the weld. Methods of overcoming these problems are discussed in Chapter 6.

Austenitic chromium–nickel steels are readily welded by four commonly used processes: shielded metal-arc, gas metal-arc, gas tungsten-arc, and submerged arc. The molten weld metal is protected from atmospheric oxidation in all four processes

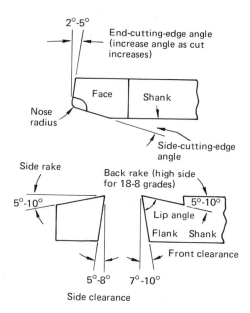

FIGURE 2-48. *The recommended tool geometry for turning stainless steel with an HSS tool.*

by flux, slag, or inert gas. The complete protection from oxidation retains all the essential elements while excluding foreign elements that effect the strength and corrosion resistance, or weld quality. Due to the low thermal conductivity, the wire-melting rate (used in gas metal-arc welding), measured in in./min/amp, is approximately 30 % higher with stainless steel than with carbon steel.

## LIGHT METALS AND ALLOYS

The age of air and mass travel has focused the attention of engineers on light metals. About 17 % of the aluminum produced in the United States is used in the manufacture of aircraft, automobiles, trucks, buses, ships, and trains.

Metals commonly classed as light metals are those having a density less than that of steel (7.8 gm/cm). Common metals in this classification are aluminum, magnesium, beryllium, and titanium.

Although aluminum is in abundant supply on the earth's crust, in all soils, clays, rocks, and minerals, it cannot be easily separated. It can only be profitably separated from bauxite. An electrolytic process is used to reduce the alumina. Commercially pure aluminum has about one-third the density of steel. It possesses excellent corrosion resistance and has good electrical, thermal, and reflectivity characteristics.

### Classification of Aluminium Alloys

Aluminum is used extensively in both the wrought and cast forms. Casting alloys are designated by a four-digit number. The first digit identifies the alloy very much the same as that shown in Table 2-12 for wrought alloys. For cast alloys,

TABLE 2-12. *Aluminum Alloy Specifications*

| Type of Aluminum Alloy | Number Group |
|---|---|
| Aluminum—99.00 % purity or greater | 1XXX |
| Copper | 2XXX |
| Manganese | 3XXX |
| Silicon | 4XXX |
| Magnesium | 5XXX |
| Magnesium and silicon | 6XXX |
| Zinc | 7XXX |
| Other element | 8XXX |
| Unused series | 9XXX |

number six is not used and number eight is tin. The second two digits, of the four-digit series, are assigned when the alloy is registered. A later modification of this alloy is indicated by a letter prefix before the numerical designation. Castings are designated by a suffix (0) which is separated from the first three digits by a decimal point. If it is an ingot of the same chemical composition, the number 1 will follow the decimal point. An example of an aluminum casting number is A514.0.

Wrought aluminum alloys are also identified by a four-digit number as shown in Table 2-12. The first digit designates the alloy type, the second digit the alloy modification, and the last two digits the aluminum purity. For alloys in use before this classification system was devised, the last two numbers are the same as the old numbers. For example, a widely used type of aluminum was 24S. In the new specification, it is 2024.

*Temper Designations.* A letter following the four-number identification, separated by a hyphen, indicates the basic temper of the metal or the treatment used in hardening, strengthening, or softening the alloy. Temper designations are shown in Table 2-13.

Annealing of a work- or strain-hardened metal is accomplished by heating to a temperature at which recrystallization takes place, followed by slow cooling. The temperatures to which it is heated will depend upon the amount of prior cold work and the alloy content.

Solution heat treatment (combined with aging) is a hardening and strengthening process that is accomplished by heating an aluminum alloy, such as 2024, to a high enough temperature to affect a solution of the hardening constituents and then quenching it rapidly. For many aluminum alloys, this is 900 to 1000°F (482–538°C). During solution heat treatment, the soluable alloying constituents are dispersed uniformly in the matrix. When subsequently quenched, there is not enough time for these constituents to precipitate out of solution, which results in a "supersaturated" solid solution. That is, the matrix has more constituents than it can normally carry in solution. This is an unstable condition and certain constituents begin to precipitate from the aluminum matrix.

Age hardening or precipitation may take place at room temperature and is known as "natural aging." Certain other alloys must be heated slightly to bring the

TABLE 2-13. *Aluminum-Alloy Temper Designations*

| | |
|---|---|
| -O | Annealed |
| -F | As fabricated |
| -H | Strain hardened |
| | -H1 Strain hardened only |
| | -H2 Strain hardened followed by partial annealing |
| | -H3 Strain hardened and stabilized |
| -T | Solution heat treated |
| | -T2 Annealed (cast shapes only) |
| | -T3 Solution treated followed by cold work |
| | -T4 Solution treated, natural aging at room temperature |
| | -T5 Artificially aged after quenching from hot-working operation such as casting or extrusion |
| | -T6 Solution treated, artificially aged |
| | -T7 Solution treated, stabilized to control growth of distortion |
| | -T8 Solution treated, cold worked, artificially aged |
| | -T9 Solution treated, artificially aged, cold worked |

precipitation to completion. This is referred to as "artificial aging." In either case, precipitation is aimed at providing the correct size, character, and distribution of the precipitate. The precipitate should be of a "gritty" type to provide the maximum resistance to slippage along the grain boundaries, and, more importantly, along the slip planes of each crystal.

## Fabricating Characteristics

*Machinability.* Pure grades of aluminum are too soft for good machinability. Work or strain hardening helps overcome some of the gumminess. The most easily machined alloy is 2011-T3. This metal is often referred to as the free-cutting aluminum alloy. It is used extensively for a broad range of screw-machine products. Small amounts of lead and bismuth help produce a better finish and easy-to-handle chips.

The softer alloys sometimes build up on the cutting edge of the tool. However, the proper use of cutting oils and a highly polished tool surface will tend to minimize this condition.

Two important physical properties that should be kept in mind when machining aluminum are modulus of elasticity and thermal expansion, as shown in Appendix 2A.

Due to the much lower modulus of elasticity (about one-third that of steel), aluminum shows much greater deflection under load. Care should be taken, in making heavy cuts and in clamping the work, to avoid distortion.

When dimensional accuracy is necessary, thermal expansion of the metal is an important consideration. Since the thermal expansion of aluminum is almost twice that of steel, care should be exercised in keeping heat down. Overheating can be reduced to a minimum by using sharp, well-designed tools, coolants, and feeds that are not excessive. It may be necessary, on larger parts, to make a thermal allowance in measuring.

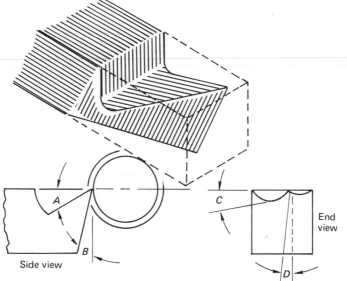

FIGURE 2-49. *A single-point HSS tool used to cut aluminum. The recommended angles are: (A) back rake, 25–50°; (B) end relief, 7–10°; (C) side rake, 10–20°; and (D) side relief, 7–10°.*

Tool angles vary widely for turning aluminum, depending on the type of material and the finish required. In general, large rake angles are employed for finishing cuts and for the softer alloys. Low rake angles are generally used for heavy cuts and for the harder, free-machining alloys. Shown in Fig. 2-49 is a sketch of a HSS lathe tool used for turning aluminum.

Surface foot speeds for aluminum range anywhere from 400 to 20,000 fpm (feet per minute). The alloys 2011-T3, 2017-T4, and 2024-T4 permit the highest surface foot speeds, long tool life, and clean cutting.

*Formability.* Most aluminum alloys can be readily formed cold. However, the wide range of alloys and tempers makes for considerable variation in the amount of forming possible.

Generally, depending upon the alloy and temper used, aluminum requires less energy and horsepower than does the forming of heavier metals. Consequently, machine life is increased, power costs are lower, and maintenance is minimized.

As the degree of temper increases, the plastic and elastic ranges decrease, which makes forming more difficult to accomplish and control. Material must be stressed beyond its yield strength in order to form it permanently, or it will tend to return to its original shape. Thus, material that is insufficiently stressed will spring back in proportion to the amount of stress applied. Of course, very little springback will be encountered in the soft (O—annealed) plate. The exact amount of springback is difficult to compute. Therefore, tools should be provided with adjustments for overbending, unless prelimary experiments are undertaken to determine the material's springback characteristic.

Each temper (degree of hardness) of an aluminum sheet alloy establishes a specific workability characteristic that must be considered in the planning of operations for bending it. Bending characteristics also vary with the thickness of the sheet.

The non–heat-treatable alloys—1100, 3003, 5005, 5050, and 5154—work-harden but have good forming qualities in the softer tempers. In the annealed state they can be bent with a very small radius up to 0.25 in. thick.

The heat-treatable alloys—2014, 2024, 6061, and 7075—can be formed readily in the annealed temper and are often worked in this state. Work can be done after solution heat treatment and quenching, but it must be done quickly before natural aging or precipitation starts. Sometimes the solution treatment is performed and the metal is placed in cold storage to delay the aging process until the parts can be formed.

### Weldability

*Physical Properties.* The physical properties that must be considered in welding aluminum are: melting point, thermal conductivity, and the coefficient of expansion, as shown in Appendix 2A.

By comparing some of the physical properties of aluminum and steel, it can be seen that the melting point of aluminum is much lower and the thermal conductivity is much higher. Therefore, more heat will be required to weld a corresponding thickness of aluminum.

The thermal expansion of aluminum is almost twice that of steel. This, coupled with the fact that aluminum weld metal shrinks about 6 % in volume upon solidification, may put the weld in tension. Some restraint is often necessary to present distortion during the welding operation. However, excessive restraint on the component sections during cooling of the weld may result in weld cracking.

The speed at which the weld is made is one of the determining factors in preventing distortion. A slow rate causes greater area heating with more expansion and subsequent contraction. For this reason, arc- and resistance-welding methods that make use of highly concentrated heat are the most used.

Inert-gas–shielded electrical-arc welding or gas-metal-arc (GMA), gas-tungsten-arc (GTA), and resistance welding (spot and seam) are the most widely used commercial methods of welding aluminum (Fig. 2-50).

Commercially pure aluminum and low-alloy aluminums such as 1100 and 3003 are easily formed and welded. These metals have little or no tendency to crack after welding, since they are quite ductile.

The heat-treatable aluminum alloys of the 6000 series (magnesium and silicon), the 2000 series (copper), and the 7000 series (zinc with magnesium and/or copper) gain their strength by solution heat treatment, quenching, and age hardening (precipitation). These metals are readily welded but the filler rod must be such that it will not produce hot-short crack sensitivity. Hot-short is a term used to describe metals that are very weak when they are near the molten state. Thus contraction cracks develop in the hot metal as it starts to solidify. Filler metals can be chosen that will overcome these difficulties. As in the case of welding the aluminum–magnesium metals, a filler rod containing 5 % magnesium will prevent weld cracking.

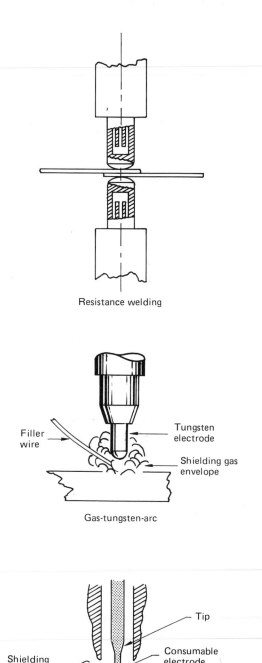

Resistance welding

Filler wire

Tungsten electrode

Shielding gas envelope

Gas-tungsten-arc

FIGURE 2-50. *Common methods of welding aluminum.*

Tip

Consumable electrode

Shielding gas envelope

Arc

Gas-metal-arc (GMA)

*Casting.* Aluminum is considered one of the more easily cast metals. It requires less energy to bring it to the molten state, about 1200°F (649°C) and is therefore easier on the refractory furnace lining. In addition to sand casting, permanent metal molds and die casting are used extensively.

The combination of low density and low melting point combine to practically eliminate most problems of sand washes (sand moving out of place) that occur when heavier metals are poured.

Special attention must be given to casting shrinkage, hot shortness, and gas absorption. Due to the large shrinkage, from 3.5 to 8.5 % by volume, it is desirable to have uniform sections wherever possible. If section changes are necessary, they should be gradual. The minimum section for sand castings is about $\frac{3}{16}$ in.

Hot shortness, the low strength of the metal immediately after solidification, makes it necessary for the casting designer to minimize shrinkage stresses.

Molten aluminum readily absorbs oxygen and hydrogen. Since most of the pinholes in aluminum castings are caused by hydrogen, precautions must be taken to minimize it. This is done by having the metal no hotter than necessary for pouring, and by keeping excess moisture from the molding sand.

Alloying elements intentionally added to facilitate aluminum casting and heat treatment are silicon, magnesium, and copper. Silicon increases the fluidity of the molten aluminum, allowing it to flow farther in thin walls of the mold cavity and produce finer detail. It also reduces internal shrinkage and reduces the coefficient of expansion.

Aluminum castings containing more than 8 % magnesium will respond to heat treatment; however, it also makes the metal more difficult to cast.

Copper is one of the principal hardening constituents in aluminum. It increases the strength of the aluminum in both the heat-treated and non–heat-treated conditions.

## COPPER AND COPPER-BASE ALLOYS

### Classification

*Coppers.* Unlike other common metals, the various types of copper are better known by name than by code number.

Commercially pure copper is available in several grades, all of which have essentially the same mechanical properties. *Electrolytic tough-pitch copper* is susceptible to embrittlement when heated in a reducing atmosphere, but it has high electrical conductivity. *Deoxidized copper* has lower electrical conductivity but improved cold-working characteristics, and it is not subject to embrittlement. It has better welding and brazing characteristics than do other grades of copper. *Oxygen-free copper* has the same electrical conductivity as the tough-pitch copper and is not prone to embrittlement when heated in a reducing atmosphere.

Modified coppers include *tellurium copper*, which contains 0.5 % tellurium for free-cutting characteristic (selenium and lead are also used for this purpose) and

TABLE 2-14. *The Main Classifications of Brasses*

| Alloy | Meaning |
|---|---|
| Alpha brasses | Contain less than 36 % zinc and are single-phase alloys. |
| Alpha-beta brasses | Contain more than 36 % zinc and have a two-phase structure. |
| Leaded brasses | Contain up to 88.5 % copper and up to 3.25 % lead. |
| Tin brasses | Known as admiralty and naval brass when the percentage of tin is low, 0.75 to 1.0 %. |
| Nickel silvers | Brasses containing high percentages of nickel. (The designation 65-18 indicates approximately 65 % copper, 18 % nickel, and the remainder zinc. Nickel is added primarily for its influence on color. When the percentage is high, a silvery white color is obtained. Nickel also improves mechanical and physical properties.) |

*tellurium-nickel copper*, an age-hardenable alloy that provides high strength, high conductivity, and excellent machinability.

**Brasses.** Brasses are principally alloys of copper and zinc. They are often referred to by the percentage of each; for example 70-30 means 70 % copper and 30 % zinc. Small amounts of other elements such as lead, tin, or aluminum are added to obtain the desired color, strength, ductility, machinability, corrosion resistance, or a combination of these properties. The main classifications of brasses are given in Table 2-14.

**Bronze.** Bronzes are usually thought of as copper–tin alloys. However, there are some types of bronze that contain little or no tin. The main classifications of bronzes are given in Table 2.15.

TABLE 2-15. *The Main Classifications of Bronzes*

| Alloy | Meaning |
|---|---|
| Phosphor bronze | Most copper–tin alloys are deoxidized with phosphorus; the small amount left in the metal may range from a trace to over 0.35 % phosphorus, increasing strength, hardness, toughness, and corrosion resistance. |
| Aluminum bronze | Contains up to 13.5 % aluminum and small amount of manganese and nickel; has good antifrictional properties. |
| Silicon bronze | Contains up to 4 % silicon; has strength similar to mild steel, and excellent corrosion resistance to brine and other nonoxidizing inorganic acids. |
| Manganese bronze | Contains up to 3.5 % manganese, which imparts high strength. |
| Beryllium bronze | Contains from 2 to 2.75 % beryllium, which makes it respond to precipitation hardening. |

## Fabricating Characteristics

*Machinability.* Because of the wide variety of copper-base alloys, some general information will be given first, followed by machinability ratings according to groups.

*General Speeds and Feeds.* For most copper alloys, it is generally good practice to use the highest practical cutting speed with a relatively light feed and moderate depth of cut. An exception is the machining of sand castings with high-speed tools, in which case low speeds and relatively coarse feeds are used to increase tool life. After the scale has been removed, the higher speeds and lighter feeds can be resumed. Tool life can be considerably increased if the castings are sandblasted, pickled in acid, or tumbled to remove the extremely hard, abrasive surface scale.

*Machinability Ratings.* Steel and steel alloys have machinability ratings based on B1112 as 100 %. Copper and copper-base alloys use either free-cutting brass—consisting of 61.50 % copper, 35.25 % zinc, and 3.25 % lead—as the base material, or B1112. The base must be designated.

To classify the relative machinability of copper-base alloys, they can be placed in three groups, based on the amount of tool life and amount of power required. The line of demarcation from one group to the other is not sharp but will act as a guide.

*Group I.* The materials in this group are the alpha-beta or two-phase brasses. The beta phase is harder and more brittle, which makes the alloy less ductile, thus permitting heavier feed rates in turning. Lead is added to both single- and double-phase brasses. It is effective in reducing the shear strength of the metal and assists in chip breakage. Alloys and ratings in this group are:

    Free-cutting brass—100 % (base)
    Selenium copper—80 %
    Leaded naval brass—80 %
    Leaded commercial bronze—90 %

*Group II.* The materials in this group are considered readily machinable, with a rating from 30 to 70 %. The principal alloys in this group are:

    Yellow brass—40 %
    Manganese bronze—30 %
    Admiralty brass—50 %
    Leaded phosphor bronze—50 %
    Leaded nickel silver—50 %
    Tin bronze—40 %

*Group III.* These materials have a machinability rating up to 30 %. This group contains the unleaded coppers, nickel silvers, some phosphor bronzes, and beryllium bronze. These alloys usually produce tough stringy chips.

As with aluminum, the main angles to consider for cutting tools are the ones on top of the tool, or the rake angles. Moderate rake is suggested for group I to reduce the tendency of the tool to grab or "hog-in." More pronounced rakes are used in groups II and III to provide free chip flow. The side rake angles in group III exceed those for steel.

Best results are obtained in drilling group I and II materials with straight-fluted drills [Fig. 2-51(*a*)]. These drills have a natural zero-degree rake angle and are particularly good for automatic machine work. Regular drills may be modified to reduce the cutting-lip angle and the tendency to grab, as shown at (*b*) of Fig. 2-51.

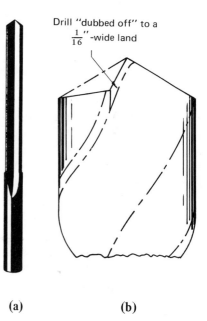

Drill "dubbed off" to a $\frac{1}{16}$"-wide land

FIGURE 2-51. *A straight-flute drill used to drill soft metals (a). A standard drill modified by grinding a $\frac{1}{16}$ in. wide flat on the cutting lip (b) to keep the drill from grabbing or "hogging-in" on soft metals.*

(a)  (b)

*Formability.* The forming qualities of copper-base alloys are dependent on three main factors: alloy composition, grain size, and temper.

*Alloy Composition.* The most-used alloys for forming are the 70-30 cartridge brasses. However, all brasses from 95-5 to 63-37 are used for pressworking oper-ations. Trouble is encountered when less than 63 % copper is used, because the beta phase is brittle and may produce fractures or waviness—surface defects.

Pure coppers such as the electrolytic tough-pitch, oxygen-free, and deoxidized coppers are very ductile, and they work-harden less rapidly than do bronzes or brasses. The deoxidized and oxygen-free coppers can withstand more severe bending and form-ing than the tough-pitch coppers.

*Grain Size.* The grain size of copper alloys is expressed as the average grain diameter in millimeters. Grain structure is related to tensile strength, elongation, hardness, and other properties.

Fine grain or ultrafine grain is material below 0.010-mm grain size. Such material is very smooth after bending, forming, or drawing and can economically be buffed to a high luster. Other grain-size recommendations for cold-working annealed copper alloys are given in Table 2-16.

TABLE 2-16.  *Grain Size for Cold-Working Annealed Copper Alloys\**.  *Courtesy American Machinist.*

| Nominal Grain Size (mm) | Typical Uses |
|---|---|
| 0.015 | Slight forming operations |
| 0.025 | Shallow drawing |
| 0.035 | For best average surface combined with drawing |
| 0.050 | Deep-drawing operations |
| 0.070 | Heavy drawing on thick gauges |

In general, ductility increases with increasing grain size.  For very severe draws, particularly of heavy-gauge strips, grain sizes as large as 0.120 mm are not uncommon.

*Temper.*  Copper and copper-base alloys may be obtained in the tempers shown in Table 2-17.

The most common tempers for rods are half-hard and hard.

The minimum bend radius varies with alloy and tempers.  Annealed copper-base alloys can be bent to a radius equal to the metal thickness, and some of them to one-half the metal thickness.

*Weldability.*  All four of the commonly employed methods—gas, metal arc, carbon arc, and gas-shielded arc—are in general use for welding copper.  Copper is also joined by brazing and by the newer method, discussed later, of ultrasonic welding. The gas method, because of the variety of atmospheres obtainable, finds wide use. Its disadvantages are cost and, in some cases, warpage and attendant high-conduction losses into the metal adjacent to the weld.

The most suitable of the coppers to use for welding is the deoxidized type. The fact that the coppers do not contain oxygen, plus the fact that there is some residual deoxidizer in the metal, reduces embrittlement to a minimum.

Resistance welds can be made on all brasses containing 15 to 20 % zinc or more.  The thermal conductivity is sufficiently low in these metals to make spot or seam welding of the lighter gauges practical.

Both soldering and low-temperature brazing are widely employed in joining coppers and brasses.

TABLE 2-17.  *Temper Designations of Copper-base Alloys*

| Temper Designation | Meaning |
|---|---|
| $\frac{1}{4}$ hard | Extra hard |
| $\frac{1}{2}$ hard | Spring |
| $\frac{3}{4}$ hard | Extra spring |
| Hard | |

*Castability.* Copper castings are made almost entirely in sand molds. The technique of casting high-conductivity copper is especially difficult, since considerable skill is required in handling the reducing agents. The molten metal does not flow well, making it difficult to fill molds that are intricate. Copper has the property of being hot-short, and will break as it solidifies if the design does not permit shrinkage.

The copper-base alloys are quite easily cast and handled in the foundry.

## ORGANIC MATERIALS

### Introduction

About 100 years ago there was no such thing as a commercial plastic material. Chemists and scientists had not yet realized that by the combination of such basic organic materials as oxygen, hydrogen, nitrogen, chlorine, and sulphur new materials could be created.

In 1868 a shortage of ivory for billiard balls developed. A determined and inventive young printer by the name of John Wesley Hyatt mixed pyroxylin, made from cotton and nitric acid, with solid camphor. The resulting material was called Celluloid. The new material was soon adapted to many uses. Colored pink it was used by dentists as a replacement for hard-rubber denture plates. Other popular uses were for toothbrush handles, combs, and motion picture film.

About 40 years passed before another major step was made in the plastics industry. In 1909 Dr. Les Henrik Baekeland introduced phenolformaldehyde resins. Although others had experimented with this combination, Dr. Baekeland was the first to achieve a controllable reaction and to develop a commercial product. This first phenolic material was given the tradename Bakelite.

Cellulose acetate was the next large-volume plastic to be developed commercially. In 1927 it was available only in rod, sheet, or tube form and in 1929 it became the first plastic material to be injection molded. Other now well-known materials began developing at an ever increasing rate, such as the vinyl resins, polystyrenes, and polyethylenes.

### Organic Structures

Early investigators began to deduce the existence of large organic molecules from the insolubility of certain substances and their resistance to melting. Chemists named them by attaching "poly" to the basic monomer unit. A high polymer was one with a very large aggregation of units, i.e., compounds of high molecular weight. Molecular weight is used to indicate the size of a polymer. It represents the cumulative weight of all elements forming the giant molecule and may range between 1000 and 1,000,000.

Another parameter that is used to designate the size of a molecule is the *degree of polymerization* (DP). It refers to the number of repeat units in the chain. In commercial plastics the DP normally falls in the range of 75 to 750 mers per molecule.

Even though the polymer molecule appears to be very large by weight, it is still below the resolving power of an optical microscope and only under certain circumstances can it be resolved by an electron microscope. Consequently, molecular weight determinations are usually made indirectly by such physical means as viscosity, osmotic pressure, or light scattering, all of which are affected by the number, size, or shape of molecules in a suspension or in a solution.

### Polymerization Mechanisms

The length that a polymer chain attains in a particular reaction is determined by such random events as: which materials are available at the end of the growing chain during a *condensation reaction*, and how soon one free radical meets another during an *addition reaction*.

*Condensation Reaction.* Where two or more reactive compounds are involved, there is a repetitive elimination of smaller molecules to form a by-product. Let us suppose, for example, that there are two monomers: one has a basic group, the amino group ($NH_2$), at each end; the other has an acid group, a carboxyl group ($COOH$), at each end. The two monomers will hook up end-to-end, the basic end of one linking up with the acid end of the other, splitting off a molecule of water in the chemical reaction. The condensation reaction takes place during the molding process. Unless the volatile by-products are completely driven off, dimensional instability and low strength will result.

The condensation method of building polymers is rather slow and tends to come to a complete stop before the molecules have attained really giant size. As the chains grow, they become less mobile and less numerous and therefore less likely to encounter free building blocks in solution. Fortunately, such products as nylon have acquired their valuable properties by the time they reach a molecular weight of 10,000 to 20,000.

*Addition Polymerization.* The addition method, in contrast with the condensation method, can build molecules of almost unlimited size. The first stage is to activate a monomer by heat and pressure as shown in Fig. 2-52. An ethylene monomer, $CH_2=CH_2$, is made by removing two hydrogens from ethane, $CH_3-CH_3$. A redistribution of electrons occurs, and a stable double bond is formed. By counting the double bonds as two, each carbon still has four energy bonds satisfied. Every atom within a molecule must have all its energy bonds completely satisfied if the compound is to be stable.

Starting with billions of molecules of a monomer in a reactor, under heat, pressure, and catalysts, one of the monomer double bonds will rearrange into two "half bonds," one at each end. These half bonds quickly combine with half bonds of other rearranged monomer molecules, forming stable "whole bonds" between them. As each monomer joins with others, the chain length grows until it meets a stray hydrogen, which combines with the reactive end, stopping the chain growth at that point.

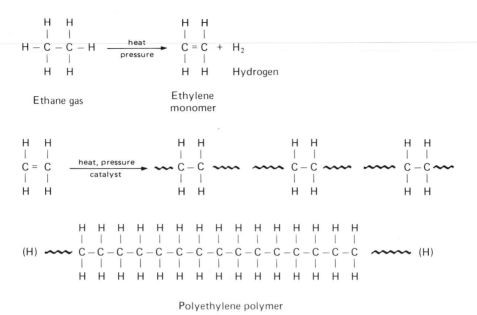

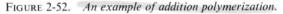

Polyethylene polymer

FIGURE 2-52.   *An example of addition polymerization.*

During the polymerization reaction, millions of separate polymer chains grow in length simultaneously, until all of the ethylene monomer is exhausted. By adding predetermined amounts of hydrogen (or other chain stoppers), the polymer chemists can manufacture a relatively consistant average chain length. Chain length (molecular weight) is important because it determines many of the properties of a plastic, and it also affects the ease of processing.

### Factors Affecting Polymer Properties

Factors that largely determine the properties of a polymer are: branching, copolymerization (random block and graft), crystal orientation and size, degree of crystallinity, molecular weight and molecular weight distribution.

*Branching.* The density of a polymer is directly related to the number and length of the side branches. The greater the branching, the less the density. The strength of intermolecular attractive forces varies inversely with the sixth power of distance between chains. Thus, as the distance is halved, the attractive force increases by a factor of 64. For this reason chain shape is as important as chain length.

Two types of branches are formed from the main stem during polymerization: short and long. Short branches (usually two carbon atoms) are formed by the action of the catalyst (which controls the rate of initiation). Increasing the catalyst and temperature yields shorter chains and lower molecular average weights (Fig. 2-53).

103

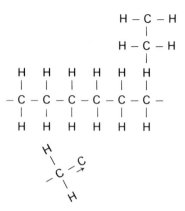

FIGURE 2-53. *An example of how short branching takes place. The growing polymer has an unsatisfied valance which attracts the monomer.*

Long branches, up to 2000 carbon atoms per branch, are formed by inter-molecular *chain transfer.* In chain transfer, the growing polymer radical attracts an atom, usually hydrogen, from an organic compound, terminating the growth of one chain and starting another. Thus, the monomer itself can be a chain transfer agent.

*Copolymerization.* Large chain molecules can be built from monomers that are not alike. The monomers need not alternate on a regular pattern, but frequently do. A polymer made from two different monomers is called a copolymer; one made from three different monomers is called a terpolymer.

Final properties of a copolymer depend on the percentage of monomer A to monomer B and how they are arranged along the chain.

*Block Copolymers.* Block copolymers have long chains of monomer A alternating with long chains of monomer B.

*Graft Copolymers.* A graft copolymer consists of a main chain of one polymer with side groups of monomers. A copolymer with a flexible polymer for the main chain but with rigid side chains is very stiff, yet has excellent resistance to impact—a combination of properties not often found in the same plastic. Shown in Fig. 2-54 is a schematic representation of various copolymer arrangements.

*Crystal Orientation and Size.* It has long been believed that single crystals could not be produced from polymer solutions because of the molecular entanglements of the chains. However, since 1953 single crystals have been reported for so many polymers that the phenomenon appears quite general. Electron-diffraction patterns show platelets about 100 Å thick in which perfect order exists. Dislocations also exist that are similar to those found in metals and low-molecular crystals.

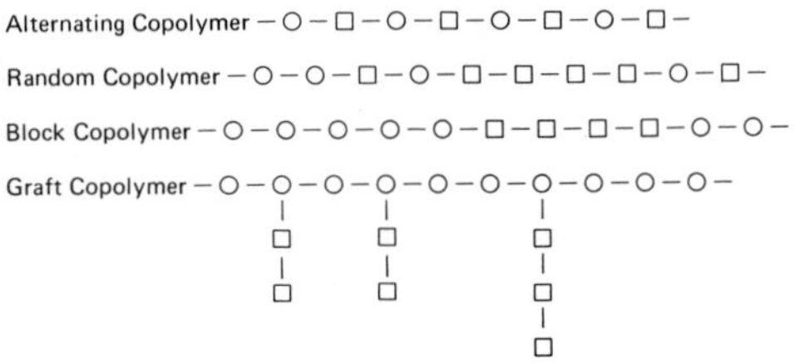

FIGURE 2-54. *A schematic of the basic arrangement of copolymers based on two monomers. Each copolymer has different properties and processing characteristics.*

Alternating Copolymer — ○ — □ — ○ — □ — ○ — □ — ○ — □ —

Random Copolymer — ○ — ○ — □ — ○ — □ — □ — □ — □ — ○ — □ —

Block Copolymer — ○ — ○ — ○ — ○ — ○ — ○ — □ — □ — □ — □ — ○ — ○ —

Graft Copolymer — ○ — ○ — ○ — ○ — ○ — ○ — ○ — ○ — ○ — ○ —
                        |              |              |
                        □              □              □
                        |              |              |
                        □              □              □
                                                      |
                                                      □

The crystallites, or spherulites as they are called, can be oriented in two different ways to produce increased strength: one-dimensional and biaxial.

An example of the one-dimensional arrangement is polypropylene monofilaments, or threadlike fibers. An example of biaxial arrangement is polypropylene film.

The size of the spherulites can be controlled by (a) rapid cooling from the melt condition, or (b) the addition of a nucleating agent. If the spherulites become too large, the polymers may become brittle. Nucleated resins normally have increased toughness.

***Degree of Crystallinity.*** Polymer structures range from partly crystalline to completely amorphous. The degree of crystallinity is largely determined by the geometry of the polymer chains. It must be remembered that the side groups attached to the main backbone of the molecular chain give it a bulkiness in three dimensions. The closer the polymer chains can be packed, the more crystalline it will become. Conversely, the farther apart the chains are held due to irregulatities, the more amorphous the structure will be. If a polymer molecule has a symmetrical shape that can pack closely, the intermolecular forces will be very large compared to those of a nonsymmetrical shape, as shown by the schematic sketch of the structures of high- and low-density polyethylene in Fig. 2-55.

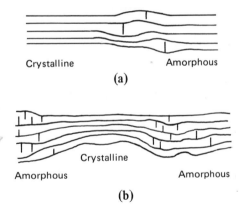

FIGURE 2-55. *The chain packing of high-density polyethylene (a) and low-density polyethylene (b). In high-density polyethylene the chains are mostly linear, with one to three branches per thousand carbon atoms.*

The shape of the polymer chain can be controlled by various modes of polymerization, as shown in Fig. 2-56. These modes are three chain arrangements and their corresponding cross-sectional views. The *isotactic* mode has very high symmetry, which allows close packing. The *syndiotactic* mode is of intermediate symmetry. The *atactic* mode is of random configuration. Considering these configurations, it is not surprising that polyethylene, with a high degree of regularity (isotactic), is closely packed and has a melting point of 338°F (170°C), while polypropylene, with a random (atactic) arrangement, flows at a temperature of 100°F (212°C).

A crystallized structure is stiffer, has higher tensile strength, has increased creep resistance, and has heat resistance, but also has a greater tendency to stress-crack than the polymers that are more noncrystallized or amorphous.

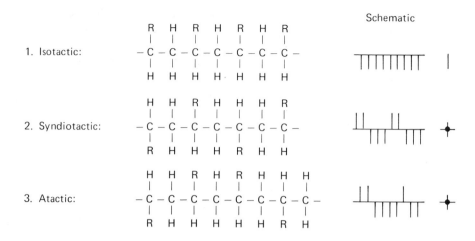

FIGURE 2-56. *Typical polymer structures where R represents a side group. The term isotactic refers to a molecule of high symmetry, syndiotactic to intermediate symmetry, and atactic to a random configuration. The modes are shown as two-dimensional; however it should be realized that even as molecules exist in three dimensions, so do the side-group locations.*

Crystalline polymers are more difficult to process, have higher melt temperatures and melt viscosities, and tend to shrink and weigh more than amorphous polymers.

The *glass transition* ($T_g$) temperature of a polymer is also related to crystallinity. Glass transition refers to the temperature at which a polymer changes from a melt to a solid. Amorphous polymers do not crystallize readily but change from a melt to a solid gradually over a range of about 80 degrees Fahrenheit (45 degrees Centigrade). The change in viscosity, when a crystalline structure goes from a melt to a solid, is more pronounced and is given as the $T_g$ temperature. If side groups attached to the main chain have different transition temperatures, the effect is to produce rubbery or leathery stages before the solid state.

Many engineering plastics such as nylon, polypropylene, and acetal share common characteristics of crystallinity and hence have similar properties.

***Molecular Weight and Distribution.*** As stated previously, polymers consist of a mixture of giant-size molecules. The size of the molecules is described in terms of *molecular weight* and the distribution of the sizes is referred to as *molecular-weight distribution* (MWD). Molecular weight of a given polymer refers to the average of all sizes of polymer chains produced during polymerization. In general, high-molecular-weight polymers have better physical properties than their low-molecular-weight counterparts. This is especially true in certain critical properties such as impact resistance (toughness), chemical resistance, and stress-crack resistance. However, because high-molecular-weight polymers also have a higher melt viscosity, they are more difficult to process.

TABLE 2-18. *Molecular Weight of Polymers is Built Up by Adding* $CH_2$ *Groups*

| Chemical Formula | Common Name | Molecular Weight | Physical State |
|---|---|---|---|
| $CH_4$ | Methane | 16 | Gas |
| $C_2H_6$ | Ethane | 30 | Gas |
| $C_3H_8$ | Propane | 44 | Gas |
| $C_4H_{10}$ | Butane | 58 | Gas |
| $C_5H_{12}$ | Pentane | 72 | Liquid |
| $C_{17}H_{36}$ | Kerosene | 240 | Liquid |
| $C_{18}H_{38}$ | Paraffin | 254 | Soft solid |
| $C_{50}H_{102}$ | Hard waxes | 702 | Brittle solid |
| $C_{100}H_{202}$ | LMW polyethylene | 1402 | Tough solid |

Molecular-weight distribution has a somewhat smaller effect on the polymer properties and processing than molecular weight, but nevertheless it is significant. In general, polymers that have a narrow MWD have higher impact strength, better resistance to environmental stress cracking, and better low-temperature toughness. On the other hand, polymers with broad MWD have better processing properties, such as less resistance to flow (important in molding and extruding) and a better drawing rate (important in producing films). Therefore it is important in selecting a polymer compound to recognize both the endproduct qualities and the ease of processing. Shown in Table 2-18 is a classification of polyethylene by molecular weight.

Both high-density polyethylene (HDPE) and ultrahigh-molecular-weight polyethylene (UHMWPE) have interesting design properties. Recent developments in copolymerization techniques have eliminated the major problem of stress cracking that has plagued polyethylene since its discovery. Recent modifications also have produced polymers with narrow molecular weight distribution. Parts molded from these new resins are more warp-resistant than are their broad-distribution counterparts, and they permit injection molding of parts that are essentially flat. An application of narrow-molecular-weight-distribution HDPE is the injection molding of 5-gallon pails and drums that have excellent resistance to impact at low temperatures and are used to replace steel and other container materials. More recently, HDPE has been found to be the best material available for hip-joint replacements.

UHMWPE is the most impact-resistant thermoplastic known. It also has excellent resistance to abrasive wear. It will outwear stainless steel by a factor of 8-to-1 under certain conditions. Abrasion resistance apparently depends more on molecular weight *distribution* than on the molecular weight itself. The effect of molecular weight on the key properties of impact strength, tensile strength, and stiffness are shown in Fig. 2-57.

At this point there is no strict definition as to what constitutes a UHMWPE; however, in this discussion it has been assumed to have a molecular weight of 2 million or higher.

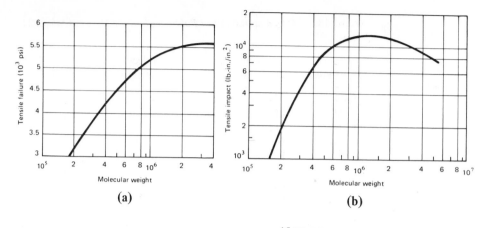

(a)   (b)

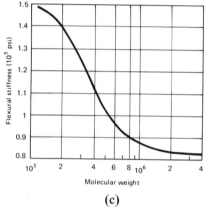

FIGURE 2-57.  *The effect of molecular weight on tensile failure (a), tensile impact (b), and flexural stiffness (c).*  *Courtesy* Machine Design.

(c)

## Thermoplastic and Thermosetting Structures

Shown in Fig. 2-58(*a*) is a sketch of an amorphous polymer with random-type attractions. The strength is limited because it depends on hit-or-miss "kissing" among the polymer chains. These secondary bonds are easily broken with heat, so such polymers are called *thermoplastics*. Common examples of this type of plastic are polystyrene and acrylics. Thermoplastics are readily molded or extruded because they melt to a viscous liquid at a temperature slightly above that of boiling water.

When the molecular chains are connected by a network of primary bonds it is very much like a three-dimensional spider web [Fig. 2-58(*b*)]. Even though heat is applied, the chains will not separate completely from one another. These polymers are termed *thermosetting*. Common examples of this type of plastic are urea-formaldehyde and phenolformaldehyde or Bakelite. In the example of phenolformaldehyde, the reaction is stopped at a point where the chains are still mostly linear. Unreacted chain portions are capable of flowing under heat and pressure. The final stage of polymerization is completed at the time of molding. The additional heat and pressure liquifies the polymer, producing a crosslinking reaction between the molecular chains.

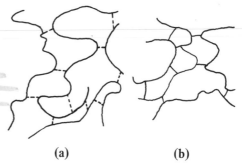

FIGURE 2-58. *An amorphous, thermoplastic polymer (a) showing weak crosslinking. Shown at (b) is a thermoset polymer with strong crosslinks. In the latter, after crosslinking has been established (usually by heat) the polymer is said to be cured.*

(a)                    (b)

Unlike the thermoplastic monomer, which has only two reactive ends for linear chain growth, a thermoset monomer must have three or more reactive ends, so the chains crosslink in three dimensions.

Once a thermoset plastic is molded, virtually all of its polymer molecules are interconnected with strong permanent bonds which are not heat reversible. If the thermoset plastic is heated too much or too long, chain breakage occurs, resulting in degradation.

### High-Temperature Plastics

High-temperature plastics have in the past been much higher in cost and presented greater processing difficulties than other materials. Now, however, some of the newer resins are processable by injection molding and the cost has been considerably reduced. The term "high temperature" in this context refers to being able to maintain mechanical properties and integrity at service temperatures above 350°F (177°C). Shown in Fig. 2-59 are precision-molded polyimide parts that are able to withstand a wide range of temperatures: from −400°F to 500°F (−240°C to 260°C) in continuous service and up to 900°F (482°C) for intermittent use. Long exposures to high temperatures causes a gradual reduction in properties, not sudden failure.

*Basic Structure.* As stated at the end of the discussion of thermosetting plastics, too much heat causes degradation. In producing high-temperature plastics, the objective is to increase the number of strong, *inert* (resistant to oxidative and chemical attack) chemical bonds in the backbone so that, at intervals, atoms are joined together by two bonds instead of one. This structure is analogous to that produced by double-joining of links in a chain. To do this, "rings" are incorporated into the polymer backbone, referred to as "aromatic rings." In this way, six carbon atoms are joined together in a hexagon structure (Fig. 2-60). This provides (a) stronger bonds (aromatic bonds are stronger than normal C—C single bonds) and (b) more bonds in a given space. Thus, more energy is required to break down the structure. The ring-type structure is typified by a polyphenylene hexagon structure or by the polyimide five-numbered rings containing carbon, oxygen, and nitrogen regularly placed around the ring.

FIGURE 2-59. *Precision polyimide parts able to withstand a wide range of temperatures: from −400°F to 500°F (−240°C to 260°C) in continuous service and up to 900°F (482°C) for intermittent use. Courtesy E. I. du Pont de Nemours & Co.*

FIGURE 2-60. *Six carbon atoms are joined together in a hexagon structure. This increases strength because there are more bonds in a given space and increases heat resistance because more energy is required to break it down.*

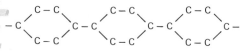

Polyphenylene

Fluorocarbons are examples of polymers whose bonds are particularly inert to chemical degradation. The polymer does not decompose chemically at high temperatures; but since it is a single-link polymer, it softens considerably at high temperatures.

A new and still experimental class of resins—called ladder polymers—are reported to have the greatest high-temperature strength yet achieved. In this structure, every atom is joined into a ladder-like structure. This type of structure is represented by the polybenzimidazoles. As might be expected, the ladder polymers are the most difficult of all to process.

### Selecting Pertinent Polymer Properties

Unlike metals, the property selection of polymers can be quite misleading. Only those properties that are clearly defined as mechanical constants are useful in comparing one polymer with another. As an example, tensile strength comparisons can be misleading because plastics tend to creep and show stress relaxation (viscoelastic effect) even at room temperature. Two ABS (acrylonitrile butadiene styrene) plastics may have identical tensile strengths of 5500 psi, but under long-term loading these may change to 2150 psi for one and 1200 psi for the other. Thus, for parts that must support loads, the long-term strength must be known.

Performance properties are those which essentially characterize the effect that mechanical loading will have at the stress encountered in the application. These include measurements of rigidity, or modulus, in various loading configurations, toughness index, Poisson's ratio, evaluation of creep under long-term conditions, impact strength, fatigue limits, and others.

The evaluation of environmental effects is considerably more critical for plastics than for metals. Too often the effect of temperature on several properties is taken as the only criteria in making a material selection. Time and space prohibit the discussion of all environmental conditions but certain key elements should be examined.

1. *Temperature exposure.* The temperature the product will operate in, including extremes and number of cycles between those extremes, must be considered.

A generalized comparison of the properties discussed in relation to temperature and modulus of elasticity of three different polystyrenes—crystalline, thermoset, and amorphous—are shown in Fig. 2-61. At room temperature, all three have about the same stiffness. At 220°F (104°C) the modulus of the crystalline material drops only slightly because of the interchain alignment. Only at well above 400°F (204°C) does the modulus drop, a signal that the crystalline regions are breaking up. The other two materials soften at 220°F (104°C). This initial loss of properties takes place at the glass transition temperature.

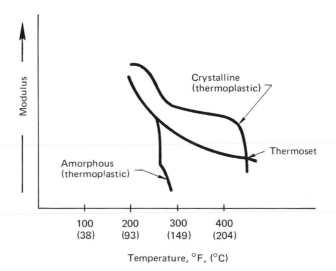

FIGURE 2-61. *A comparison of crystalline, amorphous, and thermoset properties of three different polystyrenes in relation to temperature and modulus of elasticity.*

2. *Chemical exposure.* Most polymers can be significantly damaged by many chemicals which act independently or in association with a mechanical load. Oxidation can be significant at elevated temperatures.

3. *Impact exposures.* In applications where the part is subject to frequent pounding, materials that have high impact strength or toughness must be selected. As an example, ABS plastics were chosen to replace fiber-reinforced plastic tote boxes for the post office because it would withstand the constant beating that boxes receive when traveling along conveyors.

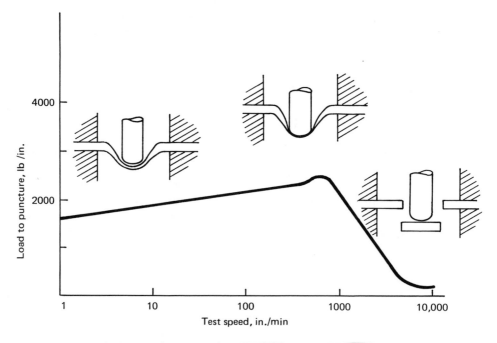

FIGURE 2-62. *A schematic and curve of a puncture test on plastics to rate comparative toughness or puncture resistance.*

Toughness varies with the rate at which the load is applied. Shown in Fig. 2-62 is a schematic of a puncture test and a corresponding curve. The plastic is clamped between two plates, exposing a 1-in. diameter surface. The rounded-end rod is then driven into the plastic at speeds up to 10,000 in. per min. As shown at low speeds, rupture is relatively independent of speed and closely related to tensile strength. As speeds increase, the load required to puncture drops off sharply. In this case, the peak for an acetal resin was reached at about 500 in. per min.

Comparison service tests show that a puncture resistance up to a speed of 2000 in./min provides a sound basis for selecting plastics that must withstand considerable abuse.

An old cliche states that the best way to evaluate a material and design is to study the finished product. Although this is fundamentally true, few companies can afford the luxury of waiting. Therefore every step from conception to final processing must be studied carefully. A thorough knowledge of materials and how they are affected by processing is essential for both the designer and the manufacturing engineer. With this knowledge, effective evaluation can be made at each step so that modifications can be made either in the design or process, or both, as required to produce the best possible product.

To design with plastics it is necessary to know not only the various materials that are available, their characteristics, and their service life, but also the various molding processes that can be used. For example, a given part may be made by injection molding, blow molding, thermoforming, or rotational molding. These processes are discussed in Chapter 5.

# ELASTOMERS

## Thermosetting Elastomers

Elastomers are a special type of thermoset polymer that have long-chain molecular structures whose glass transition temperatures are well below room temperature, usually in the range of $-20$ to $80°F$ ($-29$ to $26.7°C$). In their "green" state they are tacky, and cold-flow easily. To be useful they must be crosslinked. With rubbers this is termed vulcanization.

Elastomer molecules may be compared to a network of thin, soft coil springs. When this structure is stretched, the coils open until they are almost straight. The motion is limited by the number of junction points or crosslinks. When the stress is removed, the assembly returns to its original coiled position. The farther apart the crosslink points, the weaker the system and the greater the elongation.

Thus, an elastomer or rubber is basically a low-$T_g$ thermoset with relatively wide spacing between crosslinks in the polymer chains. This structure enables the elastomers to readily absorb and dissipate energy.

*Rubber Classification.* The standard that has been adopted for rubber classification is ASTM D2000 "A Classification System for Elastomeric Materials for Automotive Applications." While this standard was originally intended to cover automobile uses, it covers all potential rubber compositions. The following is a summary of some of the principal types.

**R Class:** R-class rubbers are those that have an unsaturated carbon chain and may be natural or synthetic.

*Natural Rubber (NR).* Because of its relatively low cost and outstanding physical properties in both reinforced and unreinforced state, natural rubber is still the only material that satisfies many applications. It may be compounded to have good abrasion and high tear resistance. Special compounding is needed if good weathering is required. It is ideal for general purpose applications where "snap" or resilience is important.

Typical applications for natural rubber are pipe gaskets, respirators, windshield wiper blades, swimming goggles, drain boards, and pharmaceutical products.

*Styrene Butadiene (SBR).* SBR was the first elastomer to find wide acceptance as a substitute for natural rubber. It is a low-cost, high-volume rubber that is available in various ratios of the two monomers. If compounded with a high-styrene and low-butadiene content, it has high tensile strength, high modulus, and high hardness without the use of large amounts of filler materials. Due to its good abrasion resistance and leather-like properties, it is used in the manufacture of shoe soles and floor tile. High-butadiene, low-styrene copolymers offer good flexibility at low temperatures.

Typical applications for SBR rubbers include tires, pipe gaskets, automotive brake components, and other mechanical items.

*Nitrile Butadiene (NBR).* NBR is a copolymer of acrylonitrile and butadiene. It is made in various ratios of these monomers. Nitrile rubbers have good oil and gasoline resistance, tensile strength, elongation properties, heat resistance, and low compression set. Special compounding can produce good weatherability.

Typical applications are seals, O-rings, gaskets, diaphragms, pipe gaskets, tank linings, boots, and bellows.

*Chloroprene (CR).* Chloroprene, also known as neoprene, is made from chlorine and butadiene in several formulations. Chloroprene rubbers have good resistance to heat, oil, chemicals, ozone, weathering, flexing, compression, stain, and flame.

Typical applications are hoses used in handling fuels, bellows, gaskets, seals, tank linings, and O-rings.

**M Class:** M-class rubbers have a saturated chain of the polymethylene type.

*Polyacrylic (ACM).* ACM elastomers are based on ethyl or butyl acrylate or a combination of the two monomers, with other comonomers.

Polyacrylic rubbers are superior to nitrile rubbers in resistance to deterioration by high-aniline-point oils, extreme-pressure lubricants, and transmission fluids. They also have excellent resistance to ozone, sunlight, and weathering, and are suitable for use down to temperatures of $-40°F$ ($-40°C$) and intermittent use at temperatures up to $300°F$ ($150°C$).

Other M-class rubbers are: chlorosulfonyl polyethylene (CSM), chloropolyethylene (CM), ethylene propylene (EPM), fluorocarbon (FKM), and epichlorohydrin (ECO). For further discussion of the various properties and uses, see the bibliography at the end of the chapter.

**Q Class:** The Q-class rubbers have a substitute group on the polymer chain and are indicated prior to the silicone (Q) designation.

*Fluorosilicone (FVMQ).* The outstanding characteristic of all types of silicone rubbers is the wide service-temperature range, $-55$ to $200°F$ ($-48$ to $93°C$). In addition, they have excellent resistance to oils and fuels.

Silicones are relatively expensive rubbers which include several types. Typical applications are for seals, gaskets, tank linings, and diaphragms. Some versions require postcure for development of optimum properties.

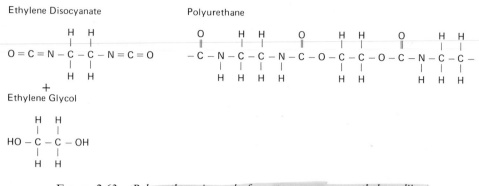

Ethylene Disocyanate

Polyurethane

Ethylene Glycol

FIGURE 2-63. *Polyurethane is made from two monomers, ethylene diiso-cyanate and ethylene glycol by the condensation process. Two molecules combine, usually with the elimination of water (H—O—H) to form the repeating units of the copolymer chain.*

**U Class:** U-class rubbers are those having carbon, oxygen, and nitrogen in the polymer chain, as shown in Fig. 2-63.

*Polyurethane Rubbers.* The two types of polyurethanes are polyester urethane (AU) and polyether urethane (EU). Both are available in viscous liquid and millable gum form. The gum forms are processed in conventional rubber equipment, and the viscous liquid form depends on the part being fabricated. Casting is frequently used for the liquid version. The properties generally exceed most other rubbers in higher hardness range, extremely high tensile strength, and abrasion resistance. In addition, these urethanes have high load-bearing characteristics, low compression set, good aging characteristics, and good oil and fuel resistance.

Typical applications include seals, bumpers, metal-forming dies (Fig. 4-12), valve seats, liners, rollers, wheels, and conveyor belts.

### Thermoplastic Elastomers

Thermoplastic elastomers are similar to the thermoset type in service properties. However, they do offer greater ease and speed of processing. Molding or extruding can be done on standard plastic-processing equipment with considerably shorter cycle times than that required for compression molding of conventional rubbers.

The principal thermoplastic elastomers are polyester copolymers (*Hytrel*, Du Pont Co.), styrene–butadiene block copolymers (*Krayton*, Shell Chemical Co.; and *Solprene*, Phillips Petroleum Co.), polyolefin (*TRR*, Uniroyal, Inc.), and polyurethane, from a number of producers.

*Polyurethane.* Polyurethane was the first major elastomer to be developed that could be processed as a thermoplastic. It does not have quite the heat resistance or the compression set resistance of the crosslinked types, but most other properties are similar.

115

## Material Specifications and Properties of Elastomers

The system of specifying elastomers as given in ASTM D2000 or SAE 5200 is based on properties required, rather than material composition. An example of a specification requirement may be as follows:

Applicable type of material: chloroprene
Original properties:
    Hardness    $50 \pm 5$ (durometer)
    Tensile    1500 psi min
    Elongation   350 % min
Air oven, 70 hr at 100°C:
    Hardness change    +15 max
    Tensile change    −15 max
    Elongation change  −40 % max
ASTM #3 oil, 70 hr at 100°C:
    Tensile change    −70 % max
    Elongation change  −55 % max
    Volume change   +120 % max
Compression set, 22 hr at 100°C:
    Set   80 % max

# THE RECLAMATION OF RUBBER AND PLASTIC

*Rubbers.* To the ecologically and energy minded world the problem of rubber reclamation has become one of prime importance. The process is not new. Goodyear patented a process for grinding waste rubber into powder and mixing it with pure rubber in 1853. Some work has been done on "desulphurizing" rubber in an attempt to break down the crosslinked structure in order to be able to reprocess and revulcanize it.

One of the most used processes is the steam process, which does not necessitate prior grinding of the waste products. The rubber is cut and put into an autoclave with 40 to 70 psi steam pressure or with superheated steam at 400 to 430°F or with saturated steam at 670 to 700 psi, after which it is dried and refined.

Somewhat the same means are employed in recovering synthetic rubbers. However, special plasticising oils or special catalysts are usually added. It is almost exclusively SBR and NBR that is recovered on an industrial scale.

Since more than 215 million old tires are discarded annually in the United States, considerable interest is being developed toward possible uses. At the present time only a small percentage (approximately 10 %) is reclaimed. A process under investigation at the University of Wisconsin utilizes liquid nitrogen for freezing the rubber and then shattering it by impact. The result is small rubber granules that can be used in athletic tracks, industrial floor mats, and landfill.

*Plastics.* Thermoplastics can be ground up and remolded, but the thermosetting resins cannot. Since collection of used plastic containers poses an environmental problem, comprehensive studies have been carried out, and are still in process, to determine the extent of plastic pollution. Newer plastics are being developed, especially for the packaging industries, that are biodegradable or susceptible to biodeterioration. Evidence has now been introduced to show that polyvinyl chloride plasticized with dodecyl esters of aliphatic dibasic acids are much more biodegradable than those formulated previously. Modifications of polyolefin polymers by the introduction of carboxyl, amide, and other substituting groups has significantly increased their biological conversion.

New mutant strains of soil microorganisms are also being developed that show increased capacity for degrading the more resistant types of synthetic polymers. The use of specialized bacteria and fungi as inoculants in waste disposal areas will do much to help the problem of degrading waste polymers.

Attention has also been given to the incineration of plastic packaging materials. A new material, Barex 210FDA, is suitable for a wide variety of food, beverage, and commercial products that can be conveniently disposed of by incineration with no hazard to the environment. The new material is described as having good transparency, shatter resistance, and lightness.

## COMPOSITE MATERIALS

The dramatic technological advances of the past few decades have, in part, been based on ever better material properties. Limitations on how far the material properties can be expanded may present equivalent limitations on further successes in aerospace, hydrospace, nuclear energy, defense, transportation, and other fields.

Composite materials that have the potential of integrating various metals, ceramics, polymers, and natural materials appear to be well on the way toward eliminating some of the obstacles that relate to strength and high-temperature performance.

### Fibers, Filaments, and Whiskers

*Fibers.* Man has only to observe nature to learn that fibers, as in plants, increase the strength of a structure considerably. The word "fiber" is generally associated with threadlike materials like those found in wool, cotton, or hemp. The main characteristic of a fiber is that its length is considerably greater, at least 200 times greater, than its effective diameter. Two well known fibers used in today's composites are glass and high-strength wire, as shown in Fig. 2-64.

*Filaments.* Filaments are defined as "endless" or continuous fibers. Filaments are used in making glass, yarn, and cloth. Glass filaments are formed by jets of air which pull glass strands from a spinneret onto a revolving drum. These filaments can be handled and processed on a mass-production basis.

*Whiskers.* Whiskers are single crystals that are grown with almost no defects. They are short, discontinuous fibers of polygonal cross sections that can be made from over

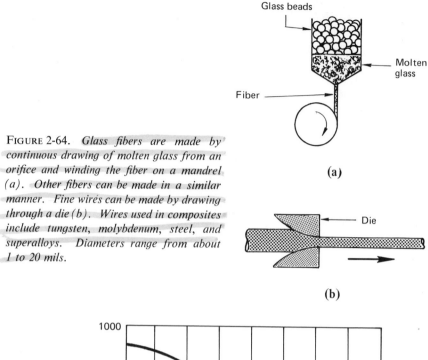

(a)

FIGURE 2-64. *Glass fibers are made by continuous drawing of molten glass from an orifice and winding the fiber on a mandrel (a). Other fibers can be made in a similar manner. Fine wires can be made by drawing through a die (b). Wires used in composites include tungsten, molybdenum, steel, and superalloys. Diameters range from about 1 to 20 mils.*

(b)

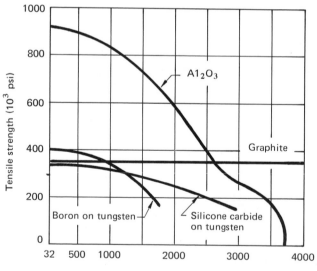

FIGURE 2-65. *Metals lose much of their tensile strength with increasing temperature. Graphite offers obvious advantages with its flat strength curve, 350,000 psi to at least 4000°F (2250°C). Graphite unfortunately is incompatible with many metal matrix materials. Silicon carbide is relatively compatible with metals and its strength is over 200,000 psi at 2000°F (1093°C). Alumina (sapphire) has the highest strength above 2500°F (1371°C).* Courtesy Machine Design.

100 materials. Common examples are copper, iron, graphite, silicon carbide, aluminum oxide, silicon nitride, boron carbide, and beryllium oxide. The tensile strength of some of these fibers in relation to heat is shown in Fig. 2-65.

Generally, whiskers grow from vapor or metallic depositions of gases or liquids on a surface, as in the case of sapphire (Fig. 2-66). Oxide, carbide, and nitride whiskers are of primary interest in composite-materials technology. For a long time, whiskers were only a laboratory curiosity, but in 1952 Herring and Galt, of the Bell Telephone Laboratories, observed that the strength of metallic tin whiskers was 1,000,000 psi.

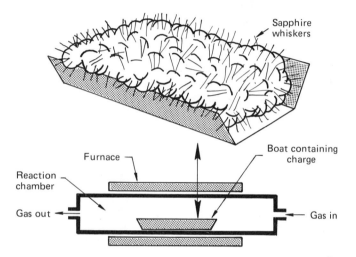

FIGURE 2-66. *Whiskers are grown from materials in a boat or dish placed in a heated, controlled-atmosphere reaction chamber. For example, aluminum in a ceramic boat, in the presence of hydrogen, forms vapors of $Al_2O$ or $AlO$. These vapors, in turn, react with oxygen (present in the form of water vapor or SiO vapor) that forms from the boat material. The result is solid, needle-like, single crystals of alumina that grow on the surfaces of the boat or its contents. Whiskers usually grow to about 0.5 in. long but sometimes may attain a length of about 2 in.*

**Multiphase Fibers.** Multiphase fibers consist of materials such as boron carbide and silicon carbide that are formed on the surface of a very fine wire substrate of tungsten (Fig. 2-67).

### Fiber–Metal Composites

The primary purpose of making a composite is to incorporate the strength or the unusual properties of the fibers into usable shapes by bonding them into a continuous matrix, as shown in Fig. 2-68. The tensile strength and stiffness of a composite

119

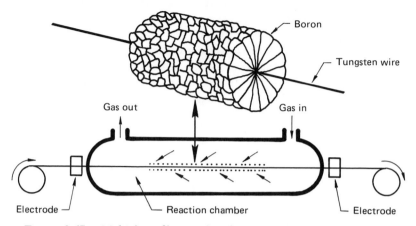

FIGURE 2-67. *Multiphase fibers such as boron on a tungsten substrate are commonly produced by vapor deposition methods. Heated tungsten wire of about 0.0005-in. diameter is moved through a chamber in which boron is deposited on the wire from gases containing a boron compound. The final diameter of the multiphase fiber is about 0.004 in.*

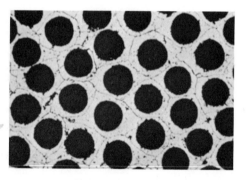

FIGURE 2-68. *The cross section of coated fibers hot-pressed in an aluminum matrix.*

are directly related to the fiber content and to the strength of the fiber, assuming the fibers are aligned in the direction of loading.

The handling, storing, and environmental conditions are important to the successful fabrication of composite structures. Bends, nicks, and scratches can seriously reduce the properties of fibers. Even long exposure to air can be detrimental.

One of the easiest methods devised of fabricating a metal-matrix composite consists of infiltrating molten metals into bundles or stacks of parallel fibers. These mats can then be further shaped by hot-rolling, forging, pressing, or other conventional forming methods. Another method of fabricating composite structures is discussed in Chapter 10 under Electroformed Composite Structures.

Fibers may also be embedded in powdered metals, which are subsequently sintered and hot-pressed to full density.

Another method used is that of plating the fibers to the desired thickness, then stacking them together and sintering to consolidate and densify the deposited matrix metal. The fibers may also be molded into a plastic matrix, usually epoxy, polyester, polyimide, or nylon.

### Present Uses and Trends

The largest expenditure of research and development money in composites has been in the area of carbon and graphite fibers. Although these fiber composites were developed for the aerospace industry, it is now found that golf shafts alone consume 70 to 75 % of the total production. The price has plummeted from $300 per pound a few years ago to $35 to $50 per pound. Volume production and better processing methods can drop this cost still further.

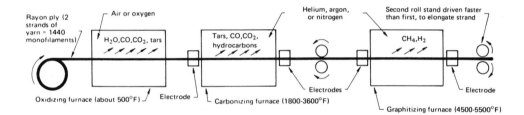

FIGURE 2-69. *Graphite fibers are made by a process that converts organic materials to carbon, then to graphite. Bundles of the rayon or polyacrylonitrile are heated to drive off most organic constituents, leaving a residue of carbon. The carbon strand is then heated to a higher temperature in controlled atmosphere chambers and a slight stress is applied to produce elongation of 40 to 60%. The strands elongate and the molecular structure is changed from carbon to graphite. The diameter of the graphite filaments is about 5 to 10 microns.*

Continuous graphite fibers (also called carbon fibers) are produced by oxidizing, carbonizing, and graphitizing polyacrylonitrile (rayon) materials at 2500°C (Fig. 2-69). Since the density of graphite composites is about half that of aluminum and one-fifth that of steel, the specific properties (strength-to-density and modulus-to-density ratios) offer substantial performance advantages over those of commonly used metals.

Graphite composites also have other properties of interest to engineers. They have low coefficients of thermal expansion, excellent wear resistance, long fatigue life, and are electrically conductive. They also have exceptional vibration dampening characteristics and their flexural strength is about 2.5 times that of steel.

Other areas being investigated for this comparatively new material are tennis racquets, skis, bicycle frames, guitars, and automobile bumpers.

## PROBLEMS

**2-1.** What are some of the major factors that design and manufacturing engineers would have to consider if management posed the problem of making an automobile hood out of aluminum that is now made out of steel? The present hood material is AISI 1010 steel, 0.035 in. thick.

**2-2.** Shown in Fig. P2-1 are three sketches of parts. Show how the design of each of these parts can be improved, assuming they will be subjected to stress loads or heat-treating stresses.

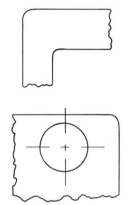

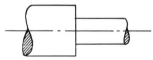

FIGURE P2-1

**2-3.** If a mild-steel butt weld in $\frac{1}{8}$ in. thick steel is normally made with $1t$ spacing between the plates, would the same rule apply to stainless steel? Why or why not?

**2-4.** A mild-steel cold-rolled rod 0.5 in. in diameter is used to support a maximum tensile load of 20,000 lb. It was found after some time that a portion of the rod had "necked down" to a minimum of 0.400 in. in diameter and the length had increased from 8 in. to 8.03 in. (a) What was the nominal

ultimate stress for this rod? (b) What was the maximum natural strain?

**2-5.** If the load at yield in problem 2-4 was 12,400 lb, what would the nominal yield stress be?

**2-6.** (a) What would the true stress be for the rod in problem 2-4? (b) What would be the true strain? (Dia. at max. load = 0.45 in.)

**2-7.** Aluminum sheets are used in roofing. Three 12-ft long sheets are used end to end. What will the total difference in length be between the temperature range of $-20°F$ to $+95°F$?

**2-8.** A steel bar is to be subjected to a load of 12,000 lbs. What diameter should it be if the stress will not be in excess of 20,000 psi?

**2-9.** An aluminum bracket with an initial area of 0.5 sq in. is used to support a maximum tensile load of 10,000 lb. After several weeks in service it fractures. At the time

of fracture it was supporting a load of 8000 lb. It was found that the area at fracture had reduced to 0.4 sq in. (a) What was the nominal stress at fracture? (b) What was the true stress at fracture?

**2-10.** The bracket, in problem 2-9, was originally 2 in. long. It was found to be 2.38 in. long at the time of fracture. What were the nominal and true strains at fracture?

**2-11.** Ten-in. diameter blanks are to be made by shearing in a standard punch and die. The material is AISI 1015 plain carbon steel $\frac{1}{8}$ in. thick. What size press, tonnage wise, would be required for this operation? Assume a 100 % safety factor. The shear strength of the material is 45 ksi.

**2-12.** A high-carbon steel rod of 0.5 in. diameter is hardened, quenched, and tempered to $R_c$ 54. How much difference would there be in the maximum tensile strength of the bar if it were tempered to $R_c$ 50?

PROBLEMS

**2-13.** (a) Why is it customary or advantageous to flame harden such parts as the bearing surfaces on top of a lathe bed or the guideways used for a large machine table to slide on? (b) What hardness, and to what depth, can be expected by this process?

**2-14.** The hardness of some razor blades needs to be checked. What types of hardness checking should be used and why?

**2-15.** (a) Given an aluminum blank 10 in. in diameter and $\frac{1}{8}$ in. thick, can it be drawn to a 6-in. diameter cup in one draw? (b) What would the percent reduction be?

**2-16.** (a) Can the 6-in. diameter cup of problem 2-15 be reduced to 3 in. in a second draw? Why or why not? (b) Would a 4-in. diameter cup be satisfactory for a second draw? Why or why not?

**2-17.** What is the difference between nominal tensile strength and true tensile strength?

**2-18.** What are the advantages of using a standard-type specimen for tensile testing over a straight rod or strip?

**2-19.** Why is a high-carbon steel so much stronger after heating above the upper critical temperature and quenching in water?

**2-20.** (a) Compare the tensile and yield strengths of 1100-O and 1100-H18 aluminum. (Use the reference material such as the Metals Handbook or the Alcoa Aluminum Handbook). (b) Explain the reason for the change in tensile and yield strengths.

**2-21.** (a) What is the main difference between 1100 aluminum and 2024 aluminum as to content. (b) Explain the mechanism by which the high tensile strength is achieved in 2024-T6. (c) What aluminum has a tensile strength of over 100,000 psi?

**2-22.** Make a graph to show by means of a curve the weldability comparison of low-, medium-, and high-carbon steels.

**2-23.** What tool steels would be used for the following applications? (a) Bolt-header dies

for cold heading. (b) Blanking dies. (c) Forging dies. (d) Milling cutters.

**2-24.** (a) What are the four types of cast irons discussed in this chapter? (b) What is the form of the carbon in each of the cast irons and what is the matrix form?

**2-25.** How does the tensile strength of an annealed ductile iron and a ferritic, malleable cast iron compare with that of annealed medium-carbon steel at a strain of $3 \times 10^{-3}$ in. per in.?

**2-26.** What would the approximate tensile strength be of a gray cast iron of the following chemical analysis: Si 1.14 %, Mn 1.07 %, P .18 %, and total carbon 2.98 %?

**2-27.** What would be the approximate tensile strength and the minimum compressive strength of ASTM class-40 gray cast iron?

**2-28.** Why is malleable cast iron generally considered nonweldable?

**2-29.** Which stainless steel would you recommend for each of the following? (a) Hardness and strength. (b) A very corrosive environment. (c) Good weldability. (d) Good machinability. (e) Good forming qualities. (f) Very high strength.

**2-30.** What modification is recommended for a standard drill if it is to be used in drilling copper or aluminum?

**2-31.** (a) Why does a metal with a large grain size form more easily than that with a small grain size? (b) What may be a disadvantage of a large grain size?

**2-32.** Why are composite materials able to achieve fantastic strengths, some even at elevated temperatures?

**2-33.** What fibers used in composites are being introduced into commercial applications?

**2-34.** What metal will compare favorably in tensile strength to a graphite composite at room temperature?

123

**2-35.** What would be the approximate tensile strength of a 0.5-in. diameter rod of graphite composite at 3000°F (1643°C)?

**2-36.** (a) Why is martempering better than the conventional quench and temper process? (b) Why isn't martempering always used instead of the conventional quench and temper process?

**2-37.** What are the main differences between martempering and austempering?

**2-38.** What are two main advantages of using cast steels in the fabrication of equipment?

**2-39.** Identify each of the following materials: 1020, 5100, 1100-O, class 20, 4150, 2024-T3, 302, 80-60-03.

**2-40.** What is the difference in structure between a plastic and an elastomer?

**2-41.** What molecular weights constitute a low range, high range, and ultrahigh range for polyethylene?

**2-42.** What type of molecular structure makes it possible to achieve a very crystalline polymer structure?

**2-43.** What is the advantage of copolymerization?

**2-44.** What type of molecular structure is needed to resist high temperatures?

**2-45.** How does the evaluation of design properties of a plastic differ from that of a metal?

**2-46.** What are some ways that rubber can be reclaimed?

## BIBLIOGRAPHY

**Books**

ALLEN, D. K., *Metallurgy Theory and Practice*, American Technical Society, Chicago, 1969.

CLARK, D. S. and W. R. VARNEY, *Physical Metallurgy for Engineers*, D. Van Nostrand Company, Inc., New York, 1962.

HEINE, R. W., C. R. LOPER and P. C. ROSENTHAL, *Principles of Metal Casting*, McGraw-Hill Book Company, New York, 1967.

LINDBERG, R. A., *Materials and Manufacturing Technology*, Allyn and Bacon, Inc., Boston, 1968.

SEHGAL, S. D., and R. A. LINDBERG, *Materials—Their Nature, Properties and Fabrication*, Kailish Publishers, Singapore, 1973.

VAN VLACK, L. H., *Elements of Material Science*, Addison-Wesley Publishing Company, Inc., Reading, Massachusetts, 1964.

WALTON, C. F. (editor), *Gray Iron Castings Handbook*, Gray and Ductile Iron Founders' Society Inc., Cleveland.

**Periodicals**

Design Guide, "Cast Iron and Steels," *Machine Design*, March 12, 1964.

JONES, C. K., S. W. MARTIN and D. J. STURGES, "Ionitriding—A Modern Casehardening Technique," *Modern Machine Shop*, Sept. 1973.

WHEETON, J. W., "Fiber Metal Matrix Composites," *Machine Design*, Feb. 20, 1969.

# 3

# *Metal-Cutting Theory and Operations*

## Introduction

Material removal by mechanical methods began in the stone age when our early ancestors discovered that materials could be shaped by a chipping process. As early as 4000 B.C., the Egyptians, to get closer control of the removal process, used a rotating bowstring device to drill holes in stone. Scientific work in metal cutting is of rather recent origin, beginning about the mid-nineteenth century. The first attempts at defining chip formation were reported in 1870.

## MECHANICS OF METAL CUTTING

In any metal-cutting operation, the unit product of metal removed is called a "chip." The chip is always thicker than the layer removed from the workpiece, thus the mechanism of plastic deformation is at work. Also the separation of the chip from the parent metal means shearing and/or cleavage is taking place. Friction develops as the chip slides over the face of the tool which results in heat and wear (Fig. 3-1).

The cutting action can be more easily analyzed if the edge of the tool is set perpendicular to the relative motion of the material, as shown in Fig. 3-2. This two-dimensional type of cutting, in which the cutting edge is perpendicular to the cut, is known as *orthogonal*, as contrasted with three-dimensional *oblique* cutting as shown in Fig. 3-1, in which the cutting edge is inclined (at an oblique angle) to the cutting direction.

FIGURE 3-1. *Schematic view of metal cutting and indication of forces.*

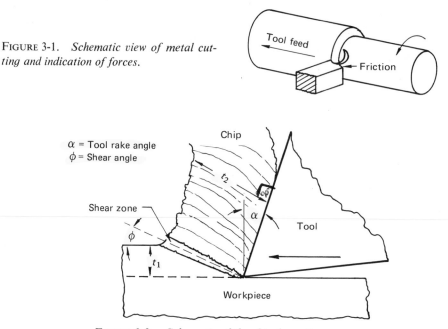

FIGURE 3-2. *Schematic of the chip-formation process.*

## Orthogonal Cutting

In orthogonal cutting, a surface layer of constant thickness ($t$) is removed by the relative movement of the tool and workpiece. Some indication of the ease or difficulty of the machining process can be obtained by comparing the chip thickness ($t_2$) to the original depth of cut ($t_1$). The ratio of $t_1$ to $t_2$ is defined as the *chip-thickness ratio* ($r$). Usually, materials that do not shear well but "pile up" in front of the tool are considered less machinable than those materials that have only a small increase in chip thickness. Thus a measure of the ease of metal removal or *machinability* is expressed by the chip-thickness ratio. This ratio may also be expressed as the shear angle, $\phi$.

$$\tan \phi = \frac{r \cos \alpha}{1 - r \sin \alpha}$$

*where:*    $\alpha$ = rake angle of the tool, as shown in Fig. 3-2.

The closer the shear angle approaches a 1 : 1 ratio, or 45°, the better the machinability is said to be.

## Chip Formation

About 95 % of the power expended for metal removal is used in the deformation taking place in the shear zone. This is the work required to form and remove the chip and the incidental plastic deformation of the surface layer of the finished workpiece.

The remaining power consumed, about 5 % of the total, is expended in stored elastic energy or *residual stresses* in the workpiece.

The chips are formed largely by shearing action and compressive stresses on the metal in front of the tool. The compressive stresses are the greatest farthest from the cutting tool and are balanced by tensile stresses in the zone nearest the tool, hence the chip curls outwardly or away from the cut surface.

### Chip Types

Chips, as formed in metal machining, have been classified into three basic types: discontinuous, continuous, and continuous with a built-up edge (Fig. 3-3).

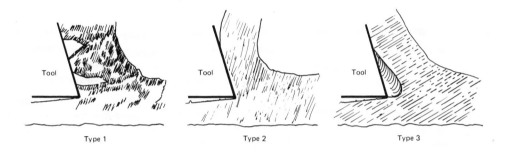

FIGURE 3-3. *The three main types of chips: (1) segmented, (2) continuous, (3) continuous with a built-up edge.*

*Discontinuous Chip.* The discontinuous chip is one that separates into short segments which may or may not adhere to each other. Severe distortion of the metal occurs adjacent to the tool face, resulting in a crack that runs ahead of the tool. Eventually the shear stress across the chip becomes equal to the shear strength of the material, resulting in fracture and separation. With this type of chip, there is little relative movement of the chip along the tool face.

*Continuous Chip.* The continuous chip is characterized by a general flow of the separated metal along the tool face. There may be some cracking of the chip, but in this case they usually do not extend far enough to cause fracture. This chip is formed at the higher cutting speeds when machining ductile materials. There is little tendency for the material to adhere to the tool. The continuous chip usually shows a good cutting ratio and tends to produce the optimum surface finish.

*Continuous With A Built-Up Edge.* This chip shows the existence of a localized, highly deformed zone of material attached or 'welded' on the tool face. Actually, analysis of photomicrographs show that this built-up edge is held in place by the static friction force until it becomes so large that the external forces acting on it cause it to dislodge, with some of it remaining on the machined surface and the rest passing off on the back side of the chip.

The built-up edge (BUE) is closely related to cutting speed with its related heat and friction. It does not appear at very low speeds since there is neither sufficient heat or friction developed. At much higher speeds, the heat developed is sufficient to cause the material in the shear zone to be annealed, eliminating the strain hardening that takes place in BUE. Studies show that the greatest BUE takes place in a temperature range of 750 to 800°F (399–427°C) for mild steel (at intermediate cutting speeds).

The BUE often results in an inferior surface finish, since the effective tool geometry is altered.

## CUTTING-TOOL MATERIALS

Relatively few materials are suitable to be used as tools for metal removal. A cutting tool must withstand high unit pressures in the order of magnitude of $10^5$ to $10^6$ psi. Chips slide across the face of the tool ranging in speeds from 10 to 10,000 fpm (30–30,000 cm/sec) at temperatures ranging from 300 to 1500°F (149–816°C). Thus a cutting tool requires a combination of properties, but principally high hardness and moderate degree of toughness at elevated temperatures. In addition, the microstructure must present wear-resistant qualities. The most common materials that meet these requirements, in varying degrees, are: high-speed steels, cast alloys, cemented carbides, cast carbides, UCON*, and ceramics.

*High-Speed Steels (HSS).* High-speed steels, which were introduced at the turn of the century, presented a vast improvement over the high-carbon steels used previously. One of the most common types of HSS contains 18 % tungsten, 4 % chromium, and 1 % vanadium. It is often referred to as tungsten-type HSS.

HSS are able to retain a high hardness up to the range of about 1000°F (538°C). The ability to hold a cutting edge at dull red heat is known as a *red hardness*. An examination of the microstructure of HSS reveals hard, brittle, refractory carbide particles held in a lower-melting-point matrix. More recently, super HSS have been introduced. A study of their microstructures reveals a more even distribution of the carbide particles, as shown in Fig. 3-4. These ultrahard HSS are essentially high-carbon, high-cobalt, molybdenum-type steels. Cutting speeds for HSS tools are given in Appendix 3A.

HSS find their best application in drills, reamers, counterbores, milling cutters, and single-point tools for lathes and shapers.

*Cast Alloys.* Cast alloys consist of cobalt, chromium, tungsten, and carbon. Since there is no iron, the materials are only obtainable in the cast condition, hence the name. After casting, the tools are subjected to heat treatment to break up the carbide network and produce high hardness throughout. Cast alloys are better known by their trade names of *Stellite, Blackalloy, Crobalt,* and *Tantung.*

* Trade name of Union Carbide Corporation.

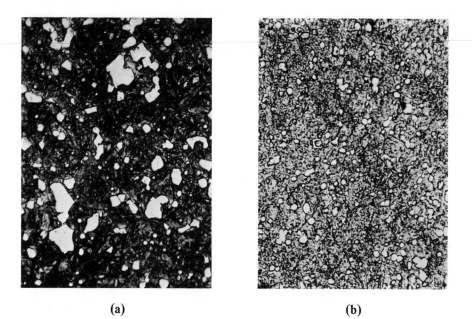

(a)                                          (b)

FIGURE 3-4. *Photomicrographs of (a) conventional HSS and (b) super HSS showing the difference in size and distribution of the carbide compounds, original magnification 1000×. Courtesy Society of Manufacturing Engineers.*

***Cemented Carbides.*** Cemented carbides have the highest ratio of carbides to matrix of the cutting-tool materials—about 80 %. Carbide tools are made by mixing pure tungsten powder under high heat with pure carbon (lampblack) in the ratio of 94 % and 6 %, respectively, by weight. The resulting compound, WC, is then mixed with cobalt until the mass is homogenous. The powder mixture is compacted at high pressure and sintered in a furnace at 2500°F (1371°C). After cooling, the carbide tools are ground and, in some cases, subjected to further finishing operations.

Although tungsten carbide tools were a boon to industry, it was found that in machining steels there was a tendency for the lower melting-point cobalt to wear first, exposing the carbide particles. This was followed by a seizure of both the high and low points of the cutting tool surface. To help prevent this condition, tantalum and titanium were added to the basic composition, hence the standard steel-cutting carbide tool contains WC–Co–TaC and TiC.

A newer development in carbide tools is a process of applying a thin 0.0002 to 0.0003 in. coating of titanium nitride (TiN) uniformly over all surfaces by means of a vapor deposition. Since this coating has no binder, it is substantially harder than the base carbide. In conventional WC grades, the chip normally welds to the cobalt binder. Test results show that, in machining alloy steels, the TiN-coated tool has a lower coefficient of friction, hence the shear angle is higher, resulting in less vibration

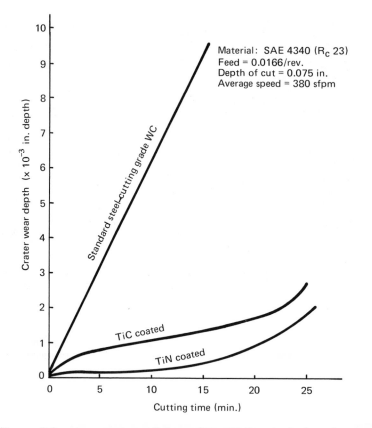

Material: SAE 4340 ($R_c$ 23)
Feed = 0.0166/rev.
Depth of cut = 0.075 in.
Average speed = 380 sfpm

FIGURE 3-5.  *A comparison of the tool wear of a standard steel-cutting grade of carbide and coated carbides.*

and better surface finish.  In addition, the coated WC tool can be used at higher cutting speeds and still maintain the same tool life (Fig. 3-5).  The crater wear for cast carbides is linear with time; while for coated carbides, the curve takes on a steeper slope, reflecting the effect of erosion of the coating.  These tools are not recommended for applications in which heavy scale or sand inclusions would erode the surface coating.  Shown in Appendix 3B is a table of carbide tools and recommended cutting speeds for given materials.  Cemented carbides used in metal cutting are classified from C-1 to C-8.  The lower numbers are of the straight tungsten-carbide–cobalt type and grade for grade are the more wear-resistant.  The higher numbers, C-5 to C-8, are the steel-cutting grades with tantalum and titanium added, as shown in Table 3-1.

***Cast Carbides.***  Cast carbides are a relatively new development and are an outgrowth of a study to find a suitable refractory alloy which would replace the low-melting binder (mainly cobalt) as used in the sintered carbide tools.  An alloy that filled this requirement was a Ti–W–C system.  When heated to 2700°C, a pseudobinary eutectic

TABLE 3-1. *Carbide Cutting-tool Classification*

| Cast Iron, Nonferrous, and Nonmetallic Materials | Steel and Steel Alloys |
|---|---|
| C-1 Roughing | C-5 Roughing |
| C-2 General purpose | C-6 General purpose |
| C-3 Finishing | C-7 Finishing |
| C-4 Precision finishing | C-8 Precision finishing |

occurs. Upon solidification, a solid solution composed of tungsten-rich alloy forms. Tungsten provides high strength. The finely distributed carbides provide wear resistance and act as a binder for the tungsten base. A comparison of the conventional W–Ti–C tool, coated and uncoated (C-5), and a cast carbide tool is shown in Fig. 3-6. At the start of the cuts, the coated carbides exhibited the greater resistance to crater wear but, as shown, the cast carbide wear is linear with time whereas the coated carbide is not.

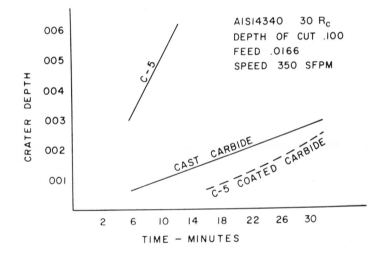

FIGURE 3-6. *A comparison of crater wear between a conventional steel-cutting grade of carbide (C-5) coated and uncoated with a cast carbide tool. Courtesy Teledyne Firth Sterling.*

***Alumina Ceramics.*** The term "ceramic tooling" is applied to tools made of aluminum oxide. The strongest tools are made by hot pressing a finely divided aluminum oxide powder in a graphite mold to obtain a solid piece free of porosity within a fraction of 1 %.

Of all tool materials, aluminum oxide best resists high temperatures. It is also hard, being surpassed only by TiC and diamonds. Its high hardness and high temperature resistance combine to provide superior wear resistance. This means excellent tool life on abrasive materials such as cast iron. Tool life is often three times that of other materials, even at three times the cutting speed. The high heat resistance also means better thermal stability with more accuracy of the machined surface.

The combination of high hardness and low toughness limits the usefulness of $Al_2O_3$ tools mainly to high-speed turning of cast irons and high-strength steels. Ceramic tools are at the point of awaiting better machine tools with greater rigidity and higher horsepower. Shown in Appendix 3C is a table of recommended cutting speeds for various materials using ceramic tools.

*UCON.* A relatively new tooling material developed by the Union Carbide Corporation is called UCON. It contains no carbides. An example of the composition, by weight, that may be used is 50 % columbium, 30 % titanium, and 20 % tungsten refractory-metal alloy. The surface hardness is on the order of 2500 to 3000 DPH (Vickers diamond-pyramid hardness). This is contrasted with tungsten carbide, whose hardness normally ranges from 1600 to 1800 DPH, and ceramics, at about 2000 DPH.

Tool inserts are made by rolling strips from the ingot. The strips are cut into roughly the size required. The material at this stage is soft, about 200 BHN. The soft inserts are placed in a special atmosphere and heated above the melting point of steel. The material reacts with nitrogen, which diffuses inwardly from the surface to the core to produce, in effect, a graded, nitrided system. The combination of high hardness and toughness along with good resistance to diffusion and adhesive wear, plus thermal shock resistance, make it an excellent cutting-tool material.

UCON inserts can be used in the speed range of 1000 to 1800 fpm for soft steels, which is two to three times faster than that recommended for tungsten carbide tools and from 10 to 20 % faster than for TiC-coated tools.

These tools are not recommended for machining stainless steel, cast irons, or exotic materials such as titanium-, nickel-, or cobalt-base alloys. If used on these materials, wear is rapid and failure occurs in much the same manner as that of conventional alloyed carbide grades.

## TOOL GEOMETRY

The principle of tool geometry is to provide a sharp cutting edge that is strongly supported. The development of basic tool geometry is shown in Fig. 3-7. The purpose of each of the angles is discussed briefly.

*Relief Angles.* As the name implies, relief angles, both side and end, provide relief for the cutting edge to keep the tool from rubbing on the cut surface. Relief angles are usually kept to a minimum to provide good support for the cutting edge. They may be increased to produce a cleaner cut on soft materials.

*Side Cutting-Edge Angle.* The side cutting-edge angle serves two purposes: it protects the point from taking the initial shock of the cut and it serves to thin out the chip by distributing the cut over a greater surface.

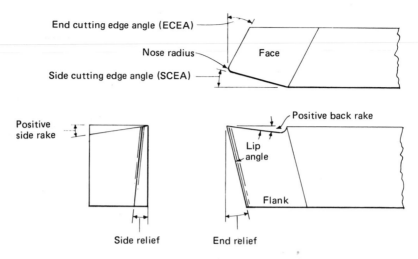

End cutting edge angle (ECEA)

Nose radius

Side cutting edge angle (SCEA)

Face

Positive side rake

Positive back rake

Lip angle

Flank

Side relief

End relief

FIGURE 3-7. *The development of basic tool geometry.*

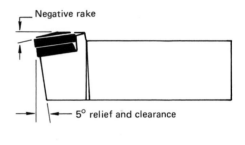

Negative rake

5° relief and clearance

FIGURE 3-8. *Negative rake angles allow a tool insert to be used on all six or eight of its cutting edges.*

***End Cutting-Edge Angle.*** The end cutting-edge angle reduces the tool contact area in respect to the metal being cut.

***Rake Angles.*** Rake angles are ground on the face of the tool and may be either positive, neutral, or negative. Positive rake angles reduce the cutting force and direct the chip flow away from the material. Negative rake angles increase the cutting force required but provide greater strength at the cutting edge. Negative angles also make it possible to use six cutting edges of a triangular insert or eight cutting edges of a square insert, as shown in Fig. 3-8.

Newer tool geometry can provide positive-rake cutting action even though the insert may be set at a negative angle (Fig. 3-9).

***Nose Radius.*** The nose radius is the rounded end that blends the side cutting-edge angle with the end cutting-edge angle. It performs the same function as the side

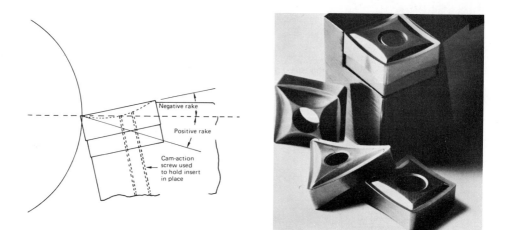

FIGURE 3-9. *Negative-rake tool holders used with effective positive-rake inserts. Thus all cutting edges can be used with the added advantage of lower cutting forces provided by the positive rake at the cutting edge. Courtesy Kennametal Inc.*

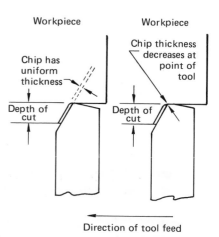

FIGURE 3-10. *The nose radius acts to thin out the chip at the point of the tool. Courtesy Adamas Carbide Corporation.*

cutting-edge angle in that it decreases the thickness of the chip as it approaches the point of the tool (Fig. 3-10). The nose radius is also important in controlling surface finish. For a given depth of cut and feed, the tool with a large nose radius will produce a smoother finish than a tool with little or no nose radius. Too large a nose radius will tend to produce chatter due to excessive contact area. Also, pressures on the cutting tool and material being machined are proportionate to the contact area. As an example, small-diameter parts require a small nose radius with light cuts to minimize tool pressures and obtain better accuracy.

## TOOL FAILURES

Carbide cutting edges lose their usefulness through wear, breakage, chipping, or deformation. Wear, which accounts for most of the tool failures, is very complex and involves chemical, physical, and mechanical processes, often in combinations.

### Wear

The study of wear between two metallic surfaces is complicated because it involves several unique mechanisms. Four of these mechanisms have been categorized as (1) abrasion, (2) adhesion, (3) diffusion, and (4) oxidation. Each of these wear mechanisms will be described briefly.

***Abrasive Wear.*** Abrasion takes place when hard constituents in the workpiece microstructure or underside of the chip " plow " into the tool face or flank as they flow over it. Highly strain-hardened fragments of a built-up edge or inclusions contribute to this type of wear. Abrasive wear is generally considered to be a mechanical wear mechanism that is predominant at low cutting speeds and at relatively low temperatures. It is generally more pronounced on the flank surface of the tool (Fig. 3-11).

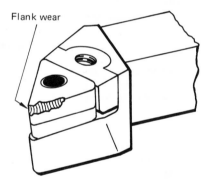

Flank wear

FIGURE 3-11. *Flank wear on cutting tool.*

***Adhesive Wear.*** The chip, as it slides over the face of the tool, is not smooth on a microscopic scale. It consists of hills and valleys, or asperities. Likewise, the cutting tool may have asperities. As these asperities slide past each other under heat and pressure, they tend to adhere. If the junction formed is weaker than both parts of the metal, a fracture occurs with little transfer of metal. However, a certain amount of plastic deformation takes place in the bonding asperities. In repeated action, work hardening takes place. After a time of deformation contact, the asperities become brittle and break off, taking with them minute amounts of tool material. Eventually a crater begins to form (Fig. 3-12). This action takes place because of the extremely high unit pressures. If, however, the temperature is near or beyond the recrystallization temperature, the bond between the chip and tool asperities is no weaker than the material adjacent to it, because work hardening has not been retained. Therefore, the rate at which the asperities are pulled off diminishes.

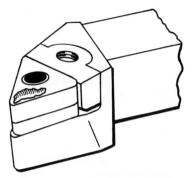

FIGURE 3-12. *A crater is formed on the face of the tool by adhesive wear.*

***Diffusion Wear.*** Diffusion wear involves the transfer of atomic particles from the tool to the workpiece and from the workpiece to the tool. Diffusion is accelerated by high temperatures caused by rapid movement of the work material near the tool surface.

***Oxidation.*** At the elevated temperature range at which a cutting tool operates, oxidation can cause rapid wear. As oxidation takes place, it weakens the tool matrix and thus the cutting-edge strength. The oxides that form are easily carried away, leading to increased wear.

### Breakage

Excessive pressures acting on the cutting edge of a tool may cause immediate failure by the loss of one or two large particles or several smaller particles. Breakage is usually attributed to mechanical shock, thermal shock, thermal cracks, fatigue, or excessive wear.

Mechanical shock, leading to breakage, is usually caused by severe interruptions of the cut on the workpiece surface or by hard inclusions in the workpiece. If this type of machining is necessary with carbide tools, tough rather than hard grades should be chosen.

Thermal shock breakage is usually due to sudden cooling of a very hot cutting edge. This often happens when a chip breaks off and the coolant strikes the tool or when the tool emerges from the cut.

Thermal cracks are caused by severe heat gradients that develop mainly on the face of the tool. The cutting edge gets hot first, then the heat penetrates into the interior of the tool by conduction. The surface is first put into compression and if it cools rapidly it contracts, putting it in tension and causing a thermal crack to appear. Repeated cycling causes more cracks, making the cutting edge weak and subject to breaking. Harder grades of carbide crack more readily than the softer grades. Although thermal cracks occur during dry machining, they are greatly accelerated by the interrupted application of cutting fluid.

***Chipping.*** Chipping is a microscopic form of breakage due to the loss of many small particles from the cutting edge and is usually caused by unhoned carbide edges. Unhoned edges lack sufficient mechanical strength to withstand the mechanical forces encountered in cutting. Chipping may also be caused by excessive vibration and chatter.

### Deformation

When a heavy load is applied close to the cutting edge of the tool, the surface at this point becomes indented while the adjacent face shows a corresponding bulge. The amount of deformation increases as the load is increased, until a critical strain is reached and fracture starts. A crack forms at the periphery of the indentation, away from the cutting edge, spreads to the edge, and then goes down the adjacent face until a small flake of carbide breaks away.

A tool's ability to withstand deformation is based on its hot hardness. The effect of temperature and hardness for various cutting-tool materials is shown in Fig. 3-13.

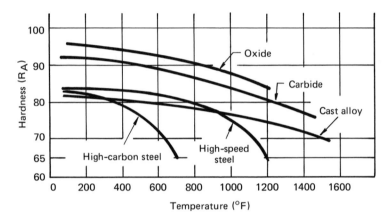

FIGURE 3-13. *A comparison of how the hardness of cutting-tool materials is affected by temperature.*

## TOOL LIFE

Tool life may be defined as that period of time that the cutting tool performs efficiently. Since it is undesirable to replace a worn tool before it becomes necessary, various quantitative criteria have been suggested to determine the maximum permissible limit of tool wear, which is in essence tool life. Recommended values of wear land are given in Table 3-2. Wear land (width of the flank wear) is shown in Fig. 3-11.

Many factors can be considered as having an effect on life, such as the microstructure of the material being cut, the desired cubic-inch removal rate, the rigidity of the setup, and whether or not cutting fluids are used.

Shown in Fig. 3-14 are four curves that have been plotted showing wear versus time. Each of the curves shows three regions generally associated with tool wear: the rapid initial wear, the relatively long period of very gradual wear, and the final period of rapid wear. Four different cutting speeds are shown, represented in order

TABLE 3-2.  *Flank-wear Tool Life Recommendations*

| Tool Material | Flank Wear (in.) | Remarks |
|---|---|---|
| Cemented carbide | 0.030 | Roughing cuts |
| Cemented carbide | 0.010–0.015 | Finishing cuts |
| HSS | 0.060 or total destruction | Roughing cuts |
| HSS | 0.010–0.015 | Finishing cuts |
| Cemented oxides | 0.010–0.015 | Roughing and finishing cuts |

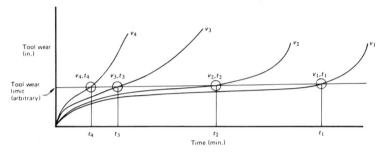

FIGURE 3-14.  *A schematic of tool-wear-time curves.*

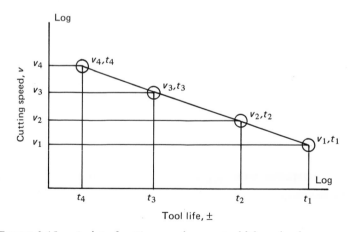

FIGURE 3-15.  *A plot of cutting speed versus tool life on log-log paper.*

of increasing speed by $V_1$, $V_2$, $V_3$, and $V_4$. The cutting times are represented from the shortest to the longest by $T_4$, $T_3$, $T_2$, and $T_1$.

If the circled points $(V_4, T_4)$, $(V_3, T_3)$, $(V_2, T_2)$, and $(V_1, T_1)$ are plotted on a special type of graph paper (log-log paper) they will fall along a straight line, as shown schematically in Fig. 3-15.

At the turn of the century, F. W. Taylor developed an equation for these curves, $VT^n = C$, where $V$ is the velocity of the work material in surface feet per minute, $T$ is the tool life in minutes, and $n$ is the slope of curve (on log-log paper). When the $V$ and $T$ are plotted, $C$ equals a constant representing the velocity of the work in surface feet per minute for one minute of tool life.

Significant changes in tool geometry, depth of cut, and feed will change the value of the constant $C$ and cause slight changes in the slope of the curve, or the exponent $n$.

Tool life curves may be constructed from the results of laboratory tests or by actual on-the-job machining operations. Shown in Fig. 3-16 is a comparison of the tool life curves as plotted for common cutting-tool materials. As an example, at a cutting speed of about 825 fpm, a carbide tool would last about 20 minutes. To get the same tool life for an HSS tool, the velocity ($V$) would have to be reduced to about 150 fpm. The values of $n$ shown in the figure may be accepted as standard.

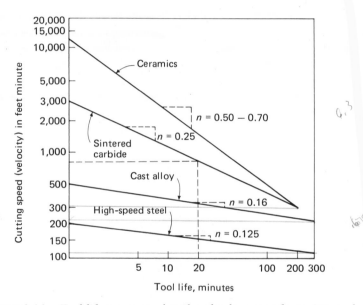

FIGURE 3-16. *Tool-life curves as plotted on log-log paper for various tool materials.*

## TOOL FORCES

The main forces acting on the cutting tool during a turning operation are tangential, longitudinal, and radial forces, as shown in Fig. 3-17*a*. The tangential force $F_t$ acts in a direction tangent to the revolving workpiece. This force is the highest of the three components. The horsepower for a turning operation is based largely on the tangential tool force and the cutting velocity.

The longitudinal force $F_l$ acts in a direction parallel to the axis of the rotating bar and is also called the feeding force. The power required to feed a turning tool is thus the product of the longitudinal tool force and the feeding velocity.

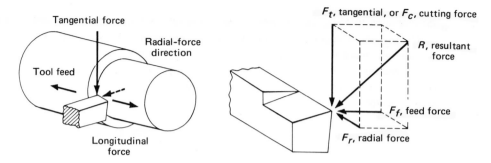

FIGURE 3-17. *Forces acting on a cutting tool during a turning operation (a). The resultant force acting on the cutting tool (b).*

The radial tool force $F_r$ is normal to the rotating work or is in a radial direction from the center of the workpiece. For turning, the radial force holds the tool at the desired depth of cut. In a turning cut, there is no velocity in the radial direction, hence there is no power required for the radial force. The longitudinal force is affected by the side rake of the tool, but it constitutes only a small percent of the total force. The radial force is much less, being about half of the longitudinal force. The three components of force may be resolved into three mutually perpendicular components, with the resultant force vector being as shown in Fig. 3-17*b*.

***Measurement of Cutting Forces.*** An indirect method of measuring cutting forces acting on the tool is with the aid of a wattmeter. A more exact method is with the aid of a tool dynamometer. Shown schematically in Fig. 3-18 is a commonly used

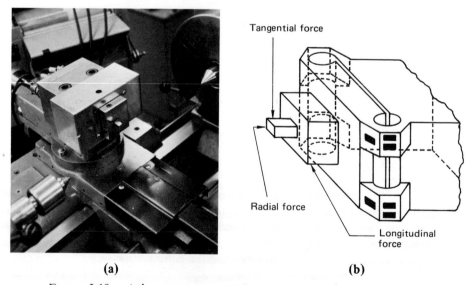

**(a)**          **(b)**

FIGURE 3-18. *A three-component, strain gage type, metal-cutting dynamometer (a). Drawing showing location of strain gages (b).*

three-force-component lathe tool dynamometer. Resistance-wire strain gages are mounted on an extended octagonal ring. The advantage of this design is that the instrument becomes independent of where the load is applied to the material.

***Effect of Speed on Force.*** In general, as speeds increase, forces increase. However, negative-rake tools that show high forces at low cutting speeds may show more than 30 % decrease in forces when the speed is increased from 50 to 300 sfpm. The increase in speed increases the temperature at the cutting edge, and as the material becomes more plastic, the shear strength is lowered.

***Effect of Feed on Force.*** An increase in feed is usually considered an increase in efficiency of metal removal, but sometimes the increase in force does not make it the best choice. An increase in the depth of cut causes a direct increase in tangential and longitudinal force but not always in the radial force.

***Cutting Forces and Tool Design.*** Cutting-force data is useful in designing tools or selecting the proper type and geometry. Shown in Fig. 3-19 is the effect of rake angle on the cutting or tangential forces versus surface speed in feet per minute. You will note the lowest cutting force is with a $+15°$ rake.

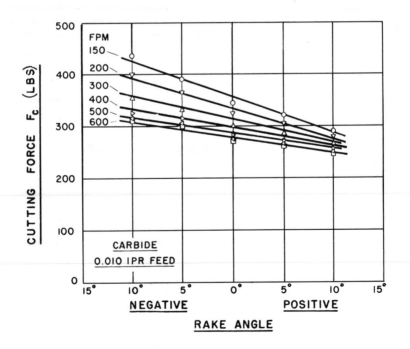

FIGURE 3-19. *The effect of rake angle on cutting force.*

## HORSEPOWER FOR MACHINING

Horsepower was first defined by James Watt as "the power exerted by a dray horse exerting a pull of 150 pounds while traveling at the rate of $2\frac{1}{2}$ miles per hour". This is equivalent to 33,000 foot-pounds per minute or 550 foot-pounds per second.

The horsepower required at the cutting edge of the tool ($HP_c$) is expressed as the summation of the individual force components and their corresponding velocities. However, the longitudinal and radial forces are considered negligible and the formula used may be expressed as:

$$HP_c = \frac{F_t V}{33,000}$$

*where:*   $F_t$ = tangential force
   $V$ = relative velocity of the work piece or tool in feet per minute

### Unit Horsepower

In metal cutting, horsepower is most often calculated in terms of *unit horsepower*. Unit horsepower is defined as the power required to cut a material at the rate of one cubic inch per minute. Unit horsepower will vary with the cutting characteristics of each material. Thus unit horsepower is given for the different hardness levels of a material as shown in Table 3-3.

Cutting speed and depth of cut in the usual shop ranges do not have a significant effect on the unit horsepower value. However, since unit horsepower is affected by changes in feed rate, feed correction factors are shown in Table 3-3. The correction factors are normally used in the calculation of turning, planing, and shaping. There has not been enough information collected to establish the accuracy of the correction factors for milling. However this is not serious since the feed per tooth for milling cutters does not vary as much as for other operations. The tables given are for calculating horsepower for sharp tools. A certain percentage (25, 50, or 100 %) is

TABLE 3-3(*a*). *Average Unit Horsepower Values of Energy Per Unit Volume*

| Material | BHN | Unit Power |
|---|---|---|
| Carbon steels | 150–200 | 1.0 |
| | 200–250 | 1.4 |
| | 250–350 | 1.6 |
| Leaded steels | 150–175 | 0.7 |
| Cast irons | 125–190 | 0.5 |
| | 190–250 | 1.6 |
| Stainless steels | 135–275 | 1.5 |
| Aluminum alloys | 50–100 | 0.3 |
| Magnesium alloys | 40–90 | 0.2 |
| Copper | 125–140 | 0.7 |
| Copper alloys | 100–150 | 0.7 |

TABLE 3-3(*b*). *Horsepower Feed Correction Factors*
*For Turning Planning and Shaping*

| Feed, ipr or ips | mm/rev or mm/stroke | Factor |
|---|---|---|
| 0.002 | 0.05 | 1.4 |
| 0.005 | 0.12 | 1.2 |
| 0.008 | 0.20 | 1.05 |
| 0.012 | 0.30 | 1.0 |
| 0.020 | 0.50 | 0.9 |
| 0.030 | 0.75 | 0.80 |
| 0.040 | 1.00 | 0.80 |
| 0.050 | 1.25 | 0.75 |

usually added to compensate for the effect of tool wear. Tool geometry does have an effect but is usually ignored in most calculations. Thus for simplicity all power computations may be based on unit horsepower ($HP_u$) which in turn is based on the cubic inch removal rate ($Q$). (See Fig. 3-20.) The gross horsepower ($HP_g$) is simply

$$HP_g = Q \cdot HP_u$$

Horsepower at the cutter may also be calculated from the product of unit horsepower and the cubic inch removal rate per minute:

$$HP_c = HP_u \cdot \frac{\text{in}^3}{\text{min}} \qquad\qquad HP_u = \frac{HP_c}{Q}$$

For turning, planing, and shaping this becomes:

$$HP_c = (HP_u)12CVfd$$

For milling:

$$HP_c = (HP_u)FWd$$

For drilling:

$$HP_c = (HP_u) \cdot \text{RPM} \cdot f \cdot \left(\frac{\pi D^2}{4}\right)$$

*where:*  $V$ = cutting speed, fpm
$C$ = feed correction factor
$f$ = feed, ipr (turning and drilling), ips (planing and shaping)
$F$ = feed, ipm
$d$ = depth of cut
$W$ = width of cut

The actual horsepower required ($HP_g$) at the cutter may be obtained by noting the reading on a wattmeter that is placed in the incoming line to the motor. The idle horsepower may be subtracted from the gross horsepower to obtain the *net horsepower*

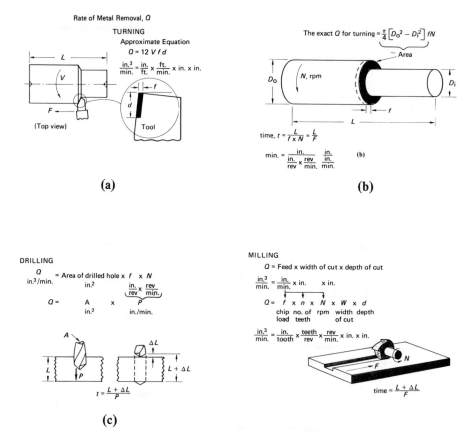

Figure 3-20. *Q defined for various machining operations. Also shown is the formula used to calculate the time required to make cuts for each type of operation.*

($HP_n$). The gross horsepower may also be estimated by applying a machine efficiency factor ($e_m$) which is usually 0.9 for a direct-belt drive and 0.7 to 0.8 for gear drives. The calculation is as follows:

$$HP_g = \frac{HP_c}{e_m}$$

## METRIC POWER CALCULATIONS

### Unit Power

Unit power ($kW_u$), which is similar to unit horsepower, is defined in terms of kilowatts required to remove one cubic centimeter of material per second.

$$kW_u = \frac{kW}{cm^3/s} \qquad \text{or} \qquad kW_u = \frac{kW}{1000\, mm^3/s}$$

The unit of time in the metric system is the second (s) because the kilowatt is defined as the rate of work on one kilonewton-meter per second. The value of unit power can be calculated by the following equation:

$$kW_u = \frac{F_t'}{f'd'}$$

*where:*   $F_t'$ = tangential cutting force in kilonewtons
$f'$ = feed rate in millimeters per revolution
$d'$ = depth of cut in millimeters

Metric units of cutting conditions are given with the prime symbol so they can be distinguished from inch values of the same term. This practice is not necessarily universal.

*Example:*  A shaft of AISI 1020 steel has a Brinell hardness of 165. The cutting speed is given as 90 m/min. The roughing depth of cut is given as 2.54 mm with a feed rate of 0.50 mm/rev. The lathe to be used is a geared-head type. Assume normal tool wear or a correction factor of 50 %. What size motor will be required?

*Solution:*

$$kW_c = kW_u \frac{C'V'f'd'}{60}$$

$$= 1.58 \frac{0.90 \cdot 90 \cdot 0.50 \cdot 2.54}{60}$$

$$= 2.71$$

$$kW_m = \frac{kW_c}{e_m}$$

$$= \frac{2.71}{0.75}$$

$$= 3.61$$

Thus 3.61 kilowatts are required at the motor for sharp cutting tool. Considering a tool wear factor of 50 %:

$$kW_m = 3.61 \cdot 1.5$$
$$= 5.41$$

Converting kilowatts to horsepower (1 kW = 1.341 hp)

$$HP_m = 1.341 \cdot 5.41$$
$$= 7.25$$

Therefore, a $7\frac{1}{2}$-hp motor would be required.

145

TABLE 3-4. *Influence of Additives on Machinability of Steels*

| Element | Helpful Effect | Element | Harmful Effect |
|---|---|---|---|
| Carbon up to 0.2 % C with low S, P, and N | Reduces gumminess of iron by forming carbide. Improves finish. | Carbon when S, P, and N are high. | Increases hardness by forming carbide. Shortens tool life. |
| Manganese up to about 1 % in carbon steels with high S. | Forms random manganese sulfide inclusions which reduce resistance to shear and reduce friction between chip and tool. | Manganese above about 0.5 % with low S. | Increases hardness by dissolving in iron. |
| Sulphur | Forms sulfide inclusions which reduce resistance to shear and reduce friction between chip and tool. | Silicon | (1) Forms hard abrasive silicates which envelop sulfides and hinder their internal lubricating action. (2) By combining with oxygen the coarser oval sulfides are changed to smaller elongated sulfides. |
| Phosphorus | Embrittles chip by promoting crack propagation when the metal is strain hardened. | | |
| Nitrogen | Similar to phosphorous but stronger. | | |
| Oxygen | Promotes oval sulfides with sufficient O as iron-manganese oxides. | Aluminum | (1) Forms very hard abrasive alumina particles. (2) Similar to item (2) for silicon. |
| Lead | Discrete particles of lead lubricate the interface between chip and tool | | |
| Bismuth | Enchances the effect of lead. | Nickel | Dissolves in and toughens iron. |
| Selenium | Forms selenide inclusion with effects similar to but stronger than sulfur. | | |
| Tellurium | Forms telluride inclusions and with effects similar to but stronger than sulfur. | | |

## MACHINABILITY

Machinability is an involved term with many ramifications, but briefly defined it is the relative ease with which the chip may be separated from the base material. The two main aspects of machinability are the properties of the material being cut and the machining conditions involved.

*Machinability Properties of the Material.* The material properties that affect machinability may be listed briefly as follows:

a. Shear strength. The shear strength can be obtained by the ratio of the chip thickness to the depth of cut or feed, depending on the operation.
b. Strain hardenability. The increase in strength and hardness with increasing plastic deformation.
c. Hardness. The characteristic of the material to resist indentation.
d. Abrasiveness of the microstructure. The abrasiveness and nature of inclusions and interfaces in the metallic matrix affect the machining qualities.
e. The coefficient of friction. The coefficient of friction varies with the type of material and the reaction of it to the tool material at the chip interface.
f. Thermal conductivity.

*Methods of Evaluating Machinability.* Common methods of evaluating the machinability are by using one or more of the following criteria:

a. Tool life for a given surface-foot speed and tool geometry. This may also be expressed in terms of cubic-inch removal rate.
b. The power required, which is also related to cutting forces.
c. Surface texture. Surface-texture requirements vary widely; however, a material that produces a better surface texture under equal conditions is considered to possess a higher degree of machinability.

Items (a) and (b) have already been discussed, and item (c) is discussed in detail in Chapter 9.

*Influence of Additives On Machinability of Steel.* Studies have been made that show the correlation between steel composition and machinability. Shown in Table 3-4 is a brief summary of elements and their effect on machinability of carbon steels. Note that elements which tend to break up the soft ferrite structure, such as carbon, manganese, and sulfur, improve machinability. However, if these elements are used in excess, they may increase the hardness and abrasiveness of the metal. See also Table 2.4.

## MACHINABILITY AND THE COMPUTER

In many manufacturing plants, machinability is a somewhat vague term with many ramifications that are too cumbersome to be placed into the calculation. Now, however, with the concept of time-shared programming of numerically controlled machines, computer technology can be used to obtain full utilization of the cutting tools.

Many of the machinability parameters have now been programmed so that relatively unskilled personnel can quickly determine precise values for the many metal cutting variables. For example, to determine the correct carbide grade for a cutting operation, an information sheet is filled out. The information is called into a computer by dialing the time-sharing network system. The printout will show a preferred grade of carbide and a second choice. Also given will be the feed rate to obtain a specified surface finish. Tool life is shown based on the input data.

Minicomputers are also being used to optimize tool life. The minicomputer monitors the cutting tools by collecting vibration frequency data comparing it with pre-established standards that have been correlated with actual wear performance. A significant improvement is made over the standard piece-count method because of the safety factor that must otherwise be added. Tools may be monitored individually, or as many as 75 or more may be monitored simultaneously.

## METAL-CUTTING ECONOMICS

The foregoing study of machinability is of little value unless it is used to obtain quality parts at minimum cost. The cutting speed and tool life that produce a minimum cost are called *economical cutting speeds* and *economical tool life*, respectively. Just what is the full significance of economical tool life? Taylor recognized some 70 years ago that a long tool life was not the goal but rather a minimum production time per workpiece.

The basic problem in determining the most economical cutting condition is that as cutting speeds increase, the efficiency of the machine tool increases but the tool life goes down. Finding the theoretical optimum point at which there is a balance of four individual costs—machining costs, tool costs, tool-changing costs, and handling costs—is the essence of metal cutting economics. The relationships between these four factors are shown in Fig. 3-21. It can be seen that machining cost decreases with increasing cutting speed. Both the tool costs and the tool changing costs are observed to increase, since they are wearing out faster at the higher cutting speeds. The handling costs are independent of cutting speeds. Adding up each of the individual costs results in the total unit-cost curve which is observed to go through the minimum point.

Each of these four basic costs will be presented individually before showing the equations used to determine the cutting speed for minimum cost or the tool life for maximum production.

*Machining Cost.* The machining cost is equal to the operating cost ($C_o$) in dollars per minute, multiplied by the time required to machine the workpiece ($t_m$):

$$\text{Machining cost} = C_o t_m$$

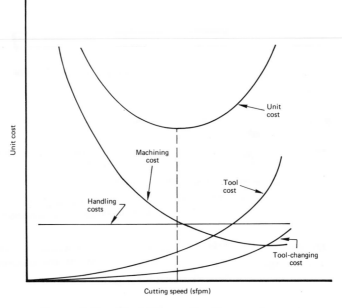

FIGURE 3-21.  *The influence of cutting speed on costs.*

***Tool Cost.***   The tool cost per operation is obtained from the product of the tool cost $(C_t)$ in dollars per cutting edge and the ratio of the machining time to the average tool life $(T)$:

$$\text{Tool cost} = C_t \frac{t_m}{T}$$

***Tool Changing Cost.***   The tool changing cost per operation is equal to the product of the cost of the operating time $(C_o)$ in dollars per minute, multiplied by the tool changing time $(t_c)$ in minutes per operation and the ratio of $t_m$ to $T$, which represents the average number of tool edges required per piece.

$$\text{Tool changing cost} = C_o t_c \frac{t_m}{T}$$

***Handling Cost.***   The handling cost per operation is simply the product of the handling time $t_h$, in minutes per workpiece, multiplied by the cost of the operating time $(C_o)$, in dollars per minute:

$$\text{Handling cost} = C_o t_h$$

***Average Unit Cost.***   The summation of these four individual costs represents the unit cost $(C_u)$ in dollars.

$$\text{Unit cost} = \text{machining cost per piece} + \text{tool cost per piece}$$
$$+ \text{ tool changing cost per piece} + \text{handling cost per piece}$$

149

or

$$C_u = C_o t_m + \frac{t_m}{T}(C_t + C_o t_c) + C_o t_h$$

There are two cost factors in this basic model ($C_o$, $C_t$) and three time factors ($t_m$, $t_c$, $t_h$), in addition to the tool life factor. Each of these factors will be discussed briefly.

*Operating Cost* ($C_o$). The operating cost equals the sum of the machine-operator's rate plus appropriate overhead.

*Tool Cost* ($C_t$). The tool cost is the cost of the insert price and a prorated cost per cutting edge of the complete tool holder and any other parts such as chip breakers, shim seats, clamps, screws, etc. If regrindable tooling is used, the tool cost is a function of the tool price, the total number of cutting edges in the life of the tool, the grinding time per cutting edge, the tool grinder's rate, and the toolroom overhead rate.

*Machining Time* ($t_m$). The machining time is the time in minutes that the tool is actually cutting. The time may be calculated by dividing the length of cut by the velocity the tool moves in the feeding direction:

$$t_m = \frac{L}{fN}$$

*where:*      $L$ = the axial length of the cut
             $f$ = the feed ipr
             $N$ = rpm.

*Tool Changing Time* ($t_c$). The tool changing time is the complete cycle in minutes to remove a tool that has failed and replace it or index it, reset for size, and be ready for the next cut.

*Handling Time* ($t_h$). The handling time is the time in minutes required to load and unload the workpiece from the machine. It includes the idle time and time necessary to advance and retract the tool.

*Tool Life Factor.* Tool life is taken from Taylor's equation, $VT^n = C$. The average tool life ($T$) in minutes per cutting edge is:

$$T = \left[\frac{C}{V}\right]^{1/n}$$

## Determination of Cutting Speed for Minimum Cost ($V_{min}$)

The total cost for an operation is made up of the four individual costs: machining cost, tool costs, tool changing costs, and the handling costs. The interaction of these factors were shown in Fig. 3-21.

The cutting speed for minimum cost of a given operation is derived by equating the total cost to the sum of the four individual costs, differentiating the costs with respect to the cutting speed, and setting the result equal to zero.

$$V_{min} = \frac{C}{\left[\left(\frac{1}{n} - 1\right)\left(\frac{C_o t_c + C_t}{C_o}\right)\right]^n} \qquad (3\text{-}1)$$

### Tool Life for Minimum Cost ($T_m$)

The minimum-cost cutting speed, $V_{min}$, is a function of the operating time costs, tool costs, and tool changing time, and is a function of the $n$ and $C$ parameters in Taylor's equation. Since the constant $C$ in Taylor's equation and in equation 3-1 are the same, and if $V$ corresponds to $V_{min}$, then the tool life that corresponds to the cutting speed for minimum cost is:

$$T_{min} = \left(\frac{1}{n} - 1\right)\left(\frac{C_o t_c + C_t}{C_o}\right) \qquad (3\text{-}2)$$

### Cutting Speed for Maximum Production

There are times when it becomes necessary to speed production beyond the point of the recommended minimum cost. In this case, it is necessary to know what the maximum production rate for the operation will be. This can be determined from the equation previously developed for the cutting speed for minimum cost, (3-1), by assuming that the tool cost is negligible, that is, by setting $C_t = 0$.

$$V_{max} = \frac{C}{\left[\left(\frac{1}{n} - 1\right)t_c\right]^n} \qquad (3\text{-}3)$$

### Tool Life for Maximum Production

By analogy to Taylor's equation, the tool life that corresponds to the maximum production rate is given by:

$$T_{max} = \left(\frac{1}{n} - 1\right)t_c \qquad (3\text{-}4)$$

The tool life at maximum production rate is a function only of $n$, the slope of the curve in Taylor's equation, and the tool changing time. Thus, for an HSS tool ($n = 0.1$) with a tool changing time of 1 min, $T_{max} = 9$ min; that is, the tool should last only 9 min. A carbide tool, where $n = 0.25$ and 1 min is needed for tool changing, should only last 3 min.

The "Hi-E" or high-efficiency range proposed by William W. Gilbert of General Electric is the cutting-speed range between minimum cost on the lower speed end and maximum production on the higher speed end (Fig. 3-22).

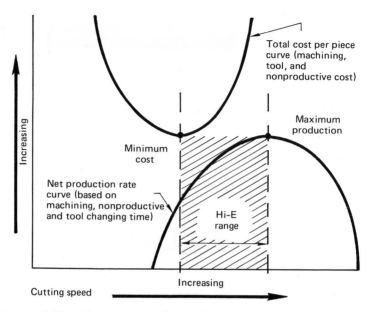

FIGURE 3-22. *The cutting-speed range between the low point on the cost curve and the high point on the productivity curve at the right is termed the " Hi-E" or high-efficiency range. Cutting speeds outside of this range sacrifice both cost and productivity.*

## HIGH-SPEED MACHINING

The ever present question—"Why can't we do it faster?"—has led to considerable research in the topic of high-speed machining. By high-speed machining is meant the removal of metal at speeds in excess of 3000 sfpm.

Detailed studies of the results in this area have shown that:

1. Aluminum oxide tools can be used successfully for high-speed machining. However, it is recommended there be a high effective negative rake, a high side-cutting edge angle, and a large nose radius. The insert must be evenly supported on a carbide pad and any tool deflection must be eliminated.
2. The machine must ensure rigidity with a minimum of deflection and vibration.

Recent advances in cutting-tool materials may stimulate a re-evaluation of the design of existing machines with the possibility of exceptional advances in metal cutting production.

## CUTTING FLUIDS

Cutting fluids are used in many metal cutting operations, as well as in grinding, to maintain optimum production rates, minimize tool wear, improve surface finish, and inhibit rust. In most cases, the cost of the fluid, which is usually recirculated, is negligible compared to the benefits obtained.

The four main functions performed by cutting fluids are: cooling both the tool and the workpiece, reducing friction at the tool chip interface (thus preventing the metal from welding to the tool), and inhibiting rust.

Cooling is necessary for the efficient operation of most metal cutting tools, ceramics being an exception. The reduction of heat increases tool life and helps stabilize the size of the finished part.

Reduction of friction decreases the power required for a given cut and also reduces the heat generated. Welding is particularly noticeable at lower speeds in carbon steels and at normal speeds in tough alloy steels.

*Types of Cutting and Grinding Fluids.* The types of cutting fluids include: water-based solutions (dispersions and emulsions), chemically inactive oils, chemically active oils, and synthetic and gaseous fluids. Where cooling is the most important function, water-base emulsions are usually recommended. If rubbing and friction problems are present, as in low-speed broaching and threading, an oil-type fluid is preferred because of its antiweld properties.

*Chemical Coolants.* Petroleum-base fluids have good lubricating characteristics, but have a heat removal rate that is often insufficient, especially with large feeds and high speeds. Chemical coolants are used mainly to cool the cutting edge of the tool. It is essential that a temperature differential exist between the tool and the chip to control welding and tool diffusion wear.

Although water has long been recognized as the most efficient cooling agent, it has a number of objectionable characteristics, such as a tendency to promote rust on the machine and workpiece and a lack of lubricating qualities. It may also contain nutrients for the growth of bacteria. Chemical coolants are modifiers of water to reduce its undesirable side effects. Where lubricity is required, soaps, wetting agents, and "chemical lubricants" are incorporated. Chemicals used are: amines and nitrites—for rust prevention; nitrates—for nitrite stabilization; phosphates and borates—for water softening; compounds of phosphorus, chlorine, and sulphur—for chemical lubrication; soaps—as wetting agents and for lubrication; glycols—as blending agents and humectants; and germicides—to control bacterial growth.

The three main classes of chemical coolants are: (1) Pure coolants with no lubricants. These are mostly water softeners and rust inhibitors. (2) Coolants with mild lubricity. These contain water softeners, rust inhibitors, and wetting agents. (3) Lubricating coolants. These have softeners, rust inhibitors, wetting agents, and chemical lubricants such as chlorine, sulphur, phosphorus, and other additives.

Because chemical coolants are primarily water modifiers, they are particularly useful for those operations in which the main function is to act as coolant. Where lubrication is important, oil-base fluids or chemical coolants with a high chemical lubricity concentration, 1 part compound to 5–10 parts water, are recommended.

## METAL-CUTTING OPERATIONS

It is not known when the first machine tool was made, but objects that appear to have been machined in a lathe have been dated by archaeologists to early 1500 B.C. Sculptured friezes show that the bow drill and a whole range of hand tools were in use hundreds of years before the Christian era.

The eighteenth century saw great advances in many engineering fields and considerable advances in machine tools. In the decade preceding 1712, Thomas Newcomen brought his steam engine to the satisfactory working state. The duplication of the engine, however, was done only with great difficulty and much of the work had to be done by hand. The same limitation applied to spinning and weaving machines. The constant search for improved methods of making wheels, shafts, and other cylindrical items led to the development of the lathe. Although the early lathes were used entirely for turning cylindrical items, they gradually became more versatile and many other tasks were assigned to them. Henry Maudsley, an Englishman, is credited with making the first screw-cutting engine lathe in about 1797.

## LATHES AND RELATED TURNING OPERATIONS

Lathe work in its most fundamental form is referred to as *turning*. Turning is used to produce cylindrical parts. The three primary factors involved are *speed*, *feed*, and *depth of cut*. The speed is often thought of as the revolutions per minute, but the important figure is the surface foot speed measured in feet per minute (sfpm). The sfpm is simply the product of the rpm times the circumference. The feed refers to the rate at which the tool advances along its cutting path and is expressed in inches of tool advance per revolution of the spindle (ipr). Some machines have a hydraulic feed, in which case the feed is expressed in inches per minute. The depth of cut is self explanatory except to note that the workpiece diameter is reduced by twice the depth of cut.

The main types of lathes are engine, turret, single-spindle automatic, multi-spindle automatic, vertical, and tracer.

### Engine Lathes

The engine lathe (Fig. 3-23) is considered to be the basic turning machine. Its main components are the headstock, tailstock, bed, carriage, and quick-change gear box.

The headstock of the engine lathe contains the gears or pulleys that are operated by an electric motor to drive the workpiece. The tailstock, mounted on the other end of the bed, can be adjusted to accommodate different lengths of stock. It may be offset laterally for turning long, slow tapers. The carriage is mounted on the ways between the headstock and tailstock. Its main function is to carry the cutting tool for the feed motions, longitudinally or across the bed on the cross slide. The compound rest on top of the cross slide can be set at any angle for accurately cutting short tapers or bevels.

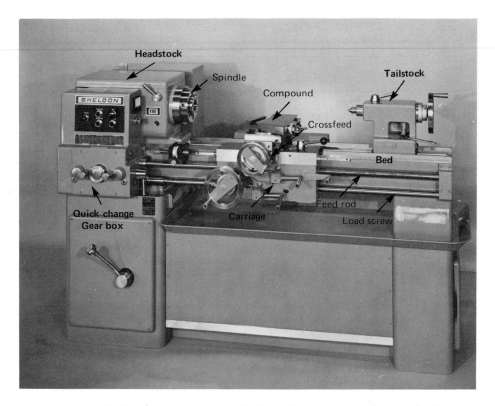

FIGURE 3-23. *An engine lathe with five main components shown in bold type. Courtesy Sheldon Machine Co., Inc.*

The size of an engine lathe is generally given by the swing over the bed, which means the maximum diameter of a part that can be machined. The swing over the cross slide is also important, since it gives you an idea of how large a diameter can be turned for its full length. Some common lathe operations are shown in Fig. 3-24.

### Turret Lathes

Turret lathes have a turret mounted on the ways in place of the tailstock. The turret generally has six faces on which tools can be mounted and can be indexed so that the appropriate tool is brought to bear on the work when required (Fig. 3-25). A square turret is mounted on the front side of the cross slide and a single tool is mounted at the rear. Once the tools are mounted for a given job, their travel is controlled by prepositioned stops, thus drastically reducing the time required to produce a part.

Automatic turret lathes have the entire cycle required to produce a part programmed. This of course means longer setup times, but it also means higher production. The automatic controls are governed by hydraulic circuits, pneumatic circuits, or a combination of these with an electrical plugboard. Plugboard control

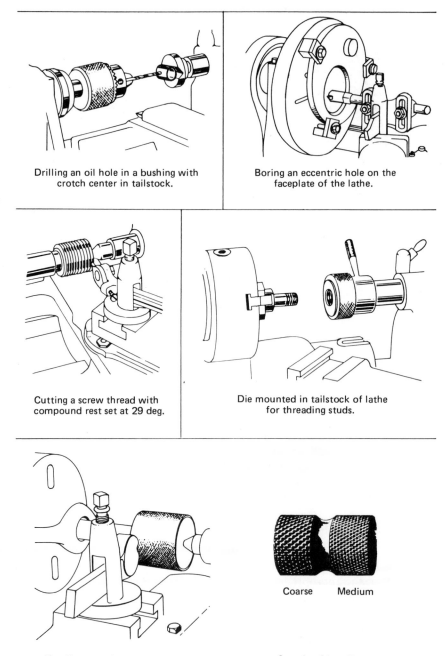

Drilling an oil hole in a bushing with crotch center in tailstock.

Boring an eccentric hole on the faceplate of the lathe.

Cutting a screw thread with compound rest set at 29 deg.

Die mounted in tailstock of lathe for threading studs.

Knurling a steel piece in the lathe.

Sample of knurling.

Coarse      Medium

FIGURE 3-24.  *A variety of operations commonly performed on the engine lathe.*

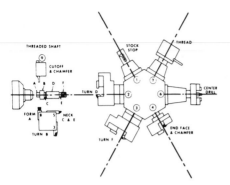

FIGURE 3-25. *A manual or semiautomatic turret lathe usually has two turrets: one at the tailstock end of the bed and one on the cross slide. A single tool is shown here at the rear cross-slide position. A fully automatic turret lathe has a tailstock turret with front, rear, vertical, and angular tool slides. Courtesy Herbert Machine Tools.*

has a board mounted on the headstock that looks something like a conventional telephone switchboard. The plugs establish which step the particular function will occupy in the sequence of operations. Some turret lathes have also been made automatic by numerical control.

### Single-Spindle Automatics

Single-spindle automatic lathes are often referred to as *automatic screw machines*. In essence, they are small cam-operated turret lathes. The distinguishing features of these machines are the positions of the turret and tool mountings. The automatic screw machine (Fig. 3-26) has front, top, and rear slides that are fed at predetermined rates by means of disk cams.

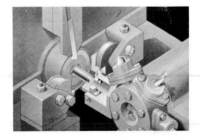

FIGURE 3-26. *A single-spindle automatic screw machine showing the turret and front, vertical, and rear cross slides. Courtesy Aluminum Company of America.*

### The Swiss-Type Automatic Screw Machine

The Swiss-type automatic screw machine has five radially mounted tools [Fig. 3-27(*a*)] that are cam-controlled. The stock can be made to feed as it is being cut, thus any desired cylindrical shape may be generated. This type of lathe is especially good for turning out small-diameter instrument parts [Fig. 3-27(*b*)]. The accuracy on small parts can be maintained to ±0.0005 in.

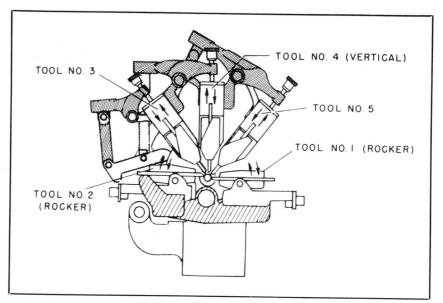

TOOL NO. 3

TOOL NO. 4 (VERTICAL)

TOOL NO. 5

TOOL NO. I (ROCKER)

TOOL NO. 2
(ROCKER)

(a)

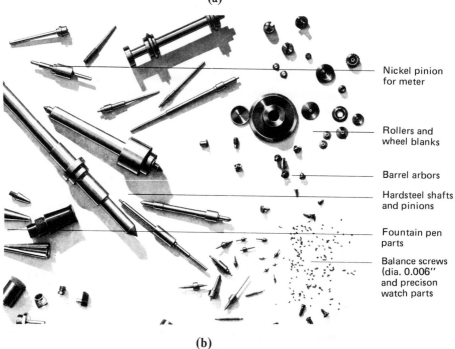

Nickel pinion
for meter

Rollers and
wheel blanks

Barrel arbors

Hardsteel shafts
and pinions

Fountain pen
parts

Balance screws
(dia. 0.006"
and precison
watch parts

(b)

FIGURE 3-27. *The tooling arrangement for a Swiss-type automatic screw machine (a) shows the relative tool locations and the arrows indicate the movement. Shown at (b) are typical parts produced.*

### Multispindle Automatic Lathes

Instead of having only one spindle, as on the engine lathe, four, six, or eight spindles may be used (Fig. 3-28). These are high-production machines. Production quantities may range from 5,000 to over 100,000 parts. Stock may be in bar or piece form. In the latter case, the lathes are called chucking machines.

### Vertical Turret Lathes

Vertical lathes have the work mounted on a rotating table (Fig. 3-29). The tools are mounted on vertical rams or turrets. These lathes are used primarily for short, heavy, large-diameter workpieces. They are also ideal for lightweight but bulky parts. Controls are usually automatic (plugboard or cam) but can also be manual or numerical control (NC).

### Tracer Lathes

Tracer lathes are basically the same as engine or turret lathes, but a two- or three-dimensional template is used to control (usually hydraulically) the path of the cutting tool (Fig. 3-30). For copying work, NC lathes can do it all without a template, but compared to tracer lathes NC machines are quite expensive.

### Machine Selection

With the wide variety of lathes available, it is sometimes a problem as to which machine to use for a given lot size. If the quantity is very small, the part would be assigned to the engine lathe. If the quantity required is 20 parts, is it still assigned to the engine lathe or would a turret lathe be better? A simple break-even formula that can be used to help with this decision is as follows:

$$\text{B.E.} = \frac{C_t}{C_A - C_B}$$

*where:*  $C_t$ = special tooling cost
$C_A$ = cost of machining per part on machine A
$C_B$ = cost of machining per part on machine B

*Example:* Assume the labor and overhead rate on a turret lathe is $15/hr and on an engine lathe $9/hr. It is estimated from standard data that it will take about 25 min/part on the engine lathe and 5 min/part on the turret lathe. Special tooling required for the turret lathe for this particular job is estimated at $45.

$$\text{B.E.} = \frac{\$45}{(25/60 \times 9) - (5/60 \times 15)} = 18$$

Therefore, under this condition, if more than 18 parts are required, it would be more economical to use the turret lathe.

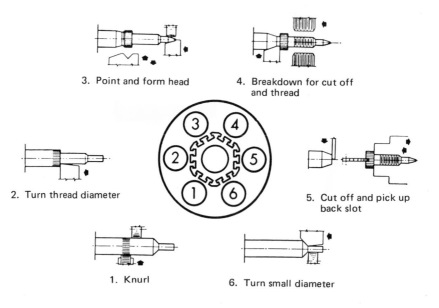

3. Point and form head

4. Breakdown for cut off and thread

2. Turn thread diameter

5. Cut off and pick up back slot

1. Knurl

6. Turn small diameter

FIGURE 3-28. *The multispindle, automatic bar machine showing the tooling used to complete a small part. Courtesy National Acme Co.*

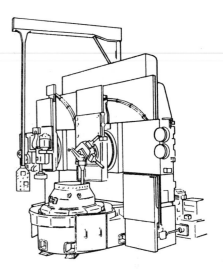

FIGURE 3-29. *A vertical turret lathe.*

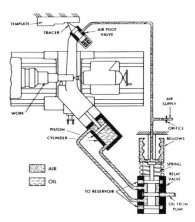

FIGURE 3-30. *An air-oil circuit as applied to the tracer lathe. Courtesy Monarch Machine Tool Co.*

## DRILLING

Drills are used as the basic method of producing holes in a wide variety of materials. The present standard drill (Fig. 3-31) has progressed through a long history of design and development and changes in geometry are being proposed regularly.

### Drill-Point Geometry

The standard drill point has a 118° included lip angle and a 12½° lip-clearance angle. In general, soft material can be drilled more efficiently with a smaller included angle; that is, bakelite, wood, or hard rubber can be drilled with a 90° drill point. The opposite is also true. Very hard materials may require a drill-point angle of about 135° and 6 to 9° lip-clearance angle, as shown in Fig. 3-32.

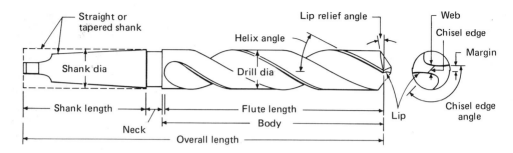

FIGURE 3-31. *The design and nomenclature of a standard twist-drill.*

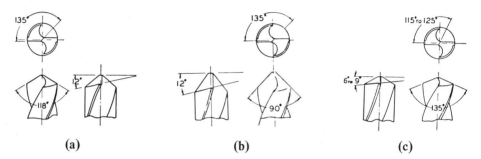

**(a)**       **(b)**       **(c)**

FIGURE 3-32. *Drill-point geometry for various materials: (a) standard, (b) for soft materials, (c) for hard materials.*

The center section of the drill or web does not produce a good cutting action; as a matter of fact, it is more an extruding action than cutting. Two means of minimizing this action are thinning the web and drilling a lead hole. Thinning the web (Fig. 3-33) may be done freehand by tool-grinding personnel or, if many drills are to be ground, with the aid of a fixture to provide accurate centering. A lead hole is a small-diameter hole drilled before the desired-size drill is used. It eliminates the section that would otherwise be subjected to the extruding action of the web.

Modifications in drill-point geometry have been developed. Shown in Fig. 3-34 is a split-point drill. This point creates a secondary cutting edge that virtually eliminates the web effect at the point. The split point was developed for drilling high-strength materials with heavy-web drills. Because the splitting cuts are rotated toward the main cutting edges, the intersection of the two cutting edges is stronger. The geometry also gives an improved centering action.

Shown in Fig. 3-35 is the action of a conventional drill as it breaks through the bottom side of the material as compared to the radial-lip grind. The radial lip produces a smooth breakthrough as compared to the shock of a conventional point. Proponents state this drill point has four to ten times longer life and that the holes produced are smooth, round, on size, and burr-free. The point is patented by the Radial Lip Machine Corporation and is based upon grinding the drill cutting lips to the shape that most evenly distributes wear over the entire cutting area.

Standard chisel point
after web thinning

FIGURE 3-33. *A standard drill point show-ing a thinned web.*

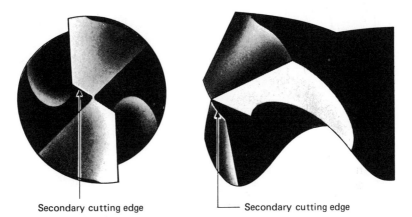

Secondary cutting edge                     Secondary cutting edge

FIGURE 3-34. *The split-point drill virtually eliminates the web effect at the cutting edge.*

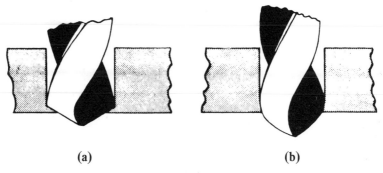

(a)                                       (b)

FIGURE 3-35. *The cutting action of a conventional drill point (a), and a radial lip (b), as it breaks through the material is shown schematically.*

### Drill-Flute Geometry

Until more recent years, the flute of the drill was thought of only as a path for the chips to emerge from the hole. However, with industry becoming more safety minded, stringy chips are now considered to be a considerable hazard to the operator. Many approaches are now being used to modify the flutes to provide a chip-breaking action. A number of flute geometries are shown in sectional views in Fig. 3-36. The main principle employed is that of causing the chip to curl so tightly that it breaks off. Although chip-breaker drills cost about 10 % more than conventional models, the time saved from not having the problem of long chips and the longer tool life easily offset the added cost. Tool life increases because of the efficient chip ejection and better penetration of the coolant to the drill tip.

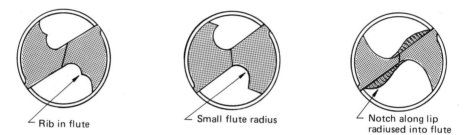

Rib in flute    Small flute radius    Notch along lip radiused into flute

FIGURE 3-36. *The flute geometry of drills modified to provide chip-breaking action.*

### Spade Drills

The spade drill is made by inserting a spade-shaped blade into a slot in a bar and fastening it securely with a screw, as shown in Fig. 3-37. The use of these drills has risen sharply in the past decade. The major reasons are their efficiency in making holes in the 1.5 to 15 in. diameter range and the lower cost. Over 3.5 in. in diameter, they are the only drills made as stock items. A standard 2 in. diameter twist drill costs about $75, whereas a 2 in. diameter spade drill, including the shank and blade, is

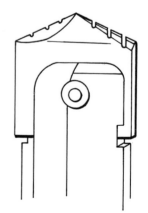

FIGURE 3-37. *A spade drill. The cutting blade shown is standard but many differently shaped points may be obtained.*

about $50. The shank can be used with more than one size tip, so it may cut the cost of a 2 in. spade drill to about $25. Standard blade material is M2 or M3 HSS and carbides. Some advantages of spade drills are: (1) rigidity—they are capable of taking heavy feeds, (2) relatively low cost—only the insert need be replaced, (3) capability of deep hole drilling (up to the length of the blade holder) with low pressure coolant, (4) chips are broken up at the cutting edge.

### Gun Drills

Gun drills (Fig. 3-38) differ from conventional drills in that they are usually made with a single flute. A hole provides a passageway for the pressurized coolant, which serves as a means of both keeping the cutting edge cool and flushing out the chips. Gun drills may be used to shallow depths, as shown in Fig. 3-39, or to depths

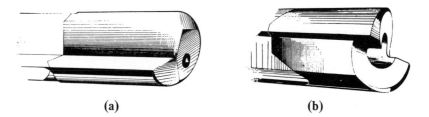

| (a) | (b) |

FIGURE 3-38. *Two types of gun drills: single-flute (a), and pin-cutting (b).*

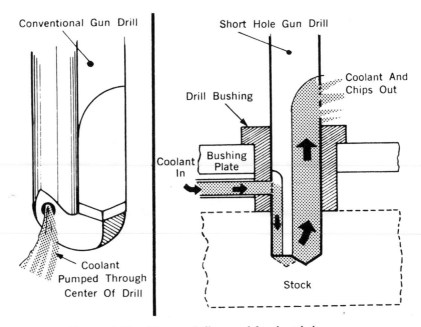

FIGURE 3-39. *The gun drill as used for short holes.*

of up to 130 diameters. Even though the drill can be made to advance rapidly (from 15 to 60 in./min), good tolerance and finish can be maintained. This is due in part to the high (60 to 1500 psi) cutting fluid pressure. Some parts that commonly require deep hole drilling are gun barrels, camshafts, crankshafts, machine tool spindles, and connecting rods.

## Drilling Thrust and Torque

Process engineers may avoid many pitfalls such as tool breakage, piece part warpage, scrap, and short tool life by knowing the relationship of thrust and torque to various drilling operations.

One might expect that the drilling torque or moment would be proportional to the amount of metal removed per revolution, which would be $(\pi/4)fd^2$, where: $f$ = feed (ipr) and $d$ = drill diameter. However, it is not. It has been found to be closely related to the empirical relationship of $(\pi/4)f^{0.8}d^{1.8}$. This means that doubling the feed per revolution will increase the torque by a factor of 1.75 instead of 2.

The torque (or moment) and thrust required for drilling mild steel without coolant is:

$$M \approx 2180 f^{0.78} d^{1.8}$$

$$T \approx 906,000 f^{0.87}\left(\frac{d}{5} + \frac{w}{d}\right)^{2.12}$$

*where:*    $M$ = torque (in./lb)
$d$ = drill diameter (in.)
$w$ = the web thickness        $f$ = feed (ipr)
$T$ = thrust

Thrust forces are important when determining drill, jig, and machine rigidity but are only 2 % or less of the total power requirement and are often disregarded.

Horsepower requirements for drilling operations can be determined even though torque measurements are not available by the use of unit power data given in Table 3-3 and the formula:

$$HP_m = (\pi d^2/4) \times f \times N \times P_u$$

## Drilling Speeds and Feeds

Due to the drill geometry, the cutting speed for drilling should be reduced to about 80 % of that used in turning. It is difficult to provide efficient cooling for the tip of the drill except in the case of gun drills or oil-hole drills. Oil-hole drills provide a passage for coolant from the shank to the tip, as shown in Fig. 3-40.

Feeds are dependent on cutting conditions and what the drill will take without deflection. A rule of thumb for feed is to make it $\frac{1}{100}$ of the drill diameter. Thus a 0.5-in. drill would have a recommended feed of 0.005 ipr.

FIGURE 3-40. *An oil-hole drill provides a passageway for coolant to the tip, where it can be used effectively. Courtesy Morse Cutting Tools.*

### Drilling Equipment

Drilling equipment ranges from the portable hand drill to large multispindle machines capable of driving hundreds of drills at one time. The main types are sensitive, radial, turret, multispindle, and individual drilling units.

***Sensitive and Upright Drill Press.*** The most common type of drill press is a hand-feed type (Fig. 3-41). The operator is actually sensitive to how fast the drill is cutting and can control it according to the conditions of the moment.

Upright drill presses are very similar to the sensitive type except they are equipped with a power feed. This machine is usually of heavier construction and is suited to a wide range of work. It may be equipped with a universal table that allows the table to be accurately positioned both longitudinally and laterally.

***Radial Drill Press.*** Large, heavy work pieces are difficult to reposition for each succeeding hole. The radial drill press (Fig. 3-42) makes it easy to reposition the drilling unit, which is independently powered. The size of a radial drill is given by

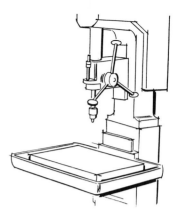

FIGURE 3-41.   *The sensitive-type drill press is characterized by hand feed.*

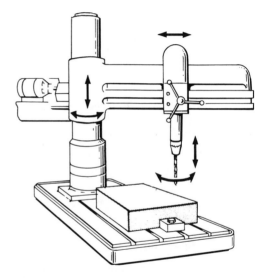

FIGURE 3-42.   *The radial drill press.*

the radius, in feet, of the largest plate in which a center hole can be drilled and the size of the supporting column in inches.

***Turret-Type Drill Presses.***   When a number of drill sizes or other tools such as reamers and taps are required to complete a part, a turret-type drill is very useful. Any one of six, eight, or ten tools can be quickly indexed into place (Fig. 3-43).   Many drill presses of this type are now numerically controlled.

***Multispindle Drilling Heads.***   A multispindle drilling head may be attached to a single-spindle machine to provide a means of drilling many holes at one time.   These heads are of two main types: adjustable, for intermediate production; and fixed, for long, high-production runs.

The adjustable head [Fig. 3-44(*a*)] is driven by a gear from the spindle through a universal-joint linkage to each drill.   Another variation is the oscillator head

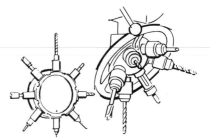

FIGURE 3-43. *The turret-type drilling machines have six, eight, or ten tools ready to use.*

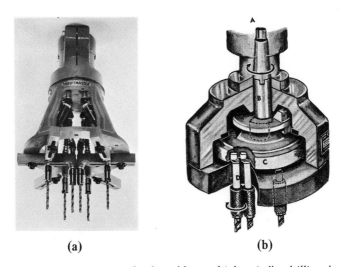

(a)                                    (b)

FIGURE 3-44. *Two types of adjustable, multiple-spindle drilling heads. The spindle at (a) has adjustable arms driven by universal joints from gears in the head. The spindle at (b) transmits power through the drill sleeve (A) which drives the crank (B) and oscillator (C). The oscillator turns the individual spindles (D) in the same direction and speed as the drive crank.*

[Fig. 3-44(*b*)], which allows drill placement anywhere within the head area and as close as the sum of the two drill-holder diameters.

Fixed-head drills are engineered for a specific job. When the production run for that pattern is over, the heads are torn down and rebuilt to the new requirements.

***Self-Contained Drilling Units.*** Self-contained drilling units consist of a power source, a feeding arrangement, and a chuck. These units come in a wide range of sizes, from small air-powered units, as shown in Fig. 3-45, to large electric-drive motor types. These units are assembled into specialized automation equipment, as shown in Fig. 3-46.

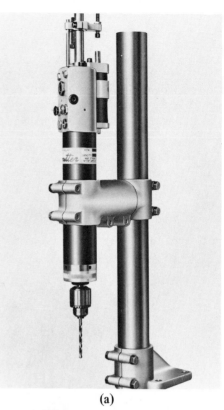

**(a)**

FIGURE 3-45. *Individual drill units range from small air-powered units to large electric-motor types. Courtesy Desoutter Corp. and Universal-Automatic Corp.*

**(b)**

FIGURE 3-46. *Individually powered drill units are mounted around a four-station indexing table for automatic drilling, counterboring, tapping, and reaming. Programmed-sequence selector switches permit single-spindle operation on any two, three, or four spindles simultaneously. Courtesy Universal-Automatic Corp.*

### Drill Jigs

Large quantity production requires quick and accurate hole location. One method of accomplishing this is with the aid of a drill jig (Fig. 3-47). The part to be drilled is placed in the jig against a number of locating pins. Clamping holds the work tightly against these pins to maintain accuracy of the part. A well designed drill jig will provide for chip clearance and removal. It also provides for quick and easy loading and unloading and foolproofing. The hardened and ground drill bushings guide the drill to the precise location.

### Reamers

Reamers are used to size a hole and provide a good finish. Hand, machine, and shell reamers are shown in Fig. 3-48. Hand reamers are recognized by their square shank, whereas machine (or chucking) reamers have a tapered shank to fit into the machine spindle. The shell or hollow reamers are used on arbors that have driving lugs. Other common types of reamers are tapered, tapered-pin, and adjustable.

FIGURE 3-47. *The drill jig provides for accurate hole location.*

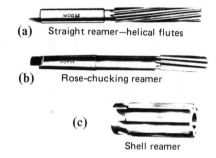

(a)  Straight reamer—helical flutes

(b)  Rose-chucking reamer

FIGURE 3-48. *Hand, machine, and shell reamers. Courtesy Morse Cutting Tools.*

(c)

Shell reamer

To work efficiently, a reamer must have all teeth cutting. A proper stock allowance is very important. For holes up to 0.5 in. in diameter, 0.015 in. is allowed for semifinish reaming on most metals. For holes over 0.5 in. in diameter, 0.030 in. is recommended. For finish reaming, the allowance is 0.005 in. for holes less than 0.5 in. and 0.015 in. for holes greater than 0.5 in. in diameter. The tolerance expected is normally ±0.0005 in.

## Drilling Accuracy

Drilling accuracy may be considered with regard to hole location, size, and geometry. The location will be determined by starting conditions. A center punch mark often serves as the drill location. This of course is subject to the skill of the layout man and the drill press operator. In production work, where drill bushings guide the drill, a high degree of accuracy is obtained. The drill bushing should be no more than 0.0005 in. larger in diameter than the drill. Excessive clearance permits the drill to wander until a hole is established. This results in oversized, tapered, or bell-mouthed holes. A drill that has unequal lip lengths will also produce oversize holes. Short-hole gun drills as previously shown can produce holes with a location tolerance of ± 0.0005 in. and a size tolerance of ± 0.0002 in.

## BORING

Up until about 1775 there was not a great demand for accuracy in bored holes. Cannons were using cannon balls in the as-cast condition, so the carefully cored holes, after being cleaned out, proved satisfactory. However, the coming of the steam

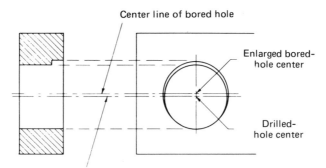

FIGURE 3-49. *A boring tool is used to enlarge and correct the hole location.*

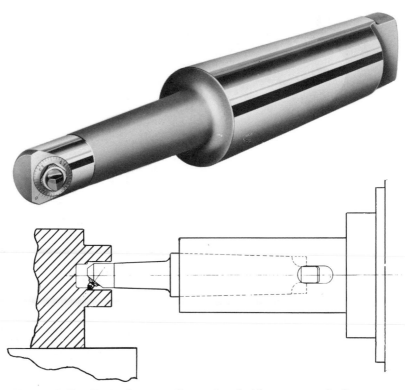

FIGURE 3-50. *Boring bars are often equipped with accurate tool-adjustment settings as shown. Boring tools may also be used for undercuts or for internal threading.*

173

engine put much higher requirements on machining capabilities. Fortunately, Wilkinson, Watt, Smeaton, and others accepted this challenge and began designing and building machines of ever increasing accuracy. Today's boring machines are able to work to within a few ten-thousandths of an inch on a production basis.

Boring is the process of using a single-point tool to enlarge and locate a previously made hole. Drills tend to wander or drift, thus, where greater accuracy is required, drilling is followed by boring and reaming. Shown in Fig. 3-49 is a sketch of how the centerline of a hole may be shifted by means of a boring operation. The single-point boring tools are usually made out of HSS or have carbide inserts, as shown in Fig. 3-50. This bar has an adjustable dial that allows it to be set to within 0.0001 in.

***Boring Machines.*** The main types of boring machines are the horizontal and the vertical, as shown in Fig. 3-51. The horizontal machine is well adapted to *in-line boring*, boring two or more holes on the same axis, as shown at Fig. 3-51(*a*). In this

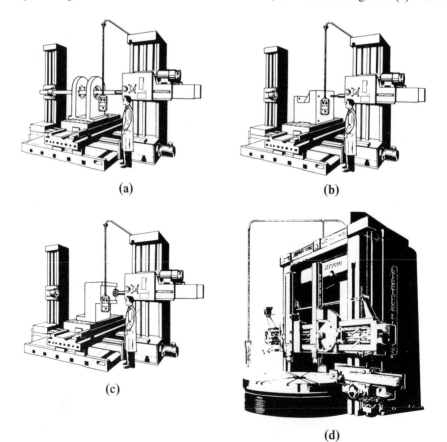

(a)

(b)

(c)

(d)

FIGURE 3-51. *Horizontal and vertical boring machines. Basic operations performed on horizontal boring machine are boring (a), drilling (b), and milling (c). Courtesy Giddings and Lewis Machine Tool Co.*

case, the boring bar is supported at the outboard end. Machines of this type also are used for milling and drilling operations, as shown at Fig. 3-51(*b*). The vertical boring mill is used to machine large, heavy parts that are difficult to mount in any other position. Vertical boring mills range in size from about 54 in. to 40 ft in diameter (table diameter).

Jig-boring machines (Fig. 3-52) get their name from the fact that they are so often used to drill and bore the holes needed for drill jigs. Outside of grinding, they are considered to be the most accurate machines of the machine shop or toolroom. They can be used to locate holes to an accuracy of $\pm$ 0.0001 in.

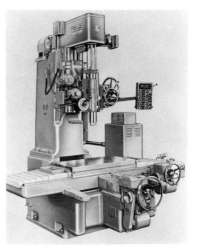

FIGURE 3-52. *A jig-boring machine. Courtesy Pratt & Whitney, Division of Colt Industries.*

## HOLE PUNCHING

Hole punching is, in many instances, much more economical than drilling. This is especially true because of the heavy-duty hydraulic punches that have been developed. Punches of the type shown in the sketches of Fig. 3-53 may be either portable or stationary and are capable of punching 1-in. diameter holes in steel ranging from 12 gauge up through $1\frac{1}{16}$ in. thick. A wide variety of hole configurations can be made.

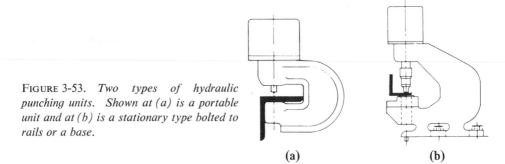

FIGURE 3-53. *Two types of hydraulic punching units. Shown at (a) is a portable unit and at (b) is a stationary type bolted to rails or a base.*

(a)                    (b)

The biggest savings in utilizing hole-punching equipment is in the time saved per hole. As an example, a 1-in. diameter hole can be punched through a 1-in. thick steel plate in approximately 2 sec, compared to about 46.8 sec required for drilling.

## MILLING

Eli Whitney, one of the foremost inventors of the early 1800s, is credited with one of the first milling machines in the United States, and with being the father of interchangeable manufacture. A contract with the U.S. government for 12,000 muskets, 4000 of which were to be delivered the first year, made it necessary for Whitney to invent ways of increasing production. Although it took eight years to complete the contract, his methods had proven to be so successful that he was awarded a contract for 30,000 muskets in 1812.

The milling machine employs a rotary multitooth cutter that can be designed to mill flat or irregularly shaped surfaces, cut gears, generate helical shapes, drill, bore, or do slotting work. It is one of the most versatile machines of either the production shop or the toolroom.

### Milling Machine Classification

Milling machines are classified broadly as standard or specials. Standard machines are further classified according to their basic structure as column-and-knee or bed type. The column-and-knee machines take their name from the two main structural components. Shown in Fig. 3-54 are two column-and-knee milling machines, one of the horizontal type and the other vertical.

A variation of the horizontal and vertical mill is a combination type (Fig. 3-55). Some machines of this type have an independently driven head mounted on an

(a)                                                    (b)

FIGURE 3-54. *A column-and-knee milling machine of the horizontal (a) and vertical (b) types. Courtesy Cincinnati Milacron Co.*

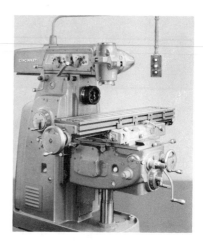

FIGURE 3-55. *A combination vertical and horizontal milling machine. The universal ram-mounted head adds range (a) and versatility (b). Courtesy Cincinnati Milacron Co.*

overarm ram. This head is of the universal type and can be swiveled for cuts at any angle between the cutter and a horizontal plane.

Shown in Fig. 3-56 are typical types of work done by cutters used on the vertical mill (*a*) and on the horizontal mill (*b*). There are some operations for which both machines are well suited, such as surfacing or slab milling and slotting. Generally, when a short-run job is to be done and a choice of machines is available, the machine that requires the least setup time is chosen. On long-run jobs, setup time is not an important factor.

***Fixed-Bed Milling Machines.*** Fixed-bed milling machines are made with a bed that does not raise or lower. All adjustments for depth of cut and lateral cutter positioning are on the spindle. This arrangement makes for rigidity and accuracy in long-run production cutting.

The distinctive feature of a fixed-bed machine is the automatic cycle. The work may be set to approach the cutter at rapid traverse, make the cut at a predetermined feed, and then automatically return to the starting position.

***Planer Mill.*** The planer mill (Fig. 3-57) is powerful, rigid, and accurate enough to produce machine finishes of good dimension and accuracy over a large surface. It also has the means of bringing several cutting heads into operation simultaneously.

### Special Types of Milling Machines

Many types of special milling machines are made to accomplish specific kinds of work more easily than the standard types. In this category are the duplicating mills, profiling machines, and pantographs. Most of these machines are vertical mills that have been adapted to reproduce accurately, by means of a tracer, the forms or contours from a master pattern (Fig. 3-58).

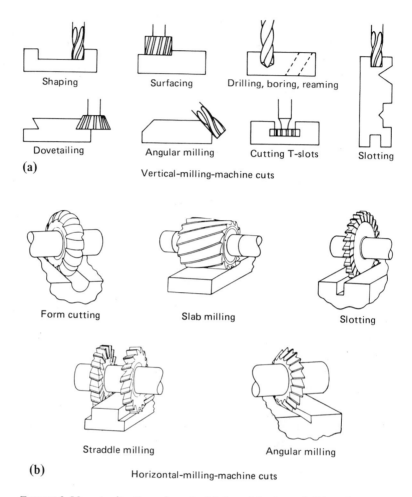

Shaping  Surfacing  Drilling, boring, reaming

Dovetailing  Angular milling  Cutting T-slots  Slotting

**(a)**  Vertical-milling-machine cuts

Form cutting  Slab milling  Slotting

Straddle milling  Angular milling

**(b)**  Horizontal-milling-machine cuts

FIGURE 3-56.  *Application of vertical (a) and horizontal (b) milling machines.*

## Milling Feeds, Speeds, and Horsepower

*Feeds.*  The conventional method of specifying feed for a milling machine is in inches per minute.  It is determined by the "chip load per tooth" or the amount of metal each tooth of the cutter should remove as it advances along the cut.  The feed in in./min is determined as:

$$F = f_t \times n \times N$$

*where:*  $f_t$ = the feed per tooth
$n$ = number of teeth in cutter
$N$ = rpm

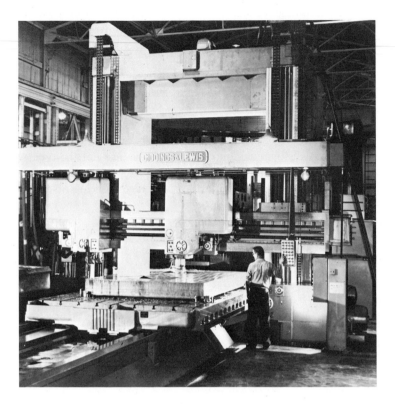

FIGURE 3-57. *A planner-type mill is capable of producing a machined surface over a large area with good accuracy and finish. Courtesy Giddings and Lewis Machine Tool Co.*

FIGURE 3-58. *A duplicating mill with twin spindles being used to profile 15 blades on two impeller rings from a single-blade master. Courtesy Kearney and Trecker, Inc.*

The recommended feed per tooth can be found in handbooks and in manu-facturers' recommendations; however, these are only starting points. The feed will also be determined by the horsepower available, the desired production rate, surface finish, and accuracy required.

Three main types of insert-tooth–type milling cutters are now available: double negative, double positive, and shear angle. The names given refer to the radial and axial rake, as shown in Fig. 3-59. The double-negative rake tool is espe-cially good for interrupted cutting or heavy roughing cuts and the removal of scale. It tends to put more pressure on the workpiece and is therefore not recommended for parts that may distort during machining. It is also not the type of cutter to use on materials that strain-harden readily. The horsepower requirements are higher than for a conventional cutter. It has the advantage of being able to utilize all edges of indexable insert-type tooling.

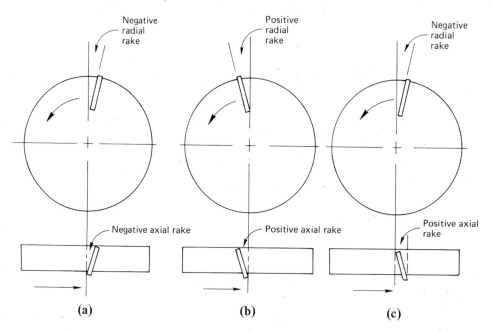

FIGURE 3-59. *Three standard tool geometries for insert-type milling cutters: double negative (a), double positive (b), and shear angle (c).*

A double-positive tool requires less horsepower and exerts the minimum of force on the workpiece. There is little tendency to chatter. As can be seen from the sketch, it is not possible to reverse the tool, so on a square insert only four cutting edges are available instead of eight. Also, the clearance angles make the cutting edges weaker.

The shear-angle cutter is of relatively new design. It combines a negative radial rake with a positive axial rake. This design causes the chips to flow away from the cutter so that higher feed rates can be used. It also produces a fine surface finish.

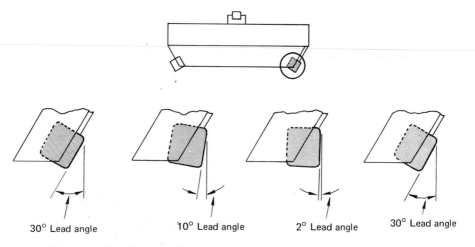

FIGURE 3-60. *Insert-milling cutters are made with various lead angles, 30° with a chamfered flat is used on the shear-angle cutters to provide an excellent finish. The 10° lead angle is used on the double-negative cutter and from 2 to 30° lead angle is used on the double-positive cutter.*

Shown in Fig. 3-60 is an enlarged view of the edge of the cutter. Shear-angle cutters are usually made with a high lead angle, whereas the double-positive angle may vary from 2 to 30°. The standard lead angle for the double-negative cutter is 10°. As with single-point tools, the lead angle serves to protect the cutting point and thin out the chip. This allows higher feed rates without overloading the insert. A 2° lead angle permits cutting up to a shoulder.

Recommended feeds per tooth for carbide milling cutters are given in Table 3-5. As a general guideline, at least two or three teeth should be in the cut at one time to reduce vibration, cutter deflection, and insert chipping.

***Speeds.*** The cutting speed is always given in terms of the number of surface feet per minute (sfpm), rather than rpm. It is related to the physical properties of the work material and the tool material. Suggested starting cutting speeds for carbide

TABLE 3-5. *Recommended feeds per Tooth for Milling Steel with Carbide and HSS Cutters*

| Type of Milling | Feed per Tooth Carbides | HSS |
|---|---|---|
| Face | .008–.015 | .010 |
| Side or straddle | .008–.012 | .006 |
| Slab | .008–.012 | .008 |
| Slotting | .006–.010 | .006 |
| Slitting saw | .003–.006 | .003 |

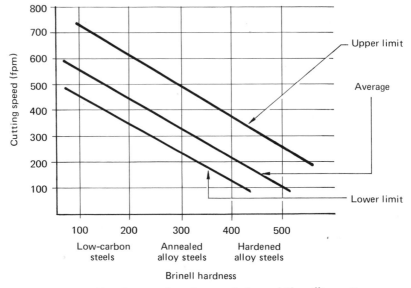

FIGURE 3-61.   *Suggested cutting speeds for carbide milling cutters.*

tools on steels of various hardness are shown in Fig. 3-61.   It may be necessary to increase or decrease these speeds to suit actual job conditions.   As mentioned previously, a built-up edge forms when machining carbon and low-alloy steels at too low a speed.   By careful observation, the speed at which the BUE is no longer formed can be found.   This may be termed the *critical speed* (CS).   The CS will be affected by hardness, chip thickness, and depth of cut.   The general recommendation for use of carbide cutters is to set a speed that is 50 to 100 % higher than the CS.

***Horsepower.***   Horsepower required for milling can be determined on the basis of the cubic inch removal rate:

$$Q = F \times w \times d \qquad \text{(See Fig. 3-20)}$$

$$HP_m = \frac{HP_u \times Q}{n} \qquad \text{(See Table 3-3)}$$

The time required to make a cut is based on the length of the cut plus cutter approach and overtravel divided by the feed rate:

$$\text{Time} = \frac{L + \Delta L}{F}$$

*where:*     $L$ = length of cut
$\Delta L$ = approach of cutter to the work (Fig. 3-62)   (An approximation of one-half the cutter diameter is sometimes used for $\Delta L$.)
$F$ = feed rate in in. per min

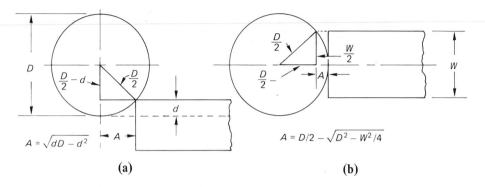

$A = \sqrt{dD - d^2}$

$A = D/2 - \sqrt{D^2 - W^2/4}$

(a)                (b)

FIGURE 3-62.  *Allowance for approach on plain or slot milling (a) and face milling (b). If the diameter of the cutter is only a little larger than the width of the work surface, the approach is D/2 for all practical purposes. For a finishing cut with a face mill, the approach is equal to D, since the cutter must be clear of the work.*

*Example:*   A milling machine is being used to make an $\frac{1}{8}$-inch deep cut on a bar of 1020 CRS that is 4 inches wide and 33 inches long.  A 10-tooth, 8-inch diameter carbide-insert face mill is used.  The chip load/tooth = 0.005 inches. The BHN hardness of the material is 280.  What is the horsepower required? The idle horsepower = .5 and the efficiency of the machine, $\eta$, = 0.9.  How long will it take to make the cut?

*Solution:*

Fig 3.61 (HP/min)

$$\text{RPM} = \frac{350(\text{CS}) \times 12 \text{ IN/FT}}{\pi \times 8 \text{ (cutter diameter)}} \approx 170$$

$$F = 170 \times 0.005 \text{ (chip load) IN} \times 10 \text{ (teeth)} = 8.5 \text{ ipm}$$

$$Q = F \times W \times d = 8.5 \times 4 \times \tfrac{1}{8} = 4.25 \text{ in.}^3/\text{min}$$

$$\text{HP}_c = \text{HP}_u \times Q = 1.46 \times 4.25 = 6.20$$

$$\text{HP}_m = \frac{\text{HP}_c}{\eta} + \text{HP}_i = \frac{6.20}{.9} + .5 = 7.38$$

$$\text{Time} = \frac{L + \Delta L}{F} = \frac{33 + 1.46}{8.5} = 4.05 \text{ min}$$

A finishing cut would take slightly longer

$$\frac{33 + 8}{8.5} = 4.83 \text{ min}$$

## BROACHING

Broaching is a method of removing metal with a tool that has a series of cutting edges in a fixed path. Each tooth removes a predetermined amount of metal in a predetermined location in the cut (Fig. 3-63). The teeth increase in size from the starting end to the rear support end. The broach can be pushed or pulled over straight or irregular surfaces, either externally or internally. The amount of stock removed per tooth varies with the type of operation and material. A general average is 0.002 to 0.005 in. per tooth for HSS broaches.

The tool is generally classified by method and type of operation (push or pull, internal or external), construction, and function.

Broaching can be used for many jobs that are done on milling, drilling, boring, shaping, planing, or keyway cutting machines. Many shapes difficult to handle by

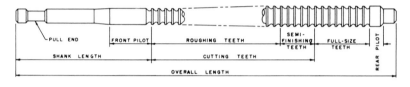

FIGURE 3-63. *An internal broach with sample workpieces. Courtesy National Broach Division, Lear Siegler, Inc.*

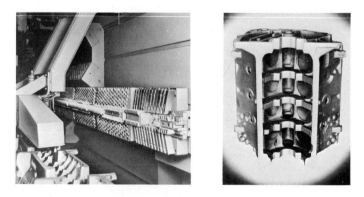

FIGURE 3-64. *A large, carbide-tipped, sectional broach is used to machine several surfaces of an engine block in one pass. Courtesy Ex-cell-o Corporation.*

other methods are ideally suited to broaching. Automobile manufacturers have replaced milling operations with surface broaching because of the combined speed, accuracy, and long tool life provided by carbide inserts. As an example, as many as 22,000 engine blocks can be broached before the tool needs resharpening (Fig. 3-64).

Broaching speeds have increased tremendously with this kind of tooling. Formerly, 20 to 40 fpm was considered average; now speeds in excess of 200 fpm are used on cast iron.

## Broaching Machines

Broaching machines are of three principal types: vertical, horizontal, and continuous-chain, as shown schematically in Fig. 3-65.

Vertical machines come in pull-up, pull-down, surface, single ram, dual ram, and press styles. These machines may be easily equipped with automatic handling features for both the workpieces and the broaches.

Horizontal machines are widely used for small-lot internal broaching. The machines come in many standard sizes, ranging from 1- to 75-ton capacities and 25- to 90-in. strokes.

The chain type or continuous broaching machines are best for large lots, 600 pieces and up per hour. These machines are used chiefly for flat surfaces or contoured

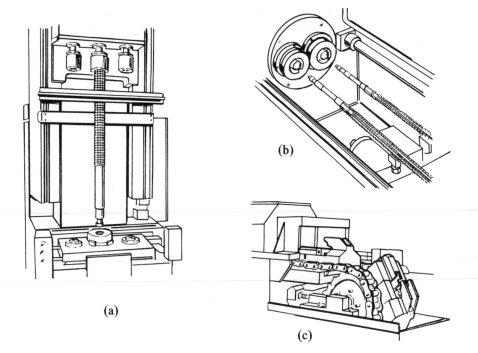

(a)

(b)

(c)

FIGURE 3-65. *Broaching machines: vertical (a), horizontal (b), and continuous (c). Courtesy of* Metalworking.

workpieces. The workpiece is mounted in a fixture that is guided by way bars and pulled past the broaches. Completed parts are automatically ejected. Hand loading of the fixture is usually required.

### Broaching Considerations

Although the broach provides the best means of producing a variety of both internal and external machined surfaces, it requires a specialized tool. That is, each broach is built for one particular job and as such should be amortized over the number of pieces to be run. There are a few general-purpose, standard-shape broaches such as square, round, rectangular, keyway, etc., that may be used on different jobs.

The horsepower required at the motor may be found as follows:

$$HP_m = \frac{Q \times HP_u}{\eta}$$

*where:* $Q = 12 \times w \times d \times v_c$ (cutting velocity in fpm)
$\eta$ = efficiency of the machine

## SHAPERS AND PLANERS

Shapers and planers are machines that employ single-point tools, similar to those used on the lathe. A reciprocating motion is used. In the case of the shaper, the tool moves back and forth across the work; while with the planer, the work moves past the tool.

### Shapers

The shaper is generally not considered a production machine; however, it is widely used in machine shops and toolrooms because it is easy to set up and operate. Also, the tools are of comparatively low cost and easily sharpened. The shaper is well suited to straight and angular cutting, as shown in Fig. 3-66.

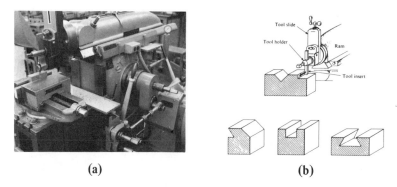

(a)                                              (b)

FIGURE 3-66. *A shaper as used in straight cutting (a) and other angular cuts commonly made (b).*

### Planers

Planers are used for much the same type of surface machining as are shapers. However, with the planer, which is usually much larger, several tools may be used at one time, as shown in Fig. 3-67. Planers are usually used to machine large castings such as machine tables and beds.

FIGURE 3-67. *A double-housing planer. Courtesy Giddings and Lewis Machine Tool Co.*

# GRINDING, ABRASIVE MACHINING, AND MECHANICAL FINISHING

Abrasives have been largely responsible for ushering in the precision capabilities of the twentieth century. Modern abrasive machines have made tolerances in the ten-thousandths range commonplace and surface finishes are measured in millionths of an inch. In recent years extensive research has changed grinding from a strictly metal-finishing operation to a competitive metal-removal method.

### Natural Abrasives

Abrasives found in nature are emery, sandstone, corundum, and diamonds. Both Turkish and American emery are natural mixtures of aluminum oxide ($Al_2O_3$) and magnetite ($FeO_4$). Due to the presence of some softer accessory minerals, both types have a milder cutting action than synthetic abrasives. Emery is often preferred for use as an abrasive on coated cloth and paper as well as in many buffing compositions. It is no longer used in making grinding wheels, thus the use of the term "emery wheel" is incorrect.

Diamonds are another type of natural abrasive, having the highest known hardness of any substance. Because of their nature, diamonds require a distinctive bond which is more specialized than that of conventional grinding wheels.

## Synthetic Abrasives

*Silicon Carbide.* The search for a method of synthesizing precious stones led Edward Acheson into the discovery of silicon carbide (SiC) in 1891. He created a small carbon-arc furnace by wrapping some wire around a pail and putting a carbon rod in the center of it. Around the carbon rod he had placed a clay mixture containing white quartz sand. After this makeshift furnace had cooled, he noted the clear crystals that adhered to the carbon rod. They were in fact silicon carbide crystals. However, Acheson termed the experiment the discovery of carborundum. Although this experiment was relatively simple, it could not have been exploited at any earlier date due to the prodigious amount of power required to make it feasible. Thus the Carborundum Company was established at Niagara Falls to be near the large hydro-electric plant.

*Aluminum Oxide.* About the time silicon carbide was synthesized, another American, Charles B. Jacobs, succeeded in producing aluminum oxide by fusing small quantities of coke and iron borings with aluminum ore (bauxite). The Norton Company acquired the rights to the process and also located at Niagara Falls because of the need for large quantities of power.

Aluminum oxide is softer than silicon carbide, making it tougher and more resistant to fracture. Thus $Al_2O_3$ wheels are not used in grinding very hard materials such as tungsten carbide because the grains would get dull prior to fracture. Aluminum oxide performs best on carbon and alloy steels, annealed malleable iron, hard bronzes, and similar metals.

*Cubic Boron Nitride (CBN).* This is a newer (1968) man-made abrasive that is harder than either $Al_2O_3$ or SiC, which have ratings of 2100 and 2480 respectively on the Knoop scale. Boron nitride (rating of 4500) is the second hardest substance ever developed by man or nature. Diamonds developed by the General Electric Company have a hardness rating of 7000.

CBN is a tight network of interlocking and alternating boron and nitrogen atoms. It is especially good for grinding hard and tough tool steels. Why CBN or Borazon (the General Electric trade name) is a more effective abrasive on these hardened materials is not fully understood. It is presently theorized that the alloying elements found in many steels have a peculiar affinity for carbon and thus prematurely erode the diamond abrasive through a chemical process. Also, diamonds revert to plain carbon at about 1500°F (816°C), while CBN can withstand temperatures above 2500°F (1371°C).

*Diamonds.* Diamonds may be classified as both natural and synthetic abrasive materials. Natural stones which are unsuitable for gems are termed *boart* and are crushed down into a series of sizes for abrasive use.

The General Electric Company was the first to manufacture diamonds on a commercial scale. In 1955 they announced the production of a commercial diamond capable of sustaining pressures up to 470,000 psi for long periods of time at very high temperatures. Other manufacturers are now using high temperatures and explosive shock waves to convert graphite into polycrystalline diamonds. Due to its polycrystalline structure, the synthetic diamond has many more cutting points than the natural diamond. Also, the random orientation of the crystals, 30 to 300 Å in size, means there are no large cleavage planes that often cause a natural diamond to split.

Commercial diamonds are now manufactured in high, medium, and low impact strengths. Shown in Fig. 3-68 is the relative performance of each type of stone on ceramics and metals. The letters refer to the type of bond used to hold the abrasive grains: (MBS)—metal bond saws; (MBG 11)—metal bond grinding; and (RVG)—resinoid and vitrified grinding. Metal bonding of the abrasive grains requires a much tougher diamond than that of the vitrified clay or resinoid material. The effectiveness of the various diamond types is directly related to the modulus of resilience (MOR) of the workpiece, which is defined as the integral under the straight line portion of the stress–strain diagram and has units of inch-pounds per cubic inch or centimeter-kilograms per cubic centimeter.

The RVG-W diamond is a modern, low impact strength, irregular, polycrystalline structure coated with metal. The metal coating strengthens the crystal and enhances the retention of the diamond in the wheel, which increases wheel life and reduces cost. It is prepared specifically for wet grinding operations and has provided significantly improved performance in the grinding of hard materials such as tungsten carbide.

## Bond Materials

The grinding wheel is made up of two materials: the abrasive grains and the bonding material. New bonding materials are being developed all the time. The most commonly used bonding materials in today's grinding wheels are vitrified materials (ceramics), resinoid materials (plastics), rubber (both natural and synthetic), shellac, and metal (sintered powdered metals).

*Vitrified Materials.* There are two forms of ceramic bond used in grinding wheels: the glass type used with $Al_2O_3$ abrasives and the porcelaneous type used with SiC abrasives. Approximately 50 % of all grinding wheels are made with the vitrified bond. The porosity and rigidity that this bond provides make it possible to obtain excellent stock-removal rates; yet it is also well suited to precision grinding.

*Resinoid Materials.* Resinoid bonding materials consist of thermosetting plastics. Because of their high strength and rigidity, wheels made of this material may be operated at speeds up to 12,000 sfpm. The resilient action of the resin bond enables

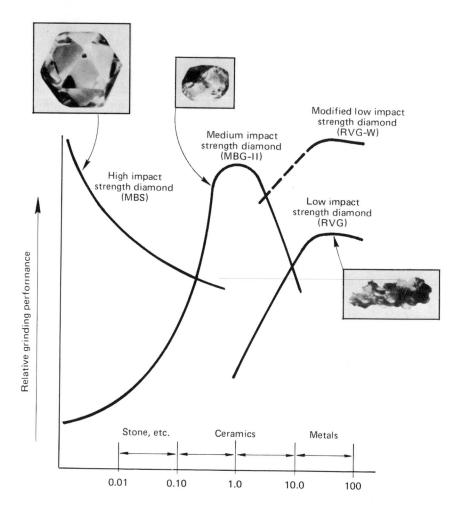

FIGURE 3-68. *Four types of synthetic diamonds used in grinding various materials. The modified, low-impact diamond (not shown) is similar to the medium-impact type but is metal-coated. Courtesy General Electric, Plastics Division.*

it to produce an unusually fine finish even when a coarse grit is used. Boron nitride crystals are given a metallic coating which increases the ability of the resinous bond to hold the crystals and to dissipate the heat generated in grinding.

***Rubber.*** The resiliency of the softer grades of rubber makes them excellent materials for polishing wheels. The harder types provide a degree of resiliency plus water-resisting qualities for safe, cool cut-off wheels, used where burr and burn must be held to a minimum.

*Shellac.* Shellac-bonded wheels are especially good for producing the high finish required in roll, camshaft, and cutlery grinding.

*Metal.* Metal bonding is used only in the manufacture of diamond wheels. The diamonds are firmly held in a shock-resistant bond and are not easily removed until fully used.

### Grain Size

Grain-size numbers used in grinding wheels are derived from the number of grains of a given size which, when laid end-to-end, would equal one inch. Thus, in interpreting the grit-size markings on wheels, simply remember that the larger the base number the smaller the grain. The American National Standards Institute has now established sieve sizes, but the old numbers are still retained. Some idea of grain size can be obtained from the fact that approximately 3.5 million grains of the 240-grain size are contained in a single gram.

Finding the correct grit size to use goes deeper than the old generalization of: "Use coarse grits for soft materials and rough finish; use fine grits for hard materials and fine finish." It is true that fine grits are generally used on hard materials. The reason for this is that they have what is termed a *higher impulse ratio;* that is, they permit a larger number of grains to contact the work per unit of time. Thus, though the grains cannot penetrate as deeply into the hard material, a greater number of cuts are being made and production rates are maintained.

It is also true that a fine-grit wheel, with a high impulse ratio, will generate more heat than a 46-grit wheel. Using a 46-grit wheel as a base, the 60-grit wheel will produce 1.8 times more heat. A 90-grit wheel has an impulse ratio of 12 : 36, a 120-grit wheel of 40 : 1.

### Wheel Structure

Structure relates to the spacing of the abrasive grain. Close grain spacings are identified by low structure numbers and are referred to as being dense. High structure numbers indicate open grain spacing. A grinding wheel may be compared to miniature teeth on a milling cutter (Fig. 3-69). The spacing of the grains allows chip room. If this space is too small for the type of material being ground, the wheel will tend to "load up." A loaded wheel heats up and is inefficient in its cutting action. Of course, too large a space is also inefficient since there are not enough cutting edges.

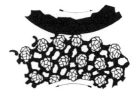

FIGURE 3-69. *The grinding wheel may be compared to a milling cutter having hundreds of tiny teeth. The voids in the wheel provide the chip room. Courtesy The Carborundum Co.*

Weak "posts"          Medium strength "posts"          Strong "posts"

FIGURE 3-70. *The grade of a grinding wheel is based on the strength of the bond posts. Courtesy The Carborundum Co.*

## Wheel Grades

The wheel grade refers to the strength of the binding posts (Fig. 3-70). As the binding posts become thicker and stronger, the wheel tends to act harder, and vice versa. Grades are designated with increasing hardness by letters from C to Z.

## Grinding-Wheel Designation

Grinding-wheel codes are easily understood since they give in logical order the complete information needed to determine the wheel's characteristics. Listed in order are the abrasive type, abrasive size, grade (hardness), and bond type, as shown in Fig. 3-71. This is a standardized code used throughout the grinding wheel industry.

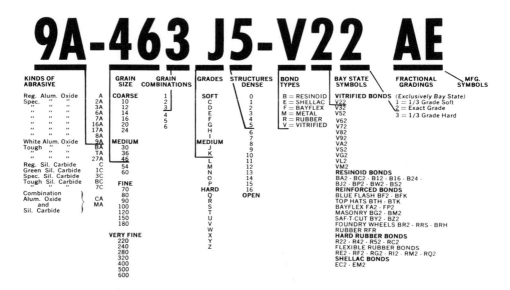

FIGURE 3-71. *The standard marking system for grinding wheels. Courtesy Bay State Abrasives Division, Dresser Industries, Inc.*

### Balancing and Dressing the Grinding Wheel

*Balancing.* When a new grinding wheel is used it should first be checked for proper balance. Large wheels are placed on arbors and set in balancing stands. Weights are shifted in the wheel flange until the wheel is in balance. Smaller wheels, 10-in. (25.4-cm) diameter or less, can be balanced on the machine spindle by shifting moveable weights in the flange. Wheels that are operated without proper balancing produce undue strains on the machine and a poor finish.

*Dressing.* After use, a grinding wheel becomes dull; that is, the sharp abrasive grains become rounded over. The wheel may also become *loaded*, a condition where the metal, or whatever is being ground, becomes inbedded in the wheel face. To restore the grinding efficiency of the wheel, it is necessary to cut away a small portion of the face to expose sharp abrasive and an open structure. The dressing can be done with a variety of tools, as shown in Fig. 3-72.

### Grinding-Wheel Selection Principles

With the extreme range of abrasive requirements and the multiplicity of wheels from which to choose, the question arises of how to select the proper grinding wheel for a specific job. The demands on grinding wheels are diverse and exacting. They may range from snagging wheels that remove as much as 60 lb (27.21 kg) of steel per hour to precision wheels that remove very little material but produce finishes as fine as 1 or 2 microinches. Often wheels must be capable of maintaining dimensional tolerances as close as 25 millionths of an inch.

To meet this range of performance specifications, all the factors discussed previously concerning abrasive types, abrasive grit sizes, hardness of the wheel, structure or grain spacing, and bond types must be considered in making a wheel selection for a given workpiece or a production run.

Experience will help narrow the selection field, but to pinpoint it to an optimum wheel may be difficult. Even an expert may have to test several varieties. A technique that has been developed and is now used in analyzing grinding-wheel efficiency is referred to as finding the optimum G-ratio.

### G-Ratio

G-ratio refers to the ratio of cubic inches of stock removed to the cubic inches of grinding wheel worn away. Shown in Fig. 3-73 is the plot of a G-ratio curve. The G-ratio is plotted as a function of wheel grade or hardness. The ratio of metal removal to wheel loss in a unit of time is also a measure of the production rate and the amount of work that a wheel is capable of performing during its useful life.

A soft wheel shows up as being relatively inefficient, but as the wheel hardness is increased, the efficiency increases until a maximum is reached. Then, despite the fact that the wheel becomes harder, efficiency drops off. Thus the curve as shown in Fig. 3-73 represents three phases of wheel action:

1. From the soft-grade, low-efficiency wheel, to where the curve begins to peak, the wheel is constantly breaking down, but at a rate that is too rapid for the most

economical use. On this part of the curve, the wheel does not load appreciably, nor does it produce excessive burn on the workpiece.

2. At the peak of the curve, the wheel is continuously breaking down at a rate that makes the most effective use of the abrasive. The abrasive grains have not become

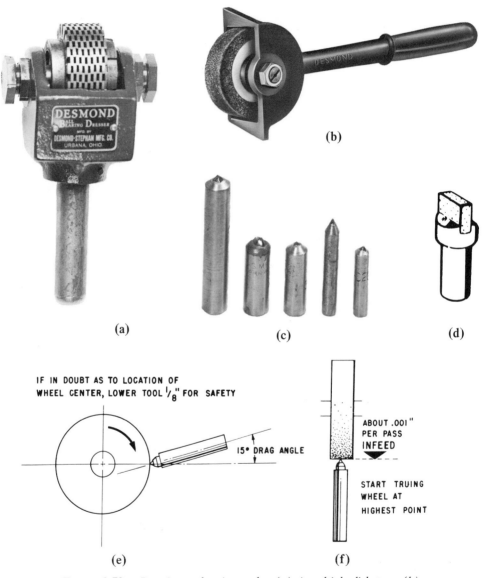

(a)

(b)

(c)

(d)

IF IN DOUBT AS TO LOCATION OF
WHEEL CENTER, LOWER TOOL $\frac{1}{8}$" FOR SAFETY

15° DRAG ANGLE

ABOUT .001"
PER PASS
INFEED

START TRUING
WHEEL AT
HIGHEST POINT

(e)

(f)

FIGURE 3-72. *Dressing and truing tools. (a) A multiple-disk type, (b) an abrasive wheel, (c) straight-shank diamond nibs, (d) diamond cluster, (e) diamond position, (f) recommended infeed per pass. Courtesy The Desmond Stephan Manufacturing Co.*

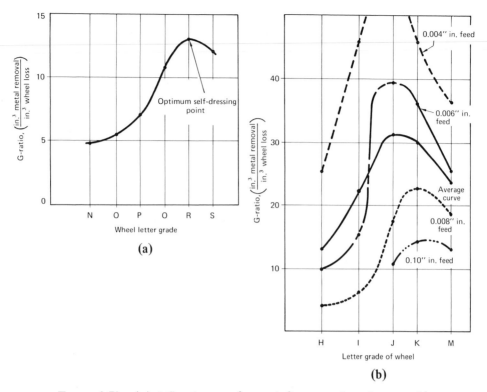

FIGURE 3-73. *(a) A G-ratio curve for a grinding operation using resenoid-bonded wheel in a snagging operation. (b) The G-ratio obtained using various infeed rates while grinding SAE 3145 steel on a centerless grinder.*

dull to the extent that the cutting rate is retarded, but they are continuously released by the bond and new sharp grains are exposed.

3. The harder wheel, as represented by J and K, show the efficiency dropping off due to a discontinuous wheel breakdown. This consists of wheel loading and unloading in cycles, accompanied by alternate pressure build-ups and releases. An increase in temperature results, the wheel burns the work, the cutting rate is decreased, and more power is consumed.

The wheel performance, as indicated by the G-ratio curve shown in Fig. 3-73(*a*), is not restricted as to bond, grit size, or any specific grinding operation. It may be used for heavy cutting, as in snag grinding, or equally well in precision grinding where the metal-removal rate as well as the wheel loss is measured in grams. Shown in Fig. 3-73(*b*) is a family of G-ratio curves based on a 60-grit wheel using increasing *infeeds* from 0.004 to 0.010 in. By infeed is meant the amount the wheel is advanced for each cut. In vertical-spindle surface grinding the cut per pass is referred to as *downfeed*. The effect of infeed on wheel performance changes the optimum grade of wheel to the harder side when heavier cuts are taken. This is logical since, as the infeed is increased, the grinding pressure becomes greater and a harder wheel is required to maintain wheel life and optimum performance.

G-ratio studies are also useful in measuring the efficiency of items related to the grinding process, such as grinding fluids, and the resistance to abrasion of various metals and nonmetallic materials.

### Grinding Operations

The wide variety of grinding operations can be classified by the main types of surfaces to be ground. These include cylindrical, surface, and internal surfaces (Fig. 3-74).

*Cylindrical Grinding.* Cylindrical grinding [Fig. 3-74(*a*)] is usually characterized by work that is mounted between centers and traverses back and forth, parallel to the grinding wheel. Work is produced that is true and concentric with the center holes.

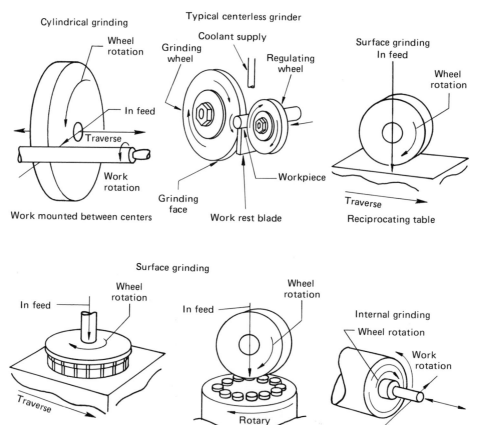

FIGURE 3-74. *Schematic view of common grinding operations.*

*Centerless Grinding.* In centerless grinding, parts are ground cylindrically but center supports are not required since the work is fed between two wheels and supported by a work rest, as shown in Fig. 3-74(*b*). Relatively high production can be attained since parts do not have to be handled individually.

*Surface Grinding.* All grinding may be construed as surface grinding. However, the term here refers to flat surfaces. The main types of machine action are shown in Fig. 3-74(*c, d,* and *e*). Face grinding has been developed to the point that it now competes with milling for efficient metal removal and comes under the term *abrasive machining.*

*Internal Grinding.* Internal grinding is used to produce a precision surface on drilled or bored holes. The work is usually held in a suitable chuck or it may be supported between rolls as in centerless grinding. Of necessity, the contact area between the wheel and the work is large and the spindle supporting the wheel is relatively small. Thus the rate of stock removal is somewhat limited. Because the wheels are usually of small diameter, spindle speeds must be very high. For example, the spindle speed for a 0.5 in. diameter wheel would be:

$$\text{RPM} = \frac{12 \times 6000}{\pi \times 0.5} \approx 46,000$$

### Grinding Speed and Feed

Traditionally, vitrified wheels have been made to operate at 5000 to 6500 sfpm. Now, however, manufacturers are taking a more critical look at all factors involved in order to increase production. If adequate power is available, an increase in grinding wheel speed will generally increase productivity, and also abrasive costs will decrease due to a decrease in the force involved. Both of these principles can be shown mathematically.

$$\text{Production ratio} = \frac{v \times b \times d}{V}$$

*where:* $v$ = work speed in sfpm
$b$ = width of grind (in.)
$d$ = depth of grind (in.)
$V$ = speed of grinding wheel in sfpm

An increase in productivity ($v \times b \times d$) and a corresponding increase in wheel speed ($V$) will allow the abrasive cost to remain fixed even though the production rate is higher. Thus, production can be increased by increasing any one of the three factors ($v, b, d$) if a corresponding increase is made in $V$.

### Grinding Force

In cylindrical grinding, the workpiece is usually the most compliant member of the machine's structural system. Thus the level of grinding force is often limited by the workpiece geometry. A diagram of the forces involved is shown in Fig. 3-75.

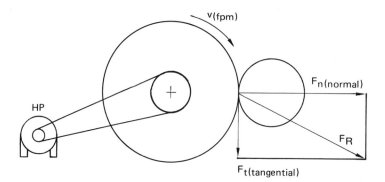

FIGURE 3-75. *Tangential force is limited by the power available for the wheel but is not changed if horsepower and speed are increased proportionately.*

The tangential force ($F_t$) or power component is limited by the power available from the wheel drive motor.

$$F_t = \frac{33,000}{V} \times HP$$

From this equation it is obvious why higher wheel speeds ($V$) are attractive, particularly for cylindrical grinding. If the wheel speed is doubled, twice as much power can be used without increasing the grinding force.

The normal grinding force ($F_n$), which tends to push the work away from the wheel, is related to $F_t$ but is dependent on the wheel specifications, the work material, coolant, type and condition of wheel dressing, etc. In general, $F_n$ is the highest when fine-grain wheels are used on hard materials and lowest with coarse-grained wheels on soft materials. For materials such as cast iron and steel, the $F_n$ ranges from two to five times the $F_t$. As an example, a grinding machine operating at 6000 sfpm with a 10-hp motor would have resultant forces ($F_r$ = vector sum of $F_n$ and $F_t$) on the work ranging from a low of 120 lb up to nearly 300 lb. A wheel running at 12,000 sfpm would cut these forces in half.

### High-Efficiency Grinding

As stated previously, the traditional speed for vitrified grinding wheels has been 6500 sfpm. Now industry is adapting to speeds of up to 16,000 sfpm and higher. Infeed rates are also comparably higher since the force on the work is reduced, as shown in the previous example.

However, high-speed grinding is not just a matter of increasing the wheel speed and/or the infeed rate on conventional machines. The interaction of the various process variables must be considered and balanced with the operating variables such as machine condition, quality of surface finish required, coolant application, wheel structure, and the hardness of the material.

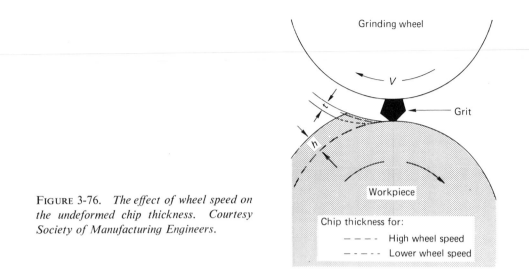

FIGURE 3-76. *The effect of wheel speed on the undeformed chip thickness. Courtesy Society of Manufacturing Engineers.*

The beneficial effects of high wheel speeds are mainly the result of a change in the thickness of the uncut chip ($t$), as shown in Fig. 3-76. If the work speed is doubled for a given depth of cut and wheel speed, the cutting force will also be doubled, as shown in Fig. 3-77. Thus there are two operating conditions to consider: high work speed that produces high cutting forces and large metal removal, or high wheel speed with low cutting forces and no change in metal removal. A marriage of these two conditions is illustrated in Fig. 3-77, where the ability to remove metal is doubled, grain loading is reduced to normal values, and the only disadvantage is a very high heat input.

Thus, to increase the metal removal rate, the infeed rate must be increased proportionately with the wheel speed to present a full layer of material for each revolution. In theory this cuts the grinding time in half. However, production is not necessarily doubled since loading and unloading times must also be considered.

### Abrasive Machining and Grinding Economics

*Abrasive Machining.* Abrasive machining is a term used to denote those grinding operations in which the metal removal rate is the main consideration, and surface finish and accuracy are secondary. It includes all operations where cost is a major factor in determining whether to grind or to use other machining methods.

Wheel speeds and feeds are important variables since they determine the maximum temperature the bond will be subjected to. In many cases, resin-bonded wheels are used. The resin loses its strength at about 600°F. If the wheel speed is too low, too much time is allowed for the heat to build up from the grain to the bond. For this reason, the higher wheel speeds are often used, up to 16,000 sfpm. Also, if the down- or infeed rate is too low, there will be a heat buildup. On the other hand, when the feed rate becomes too high, the chips are so large they no longer fit into the voids of the wheel and the grains are pried off, resulting in excessive wheel wear.

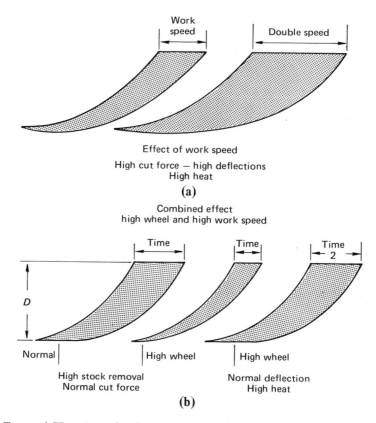

FIGURE 3-77. *A graphical comparison of the effect on cutting force in grinding, using a normal work speed and a doubled work speed (a). At (b) is shown the effect combining high work speed with high wheel speeds. Courtesy Society of Manufacturing Engineers.*

The length of the cut is also directly related to chip crowding. When the cut is relatively long, the downfeed rate must be reduced to avoid chip crowding. In order to provide a strong wheel with large chip spaces, some finer grains are often mixed with the coarse grains in making the wheel.

***Crush-Form Abrasive Machining.*** Abrasive machining is most often associated with flat surfaces; however, advances in *crush-form grinding* have made it competitive to other machining methods. Crush-form grinding involves a cylindrical or centerless grinder in which the wheel is formed by a hardened steel or carbide roller of the desired profile (Fig. 3-78). The dressing wheel is driven by the grinding wheel, which rotates at approximately 300 sfpm. The infeed rate is a few thousandths of an inch per revolution. Forming the wheel takes only a few minutes, and redressing it only a few seconds. Shown in Fig. 3-79 is an automotive front-wheel spindle that is abrasive machined in a single *plunge cut* to finish tolerances of 0.0007 in. on a crushtrue grinder.

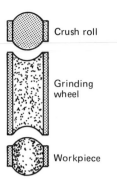

FIGURE 3-78. *The grinding wheel may be crush-formed with a hardened steel roller with almost perfect transfer of form.*

Crush roll

Grinding wheel

Workpiece

FIGURE 3-79. *An automotive spindle abrasive-machined from a rough casting to a finished dimensional tolerance of 0.0007 in. in 42 sec. Courtesy Bendix Automation and Measurement Division.*

Plunge cut refers to feeding the wheel straight into the work with no table traverse movement. In the case of the automotive spindle, the total abrasive machining time was 42 sec compared to 15 min by conventional machining.

***Grinding Economics.*** A detailed analysis of grinding costs is not within the scope of this text. The analysis can become somewhat cumbersome because there are more variables to be considered in grinding than in any other metal-working process. In addition to all the variables associated with a grinding wheel, such items as setup, load and unload, and dressing times, which in some cases may be a significant portion of the grinding cost, must be included. However, these costs are normally independent factors and can be calculated by a separate analysis. What remains is the specific grinding cost, which depends on the parameters used with a particular wheel on a specific workpiece and is associated with removing a definite amount of material from the workpiece, as measured in in.$^3$/min.

A commonly used equation for calculating specific grinding costs is

$$C = \frac{C_a}{G} + \frac{L}{tF}$$

*where:*   $C$ = specific cost of removing a cubic inch of material
$C_a$ = cost of abrasive in $/in.$^3$
$G$ = grinding ratio (volume of material removed/volume of wheel used)
$L$ = labor and overhead charge in $/hr
$F$ = machine feed rate in in.$^3$/hr
$t$ = fraction of time the wheel is in contact with the workpiece

Examining each term at the right of this equation we have:

$$\text{Specific wheel cost,} \quad \frac{C_a}{G}$$

The wheel cost in removing a cubic inch of material is a function of $G$. An aluminum oxide wheel, for example, may have an abrasive cost of $.30/in.$^3$. Thus, with a grinding ratio of 10 on a particular workpiece, the wheel cost would be $.03/in.$^3$ of material removed. $G$ may be easily determined by weighing the wheel and the work before and after grinding.

$$\text{Fixed labor and overhead cost,} \quad \frac{L}{tF}$$

This term shows the labor and overhead costs, assuming the wheel acts as a perfect cutter and the feed rate equals the material removal rate. The factor $t$, the fraction of time in which the wheel is in contact with the work, acts to increase cost as it becomes smaller by taking into consideration such items as override and approach. (This factor need not be considered in a vertical-spindle surface grinder since the wheel is in constant contact with the workpiece.)

*Example:*   What is the specific wheel cost of grinding 0.030 in. off of both sides of a gray cast iron plate 1″ × 10″ × 10″ using an aluminum oxide wheel and with the following conditions: labor and overhead = $10/hr, length of stroke = 12 in., grinding ratio = 0.5, feed rate is adjusted to 6 in.$^3$/hr, wheel cost = $.30/in.$^3$, fraction of time of wheel contact = 10/12 = 0.83 %.

*Solution:*

$$C = \frac{(.30)}{(0.5)} + \frac{(10)}{(.83)(6)} = \$2.60$$

Therefore the specific grinding cost of removing 0.030 in. from both sides of the plate would be $2.60 × 2 = $5.20.

From the foregoing discussion some generalizations can be made:

1. As $G$ becomes larger, the cost of using the wheel decreases.
2. As the feed rate $F$ increases, costs become less.
3. If the labor and overhead cost $L$ increases, the cost of using a high-wearing wheel increases.

### Abrasive-Belt Machining

In the past, cloth-coated abrasives have been regarded as being for polishing or finishing operations. In more recent years the picture has changed, so that now they are considered competitive with grinding. Some abrasive-belt machines are now built with 150-hp motors and can be used in plunge grinding operations. The metal-removal qualities of abrasive belts have been demonstrated by cutting a 1 in. square aluminum bar back 2 in. in 2 sec, and the same size gray cast iron bar the same amount in 5 sec, as shown in Fig. 3-80.

*Abrasives.* Five types of abrasive minerals are used, both natural and synthetic: flint, emery, garnet (natural), aluminum oxide, and silicon carbide (synthetic).

The grit size varies from a 12-mesh number, which is the coarsest, to a 600-mesh number, which is the finest. Simplified markings are also used, such as very fine, fine, medium, coarse, and very coarse. Just as in grinding, the metal-removal rate and surface finish will be determined by the coarseness of the abrasive, the surface foot speed of the belt, and the spacing of the abrasive grains. Closed and open coats are shown in Fig. 3-81. The open coat is used for faster cutting and on softer materials that tend to load the closed coat.

Two methods are used in applying the abrasives to the cloth backing: the gravity process and electrostatic process, as shown by the sketches in Fig. 3-82. In the gravity process, the mineral grains are dropped from an overhead hopper onto the adhesive-coated cloth. In the electrostatic process, the mineral grains pass through an electrically charged field. As the mineral and the backing pass simultaneously through the electrostatic field, the mineral grains are propelled upward and are imbedded in the adhesive on the cloth backing. This process results in having the sharpest edge of the mineral grains exposed for the best cutting action.

*Abrasive-Belt Machines.* A wide variety of abrasive-belt machines have been designed, as shown in Fig. 3-83. The conveyorized grinder [Fig. 3-83(a)] is particularly good for wide sheets (wider than 38 in.). Machines of this type are now capable of handling thin-gage sheets as well as heavy plates and lengths up to 40 ft with tolerances of ±0.001 in. or better with a finish of 3 to 5 microinches.

A good example of the use of belt grinding is in the aerospace industry. A ton of excess metal was removed from each experimental B-7 Valkyrie bomber by abrasive-belt machining. The belts were used to bring "as-rolled" precipitation-hardening

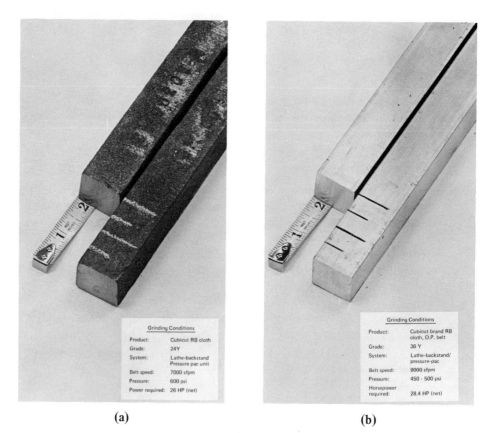

Grinding Conditions

| Product: | Cubicut RB cloth |
| --- | --- |
| Grade: | 24Y |
| System: | Lathe-backstand Pressure pac unit |
| Belt speed: | 7000 sfpm |
| Pressure: | 600 psi |
| Power required: | 26 HP (net) |

Grinding Conditions

| Product: | Cubicut brand RB cloth, O.P. belt |
| --- | --- |
| Grade: | 36 Y |
| System: | Lathe-backstand/ pressure-pac |
| Belt speed: | 9000 sfpm |
| Pressure: | 450 - 500 psi |
| Horsepower required: | 28.4 HP (net) |

(a)                                           (b)

FIGURE 3-80. *These two photographs attest to the fact that abrasive belts can be used effectively for metal removal and not just for polishing. The bars are 1 in. square. The gray iron is at 181 BHN and the 2024 aluminum at 130 BHN. The 2-in. removal (0.4 in.³/sec) on the gray iron was done in less than 5 sec. The aluminum cut was made in 2.0 sec or 1.0 in.³/sec. Courtesy 3M.*

stainless steel to a tolerance of ± 0.001 in. The process was also used to provide optimum thickness in the material where needed for structural purposes. This can be quite easily accomplished by the use of the flexible-bed sheet grinder, as shown at Fig. 3-83(*b*).

Another example of abrasive-belt grinding being competitive with conventional grinding or milling is in the machining of castings. The gray-iron casting, 170 BHN, required 0.070-in. stock removal to 0.001-in. flatness and 0.005-in. dimensional tolerance over approximately 100 sq in. of surface area. The production rate on a three-station, index milling machine was 27 sec per part. On a 3M Hi-Grind, the production rate was 210 parts per hour. The abrasive used was an aluminum/zirconium oxide mineral, electrostatically deposited on a resin-bond cloth.

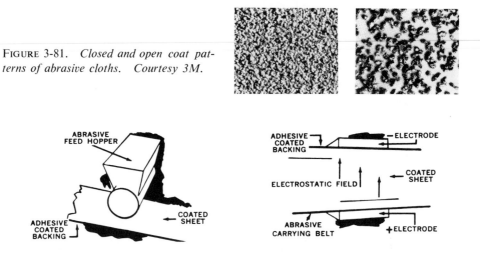

FIGURE 3-81. *Closed and open coat patterns of abrasive cloths. Courtesy 3M.*

FIGURE 3-82. *Two methods of applying minerals to the backing in the manufacture of abrasive cloth. Courtesy 3M.*

## Lapping, Honing, and Superfinishing

Designers are aware of the ever increasing demands for machines that can run faster, last longer, and operate more precisely than ever. This requires bearings, shafts, and seals that are dimensionally and geometrically accurate. Even grinding cannot meet the accuracy and surface finish required on a production basis.

As industry seeks to approach perfection, interest has focused on a family of precision-finishing processes: lapping, honing, and superfinishing. These processes are often referred to as *microfinishing*.

Each of these processes is designed to generate a particular surface finish and to correct specific irregularities in geometry. Briefly stated, microfinishing is used on parts:

1. Dimensioned with tolerances in the fourth or fifth decimal place.
2. With surface finish requirements in the 1 microinch to 8 microinch range.
3. Where an accurate geometric shape must be maintained.
4. Where surface integrity is critical.

*Lapping.* Lapping is an abrading process done either with loose-grain abrasive or with bonded-abrasive wheels or discs.

In the loose-grain process, a fine-grain grit of silicon carbide, diamond, or aluminum oxide is mixed with a carrier (usually cutting oil) and placed on the *master*. The master is a soft porous metal such as gray cast iron which holds the abrasive temporarily as it rubs or abrades against the surface to be cut. For example, gear teeth are lapped by running the newly cut gear with a master gear that is supplied with abrasive compound. Valves are often lapped to valve seats by rotating the two surfaces together with a lapping compound between. The type of abrasive and carrier

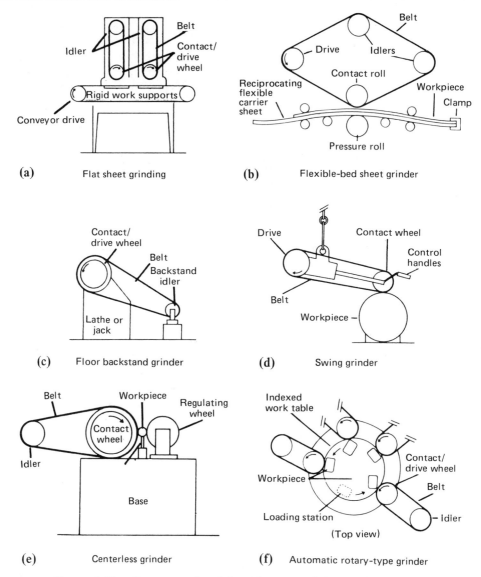

(a)    Flat sheet grinding

(b)    Flexible-bed sheet grinder

(c)    Floor backstand grinder

(d)    Swing grinder

(e)    Centerless grinder

(f)    Automatic rotary-type grinder

FIGURE 3-83.  *Some examples of the wide variety of abrasive belt grinders that are available.  Courtesy 3M.*

are dependent on the materials to be lapped.  Soft materials are lapped with aluminum oxide and hard materials with diamond or silicon carbide grit.  There is some difference of opinion as to whether the process can be called lapping if the abrasive is entirely free, that is, with no imbedding in the master.  In this case, it may be referred to as *free-abrasive machining.*

Shown in Fig. 3-84 is a lapping machine used to produce flat surfaces.  The parts to be lapped are confined to cages that impart a rotary and gyratory motion at

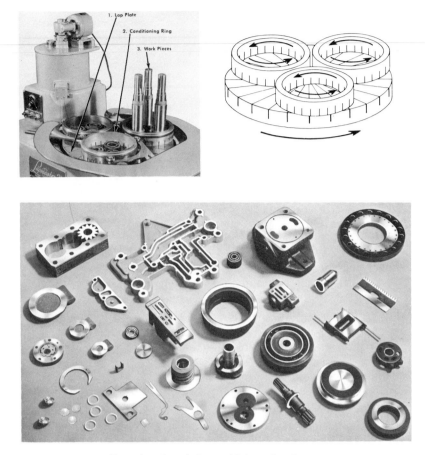

Examples of workpieces with lapped surfaces

FIGURE 3-84. *Work to be lapped is placed within the conditioning rings, which are held in place but are free to rotate. The work tends to abrade the lap plate but the rotating action of the conditioning rings causes the lap plate to wear evenly, maintaining a flat surface. Standard machines handle parts from $\frac{1}{8}$ in. to 32 in. in cross section. Steel, tool steel, bronze, cast iron, stainless steel, aluminum, magnesium, brass, quartz, ceramics, plastics, and glass can be lapped on the same lap plate. Courtesy Crane Packing Co.*

the same time, covering the entire surface of the lapping table. Parallelism is maintained by having a stationary lapping plate on top of the workpieces. Cylindrical parts are lapped by a cage arrangement as shown in Fig. 3-85.

Bonded-abrasive lapping is similar in appearance to the machine shown in Fig. 3-85. The upper section contains the bonded abrasive.

*Advantages and Limitations.* Extreme accuracy in both tolerance and geometry are the hallmarks of lapping. As an example, gage blocks used as a measurement

FIGURE 3-85. *A cage arrangement used for cylindrical lapping. Courtesy Warner & Swasey Co.*

standard are lapped to flatness of $\pm 0.000003$ in. and a parallelism of $\pm 0.000002$ in. One of the reasons this can be accomplished is that very little heat is generated in the lapping process.

Since lapping is essentially a finishing operation, parts should not be far from the expected size and geometry before lapping.

The residue of compound left on the surface of the part after lapping must be removed. This is not true of bonded lapping. Although lapping appears to be a simple process, it is considered somewhat of an art to achieve the extremely close tolerances expected.

***Honing.*** Honing is a controlled, low-velocity, abrasive-machining process that is most often applied to inside diameters or bores. One or more abrasive stones and one or more nonabrasive shoes (to equalize force throughout the workpiece) are mounted on an expanding (or contracting) mandrel (Fig. 3-86). The tool is inserted into the bore and adjusted to bear against the walls. A combination of motions gives the stones a figure-eight travel path. This motion causes the forces acting on the stones to be continually changing their direction, equalizing wear.

*Advantages and Limitations.* As with lapping, very little heat is generated, hence there is no submicroscopic damage to the workpiece surface. Tolerances can easily be maintained to within 0.0001 in. on a production basis. The surface finish produced can be very smooth (32 to 44 microinches) and part geometry can be corrected as shown in Fig. 3-87. The amount of stock removal is usually slight (0.001 to 0.0025 in.), so the cycle times are relatively short, usually 40–50 parts per hour. Materials that may be honed include soft materials like silver and brass as well as very hard materials, including the hardest alloys, carbides and ceramics.

***Superfinishing.*** Superfinishing is somewhat similar to honing but is applied primarily to outside surfaces. The process is particularly useful in finishing bearing surfaces. In ordinary use, the oil film between mating members is in danger of being punctured by any sharp points left by grinding. When the sharp points penetrate the oil film, the concentration of weight and friction cause the two metal surfaces to momentarily

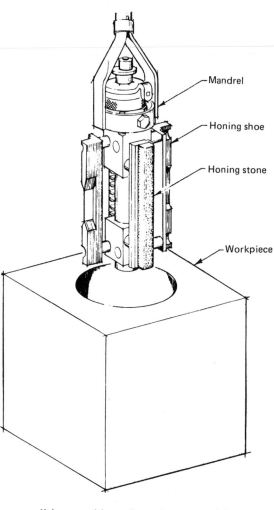

FIGURE 3-86. *The shoes and stones are uniformly spaced on the internal hone to produce even cutting forces. Honing does not distort the workpiece and produces a nearly perfect diameter.* *Courtesy* Machine Design.

weld and tear apart, causing rapid-wear conditions. Also, the minute particles sloughed off by this process further score the bearing surfaces.

The superfinishing process, as shown in Fig. 3-88(*a*), provides a gentle pressure over a wide contact area. The surface irregularities are reduced to a "near perfect" surface finish. The amount of material removed is usually small, about 0.0001 to 0.0004 in. The surface finish produced ranges from less than 1 to 80 microinches, with the average being around 3 microinches. When a surface smoother than 3 microinches is desired, the time needed to produce it goes up rapidly. As an example, a part that can be brought to a 3 microinch finish in 1 min may require 3 min more to bring it to a 2 microinch finish or better.

Flat surfaces may be superfinished as shown in Fig. 3-88(*b*). The upper spindle is a springloaded quill on which the stone is mounted. The lower spindle carries the work. Both spindles are parallel so that, when they rotate, the result is a very flat, smooth surface.

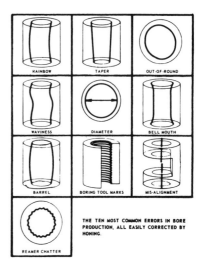

(a)

Work example

*Part:* Comparator column
*Use:* Gage
*Material:* Cast iron
*Hole diameter:* 0.813" ± 0.0002", tandem
*Length:* 1" each dia
*Finish:* Smooth base metal
*Stock removal:* 0.0015" to 0.002"
*Gage:* Ames Comparator
*Prev. operation:* Grind
*Production rate:* 50 per hr

FIGURE 3-87. *The hole geometry that can be corrected by honing. Courtesy Sunnen Products Co.*

(b)

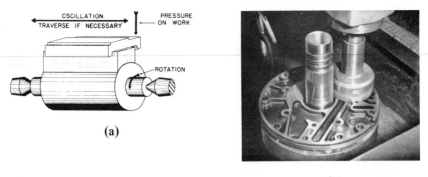

FIGURE 3-88.  *A schematic of the superfinishing process for round surfaces (a), for flat surfaces (b).*

*Advantages and Limitations.*  Superfinishing produces excellent surface finish in a minimum time after grinding, honing, or lapping.  The type of surface finish may be preselected, ranging anywhere from 1 to 80 microinches.  The process is easily adapted to high-volume production.

Superfinishing cannot be used to correct feature location such as concentricity or out-of-squareness.

### Deburring Processes

One of the most troublesome spots in the production of metal parts is the burr that remains after the cutting operation is finished.  In most cases it must be removed before another operation can be performed.  Formerly, much of deburring was a hand operation accomplished by the use of a file.  Now the manufacturing engineer has a wide choice of processes that can be used for abrasive cleaning and deburring.  Some of the more common methods used are: barrel finishing, vibratory finishing, chemical attack, and abrasive blasting.

*Barrel Finishing.*  Barrel finishing or tumbling consists of putting a number of parts in a steel drum or barrel that is usually six- or eight-sided or has shelves inside.  The abrasive medium used and the proportion in relation to the drum capacity is very important.  Some operations are done dry, but the results are usually enhanced when done wet.

There are more than 25 kinds of abrasive media used in tumbling and vibratory finishing.

The two most commonly used for today's high-speed, mass-production requirements are plastic and ceramic of various shapes, as shown in Fig. 3-89.

*Bonded Media.*  Plastic-bonded preformed media are now used extensively because they are lightweight and tough (break-resistant).  Finishes can be produced to as low as 2 microinches.  The plastic or synthetic bond mixed with a finely-grained silica-flour abrasive produces a "plater's polish," yet is relatively fast.

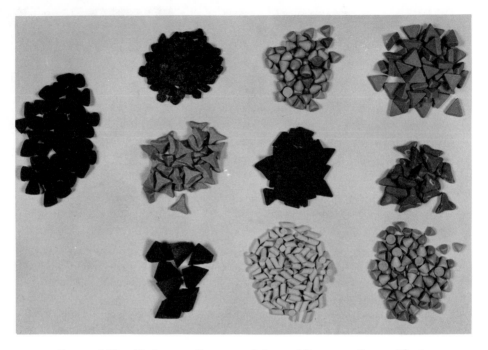

FIGURE 3-89. *Various media as used in tumbling operations. Plastic media are shown in the top row, ceramics in the middle row, and various aluminum oxide abrasive chips in the bottom row. Courtesy Almco, Queen Products Div., King-Seely Thermos Co.*

A faster cutting media is a 220-grit size, fused aluminum oxide abrasive. This aggressive media is especially good for nonferrous metals where machine lines or milling marks must be removed. It also leaves a good surface for paint adherence.

Preformed silicon carbide media are used mostly on exotic metals such as zirconium, titanium, zircoloy, and stainless steel.

High-luster finishes can be obtained by the use of abrasive, impregnated walnut shells or peach pits.

*Compounds.* Commercial compounds are composed of synthetic wetting agents, water conditioners, and various abrasives such as aluminum oxide, quartz, silicon carbide, garnet, silica flour, and cleaning chemicals. The main purpose of the compounds is to give additional cutting and cleaning power to the media.

Chemical grinding compounds consist of a blend of chemicals (no abrasives). They help maintain a clean tub and keep the abrasive media from becoming glazed.

Burnishing compounds are designed to impart high-luster finishes on all metals.

Descaling chemical compounds are used to remove heat-treat scale, forging scale, smut, oil, grease, and rust, and to " bleach " the metals. A positive step towards pollution control has been the development by various suppliers of a line of biodegradable compounds.

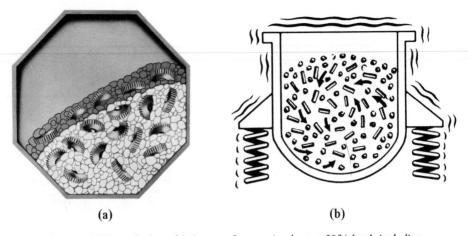

(a)　　　　　　　　　　　　　　(b)

FIGURE 3-90.　*The barrel (a) can only contain about a 50% load, including both abrasive and parts.　The rubbing and sliding action is in operation on each part only a fractional part of the cycle.　The vibratory container (b) can be loaded to within 1 or 1.5 in. of the top flange and every cubic inch of the entire load is in continuous motion.　Courtesy Materials Cleaning Systems Division, Wheelabrator-Frye, Inc.*

***Vibratory Finishing.***　The major difficulty in barrel finishing is its low efficiency [Fig. 3-90(*a*)] compared to the constant action of a vibrator (*b*).　Finishing operations include exterior and interior cleaning, descaling, deburring, grinding, radiusing, and fine finishing or burnishing.　The abrasive media and compounds are similar to those used in barrel finishing.　However, it has been found that the media do not last as long.　They wear down and fracture much faster, especially in the high-horsepower equipment.　Even with improved media, the vibrator type will expend it twice as fast as in tumbling.　However, since the complete processing is ten or twelve times faster, there is actually less media used per part.

Shown in Fig. 3-91 is an example of vibratory, abrasive, finishing equipment and sample parts.

***Thermal Deburring.***　Thermal deburring is a relatively new process that utilizes heated gases.　The part to be deburred is placed on a pallet and pushed into the thermal chamber.　The pallet acts as a seal (Fig. 3-92).　A gas mixture of 62% hydrogen and 38% oxygen is forced into the chamber.　An igniter, similar to that of a spark plug, detonates the gas mixture.　Combustion takes place in about 2 milliseconds and the temperature climbs to over 6000°F (4947°C).　The extreme temperature and the pressure, in excess of 6000 psi, vaporizes and oxidizes small objectionable burrs, but does not harm the parent metal.

***Electropolishing.***　Electropolishing is the reverse of electroplating, that is, the work is the anode instead of the cathode and metal is removed rather than added.　Many products can be efficiently and economically electropolished.

213

FIGURE 3-91. *Vibratory abrasive-finishing equipment and examples of finished parts.*

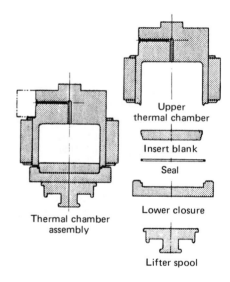

FIGURE 3-92. *A thermal chamber as used in thermal deburring. Courtesy* Modern Machine Shop.

The process is especially good on thin sheet-metal parts, intricately formed wire products, and deeply recessed complex parts. The metal loss is low, from about 0.0002 in. to 0.001 in., depending upon the original surface. Brightness and regularity of finish are dependent on uniform metal removal; therefore a smooth, dense surface, free from scratches and tool marks, is necessary.

Generally a good, bright finish may be obtained in 2 to 10 min, depending upon the original surface and range of current densities used. One ampere per square inch will remove 0.001 in. in a ten-minute period. Premixed acids may be obtained for electropolishing from E. I. du Pont de Nemours & Co., Inc.

### Contact-Wheel Finishing

Many types of contact wheels have been developed for polishing and buffing surfaces. Finishes can be varied from that of textured satin to a high luster. Harder wheels are made out of abrasive-impregnated rubber with plain, serrated, or cogtooth surfaces.

Buffing is used to produce a high luster or bright appearance. To achieve this, the abrading action is reduced to a minimum. A lubricant blended with the abrasive promotes a flowing rather than a gouging action.

Buffing is usually divided into two operations: (1) cutting down, and (2) coloring. The first operation is used to change a relatively rough surface into a smoother one. The second produces a high luster.

The abrasives used are extremely fine powders of aluminum oxide, tripoli (a porous form of silica), crushed flint or quartz, silicon carbide, and red rouge (iron oxide). The degree of ductility of the metal determines the type of abrasive to use. Soft metals, for example, do not require a cutting abrasive but a flow or blending as achieved by red rouge.

Buffing wheels are made from a variety of soft materials. The most widely used is muslin. Wheels of flannel, canvas, sisal, and heavy paper are used for special applications. A newer type made of nonwoven nylon-web material has been developed to produce a satin finish without sacrificing accuracy (Fig. 3-93).

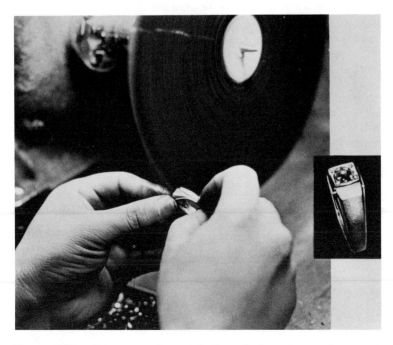

FIGURE 3-93. *Nonwoven nylon-web buffing wheels used to produce a satin finish without sacrificing accuracy. Courtesy American Buff International, Inc.*

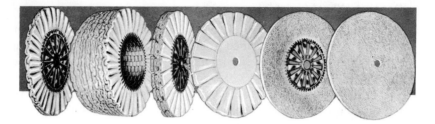

FIGURE 3-94. *A variety of buffing wheels ranging from the hard stitched to the softer pocketed type. Courtesy American Buff International, Inc.*

Muslin wheels are made in many forms (Fig. 3-94) to provide a wide range of properties. Stitched wheels are usually used for cut-down buffing. The wheels are made 0.25 in. thick and two or more are mounted on the arbor to provide the desired thickness.

Pocketed buffing wheels are used for cut-down operations on irregular contours.

Loose buffing wheels are stitched only at the hub and are used for color buffing.

The abrasive may be applied in liquid form by letting it slowly drip on the wheel or by holding a stick of the compound up to the rotating wheel. The heat of the contact is sufficient to transfer the abrasive to the wheel. Some of the sticks are made with various oils and greases and others with glues or cements. The advantage of the latter is that degreasing is not required before lacquering.

## PROBLEMS

**3-1.** A new 6-in. diameter, 8-tooth, face-milling cutter is given a performance test on a SAE 1020 CRS block (150 BHN), 4 in. square and 33 in. long. The cut is made over the long axis of the work. The following operating conditions were used:

> Sfpm = 300
> Chip load = 0.003 in.
> Depth of cut = 0.125 in.
> Idle horsepower = 0.5
> Machine mech. efficiency = 0.9

Calculate the following:
(a) RPM
(b) Feed rate in ipm
(c) Time to make the cut
(d) Cubic-inch removal rate
(e) Horsepower required at the cutter
(f) Horsepower required at the motor

**3-2.** (a) Determine the horsepower that is required to turn the surface of a 2-in. diameter soft cast iron bar under the following conditions:

> Sfpm = 200
> Depth of cut = 0.100 in.
> Feed = 0.050 ipr

(b) Is the increase in horsepower proportional to the increase in cutting speed if 300 sfpm were used?

**3-3.** (a) What causes a built-up edge to form on the tool when cutting metal? (b) How may it be eliminated? (c) Is there any disadvantage in having a built-up edge?

**3-4.** (a) How much crater wear could be expected on a standard steel-cutting grade of WC in 10 min of machining on SAE 4340 steel ($R_c$ 23) at 380 sfpm, 0.075 in. depth of cut, and 0.016 ipr feed. (b) How does this compare with the same conditions for a TiN-coated carbide tool? (c) Compare the crater wear of a cast carbide tool with that of C-5 carbide after 10 min of turning AISI 4340 steel ($R_c$ 30) at 350 fpm using a feed of 0.0166 ipr.

**3-5.** A 4-in. diameter gray cast iron bar 3-ft long is to be turned on a lathe. A roughing depth of cut of 0.100 in. is used and a 0.020-ipr feed. (a) Select a carbide grade and cutting speed. (b) Compare the machine time required to turn ten bars with a WC tool and with a ceramic tool.

**3-6.** (a) What makes UCON a distinctly different type of tool material? (b) How does it compare as to the cutting speed on mild steel with that of WC- and TiC-coated tools.

**3-7.** Make simple graphs to show the following relationships:
(a) Cutting force vs. cutting speed.
(b) Cutting force vs. side-rake angle ($-20°$ to $+20°$)
(c) Cutting force vs. feed (0.005 to .100 in.)

**3-8.** Two pieces of aluminum—1100-O and 2024-T3—are machined on the shaper with the same cutting conditions: depth of cut—0.050 in., feed—0.020 ipr, back-rake angle —$+10°$.
The measured chip thickness for 1100-O aluminum was 0.080 in., and for the 2024-T3 aluminum it was 0.060 in.
(a) Compare the cutting ratios of the metals.
(b) Compare the relative shear angles of the two metals. Make a sketch labeling the shear angle for 1100-O aluminum as A and the angle for 2024-T3 aluminum as B. Also show: the tool, $\alpha$, $t_1$, and $t_2$.
(c) Which material has the better machinability?

**3-9.** (a) What would be the recommended tool life, for the minimum cost per piece, when machining a 6-in. diameter

SAE 1020 steel bar 30 in. long under the following conditions:

Square-insert carbide tool with negative rake
Feed = 0.020 ipr
Length of cut = 25 in.
Depth of cut = 0.250 in.
Machine overhead rate = \$6.00/hr
Machine labor rate = \$3.50/hr
Carbide tool = \$1.50/insert = \$1.50/8 = \$.18/cutting edge
Tool changing time = 0.5 min
Total handling time = 5 min/part

(b) What would the cutting speed be for this tool life if the tool-life curve, as shown in Fig. 3-16, is used?
(c) How long would the cut take under the conditions given?

**3-10.** (a) If the cutting speed for one minute of tool life with a HSS tool is 200 fpm and the assumption is made that a carbide tool can cut four times faster, estimate what the tool life will be for a carbide tool cutting at 300 fpm. Use log-log paper to construct the tool-life curves similar to those shown in Fig. 3-16.
(b) What would the cutting speed be for each tool material for each of the following times:

10 min     60 min     30 min

**3-11.** In machining a new steel alloy the data indicated that $V_{10} = 316$ fpm and $V_{100} = 200$ fpm. (a) What tool life would you expect if the cutting speed were 240 fpm? (b) What is the tool material that was likely used in this operation? Why?

**3-12.** In analyzing the economics of single-point turning, what are the four cost factors that contribute to the total unit cost? Make a chart as shown for your answer.

| Cost Factors | Function of CS? | Included in Determining Max. Prod. Rate? |
|---|---|---|
|  |  |  |

**3-13.** Use Taylor's equation to find the tool life in minutes with the following conditions given:

Carbide tool
sfpm = 206
$C = 494$

**3-14.** A 0.25-in. diameter carbide drill is being used at 400 rpm for drilling soft gray cast iron. Is this speed about right? Explain.

**3-15.** Is a drill press with a 2-hp motor large enough to handle a 1-in. drill in steel using a normal feed?

**3-16.** A 0.5-in. diameter drill is being used to drill mild steel with a feed rate of 2 in./min. The drill has a web thickness of 0.1 in. (a) What is the torque and thrust? (b) What is the $HP_c$ required?

**3-17.** On 2 × 1 cycle log-log paper draw the following tool-life curves. Cutting parameters—0.125 in. depth, 0.020 in. feed. Tool carbide, +15° SCEA, and $\frac{1}{16}$ NR.

| Cast Iron sfpm | Tool Life min | SAE 1020 sfpm | Total Life min |
|---|---|---|---|
| 100 | 350 | 350 | 900 |
| 150 | 115 | 400 | 520 |
| 200 | 53 | 450 | 330 |
| 250 | 29 | 500 | 200 |
| 300 | 17.5 | 550 | 145 |
| 350 | 11.5 | 600 | 100 |

(a) What is the value of $n$ and $C$ for each of the materials?
(b) How does $n$ compare to what is considered standard for carbide tools?
(c) What is sfpm for $V_{60}$ for SAE 1020 steel?
(d) What may account for the slope of the curve being somewhat off?

**3-18.** (a) Calculate the approximate $Q$ for the turning operation shown in Fig. P3-1, $V = 100$ sfpm, $f = 0.012$ ipr.
(b) How much difference is the exact cubic inch removal rate?
(c) How long would it take to make a cut like this over a 12-in. length?

(d) Calculate the approximate $HP_c$ for the conditions given in (a) if the following forces are known:

$$F_t = 133,000f^{0.83}d^{1.0}$$
$$F_f = 33,770f^{0.48}d^{1.45}$$
$$F_r = 923f^{0.56}$$

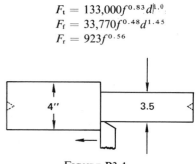

FIGURE P3-1

**3-19.** (a) Sketch typical cutting-tool vs. cutting-time curves for three different cutting speeds. (b) Label the three wear zones normally found. (c) Make a sketch to show how the data from (b) can be used to construct a tool life curve. (d) Assume the curve made from (c) was that of a carbide tool. Show the appropriate relationship to a ceramic-tool-life curve.

**3-20.** Is the cutting speed for minimum cost the optimum point at which to be cutting? Why or why not?

**3-21.** What is meant by "Hi-E" machining?

**3-22.** What is the main difference between a shear-angle milling cutter and one that is double positive? What is the advantage of each?

**3-23.** Cast iron engine blocks can be milled on the top surface by using a planer mill with a 10-in. diameter, twelve-insert carbide cutter. Each block is 2 ft long and 8 in. wide. Ten units can be set up in-line on the milling-machine table. The space between the ends of each of the blocks is 4 in. Compare the actual machine time required to face mill the blocks with broaching. Assume the following: The broach has a cutting surface as long as the blocks. It also has a 1-ft length on each end of the cutting surface. The broaching speed is 120 fpm. The carbide sfpm is based on approximately 100 min of tool life.

**3-24.** Why aren't $Al_2O_3$ wheels used for grinding WC materials?

**3-25.** What abrasive is recommended for hard, tough tool steels? Why?

**3-26.** (a) Which diamond classification is recommended for grinding WC? (b) How does it compare in impact with the regular RVG diamond?

**3-27.** (a) What are the four main factors to consider in selecting the right grinding wheel for a given job? (b) After making the grinding-wheel selection, you notice it has a tendency to "load up." What change should be made?

**3-28.** What is meant by G-ratio in grinding?

**3-29.** (a) Fifteen 2-in. rectangular flange heads are machined on a turret lathe, one at a time. The set-up time was 1.09 hr. The machine time per piece is 0.296 hr. As a comparison, the same parts were abrasive machined on a vertical grinder, where the setup time was 0.4 hr. The total amount removed from the flange was 0.030 in. The grinding chuck can accommodate eight flanges per run. The downfeed rate was 0.015 ipm. Compare the time per lot by each method. (b) What is the time saved by grinding, given in percent?

**3-30.** (a) A new 10-in. diameter, 2-in. wide wheel was placed on the grinder. A dressing operation was performed to true the wheel to the spindle. Four passes, each of 0.001 in., were made. After grinding a cast iron plate 5 in. $\times$ 15 in. from a thickness of 1.5 in. to 1.484 in., the wheel diameter measured was 9.980 in. What was the grinding ratio? (b) Comment on this grinding ratio. (c) What is the actual wheel cost for this operation if the lost abrasive is valued at 90 cents/in.³?

**3-31.** Compare the time for milling and grinding both sides of the 40 in. $\times$ 40 in. plate described in this chapter. Assume the following milling conditions: a 10-in. diameter face mill is used that has 14 indexable carbide inserts. Allow an overlap of an inch width per pass. The chip load per tooth is set at 0.008 in. The loading and unloading time is 8 min. Assume the loading and unloading time for grinding is 3 min per part. The longitudinal speed is 50 fpm. The crossfeed is 1 in. per pass, and approach and overtravel are 1 in. each.

**3-32.** A given cylindrical part can be machined either on an engine lathe or on a turret lathe. Ten operations are required. The average time for each operation on the engine lathe is 5 min. and on the turret lathe 0.3 min. The special tooling required for the turret lathe costs $60.00. The labor and overhead rate is $8.00/hr on the engine lathe and $12.00/hr on the turret lathe. It takes 30 min to set the job up on the turret lathe. Setup time per part on the engine lathe is included in each operation. How many parts would have to be made before it would pay to set up the turret lathe?

**3-33.** (a) If a cylindrical grinder with a 10-hp motor has the capability of increasing the wheel speed from 6000 sfpm to 15,000 sfpm and the infeed used was kept constant, what would be the change in tangential force on the work material? (b) What would the approximate normal force be if the operation (a) were done with a fine-grain wheel on hard material for each wheel velocity?

**3-34.** Can the wheel infeed be increased proportionately with the increase in wheel velocity? Why?

**3-35.** A hardened steel plate $1'' \times 8'' \times 16''$ is to be ground on both $8'' \times 16''$ sides with the following conditions:

Wheel diameter after dressing = 9.88 in.
Wheel width = 0.750 in.
Wheel cost = $0.90/in.³
Labor and overhead charge = $10/hr

Traverse feed rate = 0.5 in./pass
Material removed from each side
= 0.020 in.
Wheel diameter after grinding = 9.850 in.
Length of stroke = 18 in.
Longitudinal feed = 1440 in./hr

What is the specific grinding cost for this operation?

**3-36.** How does carbon content act both to increase and decrease machinability of steel?

**3-37.** How does sulphur improve machinability of low-carbon steels?

# BIBLIOGRAPHY

## Periodicals

DeVRIES, M. F., "Machining Economics—A Review of the Traditional Approaches and Introduction to New Concepts," ASTM Paper No. MR69-279, 1969.

FEINBERG, BERNARD, "Cast Carbide Tools: The Results Are In," *Manufacturing Engineering & Management*, Sept. 1971.

FEINBERG, BERNARD, "Cutting Tools, 1971" *Manufacturing Engineering & Management*, Jan. 1971.

FINDLEY, W. N. and R. M. REED, "The Influence of Extreme Speeds and Rake Angles in Metal Cutting," ASME Paper No. 62-Prod-5, May 1962.

GALIMBERTI, J. M., "Reasons for Carbide Tool Failure," *The Tool and Manufacturing Engineer*, Nov. 1968.

JUDSON, T. N., "High Velocity Turning of Semifinished Steel and Aluminum Work Pieces," ASTM Paper No. 703, 1965.

KANE, G. E. and M. P. GROOVER, "The Use of Cutting Temperature as a Measure of the Machinability of Steels," ASTM Paper No. 199, 1967.

LINDSAY, R. P., "Formula for Selecting Machining Methods," *Abrasive Engineering*, Sept. 1969.

RUDNESS, R. G., "The Technology and Applications of UCON Indexable Inserts," Press Conference Remarks at Union Carbide Corporation Office, Cincinnati, Ohio, April 1972.

TRIGGER, K. J., "Tool Chip Interface Temperatures," Trans., ASME Vol. 71, No. 2, 1949.

SPENCER, K. H., "The G-Ratio Myth," *Abrasive Engineering*, Sept. 1969.

STEWART, I. J., "How High Wheel Speeds Influence Grinding Performance," Technical Paper MR69-206, SME.

## Books

BEGEMAN, M. L. and B. H. AMSTEAD, *Manufacturing Processes*, John Wiley and Sons, Inc., New York, 1969.

DATSKO, J., *Material Properties and Manufacturing Processes*, John Wiley and Sons, Inc., New York, 1967.

DE GARMO, E. PAUL, *Materials and Processes in Manufacturing*, The Macmillan Company, New York, 1974.

DOYLE, L. E., KEYSER, C. A., *et al.*, *Manufacturing Processes and Materials for Engineers*, Prentice-Hall, Inc., Englewood Cliffs, New Jersey, 1969.

LINDBERG, R. A., *Materials and Manufacturing Technology*, Allyn and Bacon, Boston, 1968.

NIEBEL, B. W. and A. B. DRAPER, *Product Design and Process Engineering*, McGraw-Hill, New York, 1974.

ROLT, L. T. C., *A Short History of Machine Tools*, The M.I.T. Press, Cambridge, Massachusetts, 1965.

SHAW, M. C., *Metal Cutting Principles*, 3rd Edition, M.I.T. Press, Cambridge, Massachusetts, 1960.

VIDOSIC, J. P., *Metal Machining and Forming Technology*, Ronald Press, New York, 1964.

# 4

# *Metal Forming*

## Introduction

Forming takes place in a metal any time it is subjected to stresses that are greater than the yield point or when the deformation stress moves from the elastic to the plastic range. Forming in the elastic range is usually minor and temporary.

Metal forming is usually classified as being "cold" or "hot." Cold forming is done below the recrystallization temperature while hot working is done above this point but below the melting point of the metal. Hot working is preferred for primary fabrication operations because the forces required are lower and larger reductions are practical. Cold working causes more noticeable changes in mechanical properties, particularly increases in strength with a corresponding loss in ductility.

## FORMABILITY

In recent years there has been an unprecedented interest in metal-forming technology. This is due in part to the continuing demands on industry to produce lighter weight, yet stronger, more rigid components. Ever higher strength-to-weight ratios are being used in forming parts for missiles, aircraft, and more recently for automobiles. As new metals are tried, such as aluminum alloys and high-strength, low-alloy steels, they are found to be less forgiving than mild steels. Higher forming forces are required which often result in more fractures.

In the past the quality of a stamping depended on the skill of the artisan, who depended on knowledge gained by trial and error. Now, with new materials and greater demands, industry cannot wait on lengthy trial-and-error methods. Therefore considerable work has been done by both metallurgists and process engineers to provide a method of predicting formability of metals.

### Predicting Sheet-Metal Formability

Formability may be simply defined as the ease with which metal may be forced into a permanent change of shape. Several methods of predicting formability have been developed that are useful, particularly to the designer and the manufacturing engineer. These are: cup or "radial drawing," the normal anisotropy coefficient, the strain-hardening coefficient, and the forming-limit curve.

***Cup or Radial Drawing.*** Cup drawing consists of cutting a circular blank from the metal to be tested, inserting it in a die, and noting the severity of the draw it is able to withstand without tearing. By the severity of the draw is meant the ratio of the cup diameter to the blank diameter or

$$R_d = \frac{D - d}{D}$$

where:  $R_d$ = drawing ratio
$D$ = blank diameter
$d$ = punch diameter

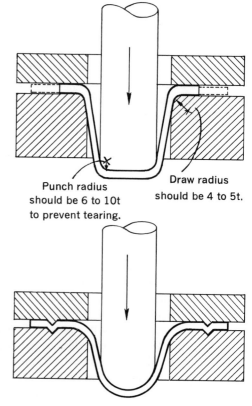

Punch radius
should be 6 to 10t
to prevent tearing.

Draw radius
should be 4 to 5t.

FIGURE 4-1. *The cup-forming test may consist of a flat-bottom punch and a lubricated blank (a), or a disk clamped and deformed by a polished, lubricated, hemispherical punch (b). In the first instance, there is drawing, bending under tension, and cylindrical stretching movements. In the second instance, there is biaxial stretching.*

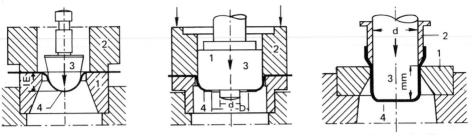

| Erichsen Cupping Test | Deep Drawing Cup Test, first draw | Deep Drawing Cup Test, second draw |

FIGURE 4-2. *Standard Erichsen cup-drawing tests designed to predict the formability of sheet-metal samples.*

A drawing ratio of 50% is considered excellent. Shown in Fig. 4-1 is a comparison of two dies used in drawing a cup. The action illustrated in the top figure is principally that of drawing, whereas the one on the right is principally that of stretching. For pure drawing, the clamping force is just sufficient to prevent buckling of the material at the draw radius as it enters the die. Thus, deformation takes place in the flange and over the draw radius. In most production parts, deformation takes place by a combination of drawing and stretching.

Standardized tests have been developed to check the drawing qualities of a metal, such as the Erichsen tests shown in Fig. 4-2.

One objection to the Erichsen cup test is the problem of determining the influence of friction on metals of differing metallurgical qualities. A test that overcomes this objection is the Fukui conical-cup test. It utilizes a hemispherical, smoothly polished punch. No blank holder is required. In each test, a drawing ratio which will result in a broken cup (Fig. 4-3) is used. Wrinkles are avoided by using a fixed

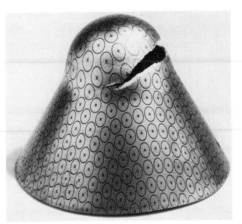

FIGURE 4-3. *A Fukui cup test used to predict the formability of sheet metal. No blankholder is needed, and wrinkles can be avoided by proper attention to the dimension ratios in the test disk. Courtesy Institut De Recherches De La Siderugie Francaise.*

ratio between the thickness of the sheet, the size of the blank, and the punch and die diameters. Under these conditions, the test produces a known amount of stretching, drawing, and bending under tension.

*The Normal Anisotropy Coefficient.* The anisotropy coefficient is derived from the ratio of the plastic width strain $\varepsilon_w$ to the thickness strain $\varepsilon_t$, as measured in a uniaxial tensile test. A material that has a high plastic anisotropy also has a greater "thinning resistance." In general, the higher the anisotropy coefficient the better the metal deforms in drawing operations.

*Strain-Hardening Coefficient.* Strains describe the actual dimensional changes in a sheet. Strain hardening refers to the fact that as a metal deforms in some area, dislocations occur in the microstructure. As these dislocations pile up, they tend to strengthen the metal against further deformations in that area. Thus the strain is spread throughout the sheet. However, at some point in the deformations, the strain suddenly localizes and necking, or localized thinning, begins. When this occurs, little further overall deformation of the sheet can be obtained without it fracturing in the necked region.

Thus the strain-hardening coefficient reflects how well the metal distributes the strain throughout the sheet, avoiding or delaying localized necking. The higher the strain-hardening coefficient, the more the material will harden as it is being stretched and the greater will be the resistance to localized necking. Necks in the metal may affect structural integrity or harm surface appearance.

For most practical stamping operations, stretching of the metal is the critical factor and is dependent on the strain-hardening coefficient. Thus stampings that entail much drawing should be made with metal having high average strain-hardening coefficients. Yield strengths should be low to avoid wrinkles or buckling.

*Forming-Limit Curve.* The forming-limit curve is one of the more recent methods developed to determine the formability of sheet metal. Essentially, it consists of drawing a curve that shows a boundary line between acceptable strain levels in forming and those that may cause failure.

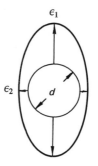

FIGURE 4-4. *The relationship of major, $\epsilon_1$, and minor, $\epsilon_2$, strains is established by measurement after forming.*

The curve is based on relating major and minor strains that are perpendicular to the plane of the sheet. To determine the major and minor strains, a grid of circles is printed on the sheet metal. This may be done by an electrolytic stencil-etching process or by using a photosensitive resist method as produced by the Eastman Kodak Co. After the metal is deformed, the circles are measured to obtain the major strain $\varepsilon_1$ and the minor strain $\varepsilon_2$, as shown in Fig. 4-4. Typically, about ten data points are obtained from a test specimen in the region of fracture. Ellipses lying both in the failed region and just outside of it are measured. The forming-limit curve is then drawn to fall below the strains in the necked and fractured zones, and above the strains found just outside these zones (Fig. 4-5).

Controlled variation in specimen size and lubrication permit the plotting of an entire forming-limit curve from one test setup. A reasonably accurate forming-limit curve may be obtained with four specimens, or 40 data points. A precision curve may be obtained with eight specimens.

Credit for recognizing the major–minor strain relationship goes to Dr. Stuart P. Keeler of the National Steel Corporation. He found the forming-limit curve to be immensely helpful in diagnosing press-shop failures, even though hundreds of production stampings were needed to produce the data for plotting one forming-limit curve. Through further study of this method at General Motors Research Laboratories (GMR), a simple procedure was developed for obtaining this curve for any metal. It was also found that "local" ductility varies for different metals, so no universal forming-limit curve could be developed. For example, two candidate metals may have peak local ductilities of 20 % and 50 % at a given minor strain. The metal with the 20 % local ductility (high strain-hardening coefficient) may turn out to be the best choice because the strain will then have a better distribution throughout, allowing the entire sheet to be stretched 20 %. If the other sheet showed little strain hardening, it might stretch by 50 % in a local area, but leave the rest of the sheet relatively unstrained.

Through the use of formability-prediction techniques, designers and fabricators are able to make a wiser choice of metals and obtain data quickly on newer metals. The essential data can be obtained before the die is designed. Also, metal suppliers will be able to establish whether a material possesses required formability before it is shipped from the plant.

### Formability, Temperature, and Recrystallization

*Temperature.* Forming operations are classified as cold or hot. Cold working is usually associated with those operations done at room temperature or below the recrystallization temperature. The properties of yield strength, strain-hardening rate, and ductility are all very much temperature-dependent. With increasing temperature, it is generally true that the yield strength and rate of strain hardening will progressively reduce and ductility will increase, as shown in Fig. 4-6. Hot-working may be defined as metalworking at a temperature above which no strain hardening takes place.

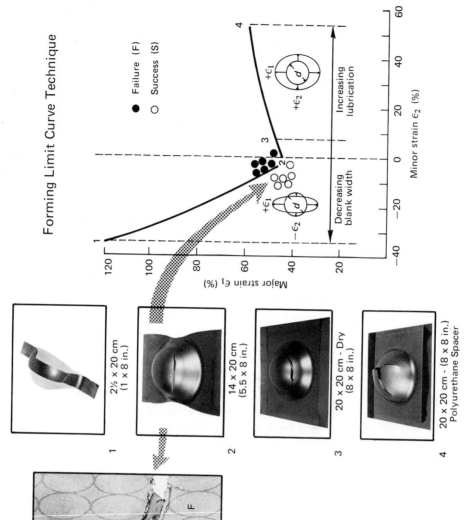

Forming Limit Curve Technique

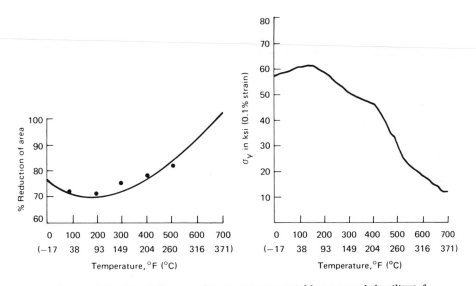

FIGURE 4-6. *The influence of temperature on yield stress and ductility of a steel containing 0.1% C, 0.5% Mn, and 0.05% Mo.*

The product designer should be well aware of the benefits of cold working, such as higher strength which permits the replacement of many components made from alloy steel with materials having a lower alloy content. Other savings are in overall weight, heat-treating costs, and reduced machining or finishing operations.

Hot working, on the other hand, allows forming operations to be done with less energy input. Not only does the metal deform more easily, but it can also accept a very large amount of deformation before cracking. As an example, billets are sheared off prior to forging. Even in moderately alloyed steels the sheared face frequently contains cracks that run into the billet and may result in forging defects. Heating the billet prior to shearing, even in the low range of 300 to 600°F (149–316°C), increases the ductility to the point where simple shear occurs and no cracks propagate into the billet.

◁ FIGURE 4-5. *A forming-limit curve may be obtained directly from four test specimens. The narrower specimens (1) and (2) provide data for negative minor strains (where original circles become "squeezed"). Data for positive minor strains (where the original circles are stretched in both directions) are obtained through controlled increase in lubrication. Detail for specimen (2) shows source of individual data points. These are obtained directly from measurements of major and minor axes of deformed ellipses in both successful (S) and failed (F) regions.*

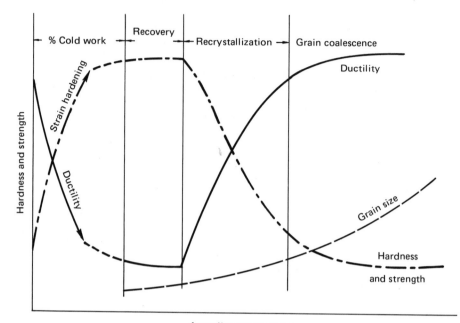

FIGURE 4-7. *A schematic representation showing the relationship of annealing on prior cold working.*

***Recrystallization.*** As discussed in Chapter 2, the strain hardening that takes place during cold working may be relieved by annealing. In the process, the grains recrystallize or new grains are formed that are strain-free. The amount of recrystallization is a function of time, temperature, and the degree of prior cold working. The whole relationship of hardness and strength, as the result of cold working, to annealing temperatures is shown schematically in Fig. 4-7. A pictorial sketch showing the effect of both cold working and hot working on the microstructure is shown in Fig. 4-8.

A newer term that has been introduced in metal forming is *warm forming*. In this process, the temperature is closely controlled such that no microstructural changes take place in the metal. As an example, AISI 5140 H steel was heated to 1225°F (655°C) to make possible roll forming the gears for an automatic transmission. This process avoided the nearly 50 % material waste inherent by the machining method. The 1225°F temperature selected provided optimum formability under the critical point of 1300°F (723°C) and a combination of the advantages of both hot and cold forming were realized (surface scale was minimized, unit loads on tooling were decreased, and dimensional control comparable to that of cold-forming was possible).

Induction heating maintained the operation within $\pm 25$ degrees Fahrenheit ($\pm 14$ degrees Celsius) of the set temperature.

The force–energy curve (Fig. 4-9) shows there is little benefit to be gained by heating SAE 5147 steel up to 400°F. From 400 to 1100°F (204–593°C) there is a proportionate reduction in force and stress. Between 1100 and 1500°F (593–816°C) the curve levels off. Above 1500°F (816°C) increases in temperature once again result in proportionate reduction in force and stress.

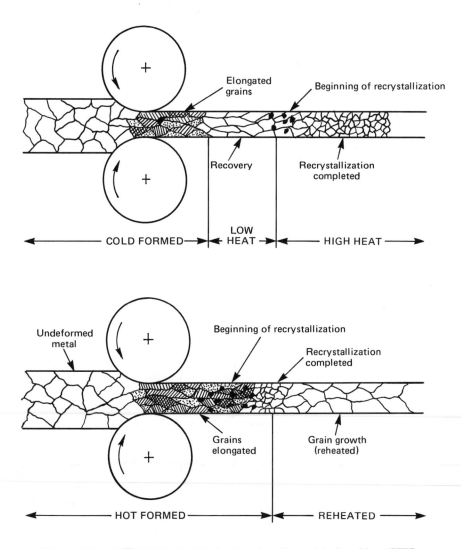

FIGURE 4-8. *A pictorial sketch showing the effect of both cold working and hot working on the microstructure of cast metals.*

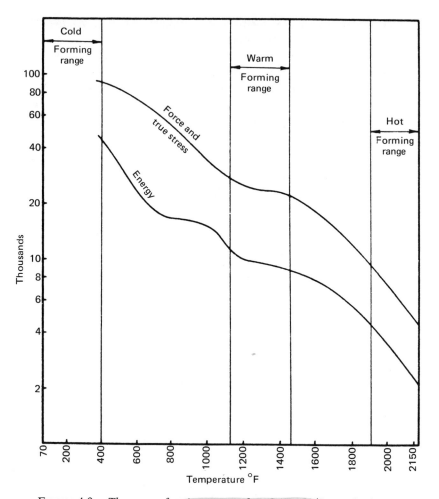

FIGURE 4-9. *The warm-forming range for SAE 5147 steel shows a leveling off of the force and energy required. Courtesy Society of Manufacturing Engineers.*

## FORMING PROCESSES

The enormous variety of plastic deformation processes for metals makes it desirable to classify them into general groups and subgroups. Traditionally, metal-forming operations have been classified as those done by "hot-working" processes or "cold-working" processes. This classification no longer seems adequate since many forming operations would fit in either classification and now there is also warm forming.

A more appropriate method of classifying metal-forming operations is to divide them into processes that intentionally alter the cross-sectional area and those that do not. With one or two exceptions, most of the metal-deformation processes can be classified as forging (either hot, cold, or warm) or as a process that merely alters the profile, as does sheet-metal forming.

### Sheet-Metal–Forming Operations

Sheet-metal–forming operations are usually associated with conventional press-working, which includes cutting the blank and then bending or drawing it to the desired shape.

### *Cutting Operations*

Common sheet-metal–cutting operations are shown in Fig. 4-10. Blanking, piercing, and notching are usually done with punches and dies that are mounted in a standard-type die set, as shown in Fig. 4-11. Die sets facilitate exchanging one setup for another, since the punch and die can be kept intact. This of course necessitates a die set for each different job; however, the time saved in getting the punch, die, stripper, etc. mounted and in alignment each time makes it economically feasible.

### *Bending*

Bending operations involve stretching the outside fibers and compressing the inside fibers, as shown in Fig. 4-12. The type of press commonly used in making bends is a press brake. The press and the type of dies used are shown in Fig. 4-13. Rubber dies are also used, as shown in Fig. 4-14.

*Bending Force.* The approximate force required to bend a piece of metal [a V-die, Fig. 4-15(a)] can be found by the use of a simple formula for a concentrated load on the center of a beam supported at two points.

$$P = 1240 \frac{t^{2.5}}{w^{1.35}}$$

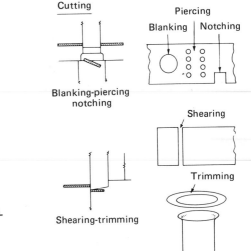

FIGURE 4-10. *Common metal-cutting operations.*

*where:*    $P$ = load, in tons/ft
            $t$ = thickness of the material, in inches
            $w$ = width of female die opening, in inches

For mild steel up to 0.5 in. thick, the recommended die opening is eight times the metal thickness. For mild steel plates thicker than 0.5 in., the recommended die opening is 10 to 12$t$.

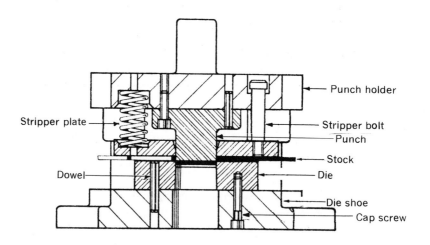

FIGURE 4-11. *A standard die set with a punch and die mounted in place.*

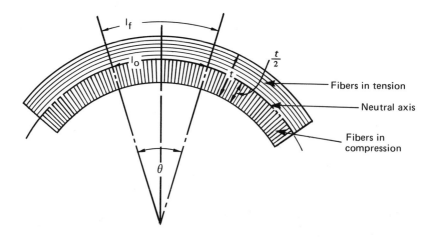

FIGURE 4-12. *Schematic of a bend specimen.*

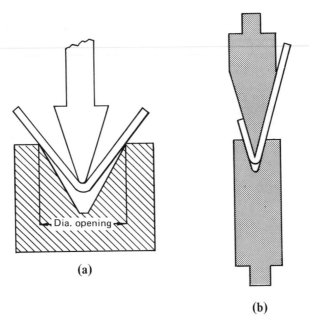

FIGURE 4-13. *Press-brake bending of metal. Air bending is shown at (a) and die bending at (b).*

***Bend Allowance.*** It is sometimes necessary for production engineers and designers to know the exact width of a strip that will contain various bends. A standard bend allowance formula is:

$$\text{B.A.} = \frac{\theta}{360}\, 2\pi(r_i + kt)$$

*where:*   $\theta$ = angle of bend, in deg.
$r_i$ = inside radius in in.
$k$ = 0.33 when $r_i$ is less than or equal to $2t$ and 0.50 when $r_i$ is greater than $2t$
$t$ = metal thickness

When $r_i \approx t$, the strip width can be found by adding the inside dimensions. For example, a channel that is $\frac{1}{8}$ in. thick, 3 in. wide, and 2 in. in depth would be $2(1.875) + 2.750 = 6.5$ in. wide before bending.

***Springback.*** One of the problems encountered in forming metal, particularly in bending, is the tendency after the force has been removed to spring back or establish internal equilibrium. As shown in Fig. 4-15(a), when a bending load is applied, tension and compression forces are set up about the neutral axis, but when the load is relieved, the outer surface shortens and the inner surface lengthens. The amount of springback is in proportion to the *sum* of the tension and compression recovery

233

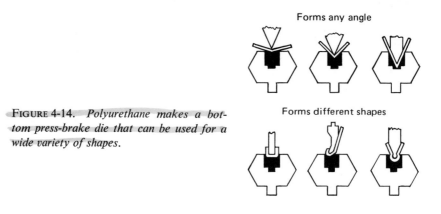

Forms any angle

Forms different shapes

FIGURE 4-14. *Polyurethane makes a bottom press-brake die that can be used for a wide variety of shapes.*

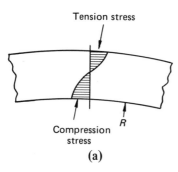

Tension stress

Compression stress

R

**(a)**

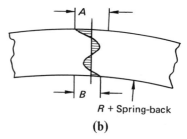

A

B

R + Spring-back

**(b)**

FIGURE 4-15. *Bending stresses are shown at (a). When the bending stress is removed, the outer surface shortens and the inner surface lengthens to relieve some of the stress (b). A method of compensating for spring-back is to apply a slightly greater force, as shown on the force–deformation diagram (c).*

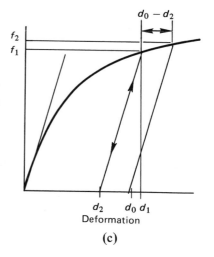

$d_0 - d_2$

$f_2$
$f_1$

$d_2$ $d_0$ $d_1$

Deformation

**(c)**

[Fig. 4-15(*b*)]. A method of compensating for springback by adding a small amount of deformation is shown in Fig. 4-15(*e*).

Springback generally increases with increasing yield strength and section thickness. In press-brake operations, as shown in Fig. 4-13, air bending (*a*) is much preferred over die bending (*b*) for metals that produce springback problems. Minimum bend radii are particularly hard to achieve on metals of very high yield strength due to substantial springback. To ensure that the proper radius is obtained, the span of the lower die opening should be 16*t*. If this span cannot be obtained, the bend can be started with a mandrel or punch of greater radius.

Polyurethane inserts are sometimes used, as in the lower die in Fig. 4-14. These save making special forms since they can assume any shape. The polyurethane forces the sheet to hug the punch. Therefore a 90° punch can be used to make a bend up to 90° by regulating the amount of punch penetration into the "rubber" where a lot of springback is encountered. The upper die can be modified so that overbending occurs to offset the springback.

### Deep Drawing

Sheet-metal–drawing operations include deep drawing, ironing, and embossing (Fig. 4-16). Deep drawing is usually done on precut blanks, but a *compound die* can be used, as shown in Fig. 4-17. A compound die is made to do both blanking and drawing in one stroke of the press. This type of die saves time and equipment. However, the tooling may become quite crowded and not only difficult to make but difficult to maintain. An answer to this problem is to spread the tooling out into a number of stages, thus cutting and forming the part progressively on what is termed

Drawing operations

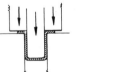

Deep double-action drawing

Reverse drawing
with cushion

Ironing

FIGURE 4-16. *Sheet-metal–drawing operations.*

Embossing

a *progressive die.* Shown in Fig. 4-18 are a number of parts that have been blanked and formed in successive stages on a progressive die.

There are times when the size or complexity of the part does not permit it to be formed on either a compound or a progressive die. In this case, the part must be transferred from one die to another until it is finished, hence the term *transfer die.* An example of a part that requires a transfer die is the deep-drawn metal cap of a

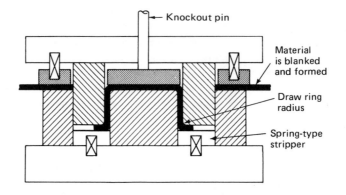

FIGURE 4-17. *A compound blank and draw die.*

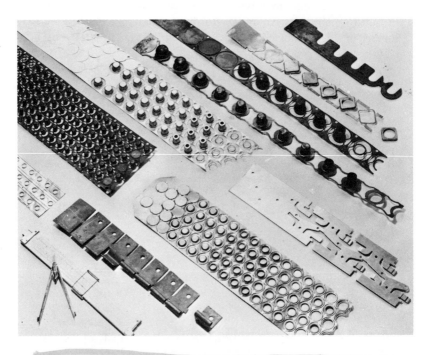

FIGURE 4-18. *Examples of progressive-die forming-and-cutting operations. Courtesy Brandes Press Div., Gougler Industries, Inc.*

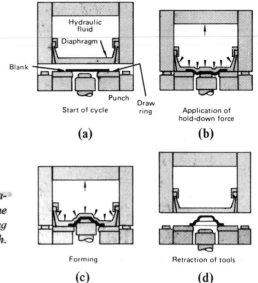

FIGURE 4-19. *The fluid-activated diaphragm applies a hold-down force to the blank, with the major portion of the forming done by the upward movement of the punch.* Courtesy Machine Design.

fountain pen. Since the depth to diameter ratio is large, the part must be transferred from one drawing die to another until the final size and shape can be achieved.

*Fluid-Activated Diaphragm Forming.* Drawing of sheet-metal parts may be accomplished by the use of fluids and a diaphragm, as shown in Fig. 4-19. The process has the advantage of closer control of the drawing operation so that parts which usually require two drawing operations can be done in one.

*Multiple-Action Dies.* A newer approach to deep drawing is by the use of multiple-action dies. In this setup (Fig. 4-20) the entire sheet is held between tooling at all times. Unsupported metal does not flow with the punch as it does in conventional tooling. The multiple, matched, male-female dies are made so that segments can be timed to move for an optimum forming sequence. The tools are mounted in a conventional press and the part is made in one stroke. The sequential power is supplied by the press, which is programmed to activate independent, external hydraulic circuits.

Although both the fluid-activated diaphragm and the multiple-action dies have good control over metal flow, which is necessary to avoid excessive thinning and buckling, it appears that the newer multiple-action die provides a more precise control.

### Metal Spinning

Production quantities often are such that it is not economically feasible to make the dies required to form the part. One approach to this problem is to make only one of the forms, usually the punch, and then form the metal over it by a spinning operation, as shown in Fig. 4-21.

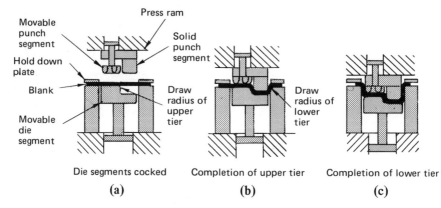

FIGURE 4-20. *Multiple-action dies are used to perform multistep forming in one operation. The punch and die segments are operated in a sequence that is coordinated with the press stroke.* Courtesy Machine Design.

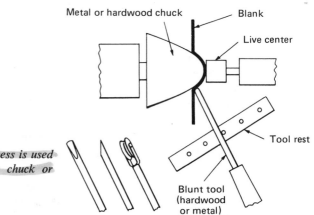

FIGURE 4-21. *The spinning process is used to form metal over a revolving chuck or punch.*

The desired form, either hardwood or steel, is mounted on the lathe headstock. A live center is brought up to hold the metal blank in place. The forming operation is accomplished by pressing the metal with wood or steel rods as it rotates. Skill and experience are required to cause the metal to flow at the proper rate, avoiding wrinkles and tears.

***Displacement Spinning or Power-Roll Forming.*** Power-roll forming is similar to hand spinning except that the part is shaped to the mandrel by rollers that produce a progressive extruding action (Fig. 4-22).

The control over workpiece contours and thickness gives the process significant advantages over drawing. Springback encountered in both pressworking and conventional spinning is eliminated. The increase in tensile strength of the finished product may be as much as 100 % or more over that of the original metal. Most

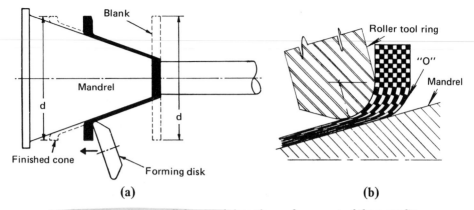

FIGURE 4-22.   *Power-roll forming (a), and an enlargement of the extruding process (b).*

parts are formed from flat blanks, but some cylinders require preformed blanks so that a sufficient volume of metal can be maintained at the desired places, such as the bottom or top flange.

### Continuous-Roll Forming

Roll forming may be defined as a process used to change the shape of coiled stock into desired contours without altering the cross-sectional area. The process utilizes a series of rolls to gradually change the shape of the metal as it passes between them (Fig. 4-23).

The intricacy of the shape, the size of the section, the thickness, and the type of material will determine the number of rolls required. A simple angle or channel with straight web and flanges can usually be made with three or four pairs of rolls, whereas complicated shapes may require ten or twelve roll passes. Usually some additional rolls are needed for idle stations and straightening.

The main advantage over other production methods is speed. For example, an item that needs no piercing or punching can be run at 300 fpm (9000 cm/min). Everyday production speeds, however, are more likely to be between 800 and 1200 fph (24,000–36,000 cm/hr) for heavy-gage steel and 1800 to 2200 fph (54,000–66,000 cm/hr) for light-gage steel.

The process has become more attractive in recent years since it can handle prepainted or electroplated surfaces without damage. As may be surmised, roll forming is for quantity production. As a rule of thumb, at least 50,000 ft (15,240 m) should be required annually to amortize the cost of the tooling. Tolerances on normal production items can usually be held within 0.005 in. (0.13 mm) and on some light-gage steel within 0.001 in. (0.039 mm).

### Stretch Forming

Stretch forming of metal consists of placing sheet material in a tensile load over a form block. The material is stressed beyond its elastic limit and into the plastic

239

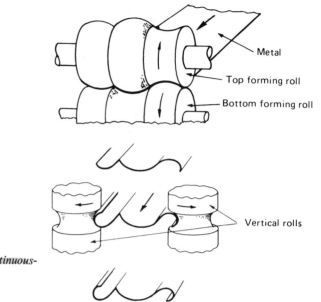

FIGURE 4-23. *An example of continuous-roll forming.*

range causing it to take a permanent set. In stretch forming, all the fibers are stretched but those on the inside of the radius of bend are stretched less than those on the outside (Fig. 4-24). Since the metal is placed under one type of load (tensile), there is no tendency to springback. However, allowances must be made for dimensional changes that occur. That is, during stretching the length increases and the width decreases.

There are two types of stretch forming. One is termed *stretch forming* and the other *stretch-wrap forming*, as shown in Fig. 4-25.

FIGURE 4-24. *In stretch forming, the outer fibers are stretched more than the inner ones.*

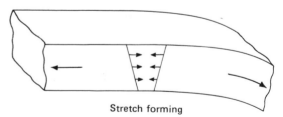

Stretch forming

Stretch forming is used to produce compound curves in sheet stock. The ends of the stock are gripped in hydraulically operated serrated jaws. Stretching is accomplished by one or more hydraulic pistons moving up under the die. Forming of this type is usually restricted to parts that do not have sharp edges, which tend to produce accelerated stretch in localized areas. The forming block or punch is usually highly polished and lubricated. The process is very useful in making prototype models of aircraft and automotive parts. It is sometimes used on a production basis for truck and trailer bodies.

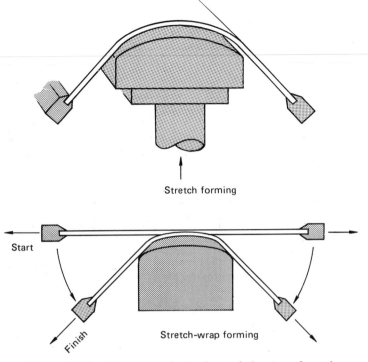

Stretch forming

Start

Finish

Stretch-wrap forming

FIGURE 4-25. *The two methods of stretch forming of metal.*

Stretch wrapping consists of first stretching the metal beyond its yield point while it is straight and then wrapping it around a form block. It is particularly suited for long sweeping bends on tubes and extruded shapes.

A big advantage of this type of forming is that only one die is needed. Also, the die may be made out of inexpensive material.

Thinning and strain hardening are inherent in the process. It is important to know the elongation values for the metal being used. Metals having a so-called "infinite gage length" elongation are well suited to this process. The thickness reduction should not exceed 5 % of the original thickness.

The production or manufacturing engineer is not only concerned with how much strain hardening takes place during a given deformation or what is the best process to use to produce a given item, but he is equally concerned that the design lends itself to production. A fundamental question is: Can any changes be made that will make it easier to produce, yet not adversely affect the quality of the product?

### Sheet-Metal-Product Design Considerations

#### *Strength and Rigidity*

Many design and fabrication techniques are available that add strength and rigidity to simple parts fabricated in sheet metal. For example, strength can be incorporated into the structure by means of flanges, ribs, corrugations, beads, etc. (Fig. 4-26).

BEADS ON LARGE PANELS

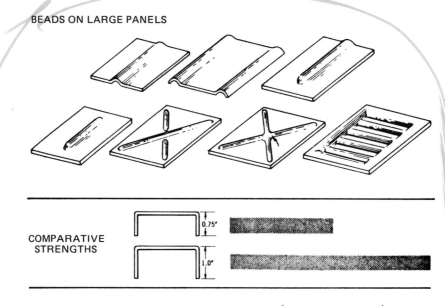

COMPARATIVE
STRENGTHS

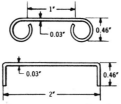

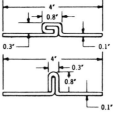

Curled edges are stronger than flanged
and present a smooth, burr-free edge.
Production, however, may require one
more operation since the curl is usually
started as a flange.

Although locking ability is sacrificed,
vertical standing seams are 3 times as
strong as flattened seams.

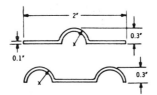

Ribs are even more efficient than
flanges. Dual-rib design yields
56.5 percent more strength for
10.8 percent more material.

Corrugated sheets are common examples of ribs, used as a continuous form.

FIGURE 4-26. *Methods of increasing strength and rigidity on sheet-metal
parts. Courtesy* Machine Design.

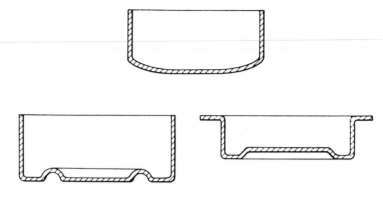

FIGURE 4-27. *Beads at the bottom of a container eliminate dishpanning and add rigidity and strength.*

Although ribs, beads, and flanges are most often thought of in connection with flat stock, they can be equally effective in supporting cylindrical shapes. Sometimes beads serve the dual purpose of adding strength and rigidity and eliminating *dishpanning*. Dishpanning, as shown in Fig. 4-27, is an oval or bulged bottom. To restore it to flatness, the excess metal is incorporated into beads.

### Press-Size Requirements

Sheet-metal parts are commonly formed on presses such as those shown in Fig. 4-28.

The press tonnage required for blanking is based on the perimeter of the cut, the thickness, and the shear strength of the material, as shown in the formula:

$$F = \frac{(LtS_s)}{2000} \times F_s$$

*where:*    $F$ = total blanking force
   $L$ = length of sheared edge
   $t$ = thickness of the material being sheared
   $S_s$ = shear strength of material in psi
   $F_s$ = safety factor (usually 2x)

The press size is designated both by the tonnage capacity that can be delivered by the ram to the press bed and by the bed area. As an example, the press shown at the left in Fig. 4-28 is termed a C-frame press or an OBI press since it has an open back and is inclinable. That is, it can be tilted backward. A standard size designation may be given as:

$$OBI-22-12\tfrac{3}{4} \times 17\tfrac{1}{2}$$

The letters stand for the type of press, 22 is the tonnage rating, $12\tfrac{3}{4}$ and $17\tfrac{1}{2}$ are bed size measurements, left to right and front to back respectively.

243

FIGURE 4-28. *Common types of presses used in blanking and forming sheet-metal parts. Courtesy South Bend Lathe, Inc.*

## Metal Forming—Altering the Cross-Sectional Shape

The two main processes used in fabrication that alter the cross-sectional shape of the metal are forging and extrusion; however, there are several variations of each of these processes.

### Forging Processes

Forgings, castings, and weldments are often used to serve the same purpose. Each means of fabrication has its own distinct advantage. If stress loads are high, forgings are preferred even though the unit cost is higher. If particularly high strengths are required, a newer process known as *ausforming*, or working the metal at tempera-

TABLE 4-1. *Typical Effects of Forging on Steel. Courtesy* Machine Design

| Steel | Deformation (%) | Tensile Strength (1000 psi) | Yield Strength* (1000 psi) | Elongation (%) |
|-------|------|------|------|------|
| H-11  | 0  | 290 | 280 | 11 |
|       | 65 | 395 | 370 | 7  |
| 4340  | 0  | 275 | 235 | .. |
|       | 72 | 310 | 280 | .. |
| D6-AC | 0  | 340 | 295 | 10 |
|       | 80 | 420 | 395 | 4  |

\* At 0.2 % elongation.

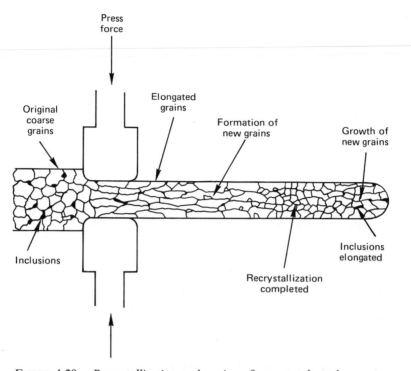

FIGURE 4-29. *Recrystallization and grain refinement takes place as a result of the forging process.*

tures below those normally used for forging, can be used on some materials to produce tensile strengths in excess of 400,000 psi (Table 4-1). Ausforming steels are variations of certain conventional alloy steels where elements such as chromium, nickel, silicon, and molybdenum have been added to allow deformation of metastable austenite. The forging process is followed by quenching and tempering. Forging also contributes to grain refinement. As the hot metal is forged, recrystallization takes place (Fig. 4-29) and the grain flow closely follows the outline of the component (Fig. 4-30). Continuous-flow lines decrease the susceptibility to fatigue and corrosion failures. Internal flaws are largely eliminated, resulting in low inspection costs and consistent results in both heat treatment and machining.

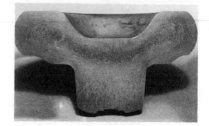

FIGURE 4-30. *The grain flow of the metal closely follows the part contour, contributing to the strength of the forging.*

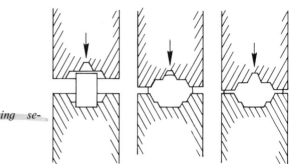

FIGURE 4-31. *A closed-die forging sequence.*

Forging processes can be divided into several groups as follows: hammer and press forging, upsetting, and extrusion forging. Some special techniques include heated-die forging, high-energy-rate forging, and swaging.

*Hammer and Press Forging.* Hammer forging is essentially the same process it has been since its early development. The quality of forgings produced under the drop hammer depends to a large degree on the skill of the operator. All hammer-forging operations, from bar to final shape, are done by repeated blows between two platens (open-die forging) or in several impressions in a die set (closed-die forging) (Fig. 4-31).

Preforming operations used to prepare the metal for the forging dies are often done by using the blocking and edging portions of a standard closed die, or on separate machine tools. The use of roll forging to make the preforms is increasing. Another approach is to use pieces cut from an extrusion (Fig. 4-32).

Press forging, although of later development than hammer forming, is becoming a major forging method. Press forging utilizes a slow squeezing action and relies less on operator skill and more on the proper design of the preform and the forging dies. At first, press forming was limited to axially symmetrical components, but now almost any shape can be forged by this method. Newer developments make it

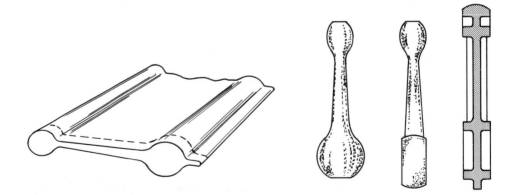

FIGURE 4-32. *An extruded preform shape that can be sliced up and forged into connecting rods.* *Courtesy* Machine Design.

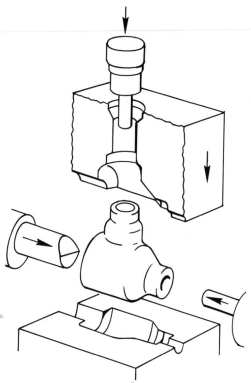

FIGURE 4-33. *Top- and side-acting auxiliary rams are used in some forging presses to produce hollow parts and undercuts.*

possible to produce bevel gears with straight and helical teeth. Rotation of the die during penetration will press bevel gears with spiral teeth. Also, hollow parts with undercut shapes can be press-forged by hydraulically operated, side-acting, auxiliary rams (Fig. 4-33).

*Flashless Forging.* Continuous work is being done in forging design to eliminate the *flash*, or excess metal that squeezes out between the dies as they close. One approach has been to force the excess metal into a compensating cavity, in the center of the forging, that will be pierced later. Flashless forgings call for close tolerance on billet weight. Some gears are now being made by this process with significant savings in material, energy, labor, and die life.

*Heated-Die Forging.* Heating the forging dies to 1600°F (871°C) improves tolerances and makes it possible to achieve thinner sections with fewer die sequences. As an example, aircraft forgings of AISI 4340 steel were made 81% deeper than that attained on conventional dies by using heated dies at 300 to 800°F (149–427°C). The process is not extensively used due to shortened die life.

**Roll Forging.** Originally developed to forge simple shapes, roll forging has expanded its field of application to many finished products. In its most simple form, a preheated billet passes between a pair of rolls which deforms it along its length (Fig. 4-34).

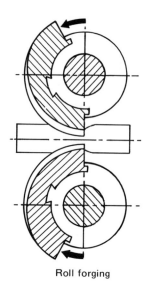

FIGURE 4-34. *Roll forging is often used in preforming for drop forging or pressing. It may also be used to produce finished forgings.*

Roll forging

**Cross Rolling.**  Cross rolling is a process of gradually deforming a shaft in the transverse direction as it passes between the rolls [Fig. 4-35(*a*)].  U-shaped tools are fastened to the rollers for unilateral or bilateral displacement of the metal, which may be hot, warm, or cold.  The part is brought to its final dimensions in one rotation.  A part such as that shown at Fig. 4-35(*b*) may be made by hot cross rolling from $1\frac{3}{8}$-in. hot bar stock in a total of 3.5 sec.  Forming proper takes 1.5 sec, while 0.2 sec is required for separation of the part from the bar and 1.8 sec for bar feeding.

The cross-rolling process is capable of showing considerable savings in material and production time over conventional closed-die forging for smaller parts.  The accuracy [0.004 in. (0.10 mm) on diameters up to 1.2 in. (3.04 cm) and 0.008 in. (0.20 cm) on diameters up to 2 in. (5.08 cm)] allows the parts to be used without further machining.  Where precision is called for, parts may be finished by grinding.

**Upset Forging.**  Upsetting is accomplished by inserting a blank of a specific length into a stationary die (Fig. 4-36).  The bore of the die is a few thousandths of an inch larger than the outside diameter of the blank.  A punch moves toward the die blank to upset that portion of the blank protruding from the die.  The maximum length that can be upset in one stroke is 2.25 to 2.50 diameters of the stock.  If this length is not enough to form the part, two or more blows may be used.  As a general rule, 4.5 diameters can be upset in two blows.  When a sliding punch is used to support part of the blank during upsetting, up to about 6.5 diameters can be formed.  A rule of thumb used to estimate impact values required for upsetting is that the impact value must be seven times the yield strength of the material.

Multiple-die machines often combine upsetting with extrusion to form a large head on small shank parts.  Contrary to upsetting, extrusion reduces the diameter of the initial stock while increasing its length (Fig. 4-37).  The reduction, expressed

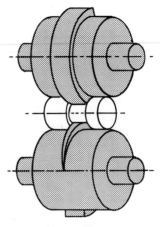

Cross rolling

(a)

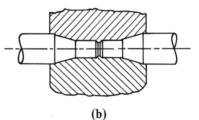

FIGURE 4-35. *Cross rolling showing U-shaped tooling (a). A typical part (b).*

(b)

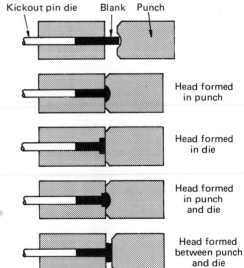

Kickout pin die    Blank   Punch

Head formed in punch

Head formed in die

Head formed in punch and die

Head formed between punch and die

FIGURE 4-36. *Upset forging may be done in several variations, as shown. Courtesy National Machinery Co.*

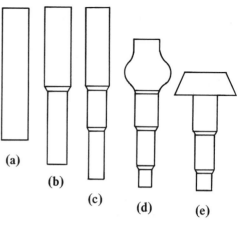

FIGURE 4-37.   *A combination of extrusion and upsetting is used to make this stempinion blank. Bar stock cut to exact required length (a), forward extrusion (b), a second forward extrusion (c), combined forward extrusion and upsetting (d), and heading (e).   Courtesy* Machine Design.

in percent reduction in area, may be 30 to 40 % or more.   The advantage of the combined process is that a typical part may have six to ten or more shank diameters.

***Extrusion Forging.***   Extrusion or impact forging is often called a cold-working process.   Three principal types of metal displacement by plastic flow are involved: backward, forward, and a combination of both backward and forward (Fig. 4-38). The process consists of hitting a slug held in the die with a punch.   The metal plastically deforms when the yield point is exceeded but extrusion does not start until the pressure becomes seven to fifteen times the initial yield strength of the alloy.

***Advantages and Limitations.***   Metals particularly suited to impact extrusion are the softer, more ductile ones such as aluminum, copper, and brass.   However, low- and medium-carbon steels are also used even though the impact factor is high.

Generally, steel impacts are limited to 2.5 times the punch diameter on reverse extrusions.   Approximate limits for aluminum extrusions are: 14 in. (35.56 cm) in diameter and 60 in. (152.4 cm) in length.   Hydraulic presses are used for loads over 2000 tons (1800 metric tons) because they have greater variation in stroke lengths, speed, and economic advantages.

Tolerances vary with materials and design, but production runs calling for 0.002- to 0.005-in. (0.05–0.13 mm) tolerance are regularly made.

Shown in Fig. 4-39 is a section of a sparkplug base exhibiting the improved grain structure and smooth flow lines obtained by cold forming.   Unsymmetrical parts can have interrupted flow lines and impaired strength if not carefully controlled.

***Applications.***   Impact extrusions compete with other press-working operations and with metal machining.   Common products are aerosol cans, cocktail shakers, lipstick cases, flashlight cases, and vacuum bottles.   Secondary operations, such as beading, thread rolling, dimpling, and machining are sometimes needed to make the complete item. Shown in Fig. 4-40 is a three-piece assembly that represents a considerable time and materials saving when manufactured as one part by cold heading and impact extrusion.

### General Design Considerations for Forgings

The following considerations will be helpful to the designer in achieving practical, economical forgings.

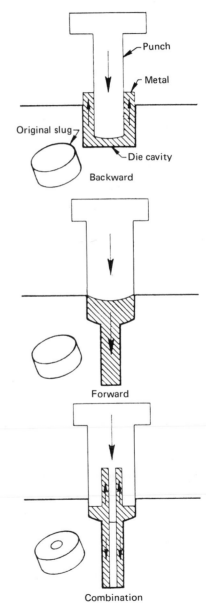

FIGURE 4-38. *The three basic methods of impact extrusion.*

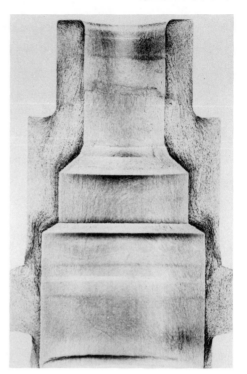

FIGURE 4-39. *A cutaway of a sparkplug base showing the smooth flow lines and improved grain structure imparted by cold forming. Courtesy National Machinery Co.*

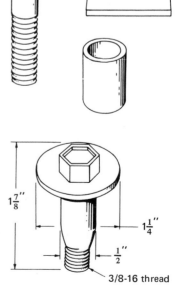

FIGURE 4-40. *A three-piece assembly redesigned for cold heading and extrusion. Courtesy Townsend Co.*

$1\frac{7}{8}''$

$1\frac{1}{4}''$

$\frac{1}{2}''$

3/8-16 thread

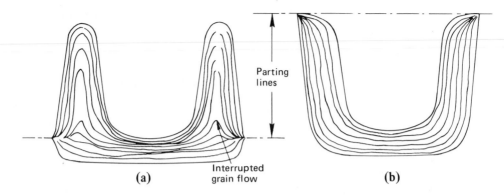

FIGURE 4-41. *Parting-line placement will have a significant effect on the grain structure of a forging as contrasted by (a) and (b).*

**Parting Line.** The parting line, where the die halves meet, helps determine grain flow and affects production rates, die costs, and die life. For maximum economy, the parting line should be kept in one plane. This makes die sinking, forging, and subsequent trimming operations simpler and therefore less costly. Also, the parting line will have a pronounced effect on the metal-flow lines, as shown in Fig. 4-41.

To avoid mismatch of the forging dies, every effort should be made to balance the forces, especially on nonsymmetrical parts. Side thrust increases as the parting line inclines away from parallel to the forging plane (Fig. 4-42).

FIGURE 4-42. *The parting line has been placed to balance die thrust forces and avoid mismatch.*

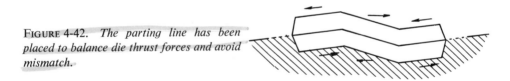

**Corner and Fillet Radii.** Small corner and fillet radii can be incorporated into modern forging designs and can be of considerable cost saving when they are used to eliminate machining operations. When this is not necessary, more generous corner and fillet radii should be used to reduce the costs and prolong die life. Also, connecting sweeps between pockets and recesses should be as generous as feasible to obtain good metal flow.

**Draft Angles.** Draftless forgings are now being produced, particularly for aircraft and missile applications. Generally the draft angle on small forgings is about 5°.

**Holes.** Forging dies can be designed with integral plugs. The plugs form recesses which can be pierced during the normal trimming operation, often at considerable savings in subsequent machining operations.

*As-Forged Surfaces.*   The as-forged surface can often be used in the finished part. However, it must be determined that the forging tolerances which apply to the finish (normally 500 microinches) are acceptable.

*Finished Size.*   The designer knows what the finished product's dimensions and tolerances should be.  The forger knows how much excess metal (the machining envelope) will be required to achieve this finished size.  The designer knows excess metal is necessary but wants it kept to a minimum, hence a problem develops.  There are many variations on this theme that arise to defeat the application of forging tolerances, such as die closure, mismatch, warpage, random surface deviations, die wear, concentricity, and out-of-roundness.  Two obvious approaches to the problem are to add enough material to take care of all contingencies or insist on tighter forging tolerances.  Both approaches will add considerably to the cost.  A more realistic approach is for the designer to establish tooling points.

*Tooling-Point Locations.*   Tooling-point locations, whether forged or machined, are locating points put on a forging so that it will "clean up" despite unpredictables.

The system is based on the fact that any part can be located and dimensioned within a framework of three planes, each located at 90° to each other, and these references, or datum planes, are used to position the parts in the machine.  Tooling points are applicable not only to forgings but also to castings and jig and fixture design.

Shown in Fig. 4-43 is a plane view of a drawing with six tooling points.  Plane B is 90° to plane A and parallel to the longest direction of the part.  It has two tooling

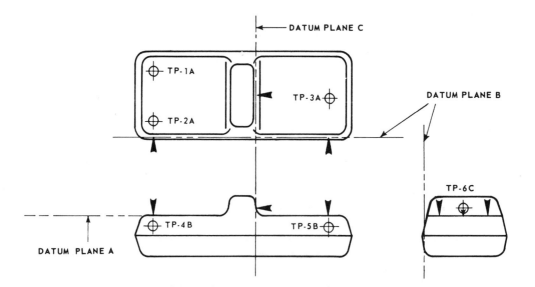

FIGURE 4-43.   *The standard location of datum planes.  Also shown are six tooling points.*

points. Plane C is 90° to each of the other planes, is the shortest dimension, and has one tooling point. Any more than three established planes would be redundant, as would more than six tooling points.

The machined surface of the tooling points will be provided by the forging manufacturer when shown on the design. When this is done, many forging companies are willing to guarantee that the part will machine out satisfactorily provided the machinist uses the proper tooling points.

Advantages of designating tooling points are:

Print tolerances are not critical, hence tolerances on forging variations are practically eliminated. An agreement can be reached between the customer and the manufacturer as to minimum and maximum machine cuts that may be desirable.

Inspection can be made more quickly from designated datum points.

The machinist always has clean, clearcut surfaces against which he can set up and in some cases on which he can clamp or chuck. These surfaces are parallel to the reference plane and will prevent problems encountered when indexing or chucking on drafted forging surfaces.

### Swaging

Swaging is a process of reducing the cross-sectional area of rods or tubes. The process is shown schematically in Fig. 4-44. In actual practice, a ring of hammers is rotated around the workpiece at a speed of about 200 rpm, delivering 1000 spm (strokes per minute). The outer surface is always circular in cross section but the inner surface of a swaged tube can be any shape from which a forming mandrel can be removed. Swaging may be started or stopped at any point along the length of the stock and is often used for pointing the ends of tubes or bars or for producing stepped diameters.

FIGURE 4-44.  *Swaging is used to reduce the cross-sectional area of rods or tubes.*

### Drawing and Extrusion

Bars or tubes that have been hot rolled are often given a cold-finishing operation to reduce the size, increase the strength, improve the finish, change the shape, or provide better accuracy. The drawing process is shown in Fig. 4-45.

Drawing is confined largely to the metal manufacturers and is not generally considered a fabricating process. Extrusions are also made by the metal manufacturer but have such a wide application that the engineer should be aware of their properties, uses, and relative economy. Shown in Fig. 4-46 is a sketch of the basic extrusion process.

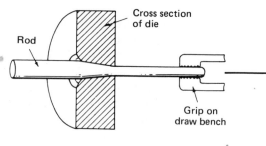

FIGURE 4-45. *Essential features of the cold-drawing process. Courtesy Steelways, published by the American Iron and Steel Institute.*

**Hydrostatic Extrusion.**   A more recent development is hydrostatic extrusion, which makes it possible to cold extrude many difficult-to-form materials such as the high-strength super alloys, arc-cast tungsten and molybdenum. In this process, the metal billet is inserted into the chamber behind the die where it is surrounded by a suitable liquid. The billet is then extruded by applying pressure to the liquid. The included angle in front of the die has a pronounced effect on the total force needed to form the metal, on the flow patterns, and on the soundness of the extrusion (Fig. 4-47). The drawing stress changes continuously with the die angle and reaches a minimum at the optimal angle.

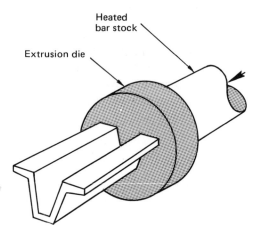

FIGURE 4-46. *Basic concept of the extrusion process.*

**Extrusion Die Angles.**   Excessive die angles cause the metal to shear within itself and form its own cone angle that does not conform to the contour of the die as shown in Fig. 4-48. The new angle that forms is known as the dead-zone cone angle. For any semicone angle between $A_1$ and $A_2$, the dead-zone angle will form. As the semicone angle goes beyond $A_2$, the die acts like a cutting tool and shaves the billet off.

Three interacting parameters—ideal deformation, shear resistance, and friction losses—vary with the die semicone angle to produce a minimum resistance curve. The ideal deformation factor involves only the power required to reduce the diameter of the billet, which varies but little with changes in the semicone angle. The shear-

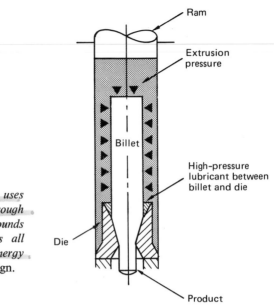

FIGURE 4-47. *Hydrostatic extrusion uses high-pressure fluid to force a billet through a die orifice. The fluid completely surrounds the billet, which virtually eliminates all billet–container friction, reducing energy requirements.* Courtesy Machine Design.

resistance parameter is determined by the force needed to produce a specific amount of distortion and increases as the semicone angle becomes larger. In either drawing or extrusion, friction is dependent upon the contact with the die wall. With a small semicone angle, the contact between the die and billet is large and therefore the friction losses are large. As the angle becomes larger, the die-billet contact area and friction loss decrease. As the curves for all three of these effects are added to make the total resistance curve, there is found, for any one set of friction and reduction conditions, a minimum point at one particular semicone angle, which is the optimal semicone angle, as shown in Fig. 4-49.

***Advantages and Limitations.*** The range of shapes that can be produced by extrusion is almost infinite. The dies required are relatively simple and low cost. Simple dies may run as low as $70, while the more intricate may cost $1500 or more. Shown in Fig. 4-50 is a multiple-hole extrusion die that can be used to produce four parts simultaneously, resulting in less die wear and higher production. Sections may be designed to reduce the number of parts needed in an assembly or to reduce machining and costs. Extrusions made in "bar" form can be sectioned for special purposes, as shown in Fig. 4-51.

Shown in Fig. 4-52 is an example of the part extrusions play in building a prototype aluminum boat.

Extrusions are usually limited to parts that can be circumscribed by a 17-in. (43.18-cm) circle for aluminum and a 6-in. (15.24-cm) circle for steel. Extremely thin sections should be avoided, as should extreme thicknesses. Hollow shapes having unsymmetrical voids are not recommended.

257

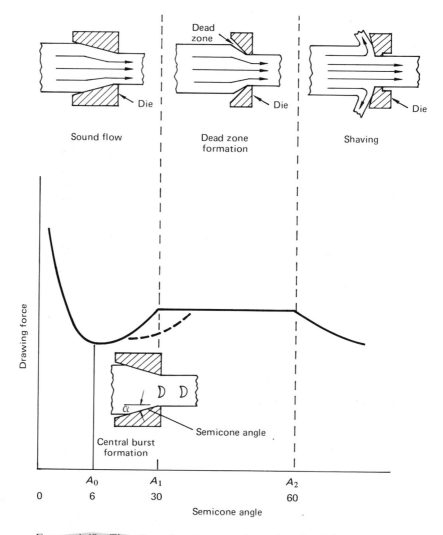

FIGURE 4-48. *The effect of semicone angles and mode of flow on drawing or extruding force. Prepared by Evans and Avitzur. Courtesy Society of Manufacturing Engineers.*

### High-Energy-Rate Forming (HERF)

High-energy-rate forming is a term that has been widely used in recent years to describe a method of efficiently applying more energy to the workpiece per unit of time than has been applied by previous techniques. HERF can be applied to both light- and heavy-gage materials, in the former it is usually thought of as another forming method and in the latter as a forging technique.

### *Theory of High-Energy Metal Forming*

The plastic flow of metal is both time and temperature dependent.

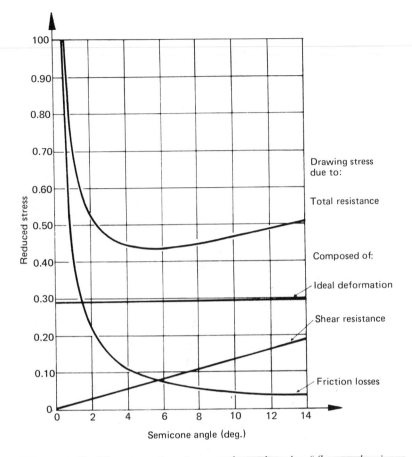

FIGURE 4-48. *The effect of semicone angles and mode of flow on drawing or extruding force. Prepared by Evans and Avitzur. Courtesy Society of Manufacturing Engineers.*

**Time Dependence.** Shown in Fig. 4-53 are basic strain-rate curves for various materials. True stress increases with strain rate. In general, the flow stress is raised considerably with increased strain rates, whereas true strain to fracture decreases a small amount.

The toughness of the material can be thought of as the area under the true stress–strain curve and is a measure of the energy-absorption ability of the metal. Replotting the curve as toughness versus strain rate gives the graph as shown in Fig. 4-54.

This graph shows that the toughness of all metals increases to some critical strain rate and then diminishes. The toughness is the ability of the metal to plastically deform under shock without fracture. Thus it can be seen that all high-velocity methods have decided limitations on the materials that can be formed. Metals that have a high toughness and strain rate should be selected.

***Temperature Dependence.*** As shown previously, increased temperatures lower the flow stress of a material. An even greater benefit is the greatly increased true strain to fracture, which can be tremendous in some metals. As an example, tungsten increases from a strain of less than 1 % at room temperature to 55 % at 1950°F (1066°C). Annealed titanium 6A1-4V increases from 12 % at room temperature to 52 % at 1200°F (649°C). Of course, not all of this increased true strain is usable in forming because of necking.

FIGURE 4-50. *A multiple-hole extrusion die provides high production and decreased die wear. Courtesy Elox Corp. of Michigan.*

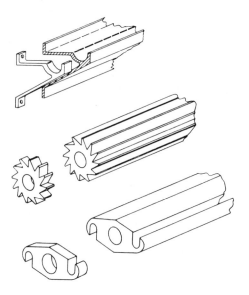

FIGURE 4-51. *Extrusions are made in bar form and can be sectioned off. Courtesy Kaiser Aluminum and Chemical Sales, Inc.*

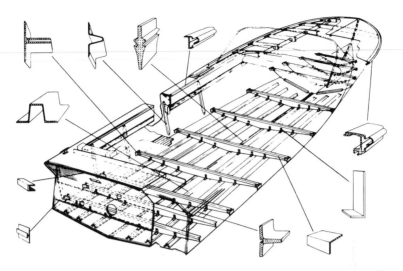

FIGURE 4-52. *A wide variety of extrusions are used to construct this prototype aluminum boat. Courtesy Kaiser Aluminum and Chemical Sales, Inc.*

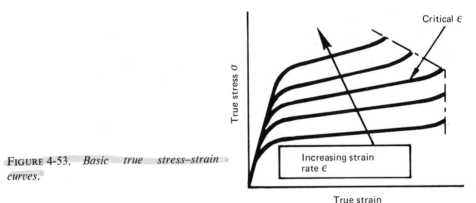

FIGURE 4-53. *Basic true stress–strain curves.*

Temperature also increases toughness. The increase in toughness does not benefit slow forming, such as hot-fluid forming, but it does benefit high-velocity–forming systems where both slip and twining occur in rapid succession.

A third benefit of increased-temperature forming is *creep*, or the plastic deformation that takes place at stresses less than those required to cause slip. It is primarily a high-temperature phenomenon and is shown schematically in Fig. 4-55(*a*).

Increasing the temperature from $T_1$ to $T_3$ greatly accelerates creep. Practical elevated-temperature creep forming is generally accomplished in a period of less than 10 min and for a temperature approximately 500°F (277°C) above the service operating temperature of the metal. The results of creep forming [Fig. 4-55(*b*)] are

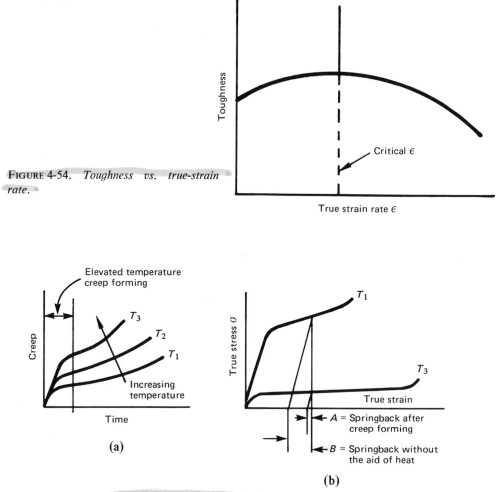

FIGURE 4-54. *Toughness vs. true-strain rate.*

FIGURE 4-55. *The effect of temperature on creep.*

the almost complete elimination of springback and the elimination of other problems in forming thin-gage metals, such as buckling and distortion.

The high-temperature systems, such as integrally heated dies and hot-fluid forming, result in deformation that is almost totally slip. Some twinning may occur in the hexagonal metals where slip is impeded by the small number of slip planes. Slip deformation in these systems has been termed elevated-temperature creep forming and is also dependent on time and the number of dislocations in the metal.

***Hydrostatic Pressures and HVF.*** Bridgman (see Bibliography at end of chapter) has shown that high transverse pressure acting on a tensile specimen during pulling will exhibit a considerable increase in a total true strain to fracture (Fig. 4-56). The true stress–strain curves show that flow stress does not change appreciably but that the true strain to fracture increases with hydrostatic pressure. Also, the mode of fracture changes from a combination of ductile-brittle failure below 170,000 psi to a totally ductile failure above this pressure, as shown by the tensile specimens.

Investigators of HVF have attempted to correlate hydrostatic pressures to formability. Transverse pressure does not benefit forming for the large-mass systems unless the pressure exceeds the compressive flow stress of the material. When this happens, as it often does in HVF, the shear stress $\tau$ greatly increases, allowing a considerable increase in plastic flow.

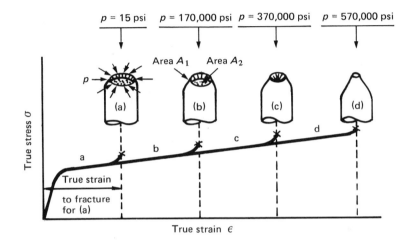

$p$ = applied transverse hydrostatic pressure
Area $A_1$ = ductile fracture = slip
Area $A_2$ = brittle fracture = separation

FIGURE 4-56.  *The effect of hydrostatic pressure on ductility.*

### Mass Versus Velocity in HERF

Energy can be increased by increasing either the mass or the velocity as shown in the kinetic energy equation:

$$KE = \tfrac{1}{2}mv^2$$

A much greater effect can be obtained by increasing the velocity since the energy is a function of the velocity squared.

Some of the early methods of HERF utilized a large mass at the expense of velocity, as shown by the drop-hammer and trapped-rubber methods (Fig. 4-57).

Another system that makes more use of the velocity factor is that of the floating piston, shown schematically in Fig. 4-58.

A comparison of the kinetic energy of the two systems can be made briefly as follows.

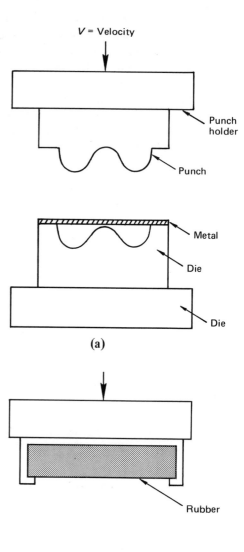

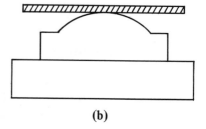

FIGURE 4-57. *Early methods of high-energy-rate forming: drop hammer (a), and trapped rubber (b).*

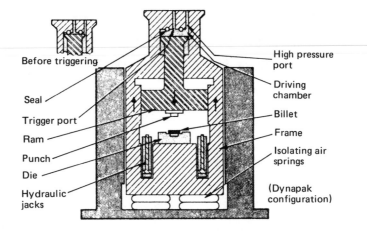

Before triggering

Seal

Trigger port

Ram

Punch

Die

Hydraulic jacks

High pressure port

Driving chamber

Billet

Frame

Isolating air springs

(Dynapak configuration)

FIGURE 4-58. *This pneumatic-mechanical system uses the rapid release of highly compressed gas to accelerate a ram to high velocity. The ram is raised against the seal by hydraulic jacks. After the work has been positioned, the jacks are lowered and a small volume of high-pressure gas is admitted through the trigger port. This forces the ram down enough to expose the top of it to the high-pressure gas, accelerating its downward movement. At the same time, the upper end of the driving chamber is exposed to the gas and the frame moves upward. Courtesy* Machine Design.

*Mass System:*

Assume a 10,000 lb punch and holder and 20 fps velocity.

$$KE = \tfrac{1}{2}mv^2 = \tfrac{1}{2}\frac{(w)}{(g)}\,v^2$$

$$= \tfrac{1}{2}\frac{(10,000)}{(32.2)}\,(20)^2 = \tfrac{1}{2}(311)(400)$$

$$= 62,100 \text{ ft lb}$$

*Velocity System:*

Assume a 325 lb mass and an impact velocity of 100 fps.

$$KE = \tfrac{1}{2}\frac{(w)}{(g)}\,v^2 = \tfrac{1}{2}\frac{(325)}{(32.2)}\,(100)^2 = \tfrac{1}{2}(10)(10,000)$$

$$= 50,500 \text{ ft lb}$$

By this simple example, we see that with an assumed velocity five times greater and a mass 30 times smaller, the total energy developed is of the same order of magnitude as with the lower-velocity, larger-mass systems. The velocity factor contributes 1000 times as much as the mass in the total energy, hence the newer term high-velocity metal working (HVM).

***High-Velocity Metal Working.*** In the high-velocity metal-forming (HVF) systems such as explosive-forming, electrohydraulic, and electromagnetic processes (shown schematically in Fig. 4-59), the mass contribution toward total energy is negligible compared to velocity. Explosive forming can be used as an example.

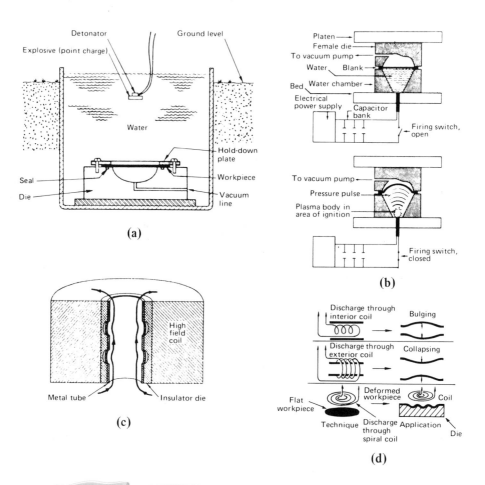

FIGURE 4-59. *High-velocity metal-forming methods: explosive forming (a), electrohydraulic forming (b), and electromagnetic forming (c). In the expansion ring or bulging concept of electromagnetic forming, the metal takes the form of the die when the field outside is removed and the unbalanced forces cause the metal to expand. The three main concepts of electromagnetic forming are shown schematically at (d). Courtesy* Machine Design.

*Explosive Forming—Electrohydraulic Forming:*
Assume 6 lb for weight of moving water front ($w$) and an impact velocity ($v$) of 700 fps.

$$\text{KE} = \tfrac{1}{2}\frac{(w)}{(g)}v^2 = \tfrac{1}{2}\frac{(6)}{(32.2)}(700)^2 \approx \tfrac{1}{2}(0.2)(5 \times 10^5)$$

$$\approx 46{,}000\,\text{ft lb}$$

It can be readily seen that the velocity to mass contributions to total energy are about 2.5 million to one for explosive–water forming. The 46,000 ft lb of energy in the example is only a small portion of the actual energy generated by the explosion as shown by the following simple calculation:

$E$ = specific energy content of explosive = $3 \times 10^6$ ft lb/lb (assumed)
$w$ = weight of average charge = 0.1 lb (assumed)

$$v = Ew = (3 \times 10^6)(0.1) = 300{,}000\,\text{ft lb}$$

$$\text{Efficiency} = \frac{\text{KE}}{v} = \frac{46{,}000}{300{,}000} = 15\,\%$$

Most explosive-forming operations have an efficiency of 15 to 20 %. The bulk of the energy, as in the example, is used in blowing the water into the air.

**_High-Temperature Metal Forming._**  Two basic systems that use heat for high-energy forming are shown schematically in Fig. 4-60. At the left is an integrally heated die with heat supplied by cartridge-type heaters. Insulation surrounds the die completely.

In hot-fluid forming, both heat and pressure are supplied by a hot, liquid-metal alloy, giving considerable advantage by lowering the flow stress of the metal.

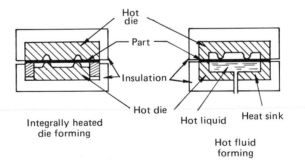

FIGURE 4-60.  *Two basic systems that use heat to aid high-velocity forming. Courtesy Society of Manufacturing Engineers.*

As an example, stainless steel has a flow stress of approximately 100,000 psi at room temperature, but reduces to 5000 psi at 1200°F (649°C). This reduces the mechanical work by a ratio of 20 : 1.

### Some Advantages and Limitations of HVF

*Advantages.*  A wide variety of materials can be formed or forged, including exotic and refractory metals, stainless steels, nonferrous alloys, and high-strength materials, some of which are not usually forgeable.

Draft allowances are reduced and in some cases eliminated.

Complex parts can be formed in one blow.

Tolerances and surface finish are improved over those obtained with conventional forging techniques.  (The surface finish usually lies in the range of 20 to 60 microinches.)

Because strength and fatigue resistance are improved, parts can be made smaller.

Repeatability is excellent.

Usually the more expensive the material, the greater the savings.

Large (20 ft or more in diameter) one of a kind domes and cylinders have been explosively formed using ice as a die.

*Limitations.*  The main limitation of HVF is in tooling materials and capacity (except for explosive forming).  Tooling materials present a serious limitation. There is a need for less expensive, long-life materials which can withstand the heavy loads involved.  The tooling materials also set limitations on the production rate since the dies can operate only so long without being allowed to cool.

Part configuration is usually limited to one-piece dies.  If two-piece or split dies are necessary, the economics of the process becomes debatable.

If several blows are required, as is sometimes done in the pneumatic-mechanical press shown in Fig. 4-58, the process again becomes economically questionable. Usually two blows are acceptable but three blows become borderline.

## PROBLEMS

**4-1.** A 10-in. diameter aluminum blank, $\frac{1}{8}$ in. thick, is to be drawn into a 4-in. diameter cup.  (a) Is this possible in one draw? Why or why not?  (b) What function does the draw radius serve?  How large should it be for this application?  (c) What punch radius should be used?  Why?

**4-2.** For drawing, what is the significance of the strain-hardening coefficient?

**4-3.** Why isn't a metal that has a large yield elongation good for forming?

**4-4.** What are some of the advantages of hot-working metal?

**4-5.** (a) What are some ways to compensate for springback?  (b) What should the die opening be for a high yield-strength material $\frac{1}{8}$ in. thick?  Make a sketch to show a sectional view of the die with the material in place and the die-opening dimension.

**4-6.** What is the force required to bend a piece of low-carbon steel plate $\frac{3}{16}$ in. thick and 8 in. wide?  Assume the die opening is $8t$.

**4-7.** A bolt blank is made by upsetting.  The material used is SAE 1018 steel.  The bar stock used is $\frac{1}{2}$ in. in diameter.  The desired hexagonal-head size is $\frac{3}{4}$ in. in diameter and 0.333 in. thick.  (a) How many strokes will be required?  (b) Estimate what impact will be required.

**4-8.** (a) What is the main advantage of forging over a cast or welded structure? (b) What is a disadvantage of using the forging process?

**4-9.** (a) Compare the impact obtained per blow from a conventional drop-hammer forging press with that from a floating-piston press under the following conditions.

Conventional drop-hammer forging press:
  Drop hammer = 1200 lb.
  Velocity of hammer = 16 fps
Floating-piston press:
  Floating piston mass = 400 lb.
  Velocity of piston = 150 fps

(b) What are the ratios of the mass and velocity to the total energy delivered between the two presses?

**4-10.** (a) How much explosive energy is required to form a $\frac{1}{8}$-in.-thick-molded steel dome, $\frac{1}{8}$ in. thick and 10 ft in diameter, under the following conditions:

$\varepsilon = 0.2$ in./in.
Average flow stress = 50,000 lb/in.$^2$

(b) What are some of the variables that may affect part (a) of this problem so that the results are not as expected?

**4-11.** What is the relationship between the flow stress of a part with increasing strain rate and true strain to fracture?

**4-12.** (a) What is meant by the toughness of a metal? (b) Should tough metals be used for HERF? Why or why not? (c) Would Silly Putty be a good material to use to illustrate the requirements for HERF? Why or why not?

**4-13.** The advantages of elevated-temperature forming are quite obvious, but what may be some disadvantages?

**4-14.** (a) What is the effect of hydrostatic pressure on true strain? (b) How does hydrostatic pressure affect fracture?

**4-15.** When are hydrostatic pressures especially beneficial for metal forming?

**4-16.** What metallurgical changes make HVF possible?

**4-17.** Why must the die angle for impact extrusions be carefully controlled?

**4-18.** State the special purpose of each of the following dies: (a) compound, (b) progressive, (c) transfer.

**4-19.** (a) Assume the forming-limit curve technique is used. The original circle diameters on the metal are 1 in. After forming $\varepsilon_1 = 1.5$ in. and $\varepsilon_2 = 1.25$ in. how would you determine if this is satisfactory or not? (b) Would it be possible to show a negative $\varepsilon_2$ strain? Explain.

**4-20.** (a) What is the main objection to the Erichsen cup test in determining the drawability of metal? (b) How does the Fukui test overcome the objection of the Erichsen test?

**4-21.** What material could be used for a die to form a dome 15 ft in diameter to a depth of 10 in. at the center? The material is mild steel $\frac{1}{4}$ in. thick. Only one is required.

**4-22.** Why should yield strength be low for a material that is used for drawing?

**4-23.** What may result if a material has a large yield elongation and is subjected to deep drawing?

**4-24.** What is the effect of temperature on the percent reduction of area that can be achieved?

**4-25.** When does hydrostatic pressure benefit forming?

**4-26.** What semicone angle produces the least total resistance in extrusion?

# BIBLIOGRAPHY

**Periodicals**

BAUER, E. G., *A Comparison of New Sheet-Forming Processes*, ASTM Paper No. MF69-519, 1969.

BLICKWEDE, D. J., "Influence of Mill Processing on Metallurgy, Formability," *Metal Progress*, Dec., 1968.

BOTROS, B. M., *Springback in Metal Forming After Bending*, ASTM Paper 67-WA/PROD-17, 1967.

BRIDGMAN, P. W., "The Effect of Hydrostatic Pressure on Ductility," *Journal of Applied Physics*, Vol. 17, 1947.

DATSKO, J. and C. T. YANG, *Correlations of Bendability of Materials With Their Tensile Properties*, ASME Transactions Series B, Vol. 82, Nov., 1960.

DRAKE, R. J. and J. W. THROOP, "How to Predict Cold Extrusion Forces," *Metal Progress*, May, 1971.

EVANS, W. M. and B. AVITZUR, *Toward Better Extrusion Die Design*, ASTM Paper No. MF67-582, 1967.

GOODWIN, G. M., *Application of Strain Analysis to Sheet Metal Forming Operations in the Press Shop*, SAE Paper No. 680093.

HALL, F., "Forged Steel is Stronger," *Product Engineering*, Oct. 25, 1965.

HECKER, S. S., "A Reproducible Cup Test for Assessing Sheet Metal Stretchability," General Motors Research Report No. 1497, Dec., 1973.

KEELER, S. P., "Understanding Sheet Metal Formability," *Machinery*, Sept., 1968.

PARKER, C. M. and F. E. CHEPKE, "Design Guide for Steel Forgings," *Machine Design*, July 22, 1965.

POMEY, G., "How Do You Test a Deep Drawing Steel?", *Metal Progress*, May, 1964.

WHITELEY, R. L., *The Importance of Directionality in Drawing Quality Sheet Steel*, ASTM Transactions, Vol. 52, P. 154, 1969.

WOOD, W. W., *A Critical Review of High-Energy Forming Methods*, ASTM Paper 292, 1960.

**Books and Pamphlets**

BURBANK, O. F., *How to Order Better Forgings*, American Machinist, Nov. 17, 1969.

*Designing for Aluminum Extrusions*, Kaiser Aluminum and Chemical Sales Inc., Kaiser Center, Oakland, Calif., 1963.

*Forming Alcoa Aluminum*, Aluminum Company of America, 1962.

LINDBERG, R. A., *Materials and Manufacturing Technology*, Allyn and Bacon, Inc., 1968.

*Principles of Forging Design and Mechanical and Physical Properties of Ferrous Forgings*, American Iron and Steel Institute, New York.

*Upsetting*, National Machining Co., Tiffin, Ohio.

# 5

## *Casting and Molding Processes*

### Introduction

There are some basic similarities in the casting and molding processes used in producing metal parts, plastic parts, and powdered-metal parts. Because of this similarity, all of these processes will be discussed in this chapter. Although most metal castings are made by pouring molten metal into a prepared cavity and allowing it to harden, castings may also be made by forcing the liquid into a mold under pressure. The same principle, of filling a mold under pressure, is used in injection molding of plastics, and in the case of powdered metallurgy, the powder is compacted into the mold under pressure.

Design and manufacturing engineers must often decide when it is advantageous to use the casting process or when it may be better to use some other method of fabrication. Given below is a list of some of the main considerations that may be used in choosing the casting process.

1. Parts that require complex internal cavities, such as asymmetric parts, or those that are quite inaccessible for machining. Also the cavities that are large and may necessitate considerable metal removal.
2. When a large number of parts are to be made out of aluminum or zinc and have rather complex structures.
3. Parts requiring heavy, formed, cross-sectional areas. Heavy sections can be fabricated if the part is relatively simple, but forming poses many problems. Fabricating and machining may be very time-consuming.
4. Castings allow bulk or metal mass to be placed advantageously, as in machine bases.
5. Damping, both sound and mechanical, is often needed in machine tools. Gray cast iron can provide this quality better than any other metal.

6. Modern foundry practices make it feasible, in some cases, to produce one of a kind. Some patterns may be made quickly out of wax or styrofoam. The part may be cast within a few hours of its conception and often with a choice of materials such as steel, cast iron, copper, brass, or aluminum.

7. Several individual parts may be quite easily integrated into one part with a savings of both material and labor.

8. Parts that are extremely difficult to machine such as the refractory material used in turbine blades may be cast to close tolerances.

9. When it is desirable to minimize directional properties of the metal. Castings have better *anisotropic* qualities than forged or wrought materials. Anisotropic refers to the directional qualities of most wrought metals that reduce ductility, impact, and fatigue properties transverse to the direction of rolling.

10. When using precious metals, since there is little or no loss of material.

There are also times when it is not advantageous to use the casting process, as follows:

1. Parts that can be stamped out on a punch press.
2. Parts that can be deep drawn.
3. Parts that can be made by extrusion.
4. Parts that can be made by cold-heading.
5. Parts made from highly reactive metals.

Sand casting of metals is an ancient process that has its beginnings in the Bronze and Iron Ages. The oldest castings yet discovered show that molten metal was poured from a clay crucible into an open mold. Ancient spears were made in two-part closed molds using cores to form sockets—a method similar to that used today. The largest bronze statue in the world today is that of the great Sun Buddha at Nara, Japan. Cast in the eighth century, it weighs 551 tons (496 metric tons) and is over 71 ft (21 m) high.

Sand casting is the most widely used and adaptable casting process. In brief, it consists of compacting sand around a metal, wood, or plastic pattern. The sand is relatively fine and is composed primarily of silica ($SiO_2$). Molding sands usually contain between 20 and 25 % clay which, with water, becomes the principal source of strength acting as the bond or binder. In some locations, clay and sand occur in the right proportions for molding and these mixtures are referred to as "natural molding sands." In other cases a clay for bonding must be added to the sand in order to develop the requisite cohesiveness and plasticity. There are several types of clays which are essentially aggregates of minute, activated, crystalline, flake-shaped particles. The clays may be southern or western bentonites, kaolinites, or fireclay. Typically, two or more of these clays may be used together in order to achieve the desired bond characteristics. Sodium silicate and organic materials are sometimes used as sand binders.

All casting processes require some type of mold. The nature of the mold may vary greatly: from a permanent type made of steel, graphite, etc., to those that are expendable, such as sand, plaster, ceramics, etc. The casting process is primarily a

TABLE 5-1. *Casting Process Classification*

| Expendable Mold Casting | Nonexpendable Mold Casting |
|---|---|
| Green sand | Permanent |
| $CO_2$ sand | Die |
| Core sand | Low pressure |
| Shell sand | Centrifugal |
| Investment | Continuous |
| Shaw process | |
| Full mold | |

function of the type of mold used, not only in terms of the mechanics of mold making and filling, but also in terms of the metallurgical results brought about by the wide range of heat-extracting qualities of the mold materials. Metal molds, for example, have a high rate of heat extraction and thus produce a tough, fine-grained casting; but they are difficult to "fill out" and usually require pressure to speed the molten metal. Plaster molds, on the other hand, extract heat very slowly and as such are easy to fill out. Even thin sections can easily be cast. However, the resulting castings tend to be soft and coarser grained. Sand molds lie between these two extremes. The principal casting processes have been classified by the type of mold used: expandable or nonexpandable (see Table 5-1).

Mold production varies widely. A die-casting mold which is capable of producing 50,000 to 100,000 pieces may cost $30,000 and take two months to produce. A sand mold can, with automated equipment, be "rammed up" in 10 to 30 sec but can be used to produce only one casting.

In some sand-casting operations, as much as 25 tons of sand may be handled for each ton of casting produced. It is obvious that in such foundries the major operation is the moving, conditioning, and molding of sand. It is for this reason that one of the primary considerations in the operation of a sand-casting foundry is the location and conditioning of the sand.

It is this great disparity in the mold-making process and in the resulting mold that dictates that the process be considered in terms of the molds rather than by melting practice, size, or metal cost.

## EXPENDABLE MOLD-CASTING PROCESSES

### Green-Sand Molding

*Green sand* refers to moist sand (from 2 to 8 % water). This method of molding is the most popular and widely used process in the foundry industry. The process is well suited to a wide variety of miscellaneous casting, in sizes of less than a pound to as large as 3 to 4 tons. This versatile process is applicable to both ferrous and nonferrous materials.

Green sand can be used to produce intricate molds since it provides for rapid collapsibility, that is to say the mold is much less resistant to the contraction of the casting as it solidifies than are other molding processes. This results in less stress and strain in the casting.

*Advantages and Limitations.* Green-sand molding is often the most economical of all the molding processes. This is particularly true when only a few castings are to be made or even for a large number of castings that are not suitable for die casting. The sand can be worked manually or mechanically and is reusable with only slight additions to correct for clay "burn-out", or deactivation, and accumulation of *fines* (fine-grain sand, finer than 270-grit).

As mentioned, the green-sand process usually gives way to die and other permanent-mold processes when high production rates are required and the tooling investment can be offset over a large number of pieces. Also, green sand tends to crush and shift under the weight of heavy sections. It is particularly weak in thin sections.

## $CO_2$-Setting Mold Sands

When sand is mixed with a sodium silicate binder, each grain is coated with a thin film of the viscous fluid. The sand is packed into the mold and carbon dioxide is passed through it. The $CO_2$ reacts with the sodium silicate, precipitating $SiO_2$, which bonds with the sand as the result of the formation of silic acid by the following reaction:

$$Na_2SiO_3 + CO_2 + H_2O \rightarrow Na_2CO_3 + SiO_2 + H_2O$$

The $CO_2$ is passed for only a few seconds at up to 20 psi pressure.

*Advantages and Limitations.* The $CO_2$-setting process makes the sand mold very hard. Therefore it can be used to support thin sections, as for example the cooling fins on a heat exchanger. Since there is little danger of breaking or crushing the mold when the pattern is taken out, less draft is necessary. Draft refers to the small amount of taper on the pattern, which allows the pattern to be removed from the sand

$SiO_2$ molds are not easily broken up and are seldom recycled. Usually the mold is scrapped after the desired number of castings have been made.

## Core-Sand Molds

Molds may be made from what is known as *core* sand. Cores, used for internal passageways in castings, will be discussed later. Core sand is a fine sand mixed with either natural or synthetic resins and water. After the mold has been made, it is baked at 350–400°F (177–204°C). Some core sands are mixed with furfuryl alcohol and alkyd resins. This mixture can be rammed into a heated mold and it will set up without baking.

*Advantages and Limitations.* Core-sand molds have more strength and can resist erosion better than some green-sand molds. They can be used for thin sections and intricate parts. The best use is in cases where the mold is made in parts that are subsequently assembled to form the complete mold. This can be especially advantageous when the mold is symmetrical and has several identical sections. The repeated parts can all be made in the same pattern section and assembled to complete the mold.

### Green-Sand Molding Process

Green-sand molding is the most frequently used casting process and will therefore be discussed in more detail than some of the other processes. The process begins with the design, which in turn is made into a three-dimensional pattern. The pattern can be made of wood, metal, plaster, styrofoam, or other materials. It is made larger than the desired casting to provide for metal shrinkage, distortion, and machining. The pattern is placed on a *molding board*, which is a little larger than the open box or *flask* in which the sand mold is to be made. The molding board is placed under the top half of the two-part flask and sand is poured around the pattern. The sand next to the pattern is always sifted or riddled to provide a better detail and is termed *facing sand*.

The sand is rammed or compacted around the pattern by a variety of methods including: hand or pneumatic-tool ramming, jolting (abrupt mechanical shaking), squeezing (compressing the top and bottom mold surfaces), and driving the sand into the mold at high velocities (sand slinging). Sand slingers are usually reserved for use in making very large castings where great volumes of sand are handled (Fig. 5-1). The jolt and squeeze method of filling the flask, usually a two-part box containing the pattern, is by far the most used. First the drag, or lower half of the flask, is filled and jolted, then the flask is turned over, the cope, or top half of the flask, is filled, and the two-part mold, with the pattern and molding board sandwiched in between is squeezed.

FIGURE 5-1. *A sandslinger can be used to fill large floor molds or a number of smaller molds in a relatively short time.*

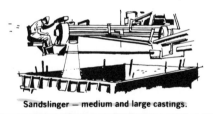

**Sandslinger — medium and large castings.**

***Patterns.*** Patterns of simple design, with one or more flat surfaces, can be molded in one piece, provided that they can be withdrawn without disturbing the compacted sand. Other patterns may be split into two or more parts to facilitate their removal from the sand when using two-part flasks. The pattern must be tapered to permit easy removal from the sand. The taper is referred to as *draft*. When a part does not have some natural draft, it must be added. A more recent innovation in patterns for sand casting has been to make them out of foamed polystyrene that is vaporized by the molten metal. This type of casting, known as the *full-mold* process, does not require pattern draft. This process will be discussed later.

*Sprues, Runners, and Gates.* Access to the mold cavity for entry of the molten metal is provided by *sprues*, *runners*, and *gates*, as shown in Fig. 5-2. A pouring basin can be carved in the sand at the top of the sprue, or a *pour box*, which provides a large opening, may be laid over the sprue to facilitate pouring. After the metal is poured, it cools most rapidly in the thin sections and along the outer surfaces where it gives up most of its heat to the sand mold. Thus the outer surface forms a shell which permits the still molten metal near the center to flow towards it. As a result the last portion of the casting to freeze will be deficient in metal and, in the absence of a supplemental metal-feed source, will result in some form of shrinkage. This shrinkage may take the form of gross shrinkage (large cavities) or the more subtle microshrinkage (finely dispersed porosity). These porous spots can be avoided by the use of risers, as shown in Fig. 5-2, which provide molten metal to make up for shrinkage losses. The science of providing the proper gate and runners to ensure sound castings is discussed later.

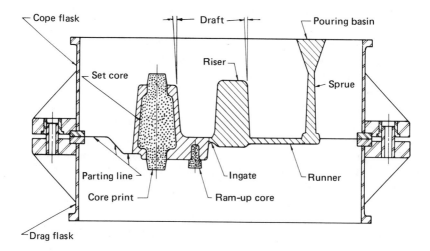

FIGURE 5-2.  *Sectional view of a casting mold.*

*Cores.* Cores are placed in molds wherever it is necessary to preserve the space it occupies in the mold as a void in the resulting castings. As shown in Fig. 5-2, the core will be put in place after the pattern is removed. To ensure its proper location, the pattern has extensions known as *core prints* which leave cavities in the mold into which the core is seated. Shown in Fig. 5-3 is an example of a core box and core placement. Sometimes the core may be molded integrally with the green sand and is then referred to as a *green-sand core*. Generally, the core is made of sand bonded with core oil, some organic bonding materials, and water. These materials are thoroughly blended and placed in a mold or core box. After forming, they are removed and baked at 350 to 450°F (177–232°C). Cores that consist of two or more parts are pasted together after baking.

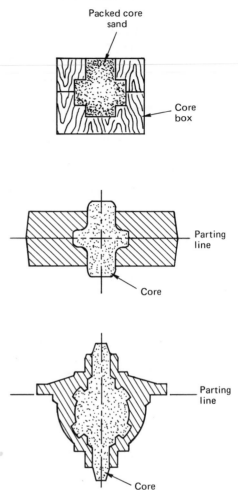

FIGURE 5-3. *Examples of a core box and core placement.*

***CO₂ Cores.*** A newer method of curing cores is with $CO_2$. Carbon dioxide reacts with the sodium silicate constituent of the binder to produce silicon dioxide, as mentioned under $CO_2$-setting molds. This method has the advantage of being able to use the cores as soon as they are removed from the core box. Also, less draft is required since high strength is developed before removal from the core box. However, a major problem with the $CO_2$-set cores is that they do not burn out and are quite difficult to remove since they do not collapse and may even cause hot cracks to occur in the casting.

***Shell-Mold Cores.*** Cores may also be made by the shell-molding process, discussed in more detail later. Only a brief description will be given here. A core box is heated and sand which contains 3 to 6 % resin is blown into it. After a dwell period, which establishes the thickness of the shell, the unheated interior sand may be drained out.

The shell wall is usually $\frac{1}{4}$ to $\frac{1}{2}$ in. thick and can be easily stripped from the box and handled directly. No further baking is required. The cores may be placed directly in the molds. Although it is possible to recover the sand from the sand–resin mixture, the procedure is questionable unless the quantity is large. Recovery consists of burning out the resin at high temperatures.

Cores should be used only when they can produce real economies by reducing later machining operations. Cores are relatively expensive to make, require expensive tooling, are a nuisance to set (place in the mold), produce scheduling problems, are often difficult to remove from the casting, and in general complicate the casting process. In some cases the cost associated with cores can constitute the greatest single cost factor in producing a casting.

*Parting-Line Placement.* The parting line is the line along which a pattern is divided for molding or along which sections of the mold separate. This surface and the pattern are on occasion sprinkled with parting dust so that the cope and drag will separate without rupturing the sand. The selection of the parting line can play an important part in the economics of production. Complex parting lines require expensive pattern equipment and are much harder to mold. On the other hand, ingenuity in the selection of a parting line can often take advantage of the natural draft found on the part, eliminate cores, allow more parts to be squeezed onto a match plate, and even eliminate some risers. It is obvious that the point where the metal enters the mold cavity, *the gate*, must lie on the parting line. While the parting line may, with sufficient justification, be allowed to wander up and down, it is most economical when it is restricted to a flat plane, thereby allowing the pattern halves to be affixed to a flat pattern board.

Generally, the flask is removed before the metal is poured to prevent it from becoming damaged and to speed up its reuse (Fig. 5-4). A pouring jacket may be put in its place if there is danger of the molten metal breaking out.

### Mechanized Molding

The preparation of molds entirely by hand is slow and costly. In most modern foundries, molds of small and medium sizes are prepared by machines. For mass production, *match-plate patterns* are normally used (Fig. 5-5). The match-plate pattern is made with the cope part of the pattern on one side of a plate, usually aluminum, and the drag part of the pattern on the other side. The plate fits between the cope and the drag. After each half of the mold has been filled with sand and compacted, the match plate is removed, leaving a nearly completed mold since runners, gates, and risers are included on the match plate. A provision for the proper location of sprue is also made.

A variation of this process is called the *cope-and-drag pattern*. This allows both the cope and drag to be "rammed up" separately but at the same time. Separate pattern plates require accurate alignment of the two halves by means of a guide and locating pins.

FIGURE 5-4. *Metal flask and mold. Courtesy Hines Flask Co.*

FIGURE 5-5. *A match-plate pattern. Courtesy Central Foundry Division, G.M.*

## Molding-Sand Preparation and Control

The selection and mixing of molding sand constitutes one of the main factors in controlling the quality of the castings. Deposits of natural molding sand are characterized by a clay content of 10 to 30 %, with about 10 % of the deposits being in the 15 to 20 % range. Natural sands are sometimes altered by additions to improve certain characteristics. Such sands are referred to as semisynthetic sands. The most common alteration is the addition of more clay to increase the green strength. The reconditioning is also done to produce certain other desirable properties as follows:

1. The sand mold should present a smooth surface. This surface is governed by grain size, grain-size distribution, and clay and fines content.

The clay content is determined by washing a sand sample through several cycles in a solution of sodium hydroxide. This will remove the clay as well as the fines and the solubles. After drying, the sand is weighed and the percentage of clay removed is defined as the American Foundry Society (AFS) clay content. It should be appreciated that the AFS clay content includes not only the active clay but also all other materials washed off, in particular the entire fines content. When casting metals of high melting temperature, such as steel, fire clays are required. A mixture of 7 to 8 % fire clay and 1 to 2 % Western bentonite may be used. Bentonite is a weathered volcanic ash.

2. The sand, after being moistened with water, should possess good moldability, produce maximum mold hardness, and be *permeable.* Permeability refers to the ability of the sand to vent steam. The moisture may be measured by weighing a sample of the moist sand and then drawing off the moisture with hot (300°F, 149°C) air and reweighing it. Perhaps a more popular method is that of "feel" by an exper-

ienced foundryman. A handful of sand is squeezed and upon releasing it the sharp edges formed between the fingers should stand up without crumbling. If they fall down the sand is too dry. If, on the other hand, it breaks at the edges and rounds off, then it is too wet. The break should be sharp and clean.

The hardness of the rammed-up mold sand may be checked with a hardness tester that has a spring-loaded steel ball. The maximum hardness that can be expected from a hand-rammed mold is about 82–85. If no penetration occurs, as on a machine-rammed mold, the hardness reading will be 100. As the mold hardness increases, it produces a casting with a better surface finish and greater dimensional accuracy. It also helps eliminate loose-sand inclusions and mold erosion.

Both the green compressive strength and the green shear strengths of the molding sands can be checked by the use of a standard AFS specimen in a universal testing machine (Fig. 5-6). However, in practice the easily taken mold hardness is usually monitored for a given sand system. This characteristic of the sand can be closely correlated to the green sand compressive and shear strength.

FIGURE 5-6. *A universal sand-strength testing machine showing a sand sample being tested for shear strength. Courtesy Harry W. Dietert Co.*

*Sand Reconditioning.* In most foundries, after the casting has been shaken out of the mold, the sand is recirculated. In mechanized foundries the sand is automatically returned by belt conveyor for reconditioning. It is aerated, lumps are broken up, magnetic particles are removed by an electromagnet, new clay is added to make up for the " burnout " or deactivated portion, and it is brought back to proper " temper " by the addition of a suitable amount of water. In addition, the sand is passed through a muller which mechanically fluffs it up and makes it suitable for remolding.

### Melting and Pouring the Metal

The metal for casting may be melted in one of several types of furnaces, as shown in Fig. 5-7. The choice of furnace is based on various factors given briefly as follows:

1. Economy, including the cost of fuel per pound of melted metal and the initial cost of the equipment plus installation.
2. The availability of the desired energy. As an example, if electric power is to be

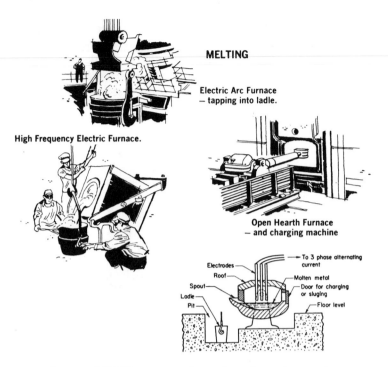

**MELTING**

High Frequency Electric Furnace.

Electric Arc Furnace
— tapping into ladle.

Open Hearth Furnace
— and charging machine

To 3 phase alternating
current

Electrodes

Roof

Molten metal

Spout

Door for charging
or sluging

Ladle

Floor level

Pit

FIGURE 5-7. *Furnaces used to melt steel or other metals.*

used it may require several years to make all the necessary arrangements needed for the heavy demands of a large foundry.

3. Temperatures required.
4. Quantity of metal required per hour or shift.
5. The compatability of the process to the metal, e.g., cupolas are good for cast iron but not for aluminum.
6. The compatibility of operation to environmental considerations. For example, in recent years foundry men have experienced that cleaning up a cupola operation can double the cost of the whole foundry. A cupola furnace is shown in Fig. 5-8.

*Crucible, or Pot-Tilting, Furnaces.* Crucible furnaces, as the name indicates, consist simply of a crucible to hold the metal while it is being melted by a gas or oil flame. It is so arranged that it can be easily tilted when the metal is ready for pouring (Fig. 5-9). The capacity is generally limited to 1000 lb. Also shown is a twin-crucible furnace that utilizes the heat that leaves the firing chamber and passes over the heating crucible to the preheat crucible.

*Cupolas.* Cupolas are designed more specifically for producing a cast-iron melt. The cupola, as shown schematically in Fig. 5-8, has several advantages:

1. Like a blast furnace used in making steel, it can be tapped at regular intervals as required under conditions of production.

2. The operating efficiency of the cupola is higher than any other foundry melting technique due to the countermovement of the heating gasses in respect to the charge. As mentioned previously, the economic advantage of the cupola has now been somewhat mitigated by the complex problem of cleaning the inherently dirty stack effluent as well as by the increased shortage of "metallugical grade" coal (low sulfur content).

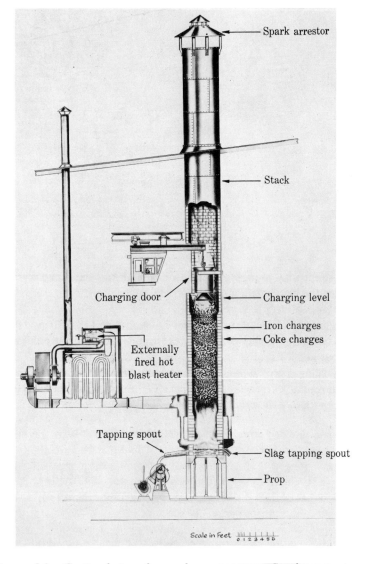

FIGURE 5-8. *Sectional view of a cupola construction. Cupolas are not as extensively used now as formerly due to the cost of cleaning them up to meet clean-air requirements. Courtesy Whiting Corporation.*

FIGURE 5-9. *A crucible or pot-tilting furnace. Courtesy Randall Foundry Corp.*

3. The chemical composition of the melt can be controlled even under conditions of continuous melting. Since the molten metal comes in contact with the carbonaceous coke, the carbon content of the melt tends to be as high as in the range of cast irons. If steels are desired, the melt must be transferred to a different type of furnace, such as an *open-hearth*, an *electric-arc*, or a *basic-oxygen* furnace, where the carbon content is lowered and the alloy content adjusted.

*Pouring the Metal.* Several types of containers are used to move the molten metal from the furnace to the pouring area.

Large castings of the floor-and-pit type are poured with a ladle that has a plug in the bottom, or as it is called, a bottom-pouring ladle. It is also employed in mechanized operations where the molds are moved along a line and each is poured as it is momentarily stopped beneath the large bottom-pour ladle.

Most ladles are of the lip-pouring type (Fig. 5-10). They are the easiest to clean out, to pour, and to reline. Ladles used for pouring ferrous metals are lined with a high alumina-content refractory. After long use and oxidation, it can be broken out and replaced. Ladles used in handling ferrous metals must be preheated with gas flames to approximately 2600–2700°F (1427–1482°C) before filling. Once the ladle is filled, it is used constantly until it has been emptied.

For nonferrous metals, simple clay-graphite crucibles are used. While they are quite susceptible to breakage, they are very resistant to the metal and will hold up

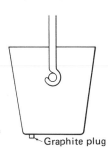
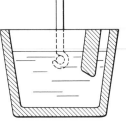

FIGURE 5-10. *Ladles used in pouring the molten metal for castings.*

Graphite plug

Bottom-pouring ladle　　　Teapot ladle

a long time under normal conditions. They usually do not require preheating, although care must be taken to avoid moisture pickup. For this reason they are sometimes baked out to assure dryness.

The pouring process must be carefully controlled since the temperature of the melt greatly affects the degree of liquid contraction before solidification, the rate of solidification which in turn affects the amount of columnar growth present at the mold wall, the extent and nature of the dendritic growth, the degree of alloy burnout, and the feeding characteristics of the risering system. These points are discussed in more detail later.

### Finishing Operations

After the castings have solidified and cooled somewhat, they are placed on a shakeout table or grating on which the sand mold is broken up, leaving the casting free to be picked out. The casting is then taken to the finishing room where the gates and risers are removed. Small gates and risers may be broken off with a hammer if the material is brittle. Larger ones require sawing, cutting with a torch, or shearing. Unwanted metal protrusions such as fins, bosses, and small portions of gates and risers need to be smoothed off to blend with the surface. Most of this work is done with a heavy-duty grinder and the process is known as *snagging* or *snag grinding*. On large castings it is easier to move the grinder than the work, so swing-type grinders are used (Fig. 5-11). Smaller castings are brought to stand- or bench-type grinders.

FIGURE 5-11. *A battery of swing grinders being used to smooth the casting surfaces where excess metal has been removed. Courtesy Lebanon Steel Foundry.*

Hand and pneumatic chisels are also used to trim castings. A more recent method of removing excess metal from ferrous castings is with a carbon-air torch. This consists of a carbon rod and high-amperage current with a stream of compressed air blowing at the base of it. This oxidizes and removes the metal as soon as it is molten. In many foundries this method has replaced nearly all chipping and grinding operations.

*Checking and Repairing.* After the castings are cleaned, they are inspected for flaws. Surface defects in steel castings may be repaired by welding.

Castings are tested both destructively and nondestructively. One destructive test consists of sawing the casting in sections to inspect for possible porosity. Other destructive tests are standard impact and tensile tests. A wide variety of nondestructive tests may be used, such as X rays, gamma rays, magnaflux, and ultrasonics. Typically, however, much of the testing is done at the time the casting is first being run, in order to prove out the gating, runnering, and risering systems. Once all the variables have been pinned down, the testing program can be drastically reduced. Usually after the initial defects have been determined and corrected, inspection is done merely on a visual basis at the time of cleanup. Of course critical parts will require more thorough testing.

*Heat Treatment of Castings.* Castings are sometimes heat treated to develop greater strength refine the grain structure, produce more isotropic properties, or produce a more homogeneous structure. Castings are typically harder and stronger in their thinner, more rapidly cooled sections. In order to improve machinability, ductility, and/or elongation, heat treatment may in some cases be incorporated into the casting process. A separate heat treatment adds considerably to the cost and should not be specified unless it is essential to the finished product.

*Sand-Casting Process Review.* The sand-casting process can become quite involved, especially for an initial presentation. Therefore a brief summary of the important steps are reviewed with a series of line drawings (Fig. 5-12).

## Shell-Mold Casting

Shell molding is a special form of sand casting which uses conventional but finer foundry sands coated with a thermosetting (heat-hardening) phenolic resin. The pattern, usually cast iron, is carefully machined and polished.

Jolting, squeezing, and ramming of the mold are all eliminated in this process. Instead, a resin–sand mixture is poured into an open-end box mounted on trunions (Fig. 5-13). The pattern plate, which has been previously heated to about 450°F (232°C), is clamped facedown on top of the box. The box is inverted and the sand mixture drops down on the hot pattern. A 20 to 35 sec exposure will result in a shell $\frac{1}{8}$ to $\frac{3}{8}$ in. (3.175–9.52 mm) thick, which is adequate for most small- to medium-sized castings. The box is returned to the upright position and the pattern, with the soft shell adhering to it, is removed and placed in an oven to cure. Instead of the dump-box approach, the newer method is that of blowing the sand mixture on the pattern. This provides a denser, more evenly distributed mold. The cure time for the shell is 30 to 40 sec, at approximately 450°F (232°C). Radiant-heat furnaces (Fig. 5-14) may also be used.

*Mold Ejection.* The cured shell fits rather snugly on the pattern, so some means must be provided for its rapid removal. This can be easily accomplished if knockout pins are incorporated into the mold board, as shown in Fig. 5-15. Other separately operated mechanical-ejection systems can be worked out, but in any case the method of removal must be an integral part of the design.

*Mold Assembly.* The shell halves are usually glued together with resin paste, which sets under pressure and the residual heat of the shells. A tongue-and-groove joint at the seam helps prevent the adhesive from entering the mold cavity and minimizes *fins* (small, thin, metal protrusions at the parting line).

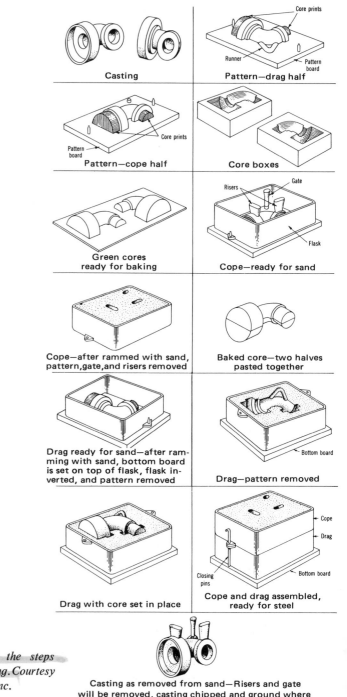

Casting

Pattern—drag half

Pattern—cope half

Core boxes

Green cores
ready for baking

Cope—ready for sand

Cope—after rammed with sand,
pattern, gate, and risers removed

Baked core—two halves
pasted together

Drag ready for sand—after ramming with sand, bottom board
is set on top of flask, flask inverted, and pattern removed

Drag—pattern removed

Drag with core set in place

Cope and drag assembled,
ready for steel

Casting as removed from sand—Risers and gate
will be removed, casting chipped and ground where
necessary, annealed, inspected, and ready for shipment

FIGURE 5-12. *A summary of the steps involved in producing a steel casting. Courtesy Adirondack Steel Casting Co., Inc.*

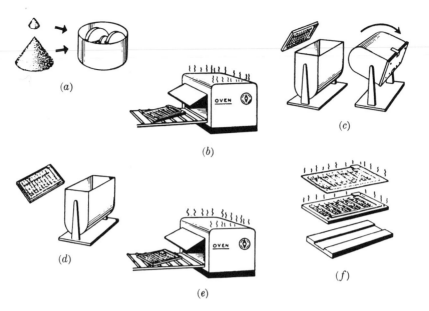

FIGURE 5-13. *The shell-molding process includes (a) mixing the resin and sand, (b) heating the pattern, (c) investing, (d) removing the invested pattern, (e) curing and (f) stripping.*

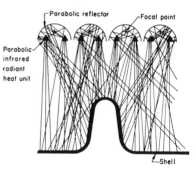

FIGURE 5-14. *Radiant-heat furnaces are used to effectively cure shell molds.*

***Pouring.*** When there is likely to be considerable pressure on the inside of the mold, it should be supported with backup material such as metal shot, gravel, or molding sand.

***Advantages and Limitations.*** Shell molding is substantially more accurate than conventional sand molding. High-speed shell-molding machines are available that can handle from 30 to 200 full molds per hour depending upon the pattern-plate size and the number of stations available.

The transmission drum shown in Fig. 5-16 was changed from a green-sand mold to a shell-mold casting. The results were longer tool life in machining and less metal removal required. Balance and drilling operations were also reduced. The savings in machining alone were enough to make the shell process feasible.

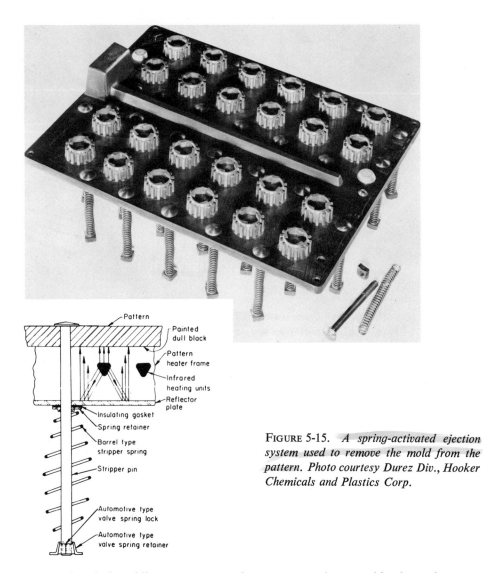

Pattern

Painted
dull black

Pattern
heater frame

Infrared
heating units

Reflector
plate

Insulating gasket

Spring retainer

Barrel type
stripper spring

Stripper pin

Automotive type
valve spring lock

Automotive type
valve spring retainer

FIGURE 5-15. *A spring-activated ejection system used to remove the mold from the pattern. Photo courtesy Durez Div., Hooker Chemicals and Plastics Corp.*

The shell-molding process requires an expensive, machined-metal pattern that makes it uneconomical unless production quantities are relatively high. Size limitations on shell-molding machines also restrict it to relatively small components.

## Plaster-Mold Casting

Plaster-mold casting is somewhat similar to sand casting in that only one casting is made and then the mold is destroyed. In this case the mold is made out of a specially formulated plaster, 70 to 80 % gypsum and 20 to 30 % fibrous strengthener. Water is added to make a creamy slurry.

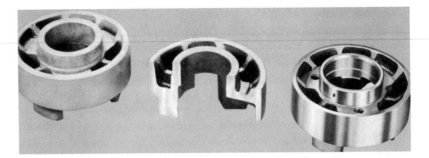

FIGURE 5-16. *The transmission drum is a good example of the use of shell molding because of the deep pockets that can be made with only a small amount of draft.*

The slurry is poured into a flask over the pattern, which is mounted on a match plate. When the plaster reaches its semiset state, the match plate is removed and the mold is put in an oven for drying. Cores, if needed, are made in the same way. The core boxes are usually made from brass, plastics, or aluminum. The process of pouring a plaster core and the assembly of its parts are shown in Fig. 5-17.

Patterns are made out of lightly lacquered plaster, sealed wood, or polished metal. Flexible molded-runner compositions and plastic compositions may also be used when undercuts are encountered. The patterns are covered with a thin film of soap, lard oil, or commercially available waxes rubbed on with a soft cloth.

Expansion plasters are available that expand as they set. The amount of expansion can be controlled by the plaster-to-water ratio. Value of $\frac{1}{16}$ to $\frac{1}{4}$ in./ft can be obtained. Thus the plaster can be made to automatically compensate for the shrink allowance of the cast metal. As an example, if a new pattern is to be made from an existing part, it can be cast right over it. When it sets, it will be larger than the part; but when the metal is cast, it will shrink and be the correct size.

(a)                                           (b)

FIGURE 5-17. *Pouring plaster into a corebox to make cores for a tire-mold tread ring (a), assembly of plaster cores into mold (b). Courtesy Aluminum Company of America.*

*Advantages and Limitations.* Plaster-mold castings provide a finish that is superior to that of the sand cast. A surface finish of 125 microinches is typical; however, 50 to 60 microinches is attainable. Excellent reproducibility of dimensions is also an advantage of the process. When working to critical tolerances such as ±0.005 in./in. (0.005 mm/mm) shrinkage, allowance must be estimated closely. Because of the low chilling rate of plaster molds, walls as thin as 0.060 in. (1.52 mm) can be cast in small parts. A draft angle of 2° is usually required, but zero draft can often be achieved.

The major disadvantage of the plaster-casting method is that chemical decomposition of the gypsum begins at 1000°F (538°C) and thus limits the process to nonferrous, low-melting alloys, mainly aluminum and aluminum-zinc alloy. It is also used for rubber, plastic vacuum, and blow-form molds discussed later in this chapter.

## Investment Molding

Casting processes in which the pattern is only used once are variously referred to as "lost-wax" or "precision-casting" processes. In any case they involve making a pattern of the desired form out of wax or plastics (usually polystyrene). Formerly, frozen mercury was also used but now OSHA safety regulations have ruled it out. The expendable pattern may be made by pressing the wax into a split mold or by the use of an injection-molding machine. The patterns may be gated together so that several parts can be made at once, as shown schematically in Fig. 5-18. A metal flask is placed around the assembled patterns and a refractory mold slurry is poured in to support the patterns and form the cavities. A vibrating table equipped with a vacuum pump is used to eliminate all the air from the mold. Formerly the standard procedure was to dip the patterns in the slurry several times until a coat was built up. This was called the *investment* process. After the mold material has set and dried, the pattern material is melted and allowed to run out of the mold.

The completed flasks are heated slowly to dry the mold and to melt out the wax, plastic, or whatever pattern material was used. When the molds have reached a temperature of 1000°F (538°C) they are ready for pouring. Vacuum may be applied to the flasks to ensure complete filling of the mold cavities.

When the metal has cooled, the investment material is removed by means of vibrating hammers or by tumbling. As with other castings, the gates and risers are cut off and ground down.

*Advantages and Limitations.* Investment casting is particularly advantageous for small precision parts of intricate design that can be made in multiple molds. Thin walls, down to 0.030 in. (0.76 mm), can be cast readily. Surfaces have a smooth matte appearance with a roughness in the range of 60 to 90 microinches. The machining allowance is about 0.010 to 0.015 in. (0.25 to 0.38 mm). Tolerances of ±0.005 in./in. (0.005 mm/mm) are normal, closer tolerances can be obtained without an excessive amount of secondary operations.

Investment molds usually do not exceed 24 in. (60.96 cm) maximum dimension. Preferably the casting should weigh 10 lb (4.53 kg) or less and be under 12 in. (30.48 cm)

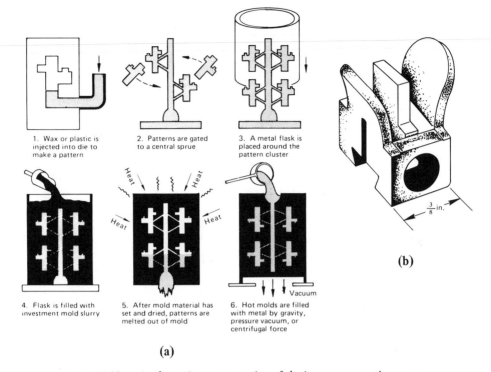

1. Wax or plastic is injected into die to make a pattern

2. Patterns are gated to a central sprue

3. A metal flask is placed around the pattern cluster

4. Flask is filled with investment mold slurry

5. After mold material has set and dried, patterns are melted out of mold

6. Hot molds are filled with metal by gravity, pressure vacuum, or centrifugal force

(a)

(b)

FIGURE 5-18. *A schematic representation of the investment-casting process (a), an example of an investment casting (b).* Courtesy Product Engineering (Copyright Morgan-Grampian, Inc., 1972) *and* Casting Engineers.

in maximum dimension. Sections thicker than 0.5 in. (1.27 cm) are not generally cast. The process has a relatively low rate of solidification. Grain growth will be more pronounced in larger sections, which may limit the toughness and fatigue life of the part.

Metals used in investment casting are aluminum, copper, nickel, cobalt, carbon and alloy steels, stainless steels, and tool steels.

### Shaw Process

The Shaw process, developed in England by Clifford and Noel Shaw, is somewhat similar to the investment-casting process in that a creamy ceramic slurry is poured over a pattern. In this case, however, the pattern, made out of plastic, plaster, wood, metal, or rubber, is reusable. The slurry hardens on the pattern almost immediately and becomes a strong green ceramic of the consistency of vulcanized rubber. It is lifted off the pattern while it is still in the rubberlike phase. The mold is ignited with a torch to burn off the volatile portion of the mix. It is then put in a furnace and baked at 1800°F (982°C), resulting in a rigid refractory mold.

*Advantages and Limitations.* The Shaw process finds its biggest advantage in being able to mold steels to very precise dimensions. Steel castings weighing over 100 lb (45.35 kg) have been made by this process. On small castings, tolerances may be held to within 0.001 in./in., but on larger parts the tolerance is generally about 0.010 in./in. The surface finish varies from 80 to 120 microinches. Pattern to finish casting can be done in 2 hr. The major tonnage of Shaw cast tooling goes into dropforging dies, but the process is also used in making plastic molds, die-casting molds, glass molds, stamping dies, and extrusion dies.

Although the Shaw process is generally quite fast, it may take as long as 5 hr to "bake out" a large ceramic mold. Parting lines may be seen on the casting where the mold halves have been put back together. The molds are quite flexible before baking but some pattern contours do not lend themselves to stripping.

Shown in Fig. 5-19 is a casting that was made by both the Shaw process and the lost-wax method. All but the curved portion of the impellor hub was made by the Shaw process, then lost wax was used to make the curved portion of the blade.

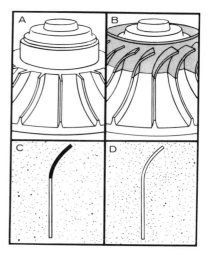

FIGURE 5-19. *A combination of the Shaw process and lost-wax process were used to produce this impellor hub. The parts shown are: metal pattern (a), wax pattern added to metal pattern (b), cross-sectional view after investment with metal pattern removed (c), cross section after wax has been melted out (d). Courtesy Lebanon Steel Foundry.*

## Full-Mold Casting

Full-mold casting may be considered a cross between conventional sand casting and the investment technique of using lost wax. In this case, instead of a conventional pattern of wood, metals, or plaster, a polystyrene foam or Styrofoam* is used. The pattern is left in the mold and is vaporized by the molten metal as it rises in the mold during pouring (Fig. 5-20). Before molding, the pattern is usually coated with a zirconite wash in an alcohol vehicle. The wash produces a relatively tough skin separating the metal from the sand during pouring and cooling. Conventional foundry sand is used in backing up the mold.

---

* Trade name of the Dow Chemical Company.

*Advantages and Limitations.* Because the pattern does not have to be withdrawn from the mold, undercuts are permitted, pattern draft is no longer required, coring is simplified, and loose pieces are done away with. The patterns can be made in a relatively short time. Thus it can be used for individualized machinery repair. Where production is high, it is possible to produce the patterns by steam expansion of the polystyrene beads in a metal mold, as used for the steam iron core shown in

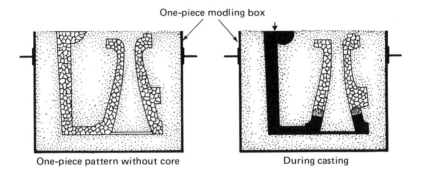

One-piece modling box

One-piece pattern without core          During casting

FIGURE 5-20. *The full-mold process utilizes a Styrofoam pattern that is vaporized by the molten metal. Courtesy Full Mold Process Inc.*

Fig. 5-21. Large patterns can be cut from Styrofoam sheets with wood-cutting tools or a hot wire. Inserts such as lubrication lines, pipes, heating elements, wear strips, cutting edges, etc. can be cast in place in the pattern or put in the pattern when it is built.

The repeatability of the full-mold process may not be as great as that where the same pattern is used over and over. This of course is dependent on the skill of those who cut and glue the patterns. Substantial savings can be obtained in making multiple patterns by the use of templates, fixtures, special patterns, tools, and jigs.

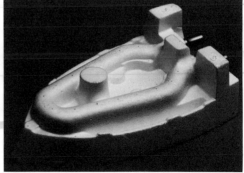

FIGURE 5-21. *Styrafoam cores, as used in a steam iron, are made in metal molds. Courtesy Full Mold Process Inc.*

## NONEXPENDABLE MOLD-CASTING PROCESSES

### Permanent-Mold Casting

One distinct disadvantage of the sand-casting process is that each time a casting is made the mold is destroyed. In the Middle Ages, iron molds were used to produce pewterware such as cups, pitchers, and other utensils. Later, tin soldiers were made by pouring metal into molds hinged together. After a time, the bulk of the still-liquid metal was poured out. The result was a thin-walled casting produced by what is now called *slush molding*.

Reusable metal molds (usually gray iron), or refractory materials, are used for nonferrous metals and cast irons. Machined graphite molds are used to a limited extent for steel castings. The molds can be used for up to about 3000 castings before they require redressing. The hot mold surface is coated with a refractory wash or acetylene soot prior to casting to protect the surface, to facilitate release of the casting, and to control cooling rates. Both metal and sand cores are used to form cavities in the cast parts. Shown in Fig. 5-22 is a typical permanent mold complete with cores.

Shown in Fig. 5-23 is a turntable used in automating the pouring of permanent molds. On this type of equipment a complete cycle can be made in 2 to 6 min, during which time the metal is poured, cooled, and ejected.

*Advantages and Limitations.* Permanent-mold castings generally have better grain structure than sand castings due to the faster cooling rate. As a rule of thumb, parts cast in metal molds show an increase of 20 % in tensile strength and elongation increase of 30 % over sand casting. The surface finish is controlled by the mold coating but it is typically in the range of 150 to 200 microinches for aluminum and between 200 and 350 microinches for ferrous metals. The castings are also less contaminated with inclusions.

Due to the high production rates (60–400 parts/hr), the high quality of the castings, and the small amount of machining required, the process is used in the production of automobile components such as aluminum pistons and also for forging dies.

The main disadvantages of permanent-mold casting is the high initial cost of the mold. The size of the casting is limited by the mold-making equipment, which is usually not over 200 lb (90.71 kg).

### Die Casting

Die casting may be classified as a permanent-mold casting system; however, it differs from the process just described in that the molten metal is forced into the mold or die under high (1000 to 30,000 psi; 6.89 to 206.8 MPa) pressure. The metal solidifies rapidly (within a fraction of a second) because the die is water cooled. Upon solidification, the die is opened and ejector pins automatically knock the casting out of the die. Most of the castings will have flash where the two die halves come together. This is usually removed in a trimming die but may be done with the aid of abrasive belts or wheels or by tumbling. If the parts are small, several of them may be made at one time in what is termed a *multicavity die*.

There are two main types of machines used: the *hot-chamber* and the *cold-chamber* types.

***Hot-Chamber Die Casting.*** The hot-chamber machine is shown in Fig. 5-24. The metal is kept in a heated holding pot. As the plunger descends, the required amount of alloy is automatically forced into the die. As the piston retracts, the cylinder is again filled with the right amount of molten metal. Metals such as aluminum, magnesium, and copper tend to alloy with the steel plunger and cannot be used in the hot chamber.

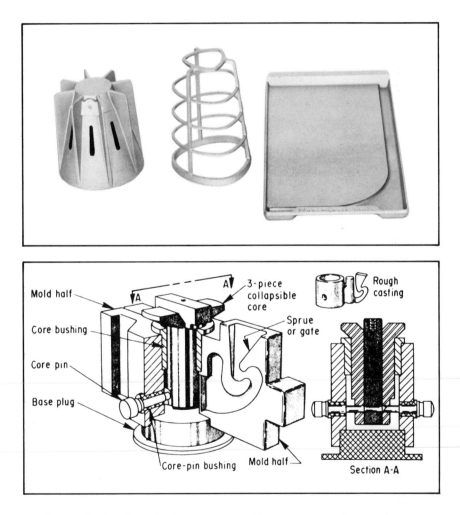

FIGURE 5-22. *A typical permanent-mold arrangement for nonferrous castings. Some examples of permanent-mold cast products. Courtesy Vacuum Die Casting Corporation.*

FIGURE 5-23.  *A turntable used to help automate the permanent-mold casting process.  Courtesy Eaton Corp.*

***Cold-Chamber Die Casting.***   This process gets its name from the fact that the metal is ladled into the cold chamber for each shot, as shown in Fig. 5-25.  This procedure is necessary to keep the molten-metal contact time with the steel cylinder to a minimum.  Iron pickup is prevented, as is freezing of the plunger in the cylinder.

***Advantages and Limitations.***   Die-casting machines can produce large quantities of parts with close tolerances and smooth surfaces.  The size is only limited by the capacity of the machine.  Most die castings are limited to about 75 lb (34 kg) zinc, 65 lb (30 kg) aluminum, and 44 lb (20 kg) of magnesium.  Die castings can provide thinner sections than any other casting process.  Wall thicknesses as thin as 0.015 in. (0.38 mm) can be achieved with aluminum in small items.  However, a more common range on larger sizes will be 0.105 to 0.180 in. (2.67 to 4.57 mm).  Examples of some of the variety of parts that can be die cast are shown in Fig. 5-26.

Some difficulty is experienced in getting sound castings in the larger capacities. Gases tend to be entrapped, which results in low strength and annoying leaks.  Of course, one way to reduce metal sections without sacrificing strength is to design in ribs and bosses.  Another approach to the porosity problem has been to operate the machine under a vacuum.  This process is now being developed.

The surface quality is dependent on that of the mold.  Parts made from new or repolished dies may have a surface roughness of 24 microinches.  The high surface finish available means that, in most cases, coatings such as chrome plating, anodizing, and painting may be applied directly.  More recently, decorative finishes of texture as obtained by photoetching have been applied.  The technique has been used to simulate woodgrain finishes, as well as textile and leather finishes, and to obtain checkering and crosshatching.

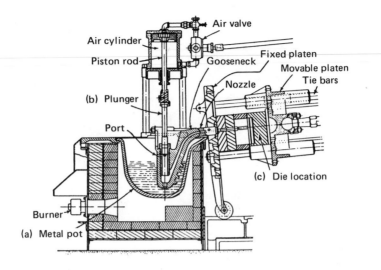

FIGURE 5-24.  *The hot-chamber die-casting process.*

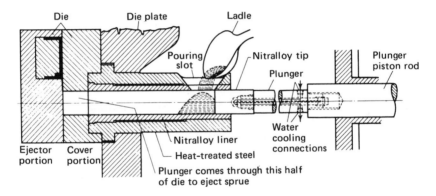

FIGURE 5-25.  *Schematic view of the cold-chamber die-casting process. Courtesy American Zinc Institute, Inc.*

The initial cost of both the die-casting machine (upwards of $40,000) and the dies (upwards of $1000 to $40,000 or more) limits this process to relatively high production quantities.  The die costs are high because of the high accuracy with which they must be made and the high polish required on the surfaces.

Perhaps the sharpest restriction on the die-casting process comes from the limited range of materials that can be cast, namely zinc-, aluminum-, magnesium-, and copper-base alloys.  Recently some progress has been made with ferrous materials. The big problem has been the die life that can be maintained.  In 1966 the General Electric Company, in an effort to create a market for the molybdenum produced by its refractory metals department, did some pioneer work in this area.  There are now two companies producing ferrous die castings on a limited basis.

FIGURE 5-26. *Examples of some of the variety of products that can be die-cast advantageously. The process has the advantage of good detail, close tolerance, good finish, and low labor cost. At the near lower left and lower right are two examples of coffee perculator wells with cast-in-place tubular heating elements. Courtesy Vacuum Die Casting Corp.*

## Low-Pressure Casting

Low-pressure casting was patented in 1910, but until recently it has not been aggressively promoted. The process may be described as a compromise between gravity-casting processes and high-pressure die casting. Because it borrows from both die casting and permanent-mold low-pressure casting, the process has a semantics problem. In Europe it is known as low-pressure permanent-mold casting and in the United States it is called low-pressure die casting.

In low-pressure casting, the metal is forced up by air pressure from a heated crucible through a feed tube or carrot to fill the die, or mold, from the bottom up (Fig. 5-27). The metal is placed under 5 to 15 lb pressure. It may take half a minute to fill the mold, so any trapped gas has time to escape. The part cools from the top down and directional solidification occurs, which is the same phenomenon that yields good part strength in gravity castings which cool from the bottom up. After the metal has solidified in the mold, the air pressure is cut off and the feed-tube metal returns to the crucible. The upper half of the mold is lifted up with the casting

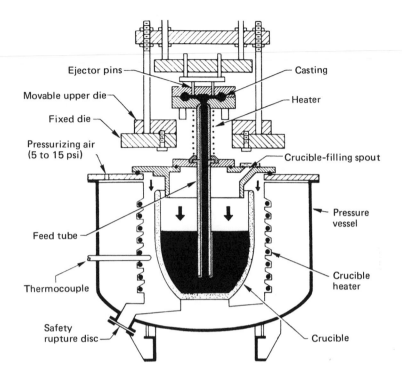

FIGURE 5-27. *In low-pressure casting, pressure is put on the molten metal of the crucible so that it flows up the feed tube to the mold. Courtesy* Machine Design.

trapped in it. After tilting or moving laterally, ejector pins drop the casting in a catcher mechanism. The whole operation is highly automated and a single operator can handle two machines.

***Advantages and Limitations.*** The closely controlled temperature of the automated low-pressure machines produces better grain structure. Oxides and impurities float on top of the crucible and are not disturbed by the low-pressure air. By maintaining pressure while the casting cools, shrinkage is fed automatically and more accurately than with risers in conventional casting. High-pressure die castings have a porosity problem which is nonexistent in the low-pressure system. Because of the pressure system, the minimum wall thickness can be reduced and the transition from heavy to light sections is more easily made. Production rates are good. As an example, 750-lb (340.19-kg) locomotive wheels can be poured at the rate of one wheel per minute. The majority of low-pressure castings are no larger than 40 lb (18.14 kg); however, some major automobile manufacturers in both the United States and Japan use the low-pressure casting system to produce aluminum engine blocks.

The economic advantages of low-pressure casting are that machines cost about one-third less than high-pressure machines and the dies are about one-third less expensive. Dies are usually cast iron, but sometimes copper, aluminum, or even plaster is used. Die life is longer than for die casting because of the low operating pressure.

Low-pressure dies are simpler in that risers can be eliminated and the runners are less complicated, but ejector pins have to be incorporated.

Tolerances must be wider for low-pressure casting than for die casting because a refractory coating is necessary inside the mold. This coating varies in thickness, depending upon its heat-transfer characteristics. Tooling costs are less than that for die casting, but more than the setup and pattern costs of sand casting. Also casting modifications are harder to make.

### Centrifugal Casting

Centrifugal casting consists of having a sand, metal, or ceramic mold which is rotated at high speeds. When the molten metal is poured into the mold it is thrown against the mold wall, where it remains until it cools and solidifies. The process is being increasingly used for such products as cast iron pipes, cylinder liners, gun barrels, pressure vessels, brake drums, gears, and fly wheels. The metals used include almost all castable alloys. Most dental tooth caps are made by a combined lost-wax process and centrifugal casting.

*Advantages and Limitations.* Because of the relatively fast cooling time, centrifugal castings have a fine grain size. There is a tendency for the lighter nonmetallic inclusions, slag particles, and dross to segregate towards the inner radius of the casting (Fig. 5-28), where it can be easily removed by machining. Due to the high purity of the outer skin, centrifugally cast pipes have a high resistance to atmospheric corrosion. Shown at Fig. 5-28(*b*) is a schematic sketch of how a pipe would be centrifugally cast in a horizontal mold.

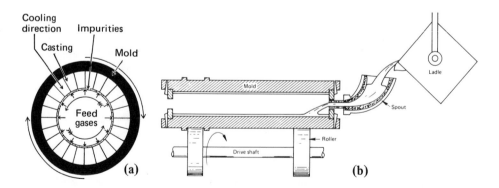

FIGURE 5-28. *The principle of centrifugal casting is to produce high-grade metal by throwing the heavier metal outward and forcing the impurities to congregate inward (a). Shown at (b) is a schematic of how a horizontal-mold centrifugal casting is made. Courtesy Janney Cylinder Co.*

Axial sections

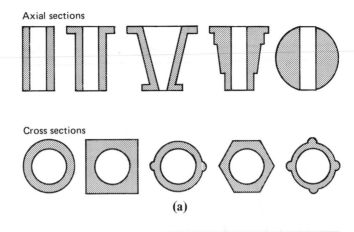

Cross sections

(a)

FIGURE 5-29. *Some examples of common cross-sectional shapes that can be made by centrifugal casting. Wall thickness is easily varied. Outside diameter may range from several inches to several feet. Also shown is a centrifugal casting used in a liquid-air converter. It is made to very close tolerances. Courtesy Sandusky Foundry and Machine Co.*

(b)

Characteristically, centrifugal castings have mechanical properties that are between those of static castings and forgings. As an example, a 3560-T6 aluminum alloy centrifugally cast has a tensile strength of 48 ksi (331.0 MPa), a yield strength of 34 ksi (234.4 MPa), and elongation of 8%. Statically cast, it has 33 ksi (227.5 MPa) tensile strength, 24 ksi (165.5 MPa) yield strength, and 3.5% elongation, a gain of about 30%.

Centrifugal casting has the advantage of being able to produce a wide variety of diameters, lengths, and wall thicknesses. The maximum castable size varies from foundry to foundry but current limits are approximately as follows: lengths of 400 in. (1016 cm), outside diameters of 68 in. (172.72 cm), and total weight about 80,000 lbs. (36,287 kg). Shown in Fig. 5-29 are common cross-sectional shapes that can be centrifugally cast. The inner surface is round but the outside can vary considerably.

Two dissimilar metal laminates can be made. By adding brazing flux to a solidified, but still spinning, casting, a dissimilar metal can be introduced, as for example a lead-rich surface to cast bearings. This method is also used to conserve more expensive metals and to provide corrosion resistance.

### Continuous Casting

Continuous casting is a relatively new process, although it was first patented by Sir Henry Bessemer in 1846. It is used to produce blooms, billets, slabs, and tubing directly from the molten metal. The process can be done vertically or horizontally,

301

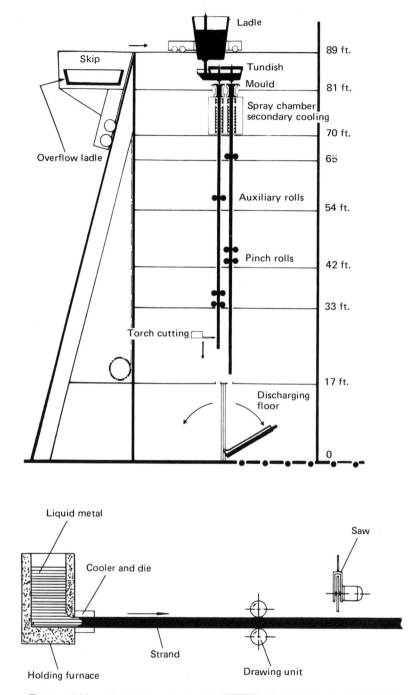

FIGURE 5-30. *The continuous-casting process may be done vertically or horizontally.* *Courtesy* Journal of Metals.

as shown in Fig. 5-30. The process is applied to copper and copper alloys, aluminum, steel, and gray and alloy-type cast irons.

The molten metal is poured into a pouring box referred to as a *tundish*. From there it flows at a controlled rate into a water-cooled mold or, if special shapes are to be made, into a graphite die as shown in Fig. 5-31. The mold is rapidly oscillated to facilitate movement. Solidification occurs within 20 to 30 sec. The solidified shape is withdrawn inch by inch by rolls which grip the section as it emerges.

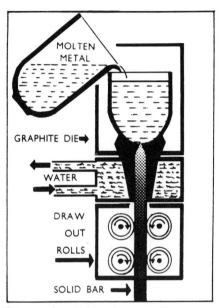

FIGURE 5-31. *A graphite die is used in the continuous-casting process to produce the desired shapes.*

Diagram of typical continuous-casting process.

Hollow rods or thick-walled tubing are made by placing a graphite core centrally in the die to a depth below the level at which solidification is complete. Contraction during solidification pulls the metal away from the die and core walls. Dies retain their dimensions for several weeks of operation, depending upon the tolerances required. Convenient lengths are cut with a torch or a circular saw.

*Advantages and Limitations.* Continuous casting can provide longer continuous lengths than can otherwise be obtained. The process eliminates some of the hot- and cold-rolling operations required in conventional production. The graphite dies used are relatively inexpensive, allowing the process to be used in small or large quantities. Graphite dies used to produce the shapes shown in Fig. 5-32 may range from $25 to $100. Because of the rapid cooling of the metal, the properties are considerably improved over sand casting.

In the case of steel, the high pouring temperature can make severe demands on the materials and design of the mold.  Also, the slow solidification rate of steel and the high casting speed make it necessary to have long dies.  This increases the chance for bulging of the cast shape due to the deep liquid core.  The general restrictions on nonferrous continuous castings are: minimum inside diameter for round stock, $\frac{7}{16}$ in. (1.111 cm), and for other shapes, 1 in. (2.54 cm); maximum outside dimensions, $9\frac{1}{4}$ in. (23.5 cm); minimum section thickness, $\frac{5}{32}$ in. (0.396 cm).

Shown in Table 5-2 is a review of the principle casting processes and many of their salient points.  Of interest here, and not discussed in the text material, are the comparative tooling costs, lead times, secondary operations and ordering quantities.

The comparative tooling cost will quickly point up which processes may be feasible for a few parts and those where only large quantities will be practical.

FIGURE 5-32. *Some of the many shapes that can be poured by the continuous-casting process.  Courtesy American Melting and Refining Co.*

By lead time is meant the amount of time required to produce the part once the design is submitted.  You will note the time is given for both sample parts and a production run.  Usually, minor changes are required in the sample parts and, when approved, additional patterns are made for the production run.

Secondary trimming operations consist of parting line smoothing and gate removal.

Ordering quantities are based on tooling costs and cycle time.  Sand castings have low tooling costs.  However, the comparatively long cycle time limits the quantities produced for a given time period.

TABLE 5-2. *Comparison of Casting Processes*

| Method | Description | Metals | Size Range | Tolerances | Comparative Tooling Cost, Units |
|---|---|---|---|---|---|
| Sand casting | Tempered sand is packed onto wood or metal pattern halves, removed from the pattern, assembled with or without cores, and metal is poured into the resultant cavities. | Most all castable metals | All sizes | $\pm\frac{1}{32}$″ up to 3 inches; $\pm\frac{3}{64}$″ from 3 inches to 6 inches; $\pm\frac{1}{16}$″ above 6 inches. Across parting line add $\pm.055$ for average casting | $2\frac{1}{2}$ |
| Permanent mold Casting | Molten metal is gravity poured into cast iron molds, coated with ceramic mold wash. Cores can be metal, sand, sand, shell, or other | Aluminum, some brass, bronze and cast iron | Aluminum: Usually $\frac{1}{2}$ lb. to 100 lb. Iron and Copper Base: 50 pounds | Aluminum: Basic $\pm.015$ to 1″. Add $\pm.002$ for each additional inch. Across parting line add $\pm.020$ for average casting | 15 |
| Investment Casting | Metal mold makes wax or plastic replica. These are sprayed, then surrounded with investment material, baked out, and metal poured in resultant cavity | Most all castable metals | Fraction of an ounce to 100 lb. | $\pm.005$″ inch | 4 |
| Plaster mold Casting | Plaster slurry is poured onto pattern halves, allowed to set. The mold is then removed from the pattern, baked, assembled, and the metal is poured into the resultant cavity | Aluminum, brass, bronze, zinc, beryllium copper | Normally up to 500 square inch area | One side of parting line $\pm.005$″ up to 2 inches. Over 2 inches add $\pm.002$″ per inch. Across parting line $\pm.010$″. Allow for parting line shift of .015″ | $3\frac{1}{2}$ |
| Die Casting | Molten metal is injected, under pressure, into hardened steel dies | Aluminum, zinc, magnesium and limited brass | Not normally over 3 feet square | $\pm.0015$″ per inch. Not less than $\pm.002$″ on any one dimension. Additional .010″ (minimum) on dimensions affected by parting line | 24 |

TABLE 5-2—*continued*

| Method | Surface Finish (RMS) | Lead Time Normal | Design Freedom 1 Lowest 5 Highest | Minimum Draft Requirement | Normal Minimum Section Thickness | Secondary Trimming Operations | Ordering Quantities |
|---|---|---|---|---|---|---|---|
| Sand Casting | Nonferrous 150–350 Ferrous 300–700 | Samples: 2–6 weeks Production: 2–4 weeks (after approval) | 4 | 1° to 5° | Nonferrous: $\frac{1}{8}"$ to $\frac{1}{4}"$ Ferrous: $\frac{1}{4}"$ to $\frac{3}{8}"$ | Grind for gate removal. Hand sand parting line | All quantities |
| Permanent Mold Casting | Aluminum: 150–200 Iron: 200–350 Copper Base: 125 | Samples: 6–14 weeks Production: 2–3 weeks (after approval) | 3 | Aluminum: 2° min Iron: External—3° Internal—7° Copper Base: External—0° Internal—$\frac{1}{2}$° to 3° | Aluminum: 9/64 for average areas Iron: 7/32 for average areas Copper base: 0.030" to .080" | Grind for gate removal | Usually 500 lbs and up |
| Investment Casting | 63–125 | Samples: 3–8 weeks Production: 3–7 weeks (after approval) | 5 | None | Aluminum: .030" Beryllium copper: .030" Stainless steel: .060" Carbon steel: .090" | Grind for gate removal | Aluminum: Usually under 2,000 lbs Other metals: all quantities |
| Plaster Mold Casting | 63–125 | Samples: 2–6 weeks Production: 2–4 weeks (after approval) | 3–4 | External: 0° to $\frac{1}{2}$° Internal: $\frac{1}{2}$° to 2° | .070" | Grind for gate removal Hand sand parting line | Aluminum: Usually under 20.0 lbs Copper base: all quantities |
| Die Casting | 32–90 | Samples: 8–16 weeks Production 2–4 weeks (after approval) | 2–3 | Aluminum: 1° to 3° Zinc: $\frac{1}{2}$° to 2° | Aluminum: 0.65" for average areas .035" for average areas | Die trim for flash and gate removal | Usually 2.0 lbs and up |

## CASTING PROCESS CONSIDERATIONS

There is much more to casting than selecting a process and making the appropriate pattern. During the past decade, research and production experiences have provided scientific principles for better casting techniques. Important considerations are: the rate at which a mold cavity is filled, gate placement, riser design, the use of chill blocks, and padding.

*Filling the Mold Cavity.* The velocity with which the molten metal fills the mold is determined by the cross-sectional area of the gating system and the mold-pouring rate. Too slow a mold-pouring rate means solidification before filling some parts, allowing surface oxidation. Too high a pouring rate caused by too large a gating system causes sand inclusions by erosion, particularly in green-sand molding, and turbulence. The minimum cross section in the gating system is called a *choke*. In the strict sense, the choke is the section in the gating system where the cross-sectional area times the potential linear velocity is at a minimum. When the gating system is choked at the bottom of the sprue, it is called a nonpressurized system. This system is somewhat less reliable than a pressurized system in which the choke is at the gate.

The first metal in the pouring basin and down the sprue usually has some turbulence which carries slag into the runner. To avoid slag in the casting, the runner should extend past the last gate to trap the initial slag (Fig. 5-33). By the time the gates become operative, the liquid level should be high enough so that no slag can enter the casting cavity. The runner should be laid out to minimize turbulence, that

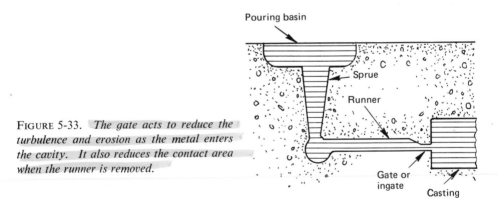

FIGURE 5-33. *The gate acts to reduce the turbulence and erosion as the metal enters the cavity. It also reduces the contact area when the runner is removed.*

is, it should be as straight and as smooth as possible. The gate that was shown in Fig. 5-2 is made to enter the cavity at the parting line. Gating arrangements may also be made at the top or bottom of the cavity. The parting line gate is the easiest for the patternmaker to make; however, the metal drops into the cavity, which may cause some erosion of the sand and some turbulence of the metal. In the case of nonferrous metals, this drop aggravates the dross and entraps air in the metal.

Top-gating is used for simple designs in gray iron but not for nonferrous alloys, since excessive dross would be formed by the agitation.

Bottom-gating provides a smooth flow of metal into the mold. However, it does have the disadvantage of an unfavorable temperature gradient. It cools as it rises, with the result of having cold metal in the riser and hot metal at the gate.

***Risers.*** Risers are designed and placed so as to ensure filling the cavity during solidification. They also act to relieve gas pressure in the mold and to reduce pressure on the lifting surfaces of the mold. The volume of metal in the riser should be sufficient to retain heat long enough to feed the shrinkage cavity and to equalize the temperature in the mold, avoiding casting strains.

The riser requirements vary with the type of metal being poured. Gray cast iron, for example, needs less feeding than some alloys because a period of graphitization occurs during the final stages of solidification, which causes an expansion that tends to counteract the metal shrinkage. Many nonferrous metals require elaborate feeding systems to obtain sound castings. Shown in Fig. 5-34 are two riser designs for the same casting.

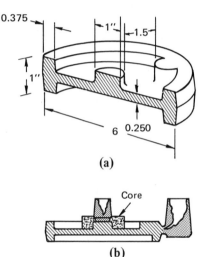

FIGURE 5-34. *Risers provide a volume of metal to feed the bosses shown at (a). Two individual risers may be used as shown at (b), or a single riser as shown at (c) with the addition of a feed pad. Courtesy of* Machine Design.

Risers are placed near the heavy sections of the casting. The volume or feed-metal requirement for gray and ductile iron can be determined as shown in Fig. 5-35. The feed metal must be located above the highest point of the casting.

***Chill Blocks.*** Chill blocks are metal blocks placed in the mold for localized heat dissipation. They may be placed at an intersection or joint where there is a comparatively large volume of metal to cool, thus relieving a hot spot or maintaining a more

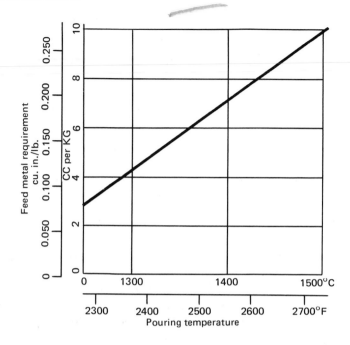

FIGURE 5-35. *A plot of the feed-metal requirement to compensate for liquid shrinkage of gray and ductile castings. As an example, if the pouring temperature is 1400°F (760°C), the FMR will be 0.190 cu in./lb of casting. Courtesy* Foundry Management and Technology.

uniform cooling rate and better microstructure. They may also be placed at the far surface of a mold, away from a riser or sprue. This will help the far end of the mold to freeze rapidly, promoting directional solidification. Chill blocks are also used at points where it is desirable to have localized hardening, as in the case of bearings or wear surfaces.

***Padding.*** Padding consists of adding to or building up a section to obtain adequate feeding of isolated sections. Shown in Fig. 5-34 were two methods of feeding the central and outside boss. Shown at (*b*) is the plan of using two risers and at (*c*) one riser with a pad. The second plan provides a yield of 45 % of the metal poured, compared to 30 % when two risers are used. The feeding distance to the central hub is $\frac{9}{2}t$, where $t$ = the thickness of the feed path. By rule of thumb, the total thickness of pad and casting (at the pad location) should not be less than one-fifth of the metal-feeding distance. This rule is not absolute but a good generalization.

***Hot Tears.*** The strength of metal near solidification or freezing point is very low. Stresses imposed at this temperature may lead to incipient flaws which later develop into well-defined cracks or *hot tears* upon cooling. Precautions must be taken to avoid stress concentrations caused by shrinkage stresses at weak points. An example

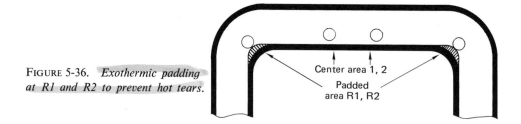

FIGURE 5-36. *Exothermic padding at R1 and R2 to prevent hot tears.*

of this type of problem is shown in Fig. 5-36. This high-quality alloy-steel casting consists of a 350-lb (158.75-kg) cradle with a 0.5-in. (12.7-cm) wall. The design is one that is conductive to hot tearing at the natural hot spots in the inside corners. Cracks occur in this radius area since this is the hottest, weakest area and is exposed to the highest stress concentration.

*Exothermic Padding.* In this case, an economical means of solving the problem is with a small amount of exothermic padding material as shown. The exothermic mixture consists of mill scale, powdered aluminum, charcoal, and refractory powder. The heat evolves as a result of the exothermic reaction and keeps the metal molten long enough so that excessive strains are not built up during final solidification and shrinkage.

Exothermic materials are also used to cover risers. The heat of the exothermic reaction keeps the riser top open to the atmospheric pressure and improves its efficiency.

## CASTING DESIGN PRINCIPLES

Casting design differs from most other fabricating design principles in that it is more intimately associated with the process itself. The effects of solidification, parting-line placement, tolerances, minimum section thickness, and draft will be examined from the design viewpoint.

### Solidification and Design

The designer of castings must be aware of how solidification is affected by or affects: part geometry, changes in cross-sectional area, mechanical properties, and heavy isolated areas.

*Solidification and Section Geometry.* After casting, and cooling has started, nucleation first takes place at the mold walls of the thinner sections. Grains form which may be dendritic, columnar, or equiaxed depending upon the cooling rate and alloys. In dendritic structures, some of the arms break off during the normal conditions of growth, thus furnishing nuclei for central, inner portions of the casting. New dendrites grow from these fragmented parts to form the equiaxed central zone. The cooling of a simple straight section presents no problem if it is not large. However,

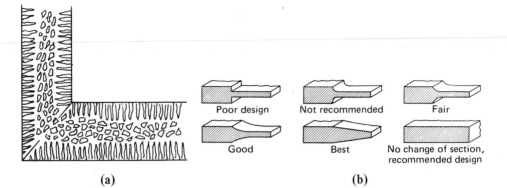

Poor design   Not recommended   Fair

Good   Best   No change of section, recommended design

(a)   (b)

FIGURE 5-37. *A weak spot is formed during solidification of a casting where two or more sections cojoin and free cooling is interrupted. A curved section avoids this problem (a). Abrupt section changes are not recommended (b). Courtesy Steel Founders' Society of America.*

when two sections cojoin or intersect, a mechanical weakness develops, as shown schematically in Fig. 5-37(*a*). Rounded sections and fillets tend to minimize these *hot spots* or solidification defects (*b*).

***Solidification and Cross-Sectional Area.*** Cooling rates are directly related to mass and surface area and may be expressed as a ratio of surface area to volume or mass. The slow cooling of a large mass with a relatively small area will produce a considerably softer material than the fast cooling of a large thin section. A means of comparing the ratio of a section to its intersection mass is that of inscribed circles (Fig. 5-38). When an inscribed circle is tangential to three major sides of a section, as shown at (1), (3), and (4) of Fig. 5-39, the cooling rate can be assumed to be equal or greater than that of a bar of the same diameter. Cooling rates may be higher than the inscribed circle indicates if small protrusions, fins, or appendages are attached to the surface.

***Solidification and Mechanical Properties.*** As discussed in Chapter 2, the solidification rate of metals has a marked influence on their mechanical properties. This is particularly noticeable in the hardness variations of heat-treatable materials. As an example, if the casting shown in Fig. 5-39 were made of malleable iron, the Brinell hardness readings may be as follows: Section 1—400 BHN, Section 3—270 BHN, and Section 4—250 BHN. The related properties are shown in Table 5-3.

***Solidification and Heavy Isolated Areas.*** Shown in Fig. 5-40 are three methods of small casting with a central hub. If the hub is solid, as shown in the top view, porosity will occur in the center of the casting because it is too large an isolated area. A step in the right direction is shown in the center section where some of the metal is removed by coring. The best solution is to make all sections as nearly uniform as possible, as shown in the bottom view.

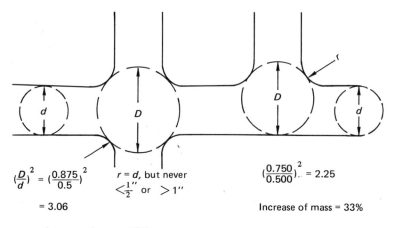

$$\left(\frac{D}{d}\right)^2 = \left(\frac{0.875}{0.5}\right)^2$$

$r = d,$ but never $< \frac{1}{2}''$ or $> 1''$

$= 3.06$

Increase of mass = 59%

$$\left(\frac{0.750}{0.500}\right)^2 = 2.25$$

Increase of mass = 33%

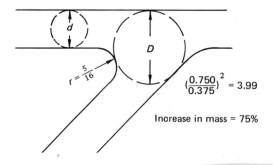

$r = \frac{5}{16}$

$$\left(\frac{0.750}{0.375}\right)^2 = 3.99$$

Increase in mass = 75%

FIGURE 5-38. *Inscribed circles are used to determine the effect of mass.*

FIGURE 5-39. *The crosshatched areas show equivalent cooling rates of various cast sections.*

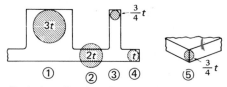

Equivalent diameter of bars which will cool at about the same rate as the particular position on the casting

**Parting-Line Placement.** With high-production molding techniques it is desirable to place the gate and feeder at the parting line. Shown in Fig. 5-41(*a*) are proper and improper parting-line placements. In the upper figure of (*a*) the parting line has been placed at the flange, causing the heavy boss at the top to become isolated from the riser, resulting in porosity. By turning the casting 90° and placing the parting line axially, the riser can feed the heavy boss directly. Parting-line placement can often

TABLE 5-3. *Properties of Ferritic and Pearlitic Malleable Irons*

|  | Structure | BHN | Tensile ksi | Yield as a % of tensile ksi | Elonga- tion %, 2 in. |
|---|---|---|---|---|---|
| Section 1 | Ferritic, maximum ductility | 200 | 80 | 69 | 10 |
| Section 3 | Pearlitic, high strength, wear resistant | 270 | 100 | 80 | 2 |
| Section 4 | Essentially pearlitic, good strength, | 250 | 80 | 75 | 3 |

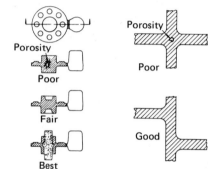

FIGURE 5-40. *Heavy isolated sections cause porosity in the casting. Courtesy Central Foundry Division, G.M.*

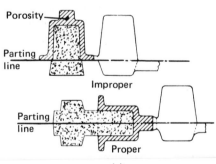

(a)

FIGURE 5-41. *Proper placement of the parting line facilitates gating and in this case helps eliminate porosity (a). Parting relocated to eliminate a core (b). Courtesy Central Foundry Division, G.M.*

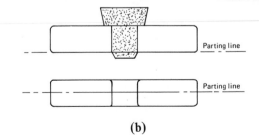

(b)

be used to eliminate a core, as shown at (*b*). Some general suggestions in regard to the parting line are:

1. Avoid putting the parting line where it may later interfere with machining or other operations.
2. Keep the parting line as even as possible to simplify pattern making and mold making, and to keep fins to a minimum.
3. Use parting lines to eliminate cores when possible.

*Tolerances.* The variety of sizes, shapes, and processes makes it difficult to give more than general guidelines as to tolerances. For high-production malleable-iron castings, the general tolerance for green-sand molding is $\pm 0.030$ in. (0.076 cm) up to 2 in. (5.08 cm) in size, with a straight-line increase in tolerance as dimension increases, as shown in Fig. 5-42. Tolerances for other processes are given in Table 5-1.

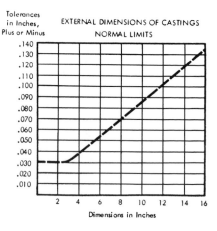

FIGURE 5-42. *General tolerances for high-production malleable castings. Courtesy Malleable Research and Development Foundation.*

*Section Thickness.* Recommended minimum section thicknesses for various casting processes are given in Table 5-1. Recommended wall thicknesses for aluminum by sand and permanent-mold casting are shown graphically in Fig. 5-43.

*Draft.* All walls perpendicular to the parting plane require draft. Normally, the casting drawing does not show draft. The standard foundry practice is to add draft to the part (Fig. 5-44). In general, green-sand molds require 2 degrees of draft. Shell molds are normally made with 1 degree of draft. Deep pockets may require from 3 to 10 degrees of draft for green sand and 1 to 2 degrees for shell molding. Draft requirements are also given in Table 5-2.

*Machining Allowances.* Drawings showing the casting prior to machining should have the finish allowance included in the dimensions. It is recommended that an undimensioned phantom line be used (Fig. 5-45) to represent surfaces after machining, thus enabling the foundry to recognize the surface for processing reference. Shown

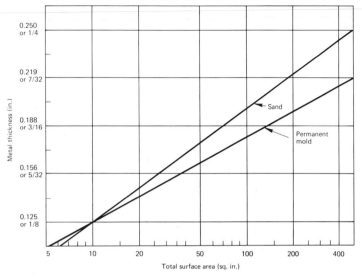

The figures in the graph below show desired minimum metal thickness that should be specified for aluminum castings of minimum complexity. When metal flow is restricted, or other factors become more important, an increase in values shown may be required.

FIGURE 5-43. *The minimum wall thickness recommended for aluminum castings. Courtesy Aluminum Association.*

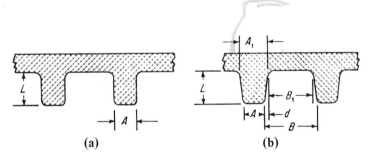

(a)                    (b)

FIGURE 5-44. *Normally a casting drawing will be made as shown at (a), standard foundry practice is to add the required draft as shown at (b).*

in Table 5-4 is the amount of metal normally added for machining aluminum and steel castings. The amount of material added relates to the overall size of the casting rather than the individual dimensions. Thus the finish allowance for any given casting is the same for all surfaces requiring machining.

## Economic Considerations

Good casting designs incorporate savings with no sacrifice of quality. The designer should consider such alternatives as multiple-cavity casting, fabrication, the use of cores, subsequent machining operations, and weight reduction.

FIGURE 5-45. *A conventional drawing showing where the finish machined surface will be.*

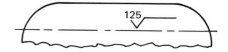

**Multiple-Cavity Casting.** Substantial savings can be realized by incorporating more than one finished part into a single casting. Molding and handling costs can be greatly reduced, since only a fraction of the number of castings need be handled. An example of multiple casting is shown in Fig. 5-46, where five bearing caps are cast together. They are machined as a unit and then cut into separate pieces.

**Fabrication.** Existing designs should be under constant evaluation. Oftentimes a part that has been fabricated by a combination of casting and welding can be made more economically by one process. Shown in Fig. 5-47 is a fabricated axle housing. The combination of casting and welding greatly simplified manufacture. Welding the heavy wall tube on each end of the housing was done by automatic submerged-arc welding at a cost of 9.36¢ per weld, exclusive of power and maintenance costs. If both welds are made simultaneously, the cost drops to 5.46¢ per weld.

Shown in Fig. 5-48 is a hub used on a rear wheel of a truck. In this case the original design called for forging and welding, as shown at (*a*). However, a reevaluation of the design proved the steel casting (*b*) to be lighter, stronger, and more economical.

**Cored Holes.** Designers are often confronted with the problem of whether to specify cored or drilled holes. The economies are best studied on an individual basis, but it is generally more economical to core larger holes, when appreciable metal can be saved and faster machining results. The universal-joint yoke shown in Fig. 5-49 was originally a solid-steel forging. A rather slow drilling operation was required to make the shaft hole. In casting, the hole was cored and was followed by a fast core-drilling operation, since a relatively small amount of metal had to be removed.

TABLE 5-4. *Machining Allowances for Aluminum and Steel Castings*

| Largest Casting Dimension (in.) | Aluminum Allowance (in.) | | | | Steel |
|---|---|---|---|---|---|
| | Sand Casting | Permanent Mold | PM with Sand Cores | PM with Shell Cores | Sand Casting |
| To 6 | 0.060 | 0.045 | 0.060 | 0.060 | $\frac{3}{16}$ |
| 6 to 12 | 0.090 | 0.060 | 0.090 | 0.060 | $\frac{3}{16}$ |
| 12 to 18 | 0.120 | 0.075 | 0.120 | 0.090 | $\frac{3}{16}$ |
| 18 to 24 | 0.150 | 0.090 | 0.180 | 0.120 | $\frac{1}{4}$ |
| Over 24 | | | Consult foundry | | |

Values shown are based on maximum build-up or safety factor for average variations in flatness, squareness, concentricity, etc., as well as linear tolerance. There are recommended minimums, and draft, when required, must be added.

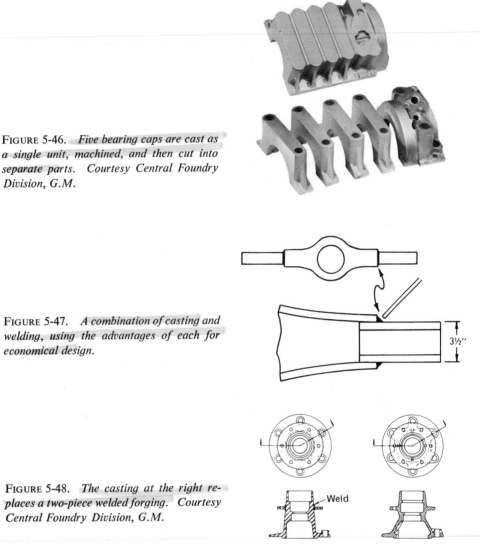

FIGURE 5-46. *Five bearing caps are cast as a single unit, machined, and then cut into separate parts. Courtesy Central Foundry Division, G.M.*

FIGURE 5-47. *A combination of casting and welding, using the advantages of each for economical design.*

FIGURE 5-48. *The casting at the right replaces a two-piece welded forging. Courtesy Central Foundry Division, G.M.*

The proper use of cores can also result in a less expensive casting. Figure 5-50 shows the original and the improved design of a truck wheel hub. The original casting could not be formed in green sand. Two cores were necessary, a *ring core* and a *body core*. By redesigning to eliminate the ribs and backdraft, the green-sand casting was made with only one core.

***Subsequent Operations.*** Good casting design, like good forging design discussed in Chapter 4, will reduce considerably the machining operations required to produce the finished product. As was shown in Fig. 4-43, six locating or tooling points are

FIGURE 5-49. *Cored holes eliminate or reduce secondary operations of drilling and boring. Courtesy Central Foundry Division, G.M.*

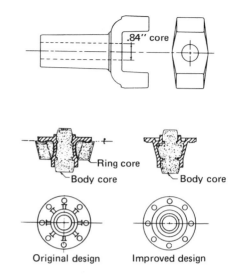

FIGURE 5-50. *The redesign of this truck-wheel hub resulted in being able to make this casting with one core instead of two. Courtesy Central Foundry Division, G.M.*

required to establish the three coordinate planes for machining. These same principles apply to castings (Fig. 5-51).

Many castings have as their principle component a round boss, as shown in Fig. 5-52. In this case, V-type locators are usually used to maintain centerline symmetry and tolerances. The V-type locator provides tooling points 1 and 2. Thus the drill hole will be concentric with the boss. The proper location of tooling point 3 is essential so that the machining of surface *P* will be in relation to the centerline of the hole.

Oftentimes it is necessary to design into the casting holding lugs or location bosses that will later be machined off. A simple example of this is shown in Fig. 5-53.

## Quality Control

Today's modern foundry is essentially a continuous chemical operation with many unique high-temperature problems. Close control of the metal poured is one of the most important functions required to ensure high-quality castings.

In large foundries, metal samples from the melting unit and holding furnaces are delivered every few minutes by a pneumatic tube system to the laboratory. In the case of ferrous materials, the samples are in the form of circular discs poured in special molds to provide a fully chilled structure. In the laboratory they are spark-tested in direct-reading vacuum spectrographs, which determine the various alloying elements except silicon and carbon. Test results are immediately sent to the control room. The control engineer can then rapidly calculate the desired weights of constituents for the next charge.

Samples of molding sand and core sand are also sent via pneumatic tube to the testing laboratory on a periodic basis. The results are returned to process control engineers and are plotted on control charts. Trends are noted and corrections made when necessary.

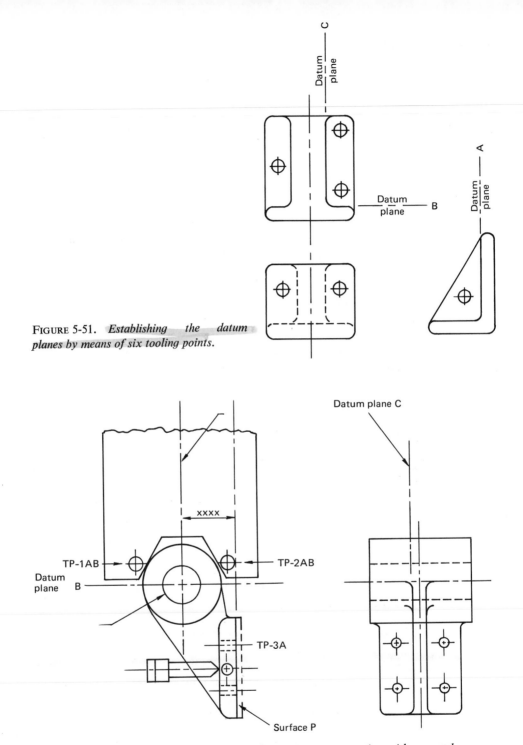

FIGURE 5-51. *Establishing the datum planes by means of six tooling points.*

FIGURE 5-52. *The location of tooling points on a casting with a round surface to maintain concentricity with the boss and alignment of the hole centerline with surface P.*

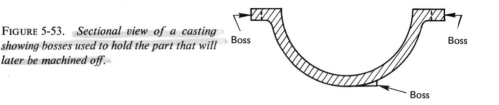

FIGURE 5-53. *Sectional view of a casting showing bosses used to hold the part that will later be machined off.*

In large, modern foundries, on-line analog and digital computers are used to control production variables. In addition to control over process variables, computers monitor plant safety, cooling-water temperatures, bearing temperatures, etc.

Although in-process control is important, it does not preclude a constant check of the final product. For example, in the manufacture of gray-iron cylinder blocks, sections are removed after each operating shift. Test specimens are surface ground and checked for hardness, graphite size, and distribution. When the hardness falls outside the specification limits, all blocks made during the suspect period are held up. Further tests show which blocks can be sent on and which scrapped.

Nondestructive testing methods such as ultrasonic testing, sonic testing, eddy current testing, and X rays are also widely used to ensure quality production.

## PLASTIC-MOLDING PROCESSES

There is a similarity between the molding of plastics and the molding of metals; and even though some of the properties are quite different, many of the same design principles are applicable.

The metal-casting process is based on heating the base material to the liquid state and then pouring or squeezing it into a mold. For most molding operations, plastics are heated to a liquid or a semifluid mass and formed in a mold under pressure. Plastics can also be poured from a liquid state into a mold using either heat or a catalyst for hardening, but this method is not used extensively except for nylon.

The basic structure of plastics was discussed in Chapter 2. You may recall that plastics are essentially giant-sized molecules that pass through a fluid to a "gel" state during the molding process. Those that form strong chemical crosslinks at this stage are termed *thermosetting*, since the process cannot be reversed. *Thermoplastics* undergo no permanent chemical change under moderate heating, and therefore the process can be repeated if a change in the product is desired. Common types of both thermoplastic and thermosetting materials and their properties are listed in Table 5-5, with a more complete listing in Appendices 5A and 5B.

Plastics may be obtained for processing in a number of different forms, as shown in Table 5-6. Also shown are the principal processing methods used and discussed in this chapter.

TABLE 5-5. *Common Plastics, Properties and Uses*

| Thermoplastic Types | Properties | Use |
|---|---|---|
| Polyethylene | Both flexible and rigid types are highly break resistant. Unaffected by food acids and household solvents. | Widely used as a coating for paper and paperboard, film packaging, disposable bags, furniture components. |
| Polypropylene | Good dimensional stability, poor weatherability, attacked by oils and organic solvents. | Appliance housings, refrigerator door liners, wall tiles, battery cases, toys, insulators, ornaments. |
| Polystyrene | Rigid, clear, transparent materials with good tensile strength. Impact type has good combination of rigidity and toughness. | Wide application in packaging, appliances, housewares, toys, and furniture. |
| Polyvinyl Chloride | Chemical and abrasion resistance, good weatherability, easily pigmented. | Sound records, blown bottles, rigid sheeting, upholstery, hose, flooring. |
| Acrylic | Outstanding optical properties. Inert to most chemicals. Excellent mechanical properties for short-term loading. | Aircraft and architectural domes, signs, machinery housings, optical components. |
| Flourocarbon (Teflon) | Low coefficient of friction. Stable up to about 400°F, good chemical resistance and weatherability. | Nonlubricated bearings, chemical-resistant pipe linings, seals, diaphrams, valves, linings of cookware. |
| Polyamide (Nylon) | High tensile strength, good impact strength, and high abrasion resistance, good chemical and electrical properties. | Gears, bushings, cams, tubing, rollers, bearings, combs, and miscellaneous household items. |

| Thermosetting Types | Properties | Use |
|---|---|---|
| Epoxy | Good flexibility, mechanical strength, adhesion, low shrinkage | Laminated plastics, printed-circuit boards, cast tooling, paints, and adhesives. |
| Phenolic | Strong, rigid, durable, and hard. Resists heat up to 300°F. Good weather resistance. | Laminates, foundry and grinding-wheel resins, gas and crude-oil pipes, appliance housings, hot-melt adhesives. |
| Silicone | Exceptional heat resistance (up to about 700°F), good moisture and chemical resistance. | Arc and thermal barriers, panel boards, terminal strips, electrical components. |
| Urethane | Rigid foam has good weatherability resistance to rot, vermin, and fungus. Flexible and semiflexible types. | Arm rests, packaging for delicate instruments, rigid types used as insulation, simulated wood beams, surfboards, and buoys. |

TABLE 5-6. *Plastic Processing Methods*

| Liquid | Powders, Granules, and Beads | Powders or Liquids | Sheet | Reinforced Plastics |
|---|---|---|---|---|
| | **Raw Material Form** | | | |
| Casting | Compression molding | Rotomolding | Thermoforming | Hand lay-up |
| | Transfer molding | | Vacuum | Vacuum bag |
| | Injection molding | | plug assist | Pressure bag |
| | Blow molding | | Pressure | Filament |
| | Extrusion molding | | plug assist | winding |
| | Expandable-bead molding | | Matched mold | |
| | Foam molding | | | |

## Casting

Casting of plastics is not used extensively except in the case of epoxies and nylons. Cast epoxies are used for making molds, dies, and bushings, as well as encapsulating electrical coils, transformers, switchgear, etc. Cast nylon, unlike other polymers, is processed directly from a liquid monomer. A catalyst, and other additives, is heated and poured into the mold at atmospheric pressure. The additives in the compound serve to "kick over" the monomer into a polymer form while it is in the mold. The nylon monomer is unique in that it is processed by direct polymerization; other thermoplastic resins cannot be processed in this manner.

*Advantages and Limitations.* The primary advantage of the casting process is the accurately contoured product, such as the gears shown in Fig. 5-54, that can be made several inches thick, yet free of voids.

Cast-nylon parts are available in two formulations, termed "Mono Cast" and "Nylatron," which contain molybdenum disulfide for improved lubrication. These materials provide an excellent balance of wear and corrosion resistance, self lubrication, impact strength, and nongalling characteristics. They are lightweight, about one-seventh that of steel, and have excellent sound- and vibration-damping qualities. Because of the slower cooling cycle in the casting process than in molding, surfaces are harder (more crystalline) and have better wear resistance. Mold cost is quite low compared to that of injection molds. A typical mold for a cast part costs about

FIGURE 5-54. *Cast nylon outlasted previous phenolic counterparts by two to one and eliminated paper contamination from lubricant throw-off. Courtesy the Polymer Corp.*

$500 to $1000. The cost of the material is competitive to that of stainless steels and brasses. Typical applications range from guides and cams for soft handling of bottles to pedestal liners for locomotive service. Other applications include large gears for all types of machinery, elevator buckets for handling foundry sand and other abrasive materials, bearings for steel-mill roll-out tables, and roll covers for paper-making machinery.

Nylons, being polymers, cannot be used where the continuous temperature exposure is 275°F (135°C) or more. Nylons absorb water on the outer surface and should not be used for applications requiring continuous immersion in water. Nylons are attacked by strong acids and bases but they are little affected by solvents, detergents, or weak acids. These materials can be machined, but care must be taken not to leave abrupt surface changes since they are notch-sensitive.

## Compression Molding

The first method used to form plastics in production quantities was by compression molding. The method has been improved and now provides several mold variations: flash type, positive, and semipositive, as shown in Fig. 5-55. The most common type used in automatic molding presses is the semipositive type. The loading chamber ensures the correct amount of material and enables accurate control of the flash or excess materials that squeeze out at the die-closing line.

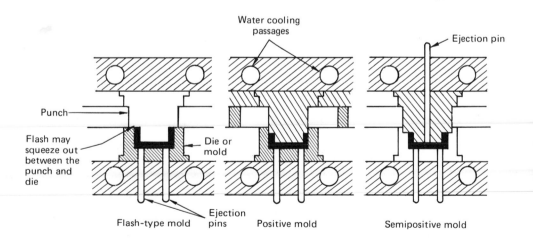

FIGURE 5-55. *Three types of compression molds. The positive mold must have a carefully measured amount of plastic to be sure it fills the cavity but not have an excess since the mold is made to prevent flash from forming.*

The mold must be heated, which is most often done with electric cartridge-type heaters placed as close to the cavities as possible but in a way that will produce uniform heat throughout the mold surface. The die area must also be cooled, which is done by circulating water through passages, as shown in the sketch.

Clamping is done by the action of the press and is usually a straight ram hydraulic arrangement or a toggle arrangement activated by either air or hydraulic cylinders. Pressure varies depending on the material being molded and the configuration of the part. As a rule of thumb, parts that are 1 in. or less deep will require a pressure of 3000 psi (20.68 MPa) as calculated on the *projected area* within the outside wall of the loading chamber. Projected area is the area of the part which is parallel to the parting line of the mold. Parts deeper than 1 in. will require 3000 psi plus 700 psi (317.5 kg/in.$^2$) for each additional inch or part of an inch.

Parts are removed from the mold with the aid of ejector pins. The shape of the part will determine whether ejection should be done from either the top or bottom of the mold half. In manual operation, the operator may remove the part from the mold. In semiautomatic operation, the operator will remove the part from the ejector pins. In a fully automatic operation, the part is made to fall free after ejection or a combing device may be added to provide a positive release.

Mold temperature varies with conditions but is generally around 320 to 380°F (160–193°C). The hotter the mold can be run, the shorter will be the cure time. However, too hot a mold may cause the plastic to set up before it has had a chance to fill all parts of the mold. As a rough guide, the cure time for phenolics (thermosetting plastics) is 60 sec for $\frac{1}{8}$ in. (.317 cm) of cross section. However, regardless of thickness, no further curing will take place after 3 min. This is because the outside layer will cure but the inside, if too thick, will still be in semigranular form.

*Transfer Molding.* Transfer molding is a variation of compression molding. It involves the use of a plunger and heated pot separate from the mold cavity. The material is loaded into the pot and, when hot, is transferred under pressure through runners into the closed mold cavity (Fig. 5-56). A three-part mold is used as shown at (*c*). As the mold is opened, the sprue and a residual disc of material is separated from the part and is removed manually. The molded part is raised from the cavity by ejection pins.

A newer development for regular compression molding and transfer molding is a screw-feeding arrangement, as shown in Fig. 5-57. The screw moves slugs of material, preheated to 290°F (143°C), to where they are dropped into the bottom of the transfer pot and then moved into the mold. This arrangement allows the cure time to be cut 25 % over that of using cold powder. Production capacity per mold has increased 400 % with this method and heavy-walled sections are cured throughout.

*Advantages and Limitations.* Compression-molded parts do not have high stresses, since the material is not forced through gates. The process is comparatively simple, which makes initial mold costs low. Transfer molding is particularly good for making complex parts where inserts are needed. In ordinary compression molding, fragile inserts may be damaged or shift before the resin reaches its plastic state. However, transfer molds are more complex and more costly to build.

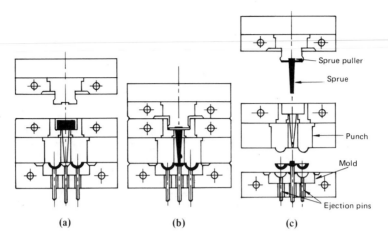

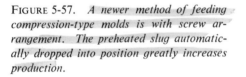

FIGURE 5-56. *The three stages of transfer molding, preform in place (a), liquification and transfer into mold (b), ejection (c).*

FIGURE 5-57. *A newer method of feeding compression-type molds is with screw arrangement. The preheated slug automatically dropped into position greatly increases production.*

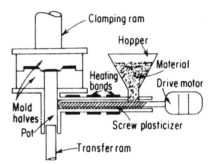

## Injection Molding

The greatest quantity of plastic parts are made by injection molding. The process consists of feeding a plastic compound in powdered or granular form from a hopper through metering and melting stages and then injecting it into a mold. After a brief cooling period, the mold is opened and the solidified part ejected. In most cases, it is ready for immediate use.

Several methods are used to force or inject the melted plastic into the mold. The most commonly used system in the larger machines is the in-line reciprocating screw, as shown in Fig. 5-58. The screw acts as a combination injection and plasticizing unit. As the plastic is fed to the rotating screw, it passes through three zones as shown: feed, compression, and metering. After the feed zone, the screw-flight depth is gradually reduced, forcing the plastic to compress. The work is converted to heat by shearing the plastic, making it a semifluid mass. In the metering zone, additional heat is applied by conduction from the barrel surface. As the chamber in front of the screw becomes filled, it forces the screw back, tripping a limit switch which activates a hydraulic cylinder that forces the screw forward and injects the

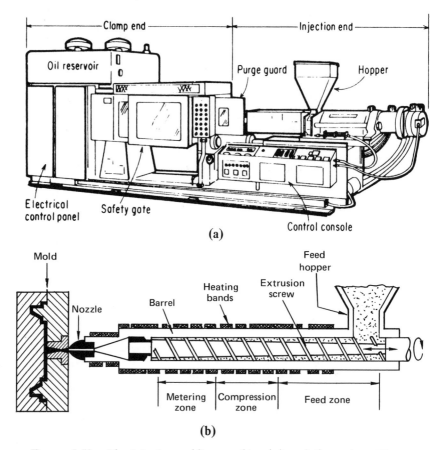

FIGURE 5-58. *The injection-molding machine (a) and the reciprocating-screw injection system (b).*

fluid plastic into the closed mold. An antiflowback valve prevents plastic under pressure from escaping back into the screw flights.

The clamping force that a machine is capable of exerting is part of the size designation and is measured in tons. A rule of thumb can be used to determine the tonnage required for a particular job. It is based on two tons of clamp force per square inch of projected area. If the flow pattern is difficult and the parts are thin, this may have to go to three or four tons.

Many reciprocating-screw machines are capable of handling thermosetting plastic materials. Previously these materials were handled by compression or transfer molding. Theremosetting materials cure or polymerize in the mold and are ejected hot in the range of 375 to 410°F (190–210°C). Thermoplastic parts must be allowed to cool in the mold in order to remove them without distortion. Thus thermosetting cycles can be faster. Of course the mold must be heated rather than chilled, as with thermoplastics.

*Advantages and Limitations.* Injection molding is one of the most economical methods of mass-producing a single item. The parts taken from the mold, in most cases, are finished products. There is very little waste of thermoplastic plastics since the runners and sprues can be ground up and reused.

Injection molds must be carefully designed by one who specializes in this type of work. They are expensive if quantities are not large. However, the quantities usually associated with injection molds makes their actual per-piece cost relatively insignificant. As a very rough figure, a one-cavity simple mold may run about $1000. Two-and three-cavity simple molds may run $1800 and $2600, respectively. Amortizing over a run of 100,000 parts would make the mold cost per piece $0.01, $0.018, or $0.026.

## Rotomolding

In rotational molding, the product is formed inside a closed mold that is rotated about two axes as heat is applied. Liquid or powdered thermoplastic or thermosetting plastic is poured into the mold, either manually or automatically. The mold halves are then clamped shut.

The loaded mold is then rolled into an oven where it spins on both axes. Heat causes the powdered materials to become semiliquid or the liquid materials to gel. (Permissible temperature ranges are considerably greater than for injection molding.) As the mold rotates, the material is distributed on mold-cavity walls solely by gravitational force; centrifugal force is not used. The mold and rotating mechanism is shown in the sketch, Fig. 5-59.

When the parts have been properly formed, the molds are cooled by a combination of cold-water spray, forced cold air, and/or cool liquid circulating inside the

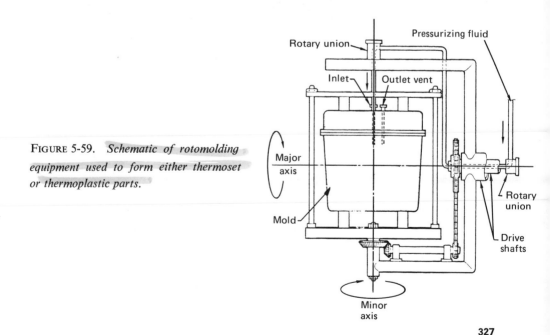

FIGURE 5-59. *Schematic of rotomolding equipment used to form either thermoset or thermoplastic parts.*

mold. The mold continues to rotate during the cooling cycle. Unloading is usually a simple manual operation, although forced air or mechanical methods are sometimes used to eject the part.

*Advantages and Limitations.* Rotomolding can be used to make parts in sizes and shapes that would be difficult by any other process. For example, the process is now being used to handle rectangular shapes $5.5 \times 5.5 \times 12$ ft ($1.65 \times 1.65 \times 3.6$ m) and cylindrical parts 15 ft (4.5 m) in diameter with 8 ft (2.4 m) straight walls. Boat hulls can be molded up to 14 ft (4.2 m) long. More common production equipment is for parts than can be contained in a 5-ft (1.5-m) sphere. Part weights may range from 0.2 oz (6 ml) to 250 lb (112.5 kg) and volumes from 1 cu in. to 1000 gallons ($16.39 \text{ cm}^3$ to 3800 l) capacity. Shown in Fig. 5-60 is a double-walled phonograph case rotomolded of high-density polyethylene. It provides shock and vibration cushioning advantages over the previous vinyl-covered wood version, and at lower cost and less weight.

FIGURE 5-60. *A double-walled phonograph case made of high-density polyethylene. It provides the added benefit of shock and vibration cushioning to the basic cost and weight reduction advantages over a prior vinyl-covered wood version. Courtesy Arvin Industries and USI Chemicals Co.*

Wall thicknesses may range from 0.030 to 0.500 in. (0.76 to 12.7 mm) by simply adjusting the amount of charge and cycle time. Heat-insulating plugs can be used to reduce the wall thickness or to eliminate the wall entirely in a given area. Wastebaskets, for example, are made by using an asbestos disc at the end that is to be open to prevent plastic formation.

Rotational molds are relatively inexpensive when compared to molds for injection and blow-molding. Some examples of typical mold costs are: a welded-steel prototype mold for making 25-gallon (95-l) refuse cans costs about $375; a single-cavity prototype mold of cast aluminum for an 8 in. (20.32 cm) diameter light globe costs about $550; a 10-cavity production mold for the same light globe in cast aluminum costs about $3500.

As with molds for other processes, the precision required is an important factor in the cost of the mold. A novelty item which has wide tolerance range can use a much cheaper mold than precision instruments. Rotomolding can often use an existing metal part as the prototype mold. For example, a metal lawn spreader became the prototype mold for a plastic lawn spreader.

In short runs, rotomolding is generally less expensive than most other molding processes. As production runs lengthen, other processes become competitive and must be carefully evaluated. However, long runs do not automatically eliminate rotomolding, as shown in the following example.

A 1-gal (3.8-l) container and lid had a unit cost for 10,000 rotomolded parts of $0.362; by injection mold, $0.942; by blow mold, $0.389. For a production run of 100,000, the costs were: rotational, $0.286; injection, $0.468; blow molding, $0.176. At a run of 1 million cans, rotational costs were $0.203; injection, $0.167; and blow molding, $0.130. For 10 million parts, the cost picture changes because of lower machine and mold factors. Here rotomolding has an advantage, at $0.093, against injection at $0.127 and blow molding at $0.108.

Changing colors during a production run is comparatively easy, whereas in injection and blow molding it is costly and time-consuming. All that is required in rotomolding is changing the color of the charge. Conceivably it can be changed with each cycle.

The basic limitations of rotomolding are: the part must be hollow and typical tolerances are $\pm 5\%$. The cycle time is slow because of the loading and cumbersome part removal unless the process is highly automated. It is difficult to get varying wall thicknesses within a part, so parts are often designed with reinforcing ribs to stiffen flat areas. The range of moldable materials is not as broad as it is for other molding processes. Basic raw materials run about 5¢ per pound higher than for injection molding, but much of this is saved because of the 100% raw-material usage without regrinding.

### Expandable-Bead Molding

The expandable-bead process consists of placing small beads of polystyrene along with a small amount of blowing agent in a tumbling container. The polystyrene beads soften under heat, which allows a blowing agent to expand them. When the beads reach a given size, depending on the density required, they are quickly cooled. This solidifies the polystyrene in its larger, foamed size. The expanded beads are then placed in a mold (usually aluminum, Fig. 5-61) until it is completely filled. The entrance port is then closed and steam is injected, resoftening the beads and fusing them together. After cooling, the finished, expanded part is removed from the mold.

*Advantages and Limitations.* Comparatively simple molds can be used to make rather large parts such as ice chests, water jugs, water float toys, shipping containers, and display figures. Expanded-bead products have an excellent strength-to-weight ratio and provide good insulation and shock-absorbing qualities.

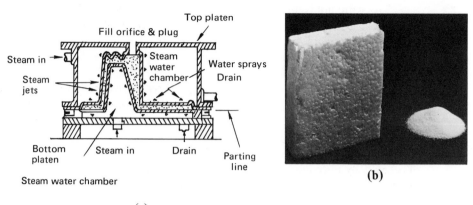

FIGURE 5-61. *Expandable-bead molding setup (a). The polystyrene beads and an expanded section (b).*

## Structural Foam Molding

Foam molding is a relatively new process that provides an economical method of making structural members with high rigidity. Basically, the process corresponds to injection molding. For the foamed effect, inert gas is dispersed throughout the polymer melt. This is done either by inducing the gas directly into the melt at the mold, called the *low-pressure system,* or by preblending the resin with a chemical blowing agent, which in the presence of heat releases inert gas. The latter is termed

FIGURE 5-62. *The chemical blowing agent is dry-blended with the resin prior to charging it into the hopper. During the processing the blowing agent decomposes with heat and releases gas to the melt. Courtesy General Electric, Plastics Division.*

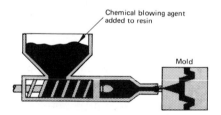

the *chemical system.* The gas–resin mixture is shot under pressure into the cavity, as shown schematically in Fig. 5-62. The gas expands within the material, filling the mold and creating the internal cellular structure. An integral skin is formed around the rigid foamed core (Fig. 5-63).

*Advantages and Limitations.* Structural foams have been used in the past as wood replacements. The detailed woodgrain finish was accurately reproduced on many molded parts such as mirror frames, coffee tables, stereo cabinets, etc. Now, however, structural foams are being used not only to replace wood and metal parts, but to make distinct structural improvements. As an example, an increase of 25 % in

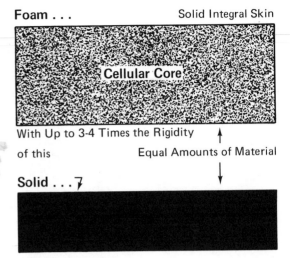

Foam . . .                                    Solid Integral Skin

Cellular Core

With Up to 3-4 Times the Rigidity

of this                          Equal Amounts of Material

Solid . . .

FIGURE 5-63. *A cross section of a foamed sample shows a rigid cellular core within a solid integral " skin." The equivalent weight of a solid injection-molded part would be thinner with $\frac{1}{3}$ to $\frac{1}{4}$ the rigidity. Courtesy General Electric, Plastics Division.*

wall thickness can result in more than twice the rigidity for a solid plastic part of equal weight. Foam offers 1.2 times the rigidity of an equal weight of aluminum and 2.4 times that of an equal weight of steel, as shown in Table 5-7.

Machines are available that can shoot well over 50 lbs (22.5 kg) per cycle, using the low-pressure system. Thus large, strong parts can be made, such as storage cabinets and office-equipment housings as shown in Fig. 5-64. Perhaps one of the most exciting uses of structural foam is an entirely new bicycle. The total weight of the new design is under 20 lbs (9 kg). It will be an all one-piece frame, or unibody construction, eliminating joints. Increased strength also results from the redesigned fork and frame members. The color is impregnated into the plastic and the serial number is imbedded and cannot be filed off. An iridescent material, to make the unit safer, is now in the developmental state.

Structural foams can be painted, woodgrained, metallized, or printed. Any type of self-tapping screws may be used for assembly. Inserts may be molded in place or installed by ultrasonic insertion. Parts may also be bonded ultrasonically or with adhesives.

Foamed parts, especially those molded in low-pressure systems, show a swirled pattern on their surface. These parts generally require painting.

Because of the thicker wall sections, foam-molding cycles are generally longer than injection-molding cycles.

TABLE 5-7. *Strength-to-Weight Ratio Comparison of Structural Foams & Metals*

| Ratio | Structural Foam | Aluminum | Steel | Zinc |
|---|---|---|---|---|
| Stiffness-to-weight | 110 | 90 | 45 | 30 |
| Flexural strength-to-weight | 133 | 45 | 20 | 5 |

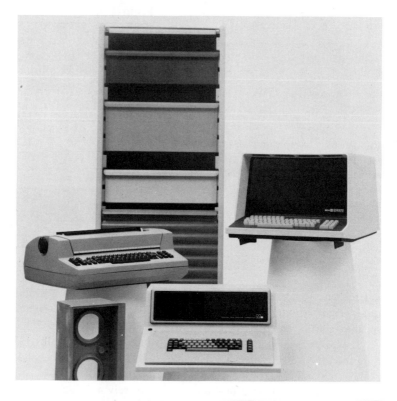

FIGURE 5-64. *Structural foams are used for a wide variety of products, particularly in appliances, business machines, air-moving systems, automotive products, industrial equipment, and furniture.* Courtesy General Electric, Plastics Division.

### Extruding

Plastic extrusion is similar to metal extrusion in that a hot material (plastic melt) is forced through a die having an opening shaped to produce a desired cross section. Depending on the material used, the barrel is heated anywhere from 250 to 600°F (121–316°C) to transform the thermoplastic from a solid to a melt. At the end of the extruder barrel is a screen pack for filtering and building back pressure (Fig. 5-65). A breaker plate serves to hold the screen pack in place and straighten the helical flow as it comes off the screen.

A special modification of the extrusion process is extruding a thin plastic film on a flat substrate such as paper, cloth, or metal. The plastic is immediately and uniformly bonded to the substrate by pressure rolls.

Metal strips, wires, or roll-formed shapes can be incorporated into the extrusion process by means of an offset die, as shown in Fig. 5-66.

Dual extrusion is a process of combining different materials in a single extrusion operation. Thus rigid and flexible vinyl extrusions that offer advantages in both assembly and sealing can be made. The rigid portion is used for shape retention or attachment and the flexible portion for sealing, cushioning, or absorbing impact, as shown in Fig. 5-67.

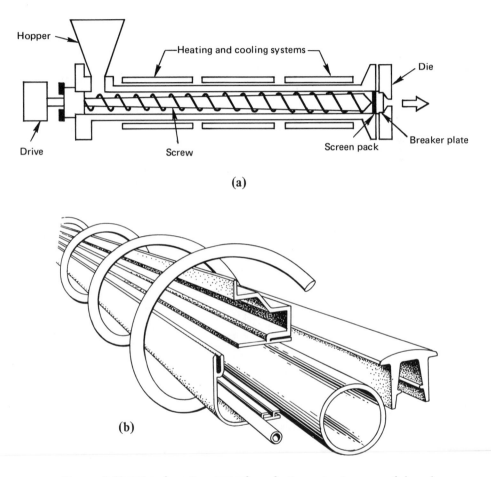

(a)

(b)

FIGURE 5-65. *A schematic view of a plastics extrusion press (a) and examples of some plastic extrusions (b).*

FIGURE 5-66. *A modification of the extrusion process allows a metal strip to be fed through the die and completely imbedded in the plastic.*

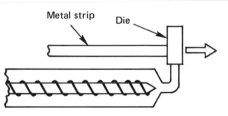

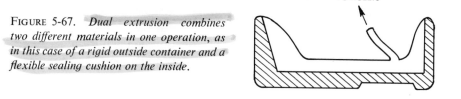

FIGURE 5-67. *Dual extrusion combines two different materials in one operation, as in this case of a rigid outside container and a flexible sealing cushion on the inside.*

## Blow Molding

Blow molding is a process that is used extensively to make bottles and other lightweight, hollow, plastic parts. Two methods are used: injection blow molding and extrusion blow molding.

Injection blow molding, as shown in Fig. 5-68, is used primarily for small containers up to about 8-oz capacity (224 g). The parison or tube is formed by the injection of plasticized material around a hollow mandrel. While the material is still molten and still on the mandrel, it is transferred into the blowing mold where air is used to inflate it. Usually no finishing operations are required. Accurate threads may be formed at the neck.

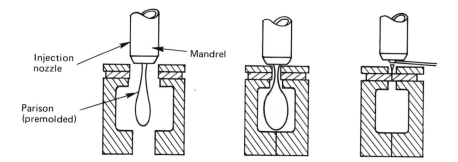

FIGURE 5-68. *In injection blow molding the parison or tube is formed around the mandrel and while it is still hot it is transferred to the blowing mold where air is injected.*

Shown in Fig. 5-69 is the extrusion-type blow molding. A molten-plastic pipe (parison) is inflated under relatively low pressure inside a split-metal mold. The die closes, pinching the end and closing the top around the mandrel. Air enters through the mandrel and inflates the tube until the plastic contacts the cold wall, where it solidifies. The mold opens, the bottle is ejected, and the tailpiece falls off.

The molds for simple, symmetrical, round shapes can be machined easily. Parts that require extensive ribbing are more easily made by casting. The parting line of the mold should be placed where it will not be aesthetically offensive. The maximum depth of undercut, as in indented handle grips, will be dependent on the flexibility of the material and the amount of wall thinning that can be tolerated.

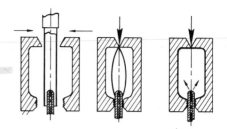

FIGURE 5-69. *The successive steps in extrusion blow molding.*

## THERMOFORMING

Thermoforming refers to heating a sheet of plastic material until it becomes soft and pliable and then forming it either by vacuum, air pressure, or between matching mold halves. There are variations of these processes, but the essential elements of each are shown in Fig. 5-70.

In vacuum forming, the plastic sheet is clamped in place and heated. The vacuum beneath the sheet causes atmospheric pressure to push the sheet down into the mold. Areas of the sheet reaching the mold last are usually the thinnest.

Molds may be made out of wood, metal, plastic, plasters, etc. The heated sheet may be clamped over the female die or draped over the male form. As the mold closes, the part is formed. Excellent reproduction of mold details, including lettering and grained surfaces, may be obtained.

In pressure-bubble, plug-assist vacuum forming, the sheet is clamped in place across the female cavity and heated. Air is introduced into the cavity and blows the sheet upward into a bubble. A photocell is often used to control the height and signals the plug to plunge into the plastic sheet. When the plug reaches its lowest position a vacuum draws the sheet against the mold, or in some cases air pressure is used.

Perhaps the most attractive feature of thermoforming is the relatively inexpensive tooling. The process is used in making a wide variety of everyday products such as disposable drinking cups, coffee-cream containers, margarine tubs, meat trays, egg cartons, and picnic plates. Output rates may reach as high as 1500 parts per minute on multiple-cavity dies. Most parts use the plug-assist pressure-forming method. Sheet thicknesses normally used range from 0.025 to 0.500 in. (0.63–12.70 mm).

### Reinforced-Plastic Molding

Reinforced plastics generally refer to polymers that have been reinforced with glass fibers. Other materials used are asbestos, sisal, synthetic fibers such as nylon and polyvinyl chloride, cotton fibers, paper, and metal filaments. Relatively new on the scene are carbon, graphite, and boron fibers, as discussed in Chapter 2 under composite materials. These high-strength composites using graphite fibers are now commercially available with moduli of 50,000,000 psi (344,700,000 MPa) and tensile strengths of about 300,000 psi (2,068,000 MPa). They are as strong as or stronger than the best alloy steels and are lighter than aluminum.

Vacuum Forming

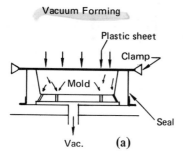

Vac.　(a)

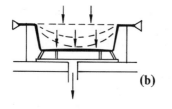

(b)

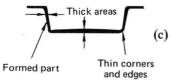

(c)

Pressure-bubble
Plug-assist Vaccum Forming

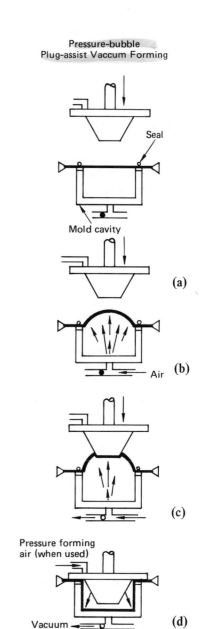

Matched-mold Forming

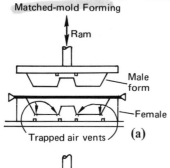

(a)

(b)

Formed part　(c)

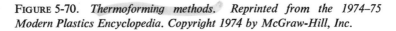

FIGURE 5-70. *Thermoforming methods.* *Reprinted from the 1974–75 Modern Plastics Encyclopedia. Copyright 1974 by McGraw-Hill, Inc.*

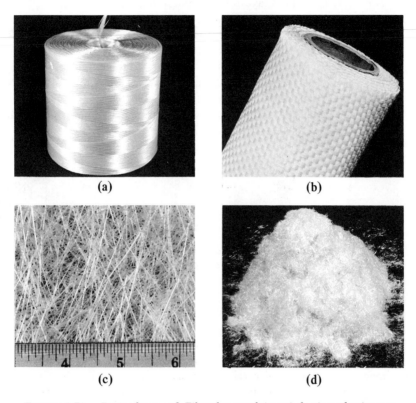

(a)

(b)

(c)

(d)

FIGURE 5-71. *Basic forms of Fiberglas used in reinforcing plastics are continuous strand (a), woven fabric (b), reinforcing mat used for medium-strength structures (c), and chopped short lengths of Fiberglas used to facilitate molding complex structures (d). Courtesy Owens-Corning Fiberglas Corporation.*

Fibrous-glass reinforcement is available in many different forms: as continuous strand, woven fabric, reinforcing mat, and chopped, short lengths (Fig. 5–71). Glass-fiber fabrics are used where rather high strengths are required. They can be of tight or open weave and have almost equal amounts of glass in each direction, or have most of their fibers running in one direction. Woven cloth is used for heavy reinforcement and mat for light and medium reinforcement. The chopped strands facilitate molding in complex structures.

### Molding Reinforced Thermosets

Size, volume, shape, strength requirements, and economics of the product will determine which of the following molding processes will be used: hand lay-up, matched metal dies, spray-up, or filament winding.

*Hand Lay-up.* Hand lay-up is the oldest and simplest method of forming reinforced thermosets. Only one mold is used and it is often referred to as the *open-mold* method. The mat or fabric reinforcement is cut and fitted to the mold and is then saturated with resin, applied by hand with the aid of a roller or brush, or sprayed on with a special gun. The resin is carefully worked into the reinforcement to ensure complete "wet-out" and to eliminate any air pockets. It is not possible to work to close tolerances, particularly in maintaining wall thicknesses.

An extension of the hand lay-up method is vacuum-bag, pressure-bag, or autoclave molding, as shown in Fig. 5-72.

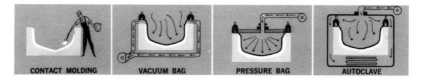

FIGURE 5-72. *Vacuum bag, pressure bag, and autoclave are extensions of the hand lay-up or contact molding. Courtesy Owens-Corning Fiberglas Corporation.*

*Matched-Die Molding.* Matched dies overcome the objections of the hand lay-up method and provide accuracy, high production, and smooth surfaces both inside and out. The quantity required, however, must be much higher to warrant the die cost. The dies are usually made of steel for small parts and cast iron for larger parts. They may also be made out of reinforced plastics or wood for short runs. Pressures used range from 200 to 300 psi (14–21 kg per cm$^2$) and temperatures from 230 to 260°F (110–127°C).

Some design limitations of this process are: there can be no undercuts, no absolutely square corners, no molded-in openings on vertical walls, and generally there must be a draft of at least 1°.

*Spray-Up.* The spray-up method is much like the hand lay-up method used in the open mold, except that glass strands or rovings are fed through a chopper that cuts them to predetermined lengths and projects them into a stream of catalyzed resin so that both resin and fiber are deposited simultaneously on the mold surface (Fig. 5-73).

*Filament Winding.* In filament winding, impregnated, fibrous, reinforcing strands or rovings are wound in continuous lengths on a suitable mandrel (Fig. 5-74). The resin used is usually an epoxy. The outstanding property of filament-wound structures is their high strength-to-weight ratio, better than that of steel or titanium.

### Forged-Plastic Parts

The forging of plastic materials is a relatively new process. It was developed to shape materials that are difficult or impossible to mold and is used as a low-cost solution for small production runs.

FIGURE 5-73. *A spray gun being used to apply both fiberglass and resin. Courtesy Spraybilt.*

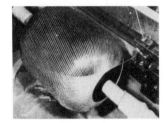

FIGURE 5-74. *Filament winding used in making a lightweight pressure vessel.*

The forging operation starts with a blank or billet of the required shape and volume for the finished part, as shown in Fig. 5-75. The blank is heated to a preselected temperature and transferred to the forging dies, which are closed to deform the work material and fill the die cavity. The dies are kept in the closed position for a definite period of time, usually 15 to 60 sec. When the dies are opened, the finished forging is removed. Since forging involves deformation of the work material in heated and softened condition, the process is applicable to thermoplastics only.

*Advantages and Limitations.* The most important advantage of the plastic-forging process is the capability of producing thick parts with relatively abrupt changes in section. Assuming such parts could be made (to an acceptable quality) by injection molding, the time element would be a factor. Injection-molded parts of this type may require about 5 min, whereas forging could be done in as little as 30 sec.

In injection molding, where the polymer is forced into the mold cavity in a fluid state and then allowed to solidify, even cooling is extremely difficult to achieve if thin and thick sections are adjacent to each other. As a result, stresses tend to distort the part. The problem is further complicated in crystalline polymers by the large reduction in volume on solidification (mold shrinkage), causing surface depressions and internal porosity. In forging, the solid material is deformed to the shape of the die cavity. The temperature is much lower than in injection molding, and no phase change is involved. Thus forging has few limitations on section thickness, which results in greater design freedom.

Extruded Blank

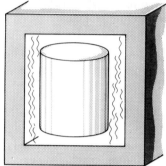

Blank is Heated

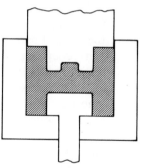

Forging Operation

FIGURE 5-75. *The sequence used to forge plastic parts. Forging dies are sometimes equipped with heating and cooling means (not shown).*

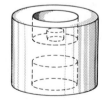

Finished Part

Only simple tooling is required for forging. The tooling pressure required is less than 1000 psi (6.89 MPa) as compared to 8000 to 30,000 psi (55.16 to 206.8 MPa) for injection molding. The temperature required is just below the melting point, 250 to 300°F (121–149°C), compared to injection molding, which requires 300 to 600°F (149–316°C). Forging dies do not require cooling passages except to increase production. There is also no need for gates or runners. Forging can be done on a small hydraulic press. For example, the parts shown in Fig. 5-76 can be made on a 5-ton hydraulic press. An injection-molding machine would cost five times as much and require from five to ten times more floor space.

Parts that are deformed significantly in the forging process have improved mechanical properties due to molecular orientation. Increases in tensile modulus, impact strength, and abrasion resistance have been observed on some materials.

As yet the number of materials used for forging is limited. Materials used commercially are polypropylene, high-density polyethylene, and ultrahigh-molecular-weight polyethylene. Amorphous polymers such as polyvinylchloride (PVC) and

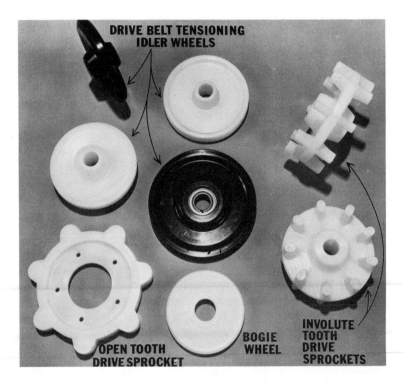

FIGURE 5-76. *Snowmobile wheels and sprockets made by plastic forging. The material is ultrahigh molecular-weight polyethylene, a polymer that cannot be injection molded. The parts have the sheen and finish of molded parts and the properties are equal or superior to machined parts. Courtesy Glasrock Products, Inc.*

acrylonitrile butadiene styrene (ABS) are less satisfactory because of their rubberlike action. Their high elastic recovery rate makes it difficult to maintain dimensional accuracy.

The forging process is generally suited to small-quantity production but it can be automated for high production rates.

Machining is the only other process besides forging and injection molding that is capable of producing complex, solid parts. Since no special tooling is involved, machining is more cost effective when the quantity is on the order of a few hundred. Machining produces scrap that may not be reusable, so with an increase in production, volume forging is favored.

### Joining Plastics

Present methods of joining plastics include hot-air welding, friction welding, heated metal plate, solvent welding, and the high-energy methods of dielectric, magnetic, and ultrasonic joining. All of these methods are applicable to thermoplastics, but only adhesive bonding and polymerization apply with adequate success to thermosetting materials. Some of these joining processes are discussed briefly.

*Hot-Air Welding.* The analogy of hot-air welding to metal welding is close, except the temperatures are much lower and there is no need to apply a flux as a shielding agent. The various types of hot-air welding are tack, hand, and high-speed hand, as shown in Fig. 5-77.

Unlike in metal welding, the rod can be made in a triangular shape so that one pass will fill large gaps. If a round rod is used, several passes have to be used, as shown in Fig. 5-78. Plastic welding can be used to produce a joint of 100 % base material tensile strength on certain types of thermoplastics. The strength depends on temperature, the amount of pressure on the rod, preparation of the material, and the skill of the operator. The temperature range should be 400 to 600°F (204–316°C) on the plastic. The filler rod should be the same as the base material.

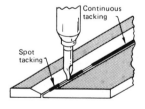
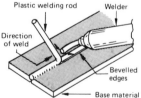
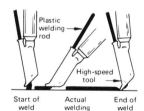
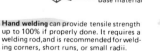

**Tack welding** is a shallow fusion of the mating surfaces of the base material and produces very little tensile strength. It requires no welding rod. Spot tacking may be sufficient, but continuous tacking may be required for added strength.

**Hand welding** can provide tensile strength up to 100% if properly done. It requires a welding rod, and is recommended for welding corners, short runs, or small radii.

**High-speed** welding is a variation of hand welding in which the rod is fed automatically. Rod is fed into the preheating tube in the welding tool by the motion of the welder as it is pulled along the joint. Welding speed ranges from 24 to 60 ipm.

FIGURE 5-77. *Hand welding of plastics using the hot-air method. Courtesy* Machine Design.

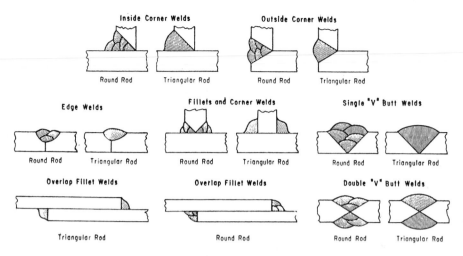

FIGURE 5-78. *Types of welds made with the round and triangular rod in plastics. Triangular rod makes it possible to make most welds in one pass.*

*Solvent Welding.* Solvent welding is a widely used plastics-joining process. The solvent may be applied to the joint by dipping, spraying, and brushing. The most common method is to hold the solvent-soaked pad against the surfaces. The edges dissolve, producing tacky surfaces which are then brought together, moved slightly to obtain mixing, and held in contact, in correct relationship, to dry and harden. Drying time ranges from a few minutes to hours.

*Friction Welding.* Spin welding is limited to round parts which can be rotated while in contact to generate heat. The frictional heat melts the interfaces and when the motion is stopped, the parts fuse together. Spinning speeds of 10 to 15 fps at the interface and pressures of 20 to 200 psi (1.4 to 14 kg per cm$^2$) must be matched to the material to produce just the right amount of heat. Melting and fusion can occur in as little as 0.25 sec. Maintaining pressure for 1 to 2 sec after braking normally completes the cycle. Shown in Fig. 5-79 is a sketch of the tooling used to spin an aerosol-bottle bottom in place.

*Ultrasonic Welding.* Ultrasonic welding is done with high-frequency sound waves (about 20 kHz). It works best on those thermoplastics that have relatively low melting points and moduli of elasticity greater than 200,000 psi (1379 MPa). A sonic tip, or tips, placed at the approximate center of the part transmits sound waves to the joint, where they are converted to frictional heat (Fig. 5-80). The operation takes 3 to 4 sec. Joint strength is close to base-material strength.

*Welding With Radio Frequencies.* Dielectric welding with certain FCC-approved frequencies is limited to plastics having a fair degree of polarity, such as saran, polyvinyl chloride, or styrene acrylonitrile. Welding is simple and rapid, but only those surfaces to be joined can be exposed to the rotating electrical field.

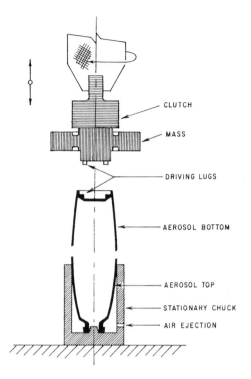

CLUTCH

MASS

DRIVING LUGS

AEROSOL BOTTOM

AEROSOL TOP

STATIONARY CHUCK

AIR EJECTION

FIGURE 5-79. *A spin weld made with an inertia tool running at a speed of 4000 rpm. The flywheel or mass comes to a stop within 0.5 sec, losing all its kinetic energy in friction and thus causing melt to form in the joint. After the flywheel stops the melt solidifies within a fraction of a second. Courtesy E. I. du pont de Nemours & Co.*

Induction welding, although applicable to virtually all thermoplastics, is specialized. Its prime use is for rapid butt welding of pipe. A metal ring inserted in the joint is induction heated and melts the surrounding plastic. A slight pressure completes the joint. The cost is high but the joint quality is good.

### Plastic Design Principles

As stated at the beginning of this section on plastics, there are many similarities in molding metals and plastics. These similarities carry over into design. The more knowledge the designer has of the manufacturing process, the better his designs will be. Often a designer will take a molder into his confidence so that the early stages of design and development will produce the best results. General considerations are material selection, process selection, tolerances, standards, and specifications.

*Material Selection.* Material selection will be based on the properties sought, such as impact strength, tensile and flexural strengths, maximum and minimum resistance temperatures, weathering qualities, flame and chemical resistance, electrical resistance, wear and scratch resistance, etc.

*Process Selection.* The part design, material, production requirements, and final price are the main factors to consider in choosing a process. Often more than one

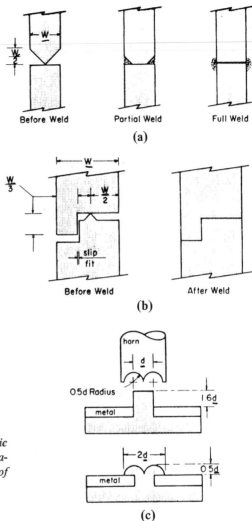

FIGURE 5-80. *Joint designs for ultrasonic welding, (a) and (b). Joint design for ultrasonic staking (c). Courtesy Society of Manufacturing Engineers.*

method can be used to produce equally acceptable parts but the final decision will be made on the basis of total cost, which includes design, material, production, tooling, etc.

The cost per part is not difficult to determine when the production rate and cost of the raw material are known. In injection molding, the cost is usually figured per thousand parts. The sprues and runners can be reground, but the cost of regrinding must not be neglected. A simple formula that can be used to determine the cost of the material is as follows:

$$C_m = \frac{c(W - R) + Rg}{N}$$

*where:*      $c$ = cost of molding power/lb
                 $W$ = total weight of part 1000 shots, in lb
                 $R$ = total weight of runners/1000 shots
                 $g$ = grinding cost/lb
                 $N$ = number of cavities in the mold

The total cost per part can then be obtained by determining the operating cost per minute and dividing by the cycles per minute.

$$\text{Operating cost/part} = \frac{C_p + C_l + C_o}{N \times 60 \times M}$$

*where:*      $C_p$ = press charge in \$/hr
                 $C_l$ = labor charge in \$/hr
                 $C_o$ = overhead cost in \$/hr
                 $M$ = cycles/minute

*Example:*   Find the cost of making 5000 television knobs that weigh $\frac{1}{8}$ oz, under the following conditions:

> Sprue and runner weight/shot = $\frac{1}{8}$ oz.
> Powder cost = 40 ¢/lb
> Grinding cost = 5 ¢/lb
> Cycle time = 20 sec
> Press charge including overhead = \$10.00/hr
> Labor rate = \$3.00/hr
> Cavities = 8

*Solution:*

$$\text{Material cost, } C_m = \frac{\$0.40(1 - 0.125)1000 + 1000(0.125)\$0.05}{8}$$

$$= \$45.53, \text{ or } \$0.044 \text{ per knob}$$

$$\text{Operating cost/part} = \frac{\$18.00 + 3.00}{8 \times 60 \times 6} = \frac{\$21.00}{2880} = \$0.0073$$

Therefore the cost per knob = 0.044 + \$0.0073, or \$0.0513.

***Tolerances and Wall Thickness.***   As with metal casting material, shrinkages must be considered in order to maintain close tolerances. In some plastics there are two shrinkages: a *mold shrinkage* that occurs upon solidification, and is shrinkage that occurs in some materials after 24 hours, *after shrinkage*. For example, the mold shrinkage for a melamine plastic may be 0.007 to 0.009 in./in. and the after shrinkage 0.006 to 0.008 in./in. Thus a total shrinkage of 0.013 to 0.017 in./in. can be figured. Minimum wall thickness recommendations for compression, transfer, and injection molding are shown in Table 5-8.

TABLE 5-8. *Recommended Minimum Wall Thicknesses for Compression, Transfer, and Injection Molding*

| Depth (in.) | Wall Thickness |
|---|---|
| Up to 2 | 0.060 |
| 2 to 4 | 0.060 to 0.080 |
| 4 to 8 | 0.080 to 0.100 |

***Standards and Specifications.*** Final drawings should be made using recognized symbols and standards. Such a system is provided in Military Standard MIL STD-8B and in Engineering and Technical Standards of Plastics and Custom Molders as found in Section 3 of the *Modern Plastics Encyclopedia.*

***General Design Considerations.*** Design points regarding castings and forgings apply, such as avoiding undercuts, heavy cross sections, adjacent thick and thin sections, sharp corners, and square holes where possible. Gate and parting-line placement are also important. These points and many others are well illustrated in several issues of *Modern Plastics Encyclopedia.*

## POWDER METALLURGY

Powder metallurgy is also a process that utilizes molds, or dies, to control the shape of the finished product. In this case finely divided metal powders are pressed into a steel or carbide die of the desired shape. The pressing is done at room temperature. The powdered particles interlock and have sufficient strength to make what is known as a "green compact." This compact is removed from the die and heated or *sintered* at a high temperature. This takes place in a neutral or reducing atmosphere at some temperature near, but below, the melting point of the metal. An exchange of atoms between the individual particles welds them together. The result is a more or less porous piece of metal (high densities can be achieved) of the approximate size and shape of the die cavity. A schematic of the basic process is shown in Fig. 5-81, including the secondary operation of coining and infiltration.

A considerable knowledge of techniques is necessary in such matters as the selection of the right combinations of metal powders, die design, sintering conditions, and any secondary operations.

### Metal-Powder Production

Metal powders are produced in several ways, such as atomization, reduction, and electrolysis.

***Atomization.*** In this process, the molten metal is forced through a nozzle, where it is atomized with compressed air and, upon solidification, a wide range of particle sizes and shapes are formed. The fineness of the powder depends on the pressure

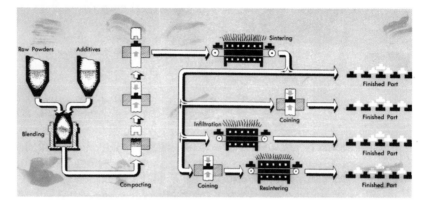

FIGURE 5-81. *Schematic diagram of the powder-metallurgy process. Courtesy* Machine Design.

of gas and the fluidity and rate of flow of the metal. The process was first applied to aluminum, zinc, lead, and tin, but now includes steels. Pure iron powder is produced by first atomizing molten cast iron. The process produces fine particles but with oxidized surfaces. The carbon and oxygen contents of the powder are controlled however, and by heating to 1740°F (950°C) decarburization occurs by mutual reaction.

*Reduction.* The reducing agents used for common metals are carbon monoxide and hydrogen. The resulting sponge metal is then crushed and ground to produce a powder.

*Electrolysis.* The electrolysis method of producing metal powders is similar to electroplating. In this process, the metal plates are placed in a tank of electrolyte. The plates act as anodes, while other metal plates are placed in the electrolyte to act as cathodes. High-amperage current produces a powdery deposit on the cathodes. After a buildup, the cathode plates are removed from the tank, scraped off, and the deposit pulverized to produce powder of the desired grain size. An annealing process follows pulverization to remove work-hardening effects of scraping.

### Superalloy-Powder Production

Demand for high-performance components operating at elevated temperatures has resulted in the development of nickel-type superalloys. The production of these alloy powders is done by three major methods: inert-gas atomization, rotating electrode, and soluble-gas atomization, as shown schematically in Fig. 5-82. The inert-gas process is usually done with argon-gas jets. The rotating-electrode process is done in a closed chamber filled with an inert gas. The molten droplets that are thrown off from the rotating electrode are collected to make up the powder. In the soluble-gas process, molten metal is pressurized in a vacuum and the liquid erupts into particles.

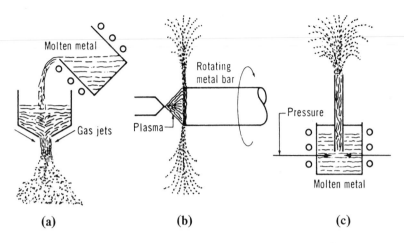

FIGURE 5-82. *The three major methods of producing superalloy powder are: (a) inert-gas atomization, (b) rotating electrode, (c) soluble-gas atomization. Courtesy Climax Molybdenum.*

## The Powder-Metallurgy Process

The concept of making parts from powdered metals is simple and straightforward, however the techniques employed can be very sophisticated, requiring a high level of technical competence and a substantial investment in capital equipment. The process consists of three basic steps: blending, compacting, and sintering.

*Blending.* Blending refers to mixing the metal powder to obtain the desired properties. Lubricants are added to the powder to reduce friction between the grains as they are being compacted as well as to reduce die wear. The blending may be done wet or dry. Wet mixing has the advantage of reducing dust and the danger of explosion which is present in some finely divided powders.

*Compacting.* Metal powders present problems of internal friction. Therefore, when placed in a die, the pressures are not distributed uniformly throughout the compact. Both hardness and density decrease as the distance from the punch increases. To rectify this condition, punches are used at both ends (Fig. 5-83) and a lubricant is mixed with the metal powder.

The use of lubricants improves the density, minimizes the load required, and increases die life. However, lubricants can create problems in feeding the powder into the die and in lubricant reaction. Lubricants must be driven off by slow heating before sintering.

For high-volume production, tungsten carbide is used as the die material. Although the cost is higher, it will outwear the normally used tool steels by a ratio of about 10 : 1. Some carbide dies can be used to produce a million parts before the tolerances are exceeded. High pressures, sometimes in excess of 50 tons per square

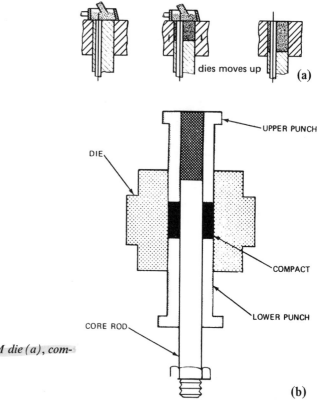

FIGURE 5-83.   *Filling the PM die (a), com-pacting a bushing (b).*

inch, (689.5 MPa) are used to cold bond the powder particles.   Some powders such as brass, bronze, and aluminum compact at relatively low pressures ranging from 12 to 25 tsi (10.8 to 22.5 metric tons per square inch).   A 95 % theoretical density can be achieved with aluminum powders at 25 tsi.

***Isostatic Pressing.***   To overcome the difficulty of variations in density due to compacting in just two directions, a method, *isostatic pressing*, has been developed to produce equal pressure on all sides of the powder.   The process can be achieved by two methods: wet bag and dry bag (Fig. 5-84).

In the wet-bag process, a plastic or rubber mold is filled with powder and is tightly sealed after being evacuated with a hypodermic-needle vacuum connection. The bag is then placed in a vessel filled with oil and pressure is increased to the desired value, usually 26 to 44 tsi (23.4 to 39.6 metric tons per square inch).

The dry-bag process employs a rigid rubber die which can maintain its shape. The bag is filled with powder and is closed on top by a punch of the press.   The pressure is then applied to the rubber die.

Due to an absence of die-wall friction, and the application of uniform pressure, isostatic compacts show almost equal density and strength in all directions, regardless of size.   Some variation in density exists from the surface to the center of the compact,

## WET-BAG TOOLING

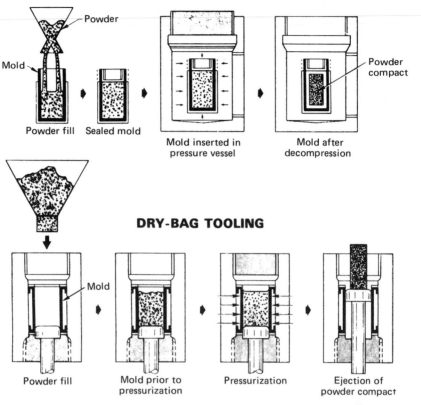

Powder

Mold

Powder fill     Sealed mold

Mold inserted in
pressure vessel

Powder
compact

Mold after
decompression

## DRY-BAG TOOLING

Mold

Powder fill

Mold prior to
pressurization

Pressurization

Ejection of
powder compact

FIGURE 5-84.  *Isostatic powder pressing may be done wet or dry as shown.*
*The wet-bag process has a separate container or mold that is loaded outside*
*of the press, whereas in the dry-bag process the mold is an integral part of*
*the press.  The wet-bag process is better suited to forming intricate parts,*
*but the dry-bag process is simpler and better suited to high production.*
*Courtesy* Machine Design.

however.  The maximum pressure in die pressing is limited for a given press by the
surface area of the compact.  For example, 10,000 psi (68.95 MPa) pressure (typical
for iron powder) requires a 500-ton (450 metric tons) press for a shape with 10 sq in.
(25.4 sq cm) of surface area.  Such equipment is rarely found in die-pressing facilities
because of the large production runs required to amortize the equipment and tooling
costs.  In isostatic compaction, pressure is not affected by part size.  Consequently,
large powder-metal parts have been made, such as blocks 24 × 32 × 26 in (60 ×
80 × 63 cm), a 1700-lb (770-kg) tungsten crucible, and an oxide insulator 24 in. (60 cm)
in diameter and 70 in. (175 cm) long.

The chief limitation of isostatic pressing is that compacts with close dimensional
tolerances cannot be made.

*Sintering.* After compacting, the *green compact*, as it is often referred to, is considered fragile but can be handled. To achieve strength and hardness, it must be sintered in a vacuum or in an atmospherically controlled furnace. Time and temperature in the furnace are closely regulated to achieve the desired properties.

*Spark Sintering.* A newer approach is that of spark sintering. In this process the powders are hot pressed and a combined ac and dc current is passed directly through the powder charge and the mold that contains it (if it is a conductor). At the same time, the powder is compacted with a programmed variation in force which minimizes the time required to bring the powder charge to a sintered density. The voltage varies, starting with 10 v and decreasing to 5 v during the 5-min cycle required to produce the part. A total current flow (ac and dc) of 1500 amp per sq in. is also typical. The use of both alternating and direct current speeds up the bonding process. The average part is compacted at a starting pressure of 400 to 500 psi (2800 to 3500 kPa), which rises to 2000 psi (13.79 MPa) at the end of the cycle.

There are several advantages in the use of spark sintering. Dies are relatively inexpensive. For example, a steel die for a given part was estimated at $1000. The same die in graphite as used in spark machining cost $50. Large parts can be made, such as a 135-lb (60-kg) iron pressing and a 32-lb (14.5-kg) beryllium pressing. The cycle time is short, in the case of the large pressings mentioned, the time was 30 to 45 min per part.

Spark sintering does not compete with traditional high-volume production of small iron, copper, aluminum, or cemented-carbide parts. It is particularly good for larger parts that must have a higher density than that produced by conventional powder-metallurgy methods.

### Secondary Operations

Many powder-metal parts may be used in the "as-sintered" condition. However, when the desired surface finish, tolerance, or metal structures cannot be obtained at this stage, additional finishing operations are used, such as sizing, coining, sinter-forging, machining, impregnation, infiltration, plating, and heat treatment.

*Sizing.* When a part must meet close tolerances, sintering is followed by a sizing operation. The sintered part is placed in a die and repressed.

*Coining.* Coining is similar to sizing except the part is repressed in a die to reduce the void space and impart greater density. After coining, the part is usually resintered for stress relief. Oftentimes sizing and coining operations are combined in the same die.

*Sinter-forging.* Sinter-forging or hot-forming is a comparatively new PM (powder-metal) process, in which the part will be placed in a furnace and heated to the upper critical temperature and then placed in a forging die where a forging blow forms the part with fine detail (Fig. 5-85). Forging may be accomplished in one blow if the preform has adequate porosity and shape.

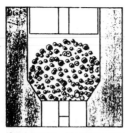

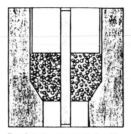

| | | |
|---|---|---|
| Metal powder—blended to any alloy depending on strength and density requirements—is loaded into compacting press. | Powder is compacted. The resulting preform is in the "green" state. | Preforms are sintered and coated with a lubricant. Up to this point, process parallels conventional PM compacting. |

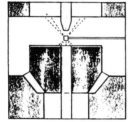

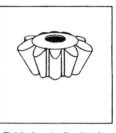

| | | |
|---|---|---|
| Preforms are reheated to hot-forming temperatures in a forming press. Lubricant is applied to the forming dies. | Preform is struck in the hot-forming press. Metal flow in the dies gives optimum strength and density throughout the finished part. | Finished part after hot forming can be as strong as parts produced by any other process. As-formed parts are accurate and uniform. |

FIGURE 5-85. *The basic steps used in hot-forming powdered-metal parts.* *Courtesy* Machine Design.

Some advantages of hot-forming of PM parts may be listed briefly as follows:

1. The number of normal forging steps are reduced; ideally only two dies are needed: one for compacting and one for forging.
2. Forging pressures are less, thus smaller presses can be used.
3. The forging temperature is lower than for conventional forging.
4. Less skill is required at the forging press or hammer.
5. Secondary operations are eliminated which are necessary in conventional forging, such as removal of flash and machining. The dies can be made with little or no draft. Shown in Fig. 5-86 are finished gears made by this process.
6. Tooling costs are lower due to lower temperature and pressure requirements.

*Machining.* The principal object of powdered metallurgy is to produce a product that is dimensionally accurate. However, certain features such as threads, reentrant angles, grooves, and side holes are usually not practical. These features are generally machined on the sintered blanks.

All the conventional metal-cutting operations are performed easily on sintered metal parts. Very sharp tools and fine feeds are necessary for PM filters to maintain the open pore structure and for self-lubricating bearings.

353

FIGURE 5-86. *Gear cutting and broaching are eliminated by hot-forming this intricate gear. Material displacement during hot-forming is planned to distribute high density and strength where it is needed most. Courtesy PM Equipment.*

**Impregnation.** When self-lubricating properties are desired, as in "lifetime bearings," the sintered parts are impregnated with oil, grease, or other lubricants. The parts are placed in tanks of specified lubricants and heated to approximately 200°F (93°C) for about 10 to 20 min. The lubricant is retained in the part due to capillary action until pressure or heat draws it to the surface.

**Infiltration.** An infiltrated part is made by first pressing and sintering the metal powder to about 77% of theoretical density. In the case of an iron-powder part that is to be infiltrated with copper, it is sent through the furnace a second time with a copper blank placed on it. The copper melts and soaks into the porous structure, producing close to 100% density. The process also provides increased strength, hardness, and corrosion resistance.

**Plating.** Prior to plating, PM parts are peened, tumbled, or given other treatments that will make the surface smooth and dense. Plating may get into the granular structure of the part and cause a galvanic action to be set up, therefore it is better to impregnate the part first. Plastic resins are usually used for impregnation as they have a low coefficient of expansion, a relatively low cost, and good filling properties, and they do not react galvanically with the metal. After impregnation, regular plating procedures are used.

**Heat Treatment.** Just as with wrought or cast metals, PM parts are heat-treated to improve grain structure, strength, and hardness. Conventional heat-treating steps can be used but care must be taken in several steps of the process. Porosity decreases heat conductivity, therefore longer heating and shorter cooling periods are required. A controlled atmosphere or vacuum furnace must be used.

TABLE 5-9. *A Comparison of Properties of Wrought and Cast Materials with Sintered Metals. Courtesy* Metal Progress

| Material | Tensile Strength (psi) | Elastic Modulus, Tension ($10^6$ psi) | Yield Strength (psi) | Elongation, % | Hardness | Maximum Service Temperature, °F | Specific Gravity | Impact Strength (Transverse Tests), (ft-lb) |
|---|---|---|---|---|---|---|---|---|
| Alloy steels, heat treated | 98,000 to 345,000 | 29 to 30 | 76,000 to 228,000 | 2 to 28 | $R_C$ 25 to 60 | 1200 | 7.75 | 15.0 to 40.0 |
| Nodular cast irons | 60,000 to 150,000 | 18 to 25 | 45,000 to 125,000 | 2 to 25 | BHN 140 to 325 | 1200 | 7.17 to 7.21 | 5.0 to 37.5 |
| Sintered alloy steels, heat treated | 50,000 to 175,000 | 14 to 22 | 40,000 to 120,000 | 0 to 5.0 | $R_C$ 15 to 48 | 1200 | 6.2 to 7.4 | 2.0 to 20.0 |
| Austenitic stainless steels, annealed | 80,000 to 115,000 | 28 to 29 | 30,000 to 55,000 | 15 to 55 | BHN 150 to 170 | 1500 | 8.02 | 200 to 412 |
| Sintered martensitic stainless steels, heat treated | 60,000 to 100,000 | 14 to 20 | — | 0.8 to 1.5 | $R_C$ 15 to 45 | 1000 | 6.2 to 6.8 | — |
| Aluminum alloy, heat treated | 35,000 to 88,000 | 10 to 10.6 | 31,000 to 78,000 | 1.5 to 15.0 | BHN 73 to 150 | 300 | 2.62 to 2.82 | 17.5 to 90.0 |
| Sintered brass | 18,000 to 27,000 | 9 to 12 (est.) | — | 9.0 to 13.0 | $R_H$ 60 to 70 | 500 | 6.8 to 7.6 | 3.0 |
| Glass-filled polystyrene | 11,000 to 17,000 | 11 to 13 | — | 1.1 to 1.3 | $R_M$ 80 to 90 | 190 to 200 | 1.25 to 1.32 | 10.2 to 15.2 |
| Molded nylon | 8,500 to 12,000 | 1.5 to 4.1 | — | 16 to 42 | $R_R$ 103 to 118 | 250 to 300 | 1.09 to 1.14 | 3.0 to 9.0 |

355

### Properties of Powdered-Metal Parts

In the past, the deficiency of engineering data for sintered metals fostered some doubts about the predictability of their mechanical properties and other design data.

Shown in Table 5-9 is a comparison of various mechanical properties of PM with wrought or cast metals. As can be seen, sintered alloys have impact strength and ductility comparable to those of ferrous castings but less than that of wrought steels. However, this is not a great problem since there are sintered steels available that have tensile strengths in excess of 150,000 psi and elongations of over 3 %. The table does not list the desirable wear and low-friction properties of sintered alloys.

The stress–strain properties of an infiltrated iron-copper carbon steel are shown in Fig. 5-87. It is essentially that of eutectoid (.80 % carbon) steel in its pearlitic condition. The copper contributes to the increased yield stength and hardenability when heat treatment is used.

### Advantages and Limitations

*Advantages.* Perhaps the most unique advantage of PM is the precise control that can be exercised over the powders. This permits variation in physical and mechanical properties while assuring consistent performance characteristics.

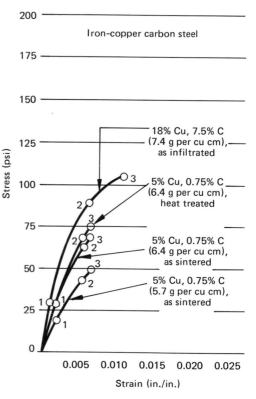

FIGURE 5-87. *The curves show the tensile properties of sintered iron-copper carbon steel: as sintered, as heat-treated, and as infiltrated.* Courtesy Metal Progress.

FIGURE 5-88. *A great variety of shapes may be produced by the powder-metallurgy process to tolerances that require no machining. Courtesy Wickes Engineering Materials.*

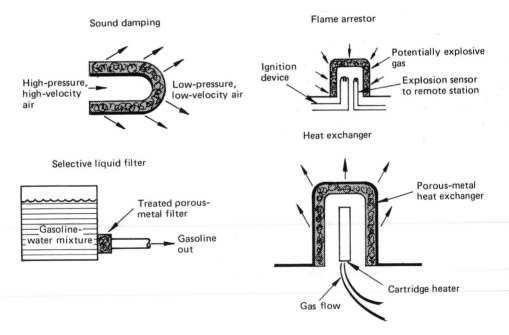

FIGURE 5-89. *A few of the uses of powdered-metallurgy structures of the porous type.*

Practically any desired alloy or mixture of metals, including those not available in wrought form such as tungsten carbide, can be produced. If need be, a single part can be made hard and dense in one area and soft and porous in another.

Parts can be produced in a variety of shapes that include irregularly shaped holes, eccentrics, splines, counterbores, gear teeth, etc., that in most cases require no machining (Fig. 5-88). As an example, a bore in a gear may be molded to a diameter between 0.5870 and 0.5875 in. and no machining is necessary.

The self lubrication through a network of small infiltrated pores can simplify design and ensure trouble-free maintenance.

The controlled porosity of PM is also important in the creation and application of filters. These filters may be used to separate or selectively diffuse the flow of

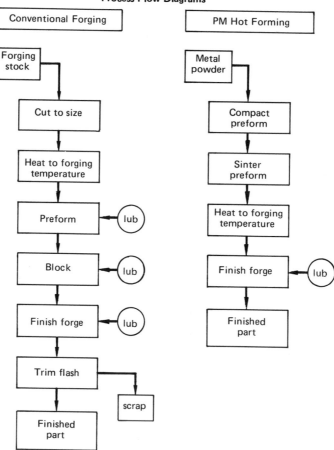

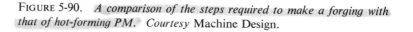

FIGURE 5-90. *A comparison of the steps required to make a forging with that of hot-forming PM.* Courtesy Machine Design.

gas or liquids, dampen sound, act as a flame arrestor, serve as a heat exchanger etc., as shown schematically in Fig. 5-89.

The excellent damping characteristics of PM parts make them important components in dictating machines, business machines air-conditioning blowers, etc.

The more recent development of hot forming has brought the PM process as a reasonable alternative to several other manufacturing processes, including shell-mold casting, lost-wax casting, and die casting (especially the newer ferrous-die casting that eliminates machining, gear cutting, and forging).

If forging requires considerable secondary machining, then PM hot-forming may offer a cost advantage. Hot forming can produce a wide range of mechanical properties all the way up to ultimate values. Tensile strength, impact resistance, and elongation can be made to match equivalent properties of forgings.

Forgings may involve four to five die setups whereas the same details can be formed with one strike in PM hot forming. A process flow diagram is used to show the steps required to produce a forging compared to PM hot forging (Fig. 5-90).

*Limitations.* Initial tooling costs are relatively high. Production volumes of less than 10,000 identical parts are normally not practical. However, there are some exceptions. There are times when even 50 pieces may prove economical, depending upon the design. The design may require characterstics easily produced by PM but costly by any other process.

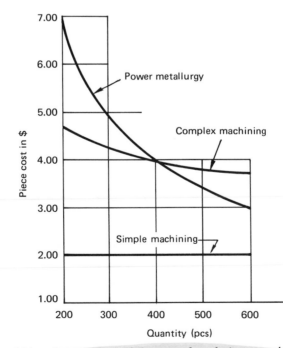

FIGURE 5-91. *A comparison of the cost of producing parts by machining or by powdered metals.*

A graphical comparison of the break-even point between PM and complex machining is shown in Fig. 5-91. Complex machining involves numerous critical dimensions, skewed surfaces, compound curves, tangent radii, intricate surface features, etc.

Weak, thin sections should be avoided, as should feather edges and deep, narrow slots.

Corrosion protection requires special attention and precautions.

Although hot forging shows great potential for certain applications, at present most parts are limited to less than 4 lb but some have been made that are twice that large. As familiarity with the process increases, so will the size and number of applications, as has been indicated by spark sintering and isostatic pressing.

## QUESTIONS

5-1. Why are investment castings made only in a relatively limited size range?

5-2. What is the main difference between the lost-wax method of casting and the Shaw process?

5-3. What is the big advantage of making castings by the full-mold process?

5-4. Why are castings made in a permanent mold generally stronger than those made by green-sand molding?

5-5. What is the main limitation of the die-casting process?

5-6. What are some of the advantages of low-pressure casting over conventional casting?

5-7. Why are the properties of a centrifugally cast part generally better than those statically cast?

5-8. (a) What is the purpose of padding in a casting design? (b) What is the purpose of exothermic padding?

5-9. What is the difference in draft requirements between sand casting and shell molding?

5-10. What are some ways a casting designer can cut costs?

5-11. Why are plastics often referred to as polymers?

5-12. What causes thermoset polymers to have different characteristics than thermoplastic polymers?

5-13. What is a unique feature of the nylon monomer?

5-14. What is the difference between compression molding and transfer molding?

5-15. What happens if a plastic mold is run too hot?

5-16. What mechanism creates the pressure for plastic injection molding?

5-17. How does one go about determining the required clamping force for a particular mold?

5-18. Why may the cost of a mold become relatively insignificant?

5-19. What is the main advantage of roto-molding?

5-20. What are structural foams being used for?

5-21. (a) What is meant by thermoforming? (b) Why are plugs used?

5-22. (a) What is the least expensive way of making a fiberglass product? (b) Why is the use of this method quite limited?

**5-23.** What advantage does the forging of plastic parts offer?

**5-24.** What methods are used to join thermosetting plastics?

**5-25.** What material is recommended for long-run PM dies?

**5-26.** What is the advantage of isostatic compacting of PM parts?

**5-27.** (a) What is the purpose of impregating PM parts? (b) How does this differ from infiltrating?

**5-28.** Can PM parts be competitive to forgings of the same size and design?

**5-29.** Why may it be necessary to machine a PM part?

**5-30.** Can the same mechanical properties be be obtained in PM parts as are obtained in wrought parts?

## PROBLEMS

**5-1.** Show by sketches how the casting designs (Fig. P5-1) can be improved.

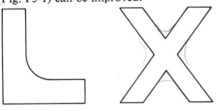

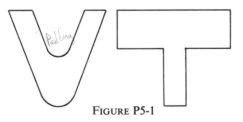

FIGURE P5-1

**5-2.** Shown in the sketch (P5-2) is a sectional view of a sand-cast steel gear housing.

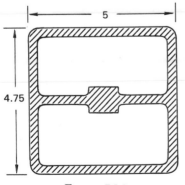

FIGURE P5-2

Problems were encountered in getting a porosity-free hub. Make a sketch to show where the riser should be placed and how you would ensure adequate metal at the hub.

**5-3.** Difficulty was encountered in this aluminum-alloy sand casting (Fig. P5-3). Even though individual risers and chills were provided for each of the four internal bosses, they were not adequately filled. Make a sketch or tracing to show how this problem might be solved.

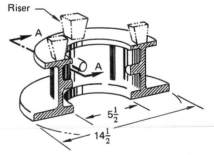

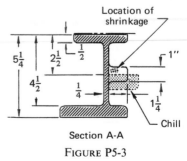

Section A-A

FIGURE P5-3

**5-4.** The sand-cast steel lever shown in Fig. P5-4 encountered excessive porosity and hot tears. Explain what the problem is and make a sketch to show how it can be overcome without adding any weight to the casting.

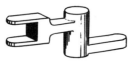

FIGURE P5-4

**5-5.** The malleable-iron gear housing shown in Fig. P5-5 was designed to be fed by six risers. Show how the part may be redesigned with less mass and three risers.

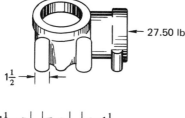

← 27.50 lb

$1\frac{1}{2}$

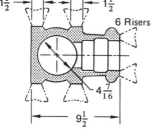

$1\frac{1}{2}$ → ← → ← $1\frac{1}{2}$

6 Risers

$4\frac{7}{16}$

$9\frac{1}{2}$

FIGURE P5-5

**5-6.** The malleable-iron gear housing as shown in Fig. P5-6 did not get adequate metal at the hub. Show how it can be corrected

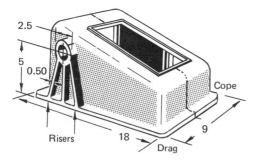

2.5

5

0.50

Cope

Risers

18

9

Drag

FIGURE P5-6

without adding any weight to the casting. Express the feeding distance of the boss in terms of $t$.

**5-7.** Shown in Fig. P5-7 is a machined-casting drawing. (a) Make a sketch and label each of the datum lines for surfaces A, B, and C. (b) Show three tooling points in the top view, two for the end view, and one for the side view. (c) Explain which surface should be machined first and why. (d) What machine should be used in machining the surfaces in each of the planes?

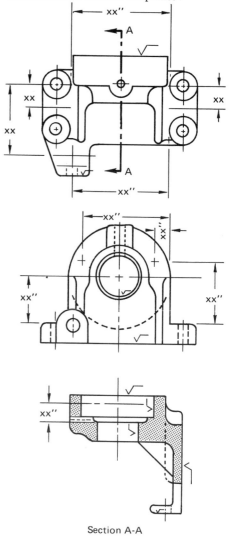

xx″

A

xx

xx

xx

A

xx″

xx″

xx″

xx″

xx″

xx″

xx″

Section A-A

FIGURE P5-7

**5-8.** Shown in Fig. P5-8 is a wheel spindle for a truck. Make a sketch to show the following: (a) parting line, (b) core, (c) riser.

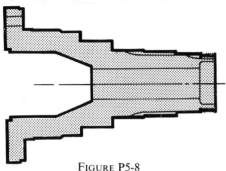

FIGURE P5-8

**5-9.** Shown in Fig. P5-9 is a casting design of a suspension arm. (a) What is the main objection to this design? (b) Make a sketch to show how it can be improved.

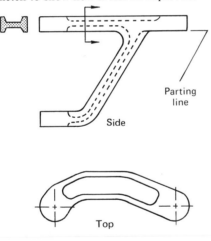

FIGURE P5-9

**5-10.** Make a sketch to show schematically how the grain structure would appear in cross section of a 3 in. × 3 in. cast-steel square bar that has had ample time to cool naturally in air.

**5-11.** Shown in Fig. P5-10 is a small 3 in. × 3 in. aluminum casting. (a) Select a process for casting this part and tell why you chose it. (b) Make a sketch to show the pattern.

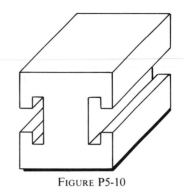

FIGURE P5-10

**5-12.** (a) State the difficulty that may be encountered in making the casting as shown in Fig. P5-11. (b) Make a sketch to show how the casting may be redesigned to alleviate the problem. (c) Assume this is an aluminum casting with overall dimensions of 6 in. × 12 in. State the following: 1. Minimum thickness of thin section if made by sand casting and by permanent-mold casting. 2. Machining allowance for the flat surface of the heavy wall section if sand cast. 3. Tolerance expected if the part is made out of malleable iron. 4. If no machining is planned and this is a high-production item, the casting process used. Why?

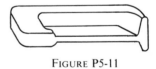

FIGURE P5-11

**5-13.** A woodworking vise jaw is made out of gray cast iron. The front plate is made with a clamping surface of 4 in. × 6 in., as shown in Fig. P5-12. Make a sketch of how this jaw should be designed for maximum strength and economy.

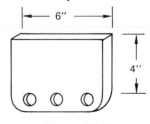

FIGURE P5-12

**5-14.** (a) Make sketches to show how each of the molded parts in Fig. P5-13 could be improved and still perform the same function. (b) State the principles by which you changed the designs.

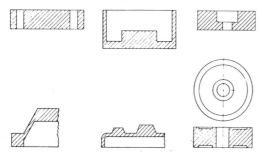

FIGURE P5-13

**5-15.** (a) Beads or ribs are used to reinforce large areas. Shown in Fig. P5-14 is a rib pattern. Show how it may be improved. (b) State your principle in making the design changes.

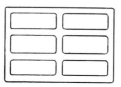

FIGURE P5-14

**5-16.** Which of the two parting line placements shown in Fig. P5-15 is the best, and why?

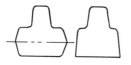

FIGURE P5-15

**5-17.** (a) Show by sketches how the designs shown in Fig. P5-16 can be improved. (b) What are the principles involved in making your design changes?

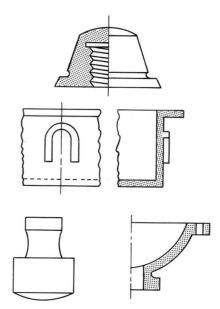

FIGURE P5-16

**5-18.** Shown in Fig. P5-17 is a full-scale drawing of a knob for a television set. (a) What molding methods could be used for a prototype run of 500? (b) What molding method would most likely be used for 10,000

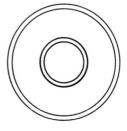

FIGURE P5-17

parts? (c) Why might polystyrene be a good choice of material? (d) Find the cost of making this part under the following conditions:

Powder cost/lb, 35¢     Press charge, $15/hr

Six-cavity mold    Labor, $3.00/hr
Total wt., 0.75 oz    Overhead,$6.00/hr
Runner wt. 0.5 oz    Cycle time,
Grinding cost/lb, 3¢     12 sec.

(e) What could be done, if anything, to improve the design? (f) Could this knob be made for high production out of a thermosetting plastic? (g) If you wanted this knob to be chromuim plated, what molding would you choose? What material would you choose? (h) What is the minimum recommended wall thickness for this part if injection molded?

**5-19.** Shown in Fig. P5-18 is a manifold blank that weighs 3.5 lb. It is used to distribute and control high-pressure oil in hydraulic motors. The tolerances are as follows: OD, ±0.010 in.; ID, ±0.005 in.; length, ±0.005 in.; slots, ±0.010 in. It should be medium hard ($R_c$ 45) in the bore location. The tensile strength should be about 50,000 psi or better. (a) Choose a material to make this part and tell why you recommended it. (b) Choose a preferred and alternate method of making this part and tell why you chose them.

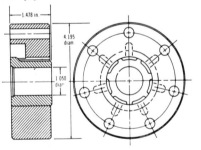

FIGURE P5-18

**5-20.** Shown in Fig. P5-19 is a transmission-synchronizer hub that weighs 1 lb. It is used to pick up and carry full driveshaft torque input for a medium-sized automotive transmission. The tolerances are as follows: spline OD, ±0.005 in.; bore spline major diameter, +0.001–0.00001 in.; bore spline minor diameter, ±0.005 in.; bore center point of die to OD, 0.003 total indicator runout (TIR); overall length +0.0001–0.005 in. The tensile strength should be about 140,000 psi with 2% elongation. The $R_c$ hardness is about 30 to 35. (a) Choose a material for this part and tell why you chose it. (b) Choose a preferred and alternate method of manufacture and explain your choice.

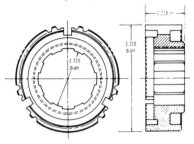

FIGURE P5-19

**5-21.** (a) What would the approximate molding pressure be for a 1 in. square plastic box with $\frac{1}{16}$ in. walls made by compression molding? (b) If the box were increased to 2 in. square, what would the compression-molding pressure be?

**5-22.** (a) Estimate the production rate per hour for the box in 5-21(a) if it is a thermosetting material $\frac{1}{16}$ in. thick and is made in a six-cavity mold. (b) What would be the approximate required clamping force?

**5-23.** What is the relative cooling rate of two castings: one is 3 in. in diameter and 3 in. long, the other is 1 in. in diameter and 16.8 in. long?

**5-24.** What material is now being used to replace metal for such items as gears, bearing caps, guides, and other wear components?

**5-25.** How does a casting nylon differ from most plastics?

**5-26.** (a) Could a product such as a typewriter case be made by injection molding, roto-molding or structural foam molding? (b) What would the consideration be in making the choice?

**5-27.** A two-cycle gasoline engine as used for a lawnmower has a 4 in. long connecting rod between the piston and the crankshaft. Which of these processes would you choose if it were (a) for a prototype engine (b) for a production run of 50,000.

1. cast nylon
2. hot-formed powdered metal
3. steel forging
4. sand-cast gray iron

## BIBLIOGRAPHY

### Books

American Society for Metals, *Metals Handbook*, 8th Edition, Vol. 1, 1961.

BIKALES, N. M., *Molding of Plastics*, Wiley-Interscience, New York, 1971.

*Casting Design Conference*, Central Foundry Division, General Motors Corp., Saginaw, Michigan.

CROSBY, E. G., and S. N. KOCHIS, *Practical Guide to Plastics Application*, Cahners Books, Boston, 1972.

DEARLE, D. A., *Plastic Molding Technique*, Chemical Publishing Co., Inc., New York, 1970.

*Designing With High-Strength Steel Castings*, Climax Molybdenum Company, New York.

HEINE, R. W., C. R. LOPER, and P. W. ROSENTHAL, *Principles of Metal Casting*, McGraw-Hill Book Company, New York, 1967.

LINDBERG, R. A., *Materials and Manufacturing Technology*, Allyn and Bacon, Boston, 1968.

Modern Plastics Encyclopedia, McGraw-Hill Inc., New York, 1971, 1972, 1973, 1974.

SEHGAL, S. D., and R. A. LINDBERG, *Materials— Their Nature, Properties and Fabrication*, Kailish Publishers, Singapore, 1973.

Steel Founders Society of America, *Steel Castings Handbook*, The Electric Printing Co., Cleveland, 1970.

The Aluminum Association, *Standards for Aluminum, Sand and Permanent Mold Castings*, New York, 1969.

### Periodicals

ADAMS, C. M., and H. F. TAYLOR, "Fundamentals of Riser Behavior," *AFS Transactions*, V. 73 P. 355–362 (1965).

ANDREOTTI, E. R. and S. W. McGEE, "Sintered Metals In Engineering Design," *Metal Progress*, April 1966.

BACKER, L. and P. GOSSELIN, "Continuous Casting: Its Metallurgical Aspects Relative to

High-Grade Alloy and Carbon Steels," *Journal of Metals*, May 1971.

BOESEL, R. W., "Spark Sintering: An Unusual Method," *Metal Progress*, April 1971.

BORDNER, P. G., G. C. FULMER, and D. MEADOWS, "Plastic Profile Extrusions," *Machine Design*, June 6, 1968.

BUFFERED, A. S. and P. U. GUMMESON, "Application Outlook for Superalloy PM Parts," *Metal Progress*, April 1971.

DREGER, D. R., "Cast Nylon," *Machine Design*, June 27, 1974.

FREDERICK, W. J., "Blow Molding Grows Up," *Machine Design*, April 2, 1970.

General Electric Company, Plastics Department, "Structural Foam Resins."

GREEN, R. G., "Assessing Injection Molding Options," *Automation*, Oct. 1970.

HALTER, R. F. and B. B. BELDEN, "PM Parts With Strength of Forgings," *Machine Design*, July 12, 1973.

HIRSCHHORN, J. S. and R. B. BARGAINNIER, "The Forging of Powder Metallurgy Preforms," *Journal of Metals*, Sept. 1970.

"How to Design for Maximum Soundness in Castings With Unequal Sections," *Machine Design*, Oct. 25, 1962.

HURSEN, H. H., "Pressure Casting of Steel," *Metal Progress*, April 1963.

JOHN, F. W., "Blow Molding Grows Up," *Machine Design*, April 2, 1970.

JOHNSON, W. R., "Porous Metal Parts," *Machine Design*, April 11, 1968.

KAMINSKY, S. J., "Welding Plastics With Hot Air," *Machine Design*, May 31, 1973.

KHOL, R., "Forged Powder Metal," *Machine Design*, April 3, 1969.

KHOL, R., "Isostatic Pressing Comes Alive," *Machine Design*, June 11, 1970.

KULKARNI, K. M., "Forged Plastic Parts," *Machine Design*, May 3, 1973.

LINSELL, A. S., "What You Should Know About Plastic Injection Molding," *American Machinist*, Aug. 14, 1967.

LOPER, C. R., R. W. HEINE, and R. A. ROBERTS, "Riser Design," *AFS Transactions*.

NICKERSON, J. A., "Rotomolding Comes of Age," *Machine Design*, Nov. 12, 1970.

OBEDA, E. G., "Ultrasonic Assembly of Thermoplastics," *The Tool and Manufacturing Engineer*, March 1968.

ORROK, N. E., "Joining Plastics," *Metal Progress*, Dec. 1966.

ROBERTS, R. A., R. P. DATE, C. R. LOPER, and D. R. POIRIER, "The Effect of Solidification Time on the Properties of Copper-Base Alloys," *AFS Transactions*, V. 76, p. 573 (1968).

SPROW, E. E., "Low-Pressure Casting for High-Performance Parts," *Machine Design*, April 1973.

VANATT, R., "Short-Run Powder Metallurgy," *Machine Design*, Dec. 21, 1967.

WILLMOT, W. J., "All About Patterns for High-Pressure Molding," *Foundry*, Oct. 1971.

# 6

## Materials Joining Processes

### Introduction

Today's manufacturing engineer must be able to assess the value of a wide variety of materials-joining processes. Tradition may dictate a process such as welding or brazing, but modern technology may show that adhesives or mechanical-joining methods would be more efficient or that a combination of adhesives and mechanical fasteners would serve more effectively.

The six main materials-joining processes used in industry today are discussed in this chapter. They are: welding, brazing, braze welding, soldering, adhesive bonding, and mechanical fasteners.

### WELDING

The English used the word *weld* long before the first torch was ever lighted. They used it in referring to a sunny flower that is used as a dye. Traced back further, the word comes from the Middle English *wellen*, meaning to pour forth or flow, which comes closer to the concept of welding.

Brazing, a process used to join metals with an alloy that melts at a temperature lower than the metals being joined, may have been developed soon after the discovery of the melting of metal, about 4000–3000 B.C. A copper panel found in Mesopotamia, believed to have been made before 3000 B.C., had separate parts joined to the main panel by brazing.

Flow welding, termed "burning-on" in the foundry, was also used in ancient times. It consists of casting molten metal in contiguity with a previously made casting to make an addition to it, or casting molten metal between two components to join them together.

TABLE 6-1. *Principal Arc Welding and Cutting Processes*

| Arc Welding | | Severing |
|---|---|---|
| Manual | Automatic | |
| Metal arc | | Metal arc |
| Submerged arc | Submerged arc | |
| Electroslag | | |
| Electrogas | | |
| Gas tungsten arc | GTA | |
| Gas metal arc | GMA | |
| $CO_2$ welding | | |
| Arc spot | | |
| Plasma arc | PA | PA |
| Carbon arc | | Air carbon arc |

Today much of welding may be considered, in effect, a relatively small-scale, high-quality casting made in a metal mold. This is termed a *nonpressure fusion process*. Fusion refers to the intimate intergranular mixing of two metals to be joined. Thus when two plates are welded together, the weld deposit is essentially a small casting that fuses to the parent metal. Here a knowledge of the metallurgical aspects of metal as it is subjected to high heat and cooling is necessary to understand the composition of the weld and the transition zone between the weld and the parent metal.

It was not until the early part of this century that welding became a common household term and several processes, including resistance welding, arc welding and cutting, and gas welding and cutting, were introduced. In recent years there has been a great proliferation of welding processes. There are now over 40 separate processes. It is not the purpose of this text to present them all but rather to cover mainly those processes that have industrial significance.

Most of the welding processes can be classified into three main types: (1) arc welding, (2) gas welding, and (3) resistance welding. Each of these main types can be further classified into a number of subtypes, as shown by the first process to be discussed, arc welding, in Table 6-1.

## Arc Welding

Arc welding is a process in which an electric arc is produced between a metal electrode or wire carrying high-amperage current and the work-piece, as shown in Fig. 6-1.

*Metal-Arc Welding.* In metal-arc welding, heat is developed between a metal electrode or wire and the workpiece. Under the intense heat of the arc, ranging from 5000 to 10,000°F (2760 to 5537°C), a small part of the base metal is brought up to the melting point. At the same time, the end of the metal electrode is melted and tiny globules or drops of molten metal pass through the arc.

369

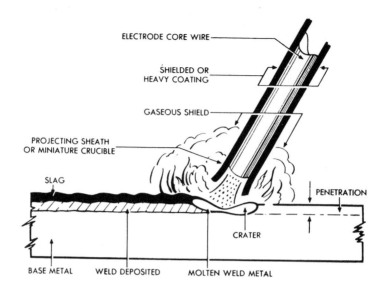

FIGURE 6-1. *The welding action of a cellulosic-coated "stick electrode"*.

The heat liberated by the arc is a function of the power used. It may be converted as follows: 1 watt = 1 joule/sec, or 1 joule = 1 watt-sec. In calculating the heat for a particular arc-welding condition, the following formula can be used:

$$W = EIt$$

*where:*    $W$ = heat in joules
$E$ = volts
$I$ = current in amperes
$t$ = time in seconds

## Arc-Welding Energy Sources

Energy for arc welding is obtained by generators, transformers, or rectifiers as shown in Fig. 6-2. The main criteria is that there be a constant current supply. In arc welding, there is a great deal of fluctuation in current requirements. When the arc is struck, the electrode is essentially in short circuit, which would immediately require a sudden surge of current unless the machine is designed to prevent this. Also, as the globules of metal travel across the arc stream, they tend to cause short circuiting. A constant current machine is designed to minimize these sudden surges.

Traditionally, power sources for metal-arc welding have a *drooping* volt-ampere characteristic as shown in Fig. 6-3(*a*). By drooping is meant the terminal voltages of the machine decrease as the welding current increases. This type of current is necessary when welding manually with covered electrodes. Since the arc length is varied, it causes the voltage to change. For example, with a long arc there is a marked change in the required voltage, as shown by letters A and B on the curve,

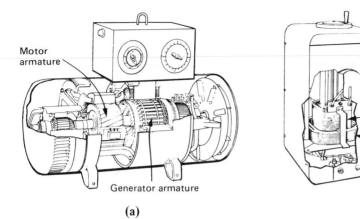

Motor armature

Generator armature

(a)

Primary coil

Secondary coil

(b)

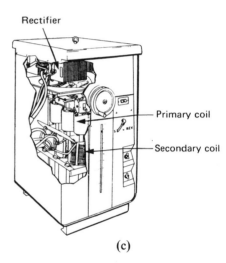

Rectifier

Primary coil

Secondary coil

FIGURE 6-2. *The energy for arc welding is obtained by generators, transformers, or rectifiers.*

(c)

but there will be relatively little change in current. On an automatic machine, increasing the wire speed will decrease the voltage, but at the same time it will increase the current somewhat. Thus the output power, volts × amps, or watts, remains relatively constant with a flat volt-ampere characteristic.

In the *flat* and *rising* power supplies as shown at (*b*) and (*c*) of Fig. 6-3, there is a much greater current change per unit voltage than that produced in the drooping-voltage characteristics. As a result, the arc has great inherent self-adjusting tendency. On automatic-control machines, the rising characteristic permits the largest variation in wire feed and current for a usable arc length without readjusting the power supply.

Originally, *rectified dc* power supplies were of the drooping characteristic but later models were available with flat or rising characteristics. A rectifier changes alternating current to direct current. In brief, the negative portion of the ac sine waves is blocked out. This can be accomplished by selenium-cell half-wave rectifying elements or with diode rectifiers. The newest model rectifiers have characteristic volt-ampere curves whose slopes may be varied.

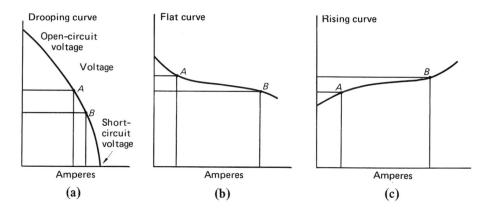

FIGURE 6-3. *Three types of volt-ampere curves. The relatively steep volt-ampere curve (a) produces only a small amount of ampere change when the arc voltage is changed. Flat and rising volt-ampere curves as used on automatic machines are designed to provide a relatively constant current by balancing the effect of volts × amps.*

Sizes of welding machines are designated according to their output rating, which may range from 150 to 1000 amperes. The output rating has usually been based on a 60 % duty cycle. This means that the power supply can deliver its rated load output for six minutes out of every ten minutes. In manual welding particularly, a power source is not required to deliver the current continuously as in other electrical equipment. Fully automatic power-supply units are usually rated at 100 % duty cycle

The size and type of energy source is dependent on the average range of work to be done. For light and medium work, and general repair and maintenance, a 150- to 200-amp machine is usually suitable. For average production work, and plant maintenance and repair, a 250- to 300-amp machine is used. For large, heavy-duty, structural-type welding, machines of 400 to 1000 amp are used. Transformers and transformer rectifiers have grown in popularity and have replaced many of the motor generator sets for general use.

*Direct Current vs Alternating Current Welding.* Direct current welding machines have motor generator sets or ac-dc transformer rectifiers. One of the big advantages of a dc energy source is that either straight or reversed polarity can be used. *Polarity* indicates the direction of the current flow. In straight polarity, the electrode is negative and the work is positive. In reverse polarity, the opposite is true. Polarity can be changed by a switch on the machine or by changing the cable connections. The significance of polarity is that there is a direct relationship between it and where the heat is liberated. Most of the heat is liberated in the positive side of the arc. That is why, ordinarily, straight polarity is used because it is preferrable to have most of the heat in the workpiece rather than in the electrode. There are times, however, when it is more advantageous not to have the heat build up in the workpiece. Thus, for overhead and vertical welding where it is desirable to have the weld pool solidify rather quickly, reversed polarity is used. Also, reversed polarity produces a greater digging action that results in better penetration (Fig. 6-4).

Alternating-current welders are theoretically both direct-current, straight-polarity (DCSP) and direct-current, reverse-polarity (DCRP) machines, as shown in Fig. 6-5. In some cases where moisture scale or oxides are present on the material welded, the flow of current in the reverse polarity direction is reduced. To prevent this from becoming a problem, it is common practice by some manufacturers of welding equipment to provide for high-frequency, low-power current, as shown in Fig. 6-6. The high-frequency current jumps the gap between the electrode and the

(a)

FIGURE 6-4. *A comparison of the weld penetration obtained by dc straight polarity (a), and dc reversed polarity (b).*

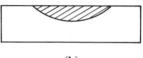

(b)

FIGURE 6-5. *An alternating-current welder is, in essence, a combination reverse- and straight-polarity machine.*

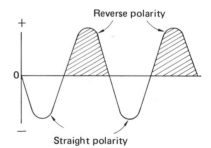

Reverse polarity

Straight polarity

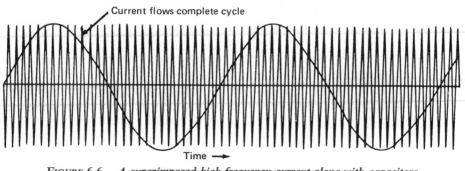

Current flows complete cycle

Time ⟶

FIGURE 6-6. *A superimposed high-frequency current along with capacitors stabilizes the welding current and produces a balanced wave form. Courtesy Airco Welding Products.*

workpiece and pierces the oxide film, making a path for the welding current. Super-imposing high-voltage, high-frequency current on the ac welding current has several advantages: (1) the arc can be started without actually touching the electrode to the workpiece; (2) a longer arc is possible, which is useful in some applications discussed later; (3) it is possible to use a wider range of welding currents for a specific diameter electrode; (4) the arc is more stable.

Another consideration in the use of welding current is that of *arc blow*. This is a phenomenon in which a strong magnetic field sets up around the electrode and tends to deflect the arc as though a strong wind were blowing, hence the term "arc blow." The arc may be deflected to the side but it is usually deflected either forward or back-ward along the direction of travel. It is especially noticeable when the electrode is used in a corner or toward the end of a joint. Arc blow is encountered principally in direct-current welding machines, since alternating current prevents the formation of strong magnetic fields.

### Electrode Types

Electrodes used in manual welding are usually covered with a flux coating that vaporizes in the heat of the arc to form a protective gas $(CO_2)$ as was shown in Fig. 6-1. This gas excludes nitrogen and oxygen from the molten metal, thus preventing the formation of undesirable oxides and promoting a smooth flow of molten metal. This form of shielding, now referred to as *shielded-arc welding* or *stick-electrode welding*, may consist of cellulosics or minerals or a combination of both.

*Cellulosic Coatings.* Cellulosic coatings derive their name from the cellulosic mater-ials from which they are made, such as wood pulp, sawdust, cotton, and various other compositions obtained in the manufacture of rayon. The expanding gas furnished by the burning cellulosic material acts to give a forceful digging action to the weld. Properly made, the electrode coating will burn just a bit more slowly than the core wire, forming a cup or crucible at the end of the rod that makes it easier to direct the arc. Cellulosic-coated electrodes can be used in any position, but are particularly useful in vertical, horizontal, and overhead positions.

*Mineral Coatings.* Mineral coatings are manufactured from natural silicates such as asbestos and clay. By adding oxides of certain refractory metals such as titanium, the harsh digging action of the arc is modified to produce one that is softer and less penetrating. This type of electrode is used to advantage when the fit-up is poor and on sheet metal where shallow penetration is desired. A large amount of slag is pro-duced, which serves to protect and control the chemistry of the deposited metal as it cools. The heavy slag retards the cooling rate of the deposited metal, allowing gas to escape and slag particles to rise to the top. Cooling stresses are reduced and a more homogeneous microstructure results. Because of the large amount of slag produced, the mineral-coated electrode is used most advantageously on *downhand* welding. Downhand welding is used in making flat welds and those inclined up to 45°.

*Iron-Powder Coatings.* The use of iron powder has been a later addition to the electrode-coating field. The use of iron powder brought several desirable effects: after the arc is struck, the metal melts away, leaving a well-defined crucible at the end of the rod which effectively concentrates the heat and gives an automatically consistent arc length. In fact, the electrode may be just slowly dragged over the work since the coating will maintain the proper arc length. The slag that forms over the weld is often self-removing and the appearance of the bead is improved. It also furnishes more inches of weld per electrode because the iron in the coating goes into the weld. However, because of the heavy coating, there are about half as many rods per pound.

*Low-Hydrogen Electrodes.* Frequently, basic mineral-coated electrodes, especially those used in welding stainless steel, manganese, and molybdenum, are referred to as low-hydrogen electrodes. The low-hydrogen electrode is especially formulated and packaged to keep the hydrogen content at a minimum. It was found that poor welds in stainless steel were due largely to the steam that formed during the welding process, which popped the mineral coating off the electrode, allowing oxides and nitrides to form in the weld. With low-hydrogen electrodes, porosity is also eliminated in welding steels of high sulfur content.

### Welding-Electrode Specifications and Selection

Not too many years ago it was easy to select an electrode because there was only one—a bare wire. It was used for all purposes and all positions. Today the situation is different. The American Welding Society–American Society for Testing Materials (AWS–ASTM) has 12 classes of mild-steel–covered electrodes, as covered by the joint AWS A5.1-64 and ASTM A233-64 standards. The characteristics of mild-steel–covered electrodes are given in Table 6-2.

The specifications are based in part of the tensile strength of the weld metal. A four- or five-digit number is used to identify each electrode type. For example:

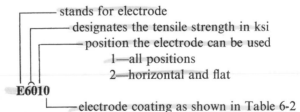

```
          ┌──── stands for electrode
          │  ┌──── designates the tensile strength in ksi
          │  │  ┌──── position the electrode can be used
          │  │  │        1—all positions
          │  │  │        2—horizontal and flat
     E6010
             └────electrode coating as shown in Table 6-2
```

Selection of electrodes is usually based on chemical composition, mechanical properties, and operating characteristics, in that order. Operating characteristics of electrodes have to do mainly with how rapidly the weld metal *fills*, (deposition rate), how fast it *freezes* (for use in various positions), and how fast the electrode *follows* (welding speed). The type and position of the joint determines whether the electrode should primarily have fill, freeze, or follow characteristics.

Fill electrodes are used primarily for easy-to-weld joints in the flat position. Examples of fill-type electrodes are E6027, E7024, and E7028. The covering of these

TABLE 6-2. *Characteristics of Mild-Steel–Covered Electrodes. Courtesy* Metal Progress

| AWS-ASTM Electrode Classification* | Welding Category | General Characteristics† | Type of Coating |
|---|---|---|---|
| **60,000 Psi Min Tensile Strength** | | | |
| E6010 | Freeze‡ | Molten weld metal from these electrodes freezes quickly. They are suitable for welding in all positions with dc reverse polarity power. They have a low deposition rate and a deeply penetrating arc. They can weld all types of joints | High-cellulose, sodium |
| E6011 | Freeze‡ | These electrodes are similar to E6010 except they may be used with ac as well as dc power | High-cellulose, potassium |
| E6012 | Follow | Compared with E6010, E6012 is characterized by a faster travel speed and smaller welds. They use ac or dc, straight polarity power and have less penetration than E6010. Their primary use is for single-pass welding of thin gage sheet metal in the flat, horizontal and vertical down positions | High-titania, sodium |
| E6013 | Follow | These electrodes are similar to E6012 except they may be used with dc (either polarity) or ac power | High-titania, potassium |
| E6027 | Fill | The deposition rate of these electrodes is high since their covering contains about 50% iron powder. Their primary use is for multipass deep groove welding in the flat position using dc (either polarity) or ac power | Iron powder, iron oxide |
| **70,000 Psi Min Tensile Strength** | | | |
| E7014 | Fill-freeze | Iron powder (30%) in the coverings of these electrodes gives tham a higher deposition rate than E6010. Usable with dc (either polarity) or ac power their primary use is for inclined and short horizontal fillet welds | Iron powder, titania |

*Continued*

TABLE 6-2.  *Continued*

| AWS-ASTM Electrode Classification* | Welding Category | General Characteristics† | Type of Coating |
|---|---|---|---|
| E7018 | Fill-freeze | Like E7014, these electrodes contain about 30% iron powder in their coverings. Their low-hydrogen coverings make them suitable for welding low and medium carbon steels (0.55% C max) in all positions and types of joints. Their weld metal quality and mechanical properties are the highest of all mild steel electrodes. Usable with dc reverse polarity or ac power | Iron powder, low-hydrogen |
| E7024 | Fill | These electrodes are similar to E7014 except that they contain more iron powder (50%) in their coverings which gives them a higher deposition rate. This makes them suitable for welding fillets in the horizontal and flat positions | Iron powder, titania |
| E7028 | Fill | Similar to type E7018, these electrodes contain about 50% iron powder in their coverings. They are used for welding horizontal fillets and grooved fillet welds in the flat position. | Iron powder, low-hydrogen |

* E6020, E7015 and E7016 are not included due to their limited usage.
† Only electrodes up to 5/32 or 3/16 in. diameter can be used in all welding positions (flat, horizontal, vertical and overhead).
‡ When welding sheet metal, these electrodes demonstrate follow-freeze characteristics.

electrodes contains as much as 50% iron powder, which increases both deposition rate and current requirements.

Freeze electrodes have relatively low deposition rates and are used for vertical and overhead or steeply inclined joints. Because their covering contains less than 10% iron powder, freeze electrodes require less current than fill electrodes. Examples are E6010 and E6011.

Electrodes such as E7014 and E7018 can be classified as fill–freeze types since they have some of the characteristics of both classes.

Follow electrodes can be used at high travel speeds with a minimum of skips and misses. Their principal use is on all types of joints in 10- to 18-gage (0.1345 to 0.0478 in. thick) sheet metal. See Appendix 6A for metric conversion factors. Typical follow electrodes are E6012 and E6013.

Shown in Fig. 6-7 are examples of joint types along with the choice of electrode. Typical operating data for stick electrodes used in making butt welds is given in Table 6-3.

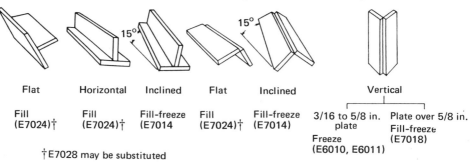

Fillet welds over 10 to 12 in. in length on 3/16 in. or thicker plate

| Flat | Horizontal | Inclined | Flat | Inclined | Vertical |
|------|-----------|----------|------|----------|----------|

Fill (E7024)†  Fill (E7024)†  Fill-freeze (E7014)  Fill (E7024)†  Fill-freeze (E7014)

3/16 to 5/8 in. plate  Plate over 5/8 in. Fill-freeze (E7018)

Freeze (E6010, E6011)

†E7028 may be substituted

FIGURE 6-7. *Examples of joints and the corresponding selection of electrodes.*

## Flux-Cored Continuous Welding Wire

As arc welding grew in popularity, particularly as a production means of fabrication, it was only natural that methods would be studied to make the process automatic or at least semiautomatic. One disadvantage of the stick electrode is the need to stop and change it as it gets short. Continuous wire seemed to be the answer, but there were different approaches as to how the arc shielding could be accomplished. A continuous covered electrode was tried and was marketed for a short time, but was not well accepted. A more feasible approach was to put the flux on the inside of a hollow wire, as shown in Fig. 6-8. This process became commercially available in 1954. At that time a $CO_2$ gas was added for auxilliary shielding. There are two processes used today, a "non-gas shield" and a "gas shield." The non-gas shield is designed to develop sufficient gas shielding from the flux itself. The welds have somewhat lower ductility and notch toughness, but it has proven satisfactory for most mild-steel applications.

The most versatile energy source for semiautomatic, flux-cored wire welding is a 500-amp, dc, constant voltage machine as shown in Fig. 6-9(*a*). Many types of wire feeders are available, but a constant-speed type as shown at (*b*) is the most satisfactory and easiest to use. Generally $\frac{3}{32}$ or $\frac{1}{8}$ in. diameter (2.36 or 3.17 mm) wire is used and the operating conditions are 375 to 575 amps with approximately 32 volts.

TABLE 6-3. *Typical Operating Data for Stick-Electrode Butt Welds*

| Plate Thickness (in.) | | Electrode Size (in.) | Amperes | Ft of Joint/Hr (100 % operating factor) |
|---|---|---|---|---|
| Sq. Butt | 1/4 | 5/16 | 325 | 45 |
| 60° V | 3/8 | 3/16 E6011 | 175 | 20.1 |
| 60° V | 1/2 | 3/16 E6027 | 280 | 17.4 |
| | | 1/4 E6011 | 275 | |
| | | 7/32 E6027 | 315 | |

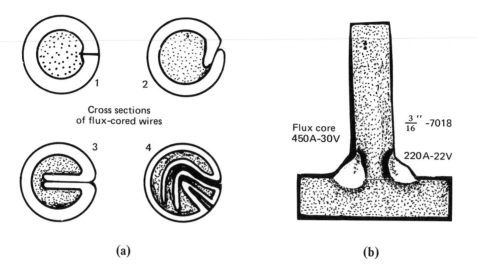

(a)

(b)

FIGURE 6-8. *Some cross-section configurations of flux-cored wire (a).*
*A fillet weld made with a $\frac{1}{8}$-in. flux-cored wire at 450 amp compared to that*
*of $\frac{3}{16}$-in. manual weld with a stick electrode at 220 amp (b). Courtesy Arcos*
*Corporation.*

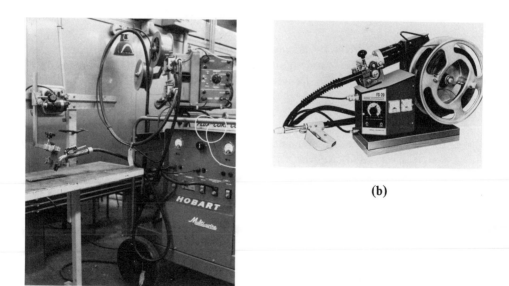

(b)

(a)

FIGURE 6-9. *A semiautomatic dc welding machine used for flux-cored wire*
*(a). A close-up view of a wire-feed unit (b). Courtesy Arcos Corporation.*

Most of the semiautomatic flux-cored welding is done with air- or gas-cooled guns and cable, even though water-cooled guns and cable are available.

Fully automatic units have been developed that will feed one, two, or three flux-cored wires simultaneously. These multihead units are frequently used to give larger fillet welds as required in heavy bridge and girder work.

*Advantages and Limitations.* The main advantages of the flux-cored wire is the speed at which the weld metal can be deposited and the depth of penetration obtained. It is especially fast in downhand welding of fillets and grooves where heavy deposits are required. A deposition rate of 35 lb/hr (15.75 kg/hr) can be maintained on automatic machines with a 0.120-in. diameter (3.05 mm) electrode. This is about three times faster than that obtained with the fastest stick electrode.

The flux-cored wire offers better control over penetration, as shown in Fig. 6-8(*b*). Not only is it deeper than with the manual electrode when desired, but also it has better control when poor fit-up is encountered. In this case, the operator allows the electrode to stick out of the gun farther. The longer "stick-out" automatically reduces the energy input into the arc, and thus into the heat that enters the base metal, and produces a higher preheat to the electrode. The immediate effect is to increase the deposition rate. Oftentimes a weld with a poor fit-up will require several passes with intermediate cleaning. With the added deposition rate, it can be made in one pass. The flux-cored wire is particularly useful where there is a large volume of work to be welded in the downhand or flat position.

The flux-cored wire has been limited to flat and horizontal welding. Without additional gas shielding, the welds are not quite as clean as with the coated electrode.

*Flux-Cored Electrode Classification.* Flux-cored wires come under the AWS 18-65 classification of electrodes as composite or "T". Thus E70T would be: E—electrode, 70—70,000 ksi tensile strength, and T—composite type.

### Solid Continuous-Welding Wires

Welding wires are made for use in many kinds of automatic and semiautomatic machines. Mild-steel wires are lightly coated with copper for good electrical contact, for corrosion resistance, and for consistent operating conditions. The wires are packaged in spools of 4 in. diameter for hand-held guns and may range up to spools weighing 750 lb for larger machines.

Mild-steel wires for gas metal-arc welding are covered by AWS specification A5.18-65 and ASTM A559-65. The classification designation consists of the letter "E" for electrode, followed by two digits, 60 or 70, representing the minimum tensile strength as welded in ksi. Next there is one of three letters—S, T, or U—signifying solid-, composite-, and emissive-coated filler metals, respectively. A typical classification is E70S-1. This is read as: an electrode of 70 ksi minimum TS, of solid construction, to be used with argon plus 1 to 5 % oxygen. The numbers following S and the dash indicate the shielding gas used as 1—AO; 2 and 3—AO and $CO_2$; 4, 5, and 6—$CO_2$.

Stainless steel wires are available in diameters ranging from 0.035 to 0.300 in. (0.90 to 7.62 mm) for gas-shielded metal-arc welding and in several sizes from 0.045 to 0.15625 in. diameter (1.14 to 3.96 mm) for submerged-arc welding. Stainless steel wires are used for joining chromium-grade stainless steels, stainless steel to mild steel, and for stainless steel overlay work.

Aluminum wires are available in many different aluminum alloys. The wire size ranges from 0.020 to 0.125 in. (0.51 to 3.17 mm) in diameter. These high-quality wires are prepared with extremely clean surfaces by a shaving process that removes the surface oxides. The wire coils are then individually foil covered to prevent exposure in shipment and storage.

### Submerged-Arc Welding

Submerged-arc welding derives its name from the fact that the arc is hidden under a heavy coating of granular mineral material or flux (Fig. 6-10). A bare wire is used and the fusible flux blanket provides protection for the molten metal from the atmosphere. The dry granular flux is fed continuously from a hopper through a pipe placed slightly ahead of the arc zone. The arc is struck in the submerged area and the weld is completed without the usual sparks, spatter, and smoke.

The current used for submerged-arc welding can be either ac or dc. DCRP is used to get deeper penetration, whereas DCSP will give a higher deposition rate with a broader bead. A flat-characteristic power source has the advantage of a fairly

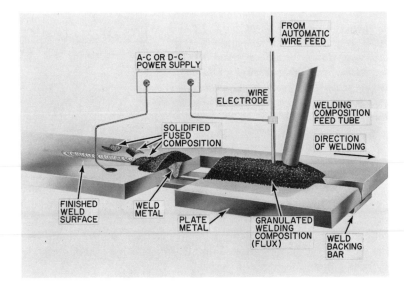

FIGURE 6-10. *A cutaway showing the components of submerged-arc welding. The arc is kept under a layer of granular flux.* Courtesy Plant Engineering.

constant voltage. Above 1000 amps, ac current is preferred to reduce arc blow. If one machine does not supply enough power, duplicate machines may be connected in parallel.

As with the flux-cored process, multiple electrodes may be used. Common arrangements are with two electrodes in tandem or in transverse (side by side or parallel) arrangement.

*Advantages and Limitations.* Submerged-arc welding has several distinct advantages. Very high currents can be used, ranging from 200 to 2000 amps. It may be noted that in conventional welding, where the arc is exposed, currents above 300 amps must be used with great care due to the intensity of high infrared and ultraviolet light rays. The ability to use high current in submerged-arc welding brings with it high deposition rates and good penetration. The process is thermally efficient since much of the heat is kept under a blanket of slag. There is a high dilution rate of base metal with weld deposit, being on the order of twice as much plate as weld metal. Each size of wire can be used over a wide range of current settings but, for any given current, the smallest wire size will give the best penetration. Because of the deep penetration possible, the use of smaller V's is practical, as shown by the example in Fig. 6-11. The weld beads are extremely smooth and most of the fused flux pops off by itself as the metal cools and contracts.

FIGURE 6-11. *A joint as prepared for manual arc welding in 0.75 in. thick plate (a), and for submerged-arc welding (b). The filler wire required per lineal foot is 1.1 and 0.7 lb respectively. A temporary backup strip is used with the submerged-arc process due to the high current density. If the joints are to be highly stressed, back gouging and running a bead on the back side is recommended.*

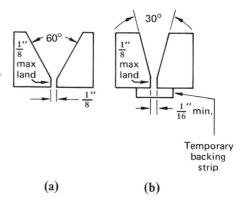

(a)  (b)

One of the disadvantages inherent in the process is that since the weld cannot be seen it is more difficult to guide it. Therefore the welding head must be accurately preset with respect to the joint or the operator must be able to adjust the head by observing an indicator, usually a pointer or light beam, focused on the joint ahead of the flux. In the case of grooved joints, rollers can be made to guide the welding head by riding along the edges of the plate.

Another disadvantage of the submerged-arc process is that it is largely limited to flat-position welding. Overhead welding is quite impractical due to the high fluidity of the weld pool and the molten flux.

Typical operating data for submerged-arc welding is shown in Table 6-4.

TABLE 6-4. *Typical Operating Data for Submerged-Arc Welding*

| Plate Thickness (in.) | Wire Diameter (in.) | Amperes | Voltage | Travel Speed (ipm) |
|---|---|---|---|---|
| 1/4 | 3/16 | 525–576 | 30–32 | 20–24 |
| 3/8 | 3/16 | 700–900 | 33–35 | 12–16 |
| 1/2 | 1/4 | 800–1000 | 33–35 | 12–15 |

*Vertical Submerged Arc.* Vertical submerged arc is done with a stationary, consumable guide tube, as shown in Fig. 6-12(*a*). The consumable, stationary wire guide is a flux-coated tube that extends the full length of the joint. A few ounces of submerged-arc composition are charged into the bottom of the joint and the weld is started by initiating an arc between the electrode, which passes through the guide to the bottom of the joint, and the joint. The flux charge becomes molten and forms a protective blanket over the weld as the electrode is continuously fed into the melt zone. The arc

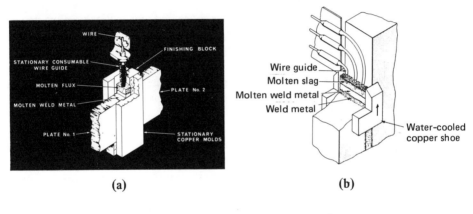

(a)  (b)

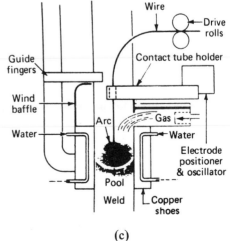

FIGURE 6-12. *Vertical submerged-arc welding (a), electroslag welding (b), and electrogas welding (c). Courtesy American Bridge Division, U.S. Steel.*

(c)

383

goes out and current is conducted directly from the electrode wire by resistance to the electrically conductive molten slag. As welding progresses, the electrode guide tube melts and enters the flux pool. Welding is completed when the molten flux reaches the top of the joint. The process is particularly good for splicing heavy, steel structural members from 0.75 to 2.5 in. (1.90 to 6.35 cm) thick.

### Electroslag Welding

A process that is quite similar to vertical submerged-arc welding is termed *electroslag welding* [Fig. 6-12(*b*)]. In this process, a granular flux is placed in the gap between the plates being welded, and, as the current is turned on, welding takes place in a U-shaped starting block, tack-welded to the bottom of the joint. As the flux melts, a slag blanket from 1 to 1.5 in. (2.54 to 3.81 cm) thick is formed. At this point the arc goes out, and current is conducted directly from the electrode wire through the slag. The high resistance of the slag causes most of the heating for the remainder of the weld.

This process is used to weld metals from 1.5 to 15 in. (3.81 to 38.1 cm) thick. The maximum reported thickness is 40 in. (101.6 cm). The molten metal and slag are retained in the joint by means of copper shoes that automatically move upward, as the weld progresses, by means of a temperature-sensitive mechanism. Two probes control the flow of granular flux from a hopper over the weld zone.

A variation of electroslag welding is *electrogas welding*. The concept is much the same. The main difference is that an inert gas is used to shield the molten metal [Fig. 6-12(*c*)]. The process is used on thinner materials than those associated with electroslag welding.

***Advantages and Limitations.*** The electroslag welding process is quite automatic, once started it will keep going until the job is completed. Heating is uniform, which keeps warpage to a minimum. Runoff tabs can be used to ensure a complete weld. Joint preparation is not required except to remove any heavy mill scale that may be present. The scale interferes with heat transfer to the copper shoes and may prevent a good fit-up, allowing some of the weld metal to run out. Each wire used deposits about 50 lb (22.5 kg) of weld metal per hour. The process can be used for welding hot-rolled carbon steels; low-alloy, high-strength steel; and quenched-and-tempered, low-alloy steels. The weld-metal properties will match the base-metal properties if the right filler metals are used.

The notch toughness of the coarse-grained, heat-affected zone of heat-treated steels may not be as good as that of the base metal. However, if the part is heat treated afterwards, the as-cast structure may be recrystallized and refined. Most electroslag systems are restricted to plates thicker than 0.5 in. (12.70 mm) and cost savings are seldom realized when they are less than 0.75 in. (19.05 mm) thick.

### Gas Shielding for Arc Welding

Inactive shielding gases such as argon and helium have long been considered ideal for protection in welding virtually all materials. However, it was not until after

World War II, when high-purity argon was produced economically on a commercial scale, that it became a reality for shielded-arc welding processes.

The function of the shielding gas, to protect the molten weld metal from the atmosphere, is quite well understood. How the shielding gas affects the arc characteristics is not so universally understood.

The chief factor influencing shielding effectiveness is the gas density. Argon is approximately one and one-third times as heavy as air and ten times as heavy as helium. Argon, after leaving the torch nozzle, forms a blanket over the weld area, whereas helium tends to rise in a turbulent fashion around the nozzle. To get the same equivalent shielding effectiveness, the flow of helium will have to be two to three times that of argon. On the other hand, helium has somewhat different electrical characteristics such that it makes for faster automatic welding and a hotter arc for wide and deep welds. Thus the higher available heat favors the use of helium over argon at the higher current densities as used on thicker plates, especially those of relatively high heat conductivity.

Both argon and helium provide excellent arc stability with dc power. However, with ac, which is almost exclusively used on aluminum and magnesium, argon yields excellent arc stability and good cleaning action while helium is slightly more erratic and the resultant weld-bead appearance is usually dirtier. Thus, with ac power argon finds wide acceptance, but helium is definitely preferred for overhead welding.

Carbon dioxide is the newest of the shielding gases. It produces a bowl-shaped weld penetration. A comparison of the type of bead obtained with argon, helium, and carbon dioxide is shown in Fig. 6-13. In welding, the carbon dioxide breaks down into carbon monoxide and oxygen. High deposition rates result when the carbon monoxide and oxygen recombine to add heat during welding. The carbon dioxide breakdown requires balancing deoxidizers and the filler metal to prevent porosity. To keep spatter within bounds and to get the best welds with carbon dioxide a short uniform arc is needed.

### Gas-Tungsten-Arc Welding (GTA)

Gas-tungsten-arc welding has popularly been called "TIG" since it is based on a *tungsten* electrode used with an *inert gas*. A wide variety of gases and gas mixtures can be used to prevent oxidation of the electrode and protect the weld.

The basic process consists of a torch with a tungsten electrode and a gas shield for both the molten weld metal and the tungsten electrode. The process is used with or without filler metal, depending upon the material, thickness, and joint configuration (Fig. 6-14). The materials may be added by hand, for manual operations, or by wire-feed mechanisms for semi-automatic and fully automatic welding.

The tungsten electrode may range in size from 0.005 in. to 0.375 in. (0.13 to 9.52 mm) and may be of three types: pure tungsten, thoriated tungsten, or zirconiated tungsten. The oxides of thorium and zirconium help the electrode hold a given configuration and improve the ease of electron emission. More electrons in the arc make for easier starting. There is very little loss of tungsten since its boiling point is 10,700°F

Argon
Spatter; narrow deep penetration

Helium
Hot arc; broad even penetration

FIGURE 6-13. *A comparison of the effect of various shielding gases on the weld bead. Courtesy* Welding Design & Fabrication.

Carbon dioxide
Stable arc; finger penetration

(5860°C). Of course, if it is allowed to touch the base metal, some tungsten will be deposited.

Gas-tungsten-arc welding is done essentially with DCSP. However, where oxides must be removed, as on aluminum and magnesium, ac current is used with a superimposed high-frequency current. In welding other metals such as stainless steel and low-carbon steel in the range of 0.015 in. to 0.030 in. (0.38 to 0.76 mm) thick, the high-frequency current may be used to help stabilize the arc and to ease starting. When a dc power source is used and ionization is well established, the high-frequency current may be shut off with a convenient foot switch or one located in the handle of the torch.

In GTA welding of aluminum, ac current is used for best results. When alternating current is used, it passes through zero twice every cycle, hence the polarity is also reversed from positive to negative. During the reverse-polarity portion (work-negative), weld cleaning takes place by electron emission and gas impingement at the work surface. The electrons leaving the work serve to disrupt and displace the oxide film so the metal can more easily flow together.

GTA welding is often used on stainless steel, nickel and cobalt alloys, aluminum and copper alloys, and titanium or other highly alloyed metals where weld purity is essential. GTA welds are noted for their clean, smooth beads. There is seldom any need for grinding and finishing, a big advantage if the welds are hard to reach.

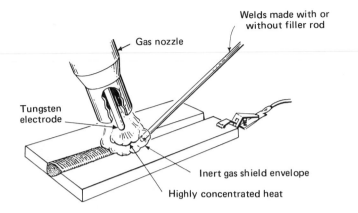

FIGURE 6-14. *Gas-tungsten-arc welding.*

Backing bars are essential for making butt welds in materials that oxidize quickly. A conductive metal bar is grooved out and placed on the underside of the weld. By flooding this groove with a shielding gas or coating the back side of the joint area with flux, it is protected from atmospheric contamination. Another approach is to have a plain strip of the base material flush with the backside of the joint to become an integral part of it. This is often referred to as a "consumable insert."

***Advantages and Limitations.*** GTA welding is especially good for welding metals that tend to oxidize rapidly, such as aluminum, magnesium, and titanium. Although it can be used to weld steel, the process is used primarily for nonferrous metals.

The main disadvantage of GTA welding is the relatively slow speed of operation.

### Gas-Metal-Arc Welding (GMA)

The success of the GTA process stimulated further research to develop a process that could deposit filler metal more easily. A natural solution was to substitute a consumable electrode for that of the tungsten. At first this did not seem practical since large globules of metal melted off and produced considerable spatter. Further research proved that much higher currents would produce a spray of tiny droplets across the arc. This type of metal spray not only produced a satisfactory arc but it also provided relatively deep penetration. The new process was termed "MIG" for metallic inert gas. Although the term is widely used, it is no longer inclusive enough, as not all gases used are inert. The basic equipment required for GMA welding is shown in Fig. 6-15.

Since its initial development, when GMA was thought of primarily as a high-current-density, small-diameter filler wire process, many variations have been produced. These variations include $CO_2$ welding, microwire* welding, and metal inert-gas welding (MIG). Variations of the arc action were also developed.

* Trade name of Hobart Brothers Company.

387

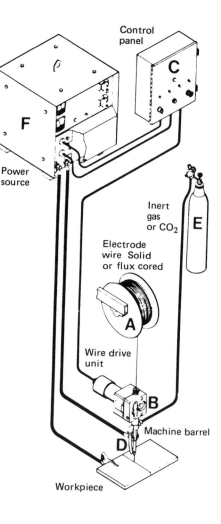

FIGURE 6-15. *Schematic representation of automatic GMA welding. The main components are: continuous consummable-wire electrode, solid or flux-cored—A, wire-drive mechanism powered by a variable-speed motor—B, control panel used to regulate wire-feed rate and all other electrical functions—C, welding gun used to guide the wire and direct gases to the arc-weld pool—D, cylinder of gas used to shield the arc and molten metal from the air—E, power source used to provide the current needed to sustain the welding arc—F. To facilitate handling for manual welding, the gun is not directly attached to the wire-drive unit.*

***CO$_2$ Welding.*** GMA welding with CO$_2$ produces smooth, high-speed welds. Carbon dioxide gas reacts with molten metal by dissociating as the temperature increases. The temperature of the welding arc is about 14,000°F (7760°C) and within the arc column complete dissociation occurs. Some of the oxygen will oxidize the molten metal. The carbon dioxide can react directly with the iron as shown in the following equation, to form iron oxide:

$$Fe(l) + CO_2(g) \rightarrow FeO(l) + CO(g)$$

In the absence of deoxidizers such as silicon and aluminum, the iron oxide can combine with carbon monoxide, which if trapped would form porosity in the weld metal.

By adding a deoxidant, the free oxygen will combine with it and form minute islands of slag that float out to the surface of the molten metal.

CO$_2$ welding is used for welding carbon and alloy steels. The filler-metal deposit is similar to the manual shielded-arc method of globular transfer.

CO$_2$ gas in cylinders is about one-ninth the cost of argon. Thus, in welding light-gauge steels, where there is good fit-up, the CO$_2$ process, because of its low shielding cost and high deposition rate, is often the most economical process available. The process, because of its good penetration, may also be used effectively for fillet and butt welds in materials 0.25 to 1.5 in. (0.63 to 38.10 mm) thick.

*Microwire Welding.* Small-diameter wires (0.020 to 0.625 in.; 0.51 to 15.87 mm) are used to advantage in GMA welding because of the high melting rate that can be achieved. The advance of the welding wire ranges from about 100 to 800 in. per minute (254 to 2032 cm/min) for all metals except magnesium, where speeds reach 1400 ipm (3556 cm/min).

At a given current value, more pounds of metal per unit can be deposited with small-diameter electrodes than with large-diameter electrodes. Also the penetration at a selected current value is greater with small-diameter electrodes than with large-diameter electrodes.

*Metal-Transfer Types.* In addition to the variations of GMA welding mentioned, there are variations in the way the metal can be transferred in the arc. These variations are termed spray transfer, globular transfer, and short-circuiting arc transfer, as shown in Fig. 6-16.

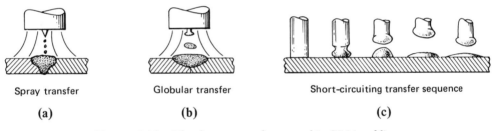

Spray transfer      Globular transfer     Short-circuiting transfer sequence

(a)        (b)        (c)

FIGURE 6-16. *The three types of arcs used in GMA welding.*

*Spray Transfer.* The spray-transfer arc is based on achieving a high enough current density for a given wire size that will cause the metal to form fine droplets and be propelled across the arc at high speed. The best spraying action is obtained when a shielding gas of argon is used. The spray transfer operates in a rather limited range for each wire size. The upper limit is set by the effects of the arc force and the lower limit by the droplets changing to globules. Small-diameter wire electrodes are used so that the current required to obtain a spray will not be excessive for lighter gage ($\frac{1}{8}$ in. thick; 3.175 cm) materials. On the other hand, high heat input with larger diameter electrodes (up to $\frac{3}{32}$ in.; 2.381 mm) can be used in welding heavy (1 in. thick; 2.54 cm) plate materials.

*Short-Circuiting Arc.* * Early research on MIG welding produced a droplet on the end of the electrode that became larger than the wire diameter and was finally forced

* "Short-arc" welding is a trade name of the Linde Company.

TABLE 6-5. *Characteristics of Shielding Gases Used for GMA Welding. Reprinted from the December, 1971, issue of* Welding Design & Fabrication. *Copyright 1971 by Industrial Publishing Co., Division of Pittway Corp.*

| Shielding Gas | Chemical Behavior | Select for: |
|---|---|---|
| 100 % Ar | Inert | All metals except steels |
| Ar + He (20 %–80 % to 50 %–50 %) | Inert | Aluminum, magnesium and copper alloys for greater heat input than straight argon, and to minimize porosity by giving better arc action than straight helium |
| Ar + 25 %–30 % $N_2$ | Inert | Greater heat input on copper; better arc action than 100 % $N_2$ |
| Ar + 1 %–2 % $O_2$ | Slightly oxidizing | Stainless and alloy steels; some deoxidized copper alloys |
| Ar + 20 %–50 % $CO_2$ | Oxidizing | Carbon and some low-alloy steels, short-circuiting arc |
| Ar + 10 % $CO_2$ + 5 % $O_2$ | Oxidizing | Various steels (European practice) |
| 90 % He + 7.5 % Ar + 2.5 % $CO_2$ | Slightly oxidizing | Stainless steels for good corrosion resistance, short-circuiting arc |
| 60 % to 70 % He + 25 % to 35 % Ar + 4 % to 5 % $CO_2$ | Oxidizing | Low-alloy steels for toughness, short-circuiting arc |

off, causing considerable spatter and lack of control. A different approach to the problem was to allow the electrode to touch the metal and short circuit momentarily. Ordinarily this short circuiting would cause the electrode to overheat. Now, however, the dc welder is built with a reactance to limit the surge of current. The limit also has a *pinch effect* on the molten metal at the tip of the electrode which is sufficient to cause separation and the arc is again established. This rapidly reoccurring sequence is repeated more than 100 times per sec. Generally, small-diameter wires are used (0.020 to 0.093 in.; 5.508 to 2.293 mm) on DCRP at a maximum of about 150 amps. The short-circuiting transfer permits welding on thinner sections with greater ease and is extremely practical for welding in all positions. A combination of argon and carbon dioxide has proven to be a very effective shielding gas. Better pinch effect. reestablishment of the arc, and less spatter are the improvements over using carbon dioxide alone. The effect of using various gases and gas combinations is shown in Table 6-5.

***Globular Transfer.*** Globular transfer for GMA welding is similar to that of manual shielded-arc welding. With this type of transfer and $CO_2$ shielding, high welding speeds can be achieved on carbon and alloy steels.

GMA welding is best done with a constant-potential power supply. It is easier to set and maintain the desired burnoff rate. The wire-feed rate control, therefore, also controls the current.

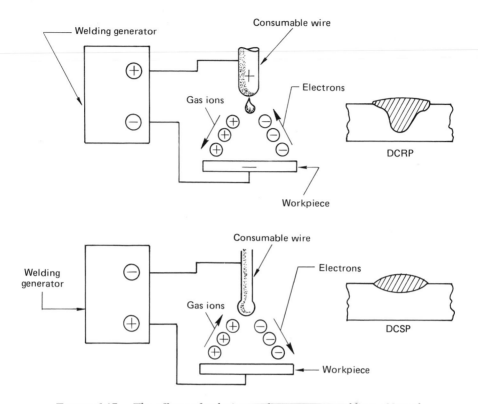

FIGURE 6-17. *The effects of polarity on direct-current welding. Note the effect on weld penetration.*

DCRP is used in GMA welding. It provides a more concentrated weld pool and deeper penetration (Fig. 6-17). Also there is a better cleaning action on the surface of the metal, which is very important when welding metals that have surface oxides, such as aluminum and magnesium. Typical operating data for GMA welding is given in Table 6-6.

***Advantages and Limitations.*** The main advantage of GMA welding is that it can produce high-quality welds at high speeds and there is no flux to remove. It is a very versatile process that is used on both light- and heavy-gauge structural plates. It is used with comparative ease on carbon steels, low-alloy steels, stainless steels, aluminum, magnesium, titanium, copper, nickel, and zirconium. With the use of the short-arc a relatively cool weld pool is produced, reducing burn-through on thin materials and also making it possible to weld in all positions.

With spray transfer, heavy wire electrodes will melt readily, making deep penetration possible. Since the individual droplets are small, they can be easily directed as required. Since the metal transfer is produced by an axial force which is stronger than gravity, the process can be effectively used on out-of-position welding.

TABLE 6-6. *Typical Operating Data for GMA Welding*

| Type of Arc | Welding Position | Wire Diameter (in.) | Current (dcrp) (amp) | Voltage | Travel Speed (ipm) |
|---|---|---|---|---|---|
| Shorting | Flat or horizontal | .030 | 130–140 | 22–24 | 5–15 depending on width of weave |
| | Vertical or overhead | .030 | 110–130 | 20–22 | 5–15 |
| Spray | Flat or horizontal | .062 | 325–375 | 25–28 | 5–15 |

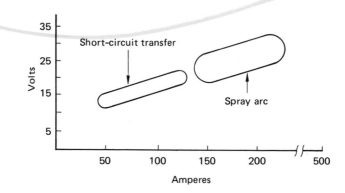

FIGURE 6-18. *The general-current areas of short-circuiting and spray-arc welding of steel.* *Courtesy* Welding Engineer.

It is best for heavy-gauge metals, since it has a tendency to burn through on light-gauge metals.

A comparison of the general range of current used for spray and short-circuiting arc are shown in Fig. 6-18. Much of the carbon and low-alloy steel welding is being done with straight carbon dioxide or a mixture of carbon dioxide and argon. Shown in Table 6-7 is an appropriate cost comparison of welding a 0.25-in. (0.63 mm) thick square-butt joint in mild steel with manual stick electrode, GMA, and submerged-arc welding.

**Gas-Metal-Arc Spot Welding**

A variation of the GMA process is spot welding. The same power sources and wire feeders are used, but a timing device and special gun are added. Shown schematically in Fig. 6-19 are the steps required to make an arc spot weld. The arc actually penetrates the top sheet and welds into the bottom sheet with a weld nugget. This type of spot welding has the advantage of being portable and requires access to only one side of the workpiece. Only a small amount of surface cleaning is required.

TABLE 6-7. *Approximate Cost Comparisons of Shielded Arc, GMA, and Submerged Arc*

| Cost Data for $\frac{1}{4}$ in.$^2$ Steel Butt Joint | Shielded Electrode | GMA $CO_2$ | Submerged Arc |
|---|---|---|---|
| Operating factor | 40% | 50% | 35% |
| Welding speed, ipm | 8 | 15 | 24 |
| Welding current | 300 | 420 | 800 |
| Arc time, h/100 ft of weld | 2.5 | 1.33 | .84 |
| Electrode, lbs/100 ft | 30 | 15 | 17 |
| Welding flux, lb/100 ft | — | .92 | 2.10 |
| Labor cost/100 ft at 3.50/h | 18.75 | 7.60 | 8.40 |
| Electrode cost/100 ft | 4.50 | 6.40 | 2.21 |
| Power cost/100 ft | .50 | .50 | .50 |
| Direct costs | $23.75 | $15.42 | $13.21 |
| Overhead | 18.75 | 7.60 | 8.40 |
| Total/100 ft of weld | $42.50 | $23.02 | $21.60 |

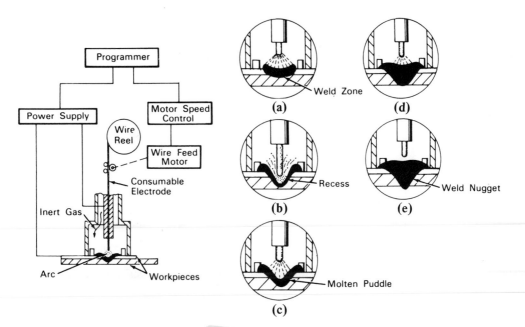

FIGURE 6-19. *GMA spot welding. After the arc is struck, the temperature and length of the arc are varied by a predetermined program of (a) preheat, (b) blowing the metal out of the weld zone to make a recess, (c) melting the sides of the recess, (d) forming the weld nugget, and (e) interruption of the arc.*

### Plasma-Arc Welding

Plasma is often referred to as the fourth state of matter: liquids, solids, gases, and plasma. Plasma is not new, it has always existed in varying degrees whenever an arc is formed, like the air in the path of a lightning flash. Basically, plasma is a partially ionized gas produced by the passage of gas through an electrical field which separates into free electrons, neutrons, and ions. The energy required to effect this dissociation of the gas is very high. However, upon contact with some exterior surface, the latent heat is released as the atoms recombine. The result is a plasma jet driven

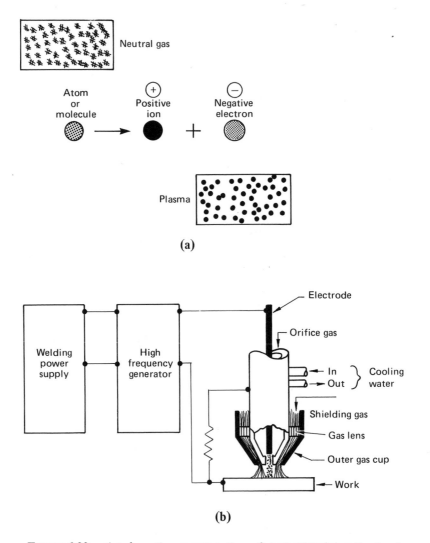

FIGURE 6-20. *A schematic representation of ionization (a). Sectional view of the plasma-welding torch (b).*

at a supersonic speed and yielding temperatures in excess of 50,000°F (27,500°C). A schematic representation of ionization is shown in Fig. 6-20(a).

A method commonly used to initiate plasma flow is to generate a high-frequency spark between the electrode and the nozzle [Fig. 6-20(b)]. After the start-up, direct current from the main power supply takes over to maintain ionization. The torch shown has a gas lens, which is simply a wire-mesh grating designed to eliminate turbulence in the flow of the shielding gas. The nozzle is commonly made of copper through which tap water is circulated for cooling purposes. The flow of inert gases also serve in the cooling operation, although the main function is that of shielding the workpiece from the atmosphere. The boundary layer of gas near the walls of the torch remains relatively cool with the discharge current concentrated at the more conductive central region of the plasma. This effect is often referred to as "thermal pinch" (Fig. 6-21).

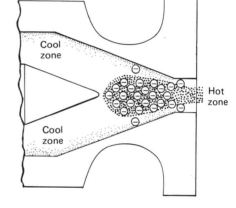

FIGURE 6-21. *Electrons move more freely in the center hot zone of the plasma than in the cooler boundary layers. Therefore the majority of the electrons and the discharge current is concentrated in the hot zone known as the "thermal pinch."*

There are two main types of plasma torches: transferred arc and nontransferred arc. The nontransferred arc has both electrodes in the torch. The transferred arc, in which the workpiece is positively charged, is far superior since it transfers more energy and is less susceptible to magnetic deflection. It is of course limited to conductive materials. The two types of plasma torches are shown, and compared to the GTA torch, in Fig. 6-22.

Argon is suitable as an orifice and shielding gas for virtually all metals but it does not always produce optimum results. For example, a hotter flame may be required, in which case hydrogen is added. Stainless steel, Iconel, nickel, and copper-nickel alloys are metals for which 5 to 15 % helium is added to the orifice and shielding gases.

Plasma torch welding can be divided into two basic areas: low and high operating currents. The low operating current is known as the *plasma needle arc.**

---

* Trade name of the Linde Company.

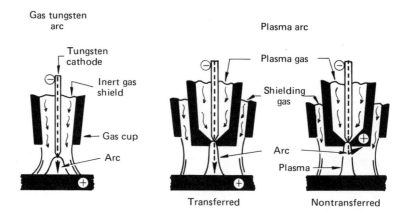

FIGURE 6-22. *The two basic types of plasma torches compared with the GTA torch. Note that the arc shape in the GTA torch is conical and expands rapidly. Courtesy Union Carbide Corp., Linde Division.*

***The Plasma Needle Arc.*** The needle arc operates with a transferred current in the range of 0.1 to 10 amp with 1 cubic foot per hour gas-flow rate. The collimated shape of the arc is needle-like in appearance. Its total divergence is only 6°, as compared to 45° for the GTA process. This collimated arc can stand relatively large variations in arc length before welding behavior will be seriously affected. The less critical the length the more easily it is adapted to automated equipment.

***High-Current Plasma Welding.*** The high-current plasma torch has several holes in the nozzle for welding materials $\frac{3}{32}$ in. (2.38 cm) thick or more, as compared to a single port for the needle arc (Fig. 6-23). The effect of the side ports is to squeeze the plasma jet's normally circular heat pattern into a narrow, elongated shape on the workpiece. When the multiports of the nozzle are aligned with the centerline of the weld, greater welding speeds are attained and the heat-affected zone is narrower.

The low-current, needle-arc weld puddle is narrow and slightly depressed but otherwise identical to that of GTA welding. With the high-current plasma process, the weld puddle area has a hole pierced through the entire thickness of the weldment [Fig. 6-24(a)]. This void is known as the "keyhole" effect. The keyhole is formed at the leading edge of the weld puddle, where the forces of the plasma jet displace the molten metal to permit the arc to pass completely through the workpiece. As the torch progresses, the molten metal, supported by surface tension, flows in behind the keyhole to form the bead. The keyhole ensures complete penetration and weld uniformity. Since the weld puddle is supported by surface tension, close-fitting backing bars are not required but underbead protection should be provided by a channel filled with shielding gas.

***Advantages and Limitations.*** The low-current needle arc is well suited for welding lightweight, thin-wall structures up to $\frac{1}{32}$ in. (0.793 cm) thick. It has been used to good advantage in welding complicated structures such as radiators, combustion

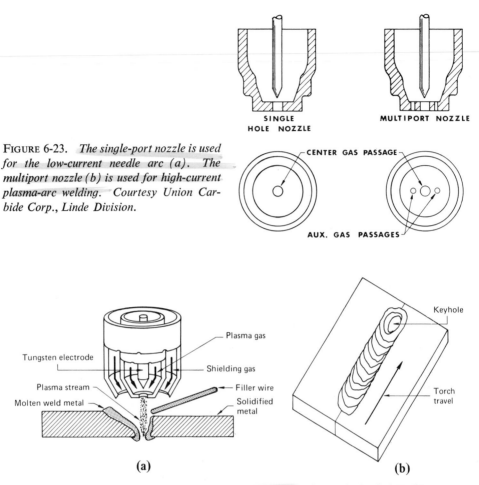

FIGURE 6-23. *The single-port nozzle is used for the low-current needle arc (a). The multiport nozzle (b) is used for high-current plasma-arc welding.* Courtesy Union Carbide Corp., Linde Division.

FIGURE 6-24. *Schematic of the plasma-arc torch utilizing the keyhole technique (a). The keyhole effect seen throughout the plasma-welding operation indicates complete penetration (b). This type of weld can be made on thin gages of titanium, stainless steel, or aluminum at high speeds.*

chambers, exhaust manifolds, and air ducting. The concentrated heat makes the process ideal for welding small instrument parts, fine wires, metal mesh, and similar components.

High-current plasma-arc welding shows its greatest advantage in making square-butt joints in metals ranging from 0.090 to 0.250 in. (2.29 to 6.35 mm) where keyholing is appropriate. Keyholing can be used on thick sections but the speed is normally lower than that obtained with GTA welding.

Plasma welding is less sensitive to edge mismatch than GTA welding. More variations in standoff distance can be tolerated so sections of different thickness can be welded without trouble. Plasma torches can be operated at 400 amp continuously and at 1000 amp for short periods. With this added power, the plasma torch produces a bead about one-third as wide per unit depth as GTA welding.

Perhaps the main disadvantages of the plasma arc are that it is noisy and the original cost of the equipment is high, being about four or five times that of GTA or GMA welding.

### Plasma-Mig Welding

A newer welding process has been developed by combining a plasma-arc torch with that of the MIG torch. The process has been independently but simultaneously developed by Philips Research Laboratories in the Netherlands and by NASA in the United States.

In the combination arrangement, the current is passed through the tungsten to establish the arc (Fig. 6-25). The gas, a standard argon-helium mix, is routed through the inside of the tungsten cylinder to form a plasma. The wire is moved into the plasma as in the GMA process. The combined high current density and plasma melt the feed wire rapidly while providing an intense heat flux at the workpiece.

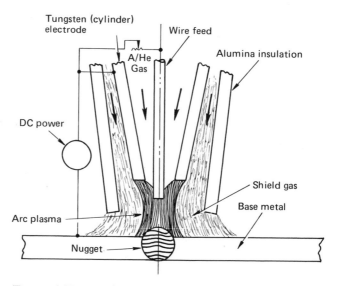

FIGURE 6-25. *A schematic drawing of the plasma-MIG torch.*

Using straight polarity, a narrow arc is formed at comparatively low welding currents through the wire. The arc can be used for deep penetration of thick material and for high-speed welding of thin plate. The process is highly suitable for pre-programmed control.

### Plasma-Arc Cutting

Plasma-arc cutting is particularly advantageous for severing hard, flame-resistant metals. The torch used is a transferred-type arc with a nonconsumable electrode. The principal difference between this torch and the torch used in welding is the absence of the shielding gas. The plasma flow is initiated through a pilot arc established by a high-frequency generator. Current to maintain the jet is provided by a direct-current power supply.

The orifice gas, fed to the torch at the rate of 50 to 350 cubic feet per hour, is determined by the particular application. Aluminum and magnesium are cut with nitrogen, nitrogen-hydrogen, and argon-hydrogen. Stainless steel thicker than 2 in. (50.8 mm) is best cut with argon and hydrogen.

Oxygen is an excellent cutting gas although its corrosive effect on the electrode makes it expensive. The best way of using oxygen is to introduce it downstream in the flow orifice, as shown in Fig. 6-26.

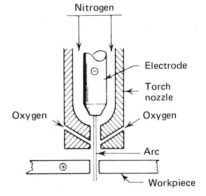

FIGURE 6-26. *Oxygen introduced downstream in the flow orifice eliminates oxidation and erosion of the electrode.*

***Advantages and Limitations.*** The plasma-arc torch has several advantages over the conventional oxyacetylene cutting torch:

1. It can cut faster and leave a narrower kerf (width of cut). Mild steel, 0.75 in. (19.05 mm) thick, can be cut at 70 ipm (1778 mm/min), compared to oxyacetylene cutting at 20 ipm (508 mm/min.).
2. Operating costs are less, as shown in Fig. 6-27.
3. The plasma arc can be used to cut any metal, since it is primarily a melting process.

The main disadvantage of plasma-arc cutting is the high initial cost of the equipment. Thus, only those who do quantity cutting will be able to amortize the cost in a reasonable time.

Shown in Table 6-8 is a brief summary of the arc-welding processes that have been presented and their use on common metals.

### Plasma-Flame Spraying

The plasma-arc metallizing torch is essentially of the nontransferred-arc type. A simplified schematic sketch is shown in Fig. 6-28. The metallic or ceramic materials are introduced into the flame and reach the molten or plastic state in a fraction of a second. The high velocity of the stream leaving the gun propels the particles to the

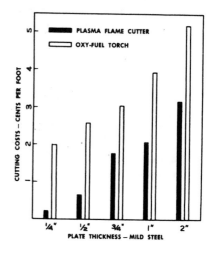

FIGURE 6-27. *The comparative costs of oxyacetylene cutting of mild steel and plasma cutting. Courtesy* Thermal Dynamics.

part at approximately 1000 ft/sec (30,000 cm/sec). Upon impact with the base materials, these particles resolidify to form the coating.

The process develops temperatures as high as 30,000°F (5380°C), so it is necessary to use chilled water for cooling. The sprayed parts, only inches away from the gun, remain at temperatures ranging from 200 to 400°F (93–204°C).

Although some plasma-arc guns are made to be hand-held, it is best in production work to have them machine-mounted and equipped with traversing mechanisms. Some coating is done in enclosed, soundproof rooms with windows of darkened glass, which permits monitoring.

All metals, ceramics, and certain polymer substrates can be coated by the plasma-spray process. Most form a mechanical bond with the substrate, but certain metallic, exothermic coatings achieve true metallurgical bonds. The general range of bond strength is from 3000 to 8000 psi (2.11 to 5.63 kg/mm$^2$). Due to stresses and differences in the coefficient of expansion, the thickness of the coating is limited to prevent chipping and spalling. The coatings have a porosity rating ranging from 0.5 to 10 %, depending on the material.

Plasma-flame–sprayed coatings serve many purposes. One of the main applications is in providing wear-resistant surfaces. As an example, a hardened steel part used on a multifunction IBM card showed excessive wear after less than 40,000 punched cards. After plasma-flame spraying, no wear could be measured after 1.5 million cards.

In addition to wear applications, plasma-sprayed coatings are used to impart electrical conductivity, as in spraying copper on glass. Other applications are those of providing thermal barriers on high-friction-coefficient surfaces (greater than 1.0) or on low-coefficient surfaces (less than 0.15).

Plasma plating also serves in the fabrication of tungsten cermets, metal oxides, and other materials by applying successive coatings to a mandrel, which is later removed. These parts have densities up to 90 % of theoretical. The process is shown schematically in Fig. 6-29.

TABLE 6-8. *Welding Processes Summary Chart for Common Metals*

| Material | Oxyacetylene | Stick Electrode | GTA | GMA | Plasma Arc (PA) |
|---|---|---|---|---|---|
| Aluminum & alloys | Weld or braze | Satisfactory, not normally used | Best for thin gage | Best for medium & thick gage | Foil & thin gage. Medium gage** |
| Brass, naval | Weld or braze | * | GTA braze | * | PA braze |
| Cu, deoxidized | Weld or braze | * | Best for thin gage | Best for medium & thick gage | Best on foil & thin gage. Medium gage** |
| Iron, cast | Weld or braze | Satisfactory | * | * | * |
| Iron, wrought | Weld or braze | Most popular | * | * | * |
| Lead | Weld | * | * | * | * |
| Magnesium | * | * | Best for thin gage | Best for medium & thick gage | Best on foil & thin gage. Medium gage** |
| Steel, low-carbon | Used for thin material | Most popular | Satisfactory | Satisfactory | Used on foil & thin gage. Medium gage** |
| Steel, high-carbon | Used for thin material | Most popular | Satisfactory, not popular | Satisfactory | Used on foil & thin gage. Medium gage** |
| 304 Stainless | Weld or braze | Satisfactory | Best for thin gage | Best for medium & thick gage | Best on foil & thin gage. Medium gage** |
| Titanium | * | Most popular | Satisfactory, not popular | Best for medium & thick gage | Best on foil to medium gage |

* Signifies the process is not normally used.
** High-current plasma arc.

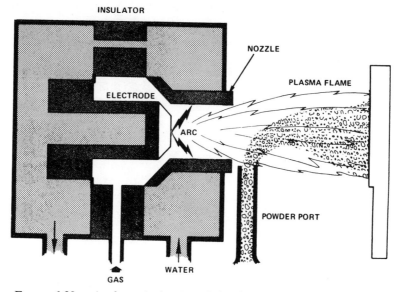

FIGURE 6-28. *A schematic drawing of the plasma-flame spraying process. The plasma flame is developed through dissociation of argon-hydrogen or nitrogen-hydrogen gas as it passes through the electrical field. Courtesy Society of Manufacturing Engineers.*

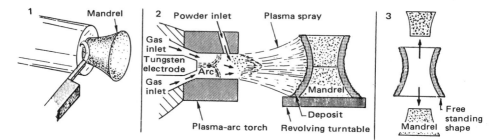

FIGURE 6-29. *Fabricating a tungsten part by plasma-plating a mandrel. The mandrel is designed and made (a). The mandrel is sprayed with the plasma-arc torch. The process can be done in an atmosphere controlled chamber (b). The mandrel is removed and the free standing shape is the part (c). Spraying is often followed by sintering and light machining on the outside if necessary. Courtesy Machine Design.*

## GAS WELDING

Gas welding is a process of joining metals with a high-temperature flame. The most common gases used are oxygen and acetylene. Acetylene is referred to as a fuel gas and oxygen as a gas that supports combustion. Other fuel gases used in metal cutting are propane and natural gas. A newer gas that has been used for a few years is stabilized methylacetylene propadiene, usually referred to as MAPP*. Occasionally other "new" fuels will appear but these are mostly modified propane-based fuels.

In recent years the Occupational Safety and Health Act (OSHA) has forced fabricators to search for fuels that are both safe and productive. This search has shown that fuel gases can have a major effect on fabricating costs and quality as well as worker safety.

*Fuel-Gas Properties.* The three properties of a fuel gas that most directly affect production are flame temperature, flame-propagation rate, and gross heat of combustion. Of these three, the latter is the most important.

Flame temperature, either measured or calculated, refers to the maximum figure for a given flame under ideal conditions and with a specific oxygen/fuel ratio. It does not indicate either the temperature or heat distribution throughout the flame.

The flame-propagation rate refers to the rate at which a flame front progagates through a column of a fuel gas mixed with oxygen. A fuel gas such as acetylene, which has a high flame-propagation rate (17.7 ft/sec; 45 cm/sec), burns stably when leaving a cutting-torch tip at port velocities of 350 to 750 in./sec (889 to 1905 cm/sec). A flame of low-propagation-rate fuel gas such as propane (5.5 ft/sec; 14 cm/sec) tends to be blown off the tip at high port velocities, resulting in unstable burning conditions and poor heat transfer.

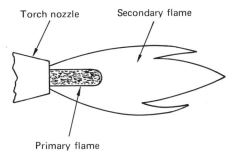

FIGURE 6-30. *Both acetylene and MAPP, when mixed with oxygen, have most of the heat concentrated in the primary cone.*

Gross heat of combustion is given in Btu's per cubic foot of fuel since that is the way it is used. Gross heat considered by itself is misleading. For example, propane, which is considered a "cool" gas, is hot in terms of Btu's per cubic foot but cool in terms of temperature. Its use is generally limited to brazing or soldering. Two gases with high Btu per cubic foot and high flame temperature are MAPP and acetylene. They also have good heat distribution in both the primary and secondary flame cone, as shown in Fig. 6-30. Propane, natural gas, and propylene have most of the heat concentrated in the secondary cone. All fuel gases have a critical distance or, as it is sometimes referred to, a "coupling" distance. For the oxyacetylene flame it is

* Trademark of Dow Chemical Company.

$\frac{1}{16}$ in. (1.587 mm) out from the blue, primary flame. If the distance is varied too much, the cool part of the flame comes into play and welding and brazing difficulties result. The coupling distance for MAPP is less critical, which translates into lower, less expensive operator skills.

*Fuel-Gas Safety.* All fuel gases are potentially dangerous and should be treated with respect. However, the safety hazards can be minimized by a review of fuel-gas safety properties. The first property to consider is the inherent stability of the chemical bond in the fuel-gas molecule. Acetylene, with its triple bond [Fig. 6-31(*a*)], is the most unstable of all fuel gases. It will decompose under mechanical shock loading; that is, a dropped acetylene cylinder can explode. MAPP, on the other hand, which has a triple bond in the methylacetylene portion and a double bond in the propadiene portion [Fig. 6-31(*b*)], is relatively stable.

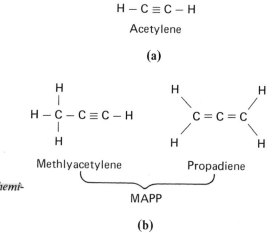

$$H - C \equiv C - H$$

Acetylene

**(a)**

Methlyacetylene        Propadiene

MAPP

**(b)**

FIGURE 6-31. *A comparison of the chemical structure of acetylene and MAPP.*

Common practice limits acetylene line pressure to a maximum of 15 psi (1.05 kg/cm²). However, tests by the U.S. Bureau of Mines reveal that acetylene can explode at pressures as low as 5.9 psi (0.41 kg/cm²). That is the reason why acetylene in storage tanks must be absorbed in a solution of acetone which is packed in inert, porous filler material such as Fuller's or diatomaceous earth. As a result, a standard 240-lb (108-kg) acetylene cylinder actually contains only 20 lb (9 kg) of acetylene.

All fuel gases will explode when mixed with oxygen or air under suitable conditions. Acetylene forms an explosive vapor with as little as 2.5 % and up to 80 % of the volume. The explosive limits of MAPP and acetylene are compared in Fig. 6-32.

*Fuel-Gas Economics.* Many variables make it difficult to assess which fuel gas can do a job most economically. Perhaps the best way is by actual trial runs. However, to be accurate these trials require close monitoring of both the fuel gas and oxygen flow rates as well as developing some method of determining that equal amounts of work are performed. The amount of oxygen consumed per cubic foot of fuel burned is given in Table 6-9. Fuel gases must not be equated cylinder-by-cylinder since a 240-lb (108-kg) acetylene cylinder contains 20 lb (9 kg) of acetylene whereas a 120-lb (54-kg) MAPP cylinder contains 70 lb (31.5 kg) of MAPP.

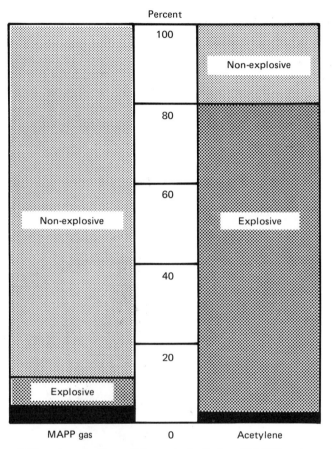

FIGURE 6-32.  *A comparison of the explosive limits of MAPP and acetylene in air.*

TABLE 6-9.  *Oxygen Consumed Per Cubic Foot of Fuel Burned. Courtesy Modern Machine Shop*

| Fuel | Flame Temperature | Total Heat (Btu/cu ft) | Cu Ft of Oxygen Per Cu Ft of Fuel |
|---|---|---|---|
| Acetylene | 5589 | 1470 | 1.4 |
| MAPP gas | 5301 | 2406 | 2.5 |
| Propylene | 5193 | 2371 | 3.5 |
| Natural gas | 4600 | 1000 | 1.9 |
| Propane | 4570 | 2498 | 4.3 |

In terms of comparing actual work done by various gases, it appears that flame cutting may be the easiest to assess. Data indicates that MAPP flame cutting is about 15 % faster than acetylene, 17 % faster than propylene, and 21 % faster than propane or natural gas. Actual production figures may very widely due to the many operating variables.

Advantages of gas welding:

1. The gas flame is generally more easily controlled and is not as piercing as shielded-arc welding. Therefore it is used extensively for sheet-metal fabrication and repairs.
2. The oxyacetylene or oxy-MAPP flame is versatile. It may be used for brazing, preheating, postheating, heating for forming, heat treatment, flame cutting, etc.
3. The gas welding outfit is very portable (see Fig. 6-33).

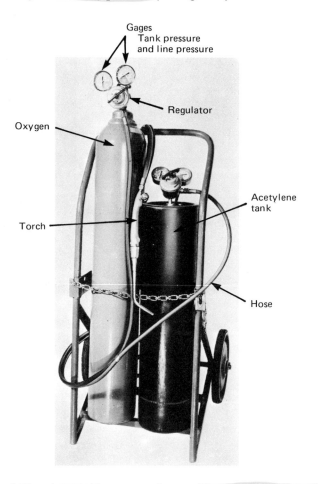

FIGURE 6-33. *A portable oxyacetylene welding outfit, complete with carrying cart. Courtesy Union Carbide Corp., Linde Division.*

Limitations of gas welding:

1. The process is much slower than arc welding.
2. Harmful thermal effects are aggravated by prolonged heat. This often results in increased grain growth, more distortion and, in some cases, a loss of corrosion resistance.
3. Most oxyacetylene tanks are leased and, when kept beyond the specified time, demurrage charges are incurred.
4. There are safety problems in handling gases.
5. Flux applications and the shielding provided by the secondary envelope of the flame is not nearly as positive as that supplied by the gases used for GTA and GMA welding.

## BRAZING

Brazing is a process of joining metals with a nonferrous filler metal that has a melting point below that of the metals being joined. By AWS definition the melting point of the filler metal will be above 800°F (427°C). Below this temperature are the solders. The filler metal must *wet* the surfaces to be joined, that is, there must be a molecular attraction between the molten filler material and the materials being joined. The brazing alloy, when heated to the proper temperature, flows into the small joint clearances by capillary action. A limited amount of alloying occurs between the filler metal and the base metal at elevated temperatures (Fig. 6-34). As a result, the strength of the joint when properly made may exceed that of the base material. The strength is attributed to three sources: atomic forces between the metals at the interface; alloying, which comes from diffusion of the metals at the interface; and intergranular penetration.

The heat for brazing may be provided in many different ways, the most common of which are by torch, induction, furnace, and hot dipping.

*Torch Brazing.* Torch brazing is one of the oldest and most widely used methods of heating for brazing, being versatile and adaptable to most jobs. It is especially good for repair work in the field and for small-lot jobs in the shop. Being a manual operation, torch brazing may have high labor costs and the skill and judgment of the torch operator will determine the quality of the work.

Torch brazing may be done with the use of natural gas and oxygen, air and acetylene, butane, propane, MAPP, and the regular oxyacetylene. If the latter is used, a neutral flame is recommended that is a balanced condition of acetylene and oxygen. A slightly reducing flame (excess acetylene) may be used to avoid oxidation of the metal. The filler metal is touched to the joint when the flux becomes a clear liquid. It is the parent metal, not the flame, that transfers the heat to the brazing alloy.

*Induction Brazing.* The heat for induction brazing is furnished by an ac coil placed in close proximity to the joint (Fig. 6-35). The high-frequency current is usually provided by a solid-state oscillator which produces a frequency of 200,000 to 5,000,000

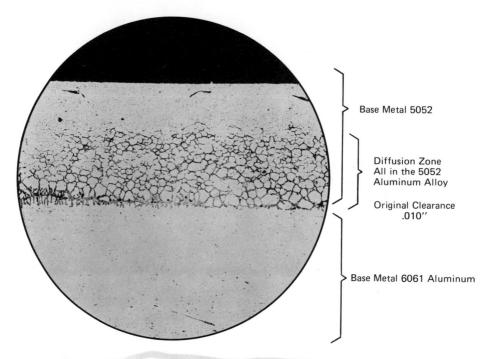

Base Metal 5052

Diffusion Zone
All in the 5052
Aluminum Alloy

Original Clearance
.010"

Base Metal 6061 Aluminum

FIGURE 6-34. *In brazing, a limited amount of alloying occurs between the filler metal and the base metal at elevated temperatures. Courtesy Wall Colmonoy.*

Hz. These alternating high-frequency currents induce opposing currents in the work, which, by electrical resistance, develop the heat. This method has the advantage of providing: (1) good heat distribution, (2) accurate heat control, (3) uniformity of results, and (4) speed. It is especially good for certain types of repetitive work that require close control.

*Furnace Brazing.* Parts that are to be furnace brazed have the flux and brazing, material preplaced in the joint. If the furnace has a neutral or shielding atmosphere the flux may not be necessary. The furnace may be batch or conveyor type. Automatic controls regulate both time and temperature and, where applicable, atmosphere. Shown in Fig. 6-36 are the three steps used in copper-hydrogen furnace brazing.

*Dip Brazing.* Dip brazing gets its name from the fact that parts are jigged (or in some cases the parts are self jigging, as shown in Fig. 6-37) and are placed in a chemical or molten-metal bath maintained at the correct brazing temperature. In a chemical bath the parts are first thoroughly cleaned, then assembled with a filler-metal preform placed on or in the joint. The bath, usually molten salts, is maintained at a higher temperature than the filler metal being used. After dipping, the parts are removed and immediately cleaned to remove the flux. The number and size of the parts to be brazed is limited only by the capacity of the bath and the handling facilities.

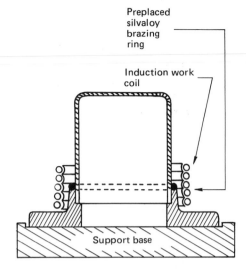

Preplaced
silvaloy
brazing
ring

Induction work
coil

Support base

FIGURE 6-35. *An example of external induction brazing. The preform brazing filler material is preplaced in close contact to the work to assure melting at the proper temperature. Courtesy Engelhard Industries.*

Staked and pasted

Heat to 2000°F
under hydrogen atmosphere

FIGURE 6-36. *The three phases of furnace brazing. The hydrogen acts as a reducing agent removing any light oxide film from the surfaces of the components. This technique eliminates the need for fluxing agents. In addition, the hydrogen protects the surface from oxidation during the heating and cooling cycle. Thus clean, bright assemblies having high joint strength are produced. Courtesy Society of Manufacturing Engineers.*

Brazed joint

## Brazing-Joint Design

Two factors are especially important in brazing-joint design—area and clearance. An example of the area principle can be shown in designing a butt or lap joint (Fig. 6-38). Lap joints are used in preference to the butt or scarf joint. An overlap of three to four times the thinnest member will usually give the maximum efficiency.

Overlaps greater than this lead to insufficient penetration, inclusions, etc. This joint is also recommended for the leak tightness and good electrical conductivity. Some principles of joint design are illustrated in Fig. 6-39.

***Brazed-Joint Clearance.*** The joint clearance determines the thickness of the alloy film that will be formed between the parts. This has an important influence on the strength of the joint. Maximum strength is obtained when there is just enough room

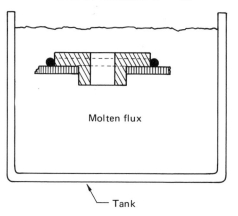

Molten flux

FIGURE 6-37. *Parts to be dip brazed are first placed in jigs or, as shown here, the parts may be self-jigging.*

Tank

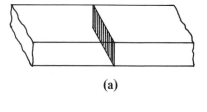

(a)

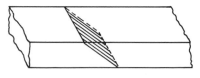

(b)

FIGURE 6-38. *The area principle in braze joints. The butt joint (a) is very poor. A scarf butt joint (b) is much better. A lap joint (c) should provide an overlap of 3t.*

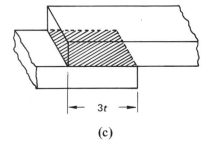

3t

(c)

for the alloy to flow through the joint when heated and the flux and gases are allowed to escape. The ideal clearance for production work ranges between 0.002 and 0.005 in. (0.050 to 0.127 mm). Joints below 0.001 in. and above 0.008 in. (0.025 and 0.203 mm) should be avoided if at all possible.

In determining the clearance, the thermal expansion of the parts must be considered and sufficient clearance allowed so that at brazing temperatures there will still be sufficient room for the entrance of the alloy. In Appendix 2A is a table showing the thermal expansion of the various metals which can be used to determine brazing-joint clearance.

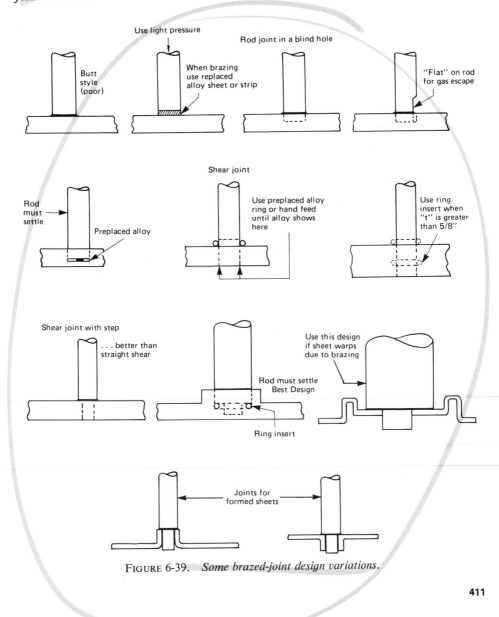

FIGURE 6-39. *Some brazed-joint design variations.*

*Example.* A 3-in. diameter stainless steel plug is to be silver brazed into a recessed hole in a steel plate. Determine the clearance required. Assume a temperature increase of 1250°F.

*Solution:* Coefficient of expansion for stainless steel and steel respectively: 9.3 in. $\times 10^{-6}$/in./°F $- 6.7$ in. $\times 10^{-6}$/in./°F $= 3(0.0000093 - 0.0000067)1250 = 0.00975$. Therefore the total brazing clearance before heating is 0.0035 in. + 0.00975 in. = 0.01325 in.

**Brazed-Joint Strength.** There is unfortunately no industrywide standard-test specimen used for brazed joints. The published data available is applicable only to specific test specimens and are not comparable to other designs. However, the following simple formula is useful in determining the lap length (depth of shear) on tubular or flat joints.

$$L_l = \frac{(S_t)\, t\, (F_s)}{S_s(D)}$$

*where:*    $L_l$ = lap length
$S_t$ = tensile strength of weakest member
$t$ = thickness of weakest member
$F_s$ = factor of safety
$S_s$ = shear strength of brazing alloy
$D$ = diameter of shear area. *Note:* $D$ is omitted on flat joints.

*Example.* A $\frac{1}{4}$ in. $\times$ 2 in. strap is brazed to a $\frac{1}{2}$ in. thick steel plate. The brazing alloy used has an $S_s$ of 36,000 psi. The desired safety factor is 3. The tensile strength of the steel is 50,000 psi.

$$L_1 = \frac{50,000 \times 0.250 \times 3}{36,000}$$

$$= 1.041$$

Therefore the lap length should be about $4t$.

**Brazing Fluxes.** The primary purpose of brazing fluxes is to dissolve and absorb oxides which heating tends to form. Therefore the parts to be brazed must first be cleaned of all existing oxides, oil, grease, and other foreign matter either by degreasing, grinding, pickling, or a combination of these cleaning processes. When the metal is heated to the brazing temperature, the flux becomes a clear liquid that wets the surface and aids in the flow of the filler metal. Some of the most common ingredients of fluxes are sodium, potassium, and lithium. They are used in making up chemical compounds such as borates, fluorides, chlorides, boric acids, alkalies, and wetting agents.

Generally, fluxes are available in three forms: powder, paste, or liquid. Of the three, paste is most commonly used, although powdered flux is frequently mixed with water or alcohol to give a pastelike consistency.

*Brazing Filler Metals.* Brazing filler metals are divided into seven classifications. In order of popularity, these are: silver, copper, copper zinc, copper phosphorus, aluminum silicon, copper gold, and magnesium. These alloys are produced in many forms such as wire, rods, coated sheets, and powder. Frequently, brazing alloys are preformed into rings or special shapes to simplify placement of the correct amount at the joint.

The service temperature of brazing varies widely. Most are for room temperature but some have been developed for stainless steel that have a service temperature of up to 2200°F (1204°C).

*Advantages and Limitations.* Brazing is well suited to mass-production techniques for joining both ferrous and nonferrous metals. Some of the principal advantages are:

1. Dissimilar metals can be joined easily.
2. Assemblies can be joined in virtually a stress-free condition.
3. Complex assemblies can be joined in several steps by using filler metals with progressively lower melting temperatures.
4. Materials of different thicknesses can be joined easily.
5. Brazed joints require little or no finishing other than flux removal.

Limitations:

1. Joint design is somewhat limited if strength is a factor.
2. Joining is generally limited to sheet-metal thicknesses and relatively small assemblies.
3. Cost of joint preparation can be high.

### Braze Welding

Braze welding is similar to brazing in that the base metal is not melted but joined by an alloy of lower melting point. The main difference is that in braze welding the alloy is not drawn into the joint by capillary action. A braze-welded joint is prepared very much like one for welding except that an effort should be made to avoid sharp corners that are easily overheated and may also be points of stress concentration.

Braze welding is used extensively for repair work and some fabrication on such metals as cast iron, malleable iron, wrought iron, and steel. It is used, but to a lesser extent, on copper, nickel, and high-melting-point brasses.

## SOLDERING

Soldering is a means of joining two or more pieces of metal by means of a fusible alloy or metal, called solder, applied in the molten state. Soldering can be distinguished from brazing in that it is done at a lower temperature (below 800°F, 427°C) and there is less alloying of the filler metal with the base metal.

*Solder Alloys.* The four metals used as the principal alloying elements of all nonproprietary solder formulations are tin, lead, cadmium, and zinc. These alloying elements fall into three thermal groups. Those that melt below 596°F (313°C) are called low-melting-point solders. Those that melt between 596 and 700°F (313–371°C) are called intermediate solders. Those that melt between 700 and 800°F (371–427°C) are termed high-temperature solders. The strongest and most corrosion-resistant are the high-temperature solders. The strength of low-temperature solders in shear is a bit in excess of 5000 psi (3.52 kg/mm$^2$).

The low-melting-point solders (378–596°F, 190–313°C) are the most widely used and are of the tin-lead or tin-antimony type. The tin-lead alloys provide good strength at low temperatures. A combination of 62 % lead and 38 % tin produces the lowest melting point and in practice is referred to as 60–40 solder. Being a eutectic alloy, it behaves like a pure metal and has only one melting temperature, 362°F (183°C), at which it is completely liquid, rather than a range of temperatures. Increasing the tin content produces better wetting and flow characteristics.

The tin-lead-antimony solders are generally used for the same types of applications as the tin-lead alloys except that they are not recommended for use on aluminum, zinc, or galvanized steel.

The tin-zinc group of alloys is used mainly for soldering aluminum, primarily where a lower soldering temperature than that of a zinc-aluminum solder is required. Zinc-aluminum solders are designed specifically for soldering aluminum.

Indium solder, 50 % indium and 50 % tin, is particularly suitable for products subjected to cryogenic temperatures and for glass-to-metal bonds.

*Solder Design Considerations.* Soldered joints should not be placed under great stress. In sheet metal, the joint should take advantage of the mechanical properties of the base metal by using interlocking seams, edge reinforcing, or rivets. Butt joints should be avoided. Lap joints loaded in pure shear may be used if the proper clearance is allowed. Clearances of 0.003–0.010 in. (0.076–10.254 mm) are permissible but the best strength and ease of soldering is obtained with about 0.005 in. (0.127 mm) clearance.

*Soldering Methods.* There are many ways of transmitting heat to the joint area. These include the well-known soldering "irons" or "coppers," torches, resistance heating, hot plates, ovens, induction heating, dip soldering, and wave soldering.

*Aluminum Soldering.* Aluminum is often thought of as being very difficult to solder. One reason is that an oxide film forms quickly over the surface after cleaning. Now, however, there are several methods of removing the oxide as the soldering is being done. These methods are flux soldering, friction soldering, and ultrasonic soldering, as shown in Fig. 6-40. Ultrasonic soldering is now being used to solder aluminum tubing in heat-exchanger coils without flux. The return bends, as shown in Fig. 6-41, are pretinned (coated with solder) and placed in the untinned bell. The assembled joint is then immersed in the ultrasonically activated solder. The sonic energy removes the oxides. The process lends itself to automation.

Of the low-temperature solders used for aluminum, a zinc-lead eutectic, which melts at 400°F (204°C), has an advantage over brazing because less heat is required, which results in a minimum loss of strength and hardness and the least distortion. This is particularly important in complex structures. If greater strength is required, the intermediate or high-temperature solders may be used. The high-temperature

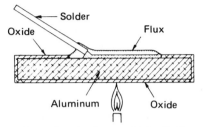

Removing aluminum oxide by fluxing. Chemical action of the flux removes the oxide.

**(a)**

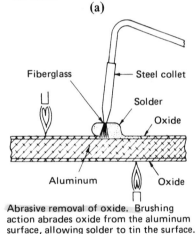

Abrasive removal of oxide. Brushing action abrades oxide from the aluminum surface, allowing solder to tin the surface.

**(b)**

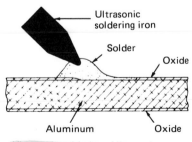

FIGURE 6-40. *Three methods of removing the oxide film from aluminum during the soldering process. Courtesy of Reynolds Metals Company.*

Ultrasonic soldering. Ultrasonic waves cause cavitation, breaking up the oxide and floating it to the top of the solder puddle. A frequency of 20 kc is often used to cause active cavitation.

**(c)**

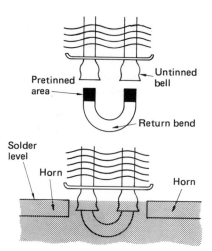

FIGURE 6-41. *Ultrasonic energy is used to remove the oxide as the aluminum tubing is immersed in the liquid solder.* *Courtesy* Product Engineering.

solders contain 90 to 100 % zinc and are the least expensive and the strongest of all aluminum solders.

Corrosion in aluminum-soldered joints can be particularly troublesome. It may be either chemical or galvanic in nature. In any case, moisture must be present. By coating the joint with a waterproof coating, the problem can be eliminated.

## FLAME CUTTING

Flame cutting is a material-separation process, but since it is a necesssary step in fabrication and is accomplished with welding equipment, it is included in this chapter.

The burning of red-hot iron or steel in an atmosphere of pure oxygen was a common laboratory experiment in the early 1800s. Theoretically, 4.6 cu ft (0.085 m$^3$)

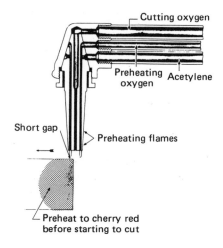

FIGURE 6-42. *The cutting torch has an oxygen jet surrounded by heating flames.*

of oxygen is required to oxidize 1 lb (0.45 kg) of iron. This principal has been built into a cutting torch (Fig. 6-42). The torch somewhat resembles the welding torch but it has a provision for several neutral flames and a jet of high-pressure oxygen in the center. The preheating flames are brought close to the metal as in welding and when the metal becomes cherry red in color, about 1500°F (816°C), the oxygen jet is released by a quick acting valve.

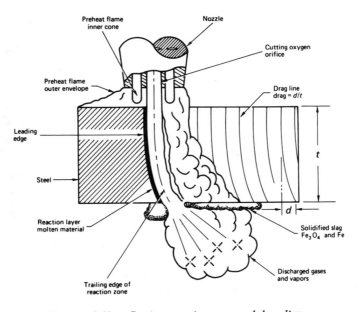

FIGURE 6-43. *Cutting reaction zone and drag lines.*

The quality of the cut is governed by tip size, oxygen pressure, cutting speed, and preheating flames. Cuts that are improperly made will produce ragged and irregular edges, with slag adhering at the bottom of the plates. One indication of the proper speed is the drag lines, as shown in Fig. 6-43. Drag is the horizontal lag that exists between the entrance and exit points of the cutting oxygen stream. It is generally specified as a ratio or as a percentage as follows:

$$\text{Drag} = \frac{d \text{ in.}}{t \text{ in.}}$$

$$\text{Drag \%} = \frac{d}{t} \times 100$$

*where:*   $d$ = amount of horizontal lag
          $t$ = the thickness of the metal being cut

The amount of drag that is permissible is dependent upon the quality of the part required. In the case of precision machine cutting, nearly vertical drag lines are

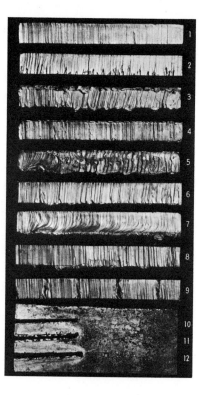

FIGURE 6-44. *Faults to avoid in torch cutting. Courtesy Union Carbide Corp., Linde Division.*

used. Generally a drag of 10 % is the uppermost limit, but for straight cutting it may go as high as 20 %. Defects that arise in metal cutting are shown and explained in Fig. 6-44.

Although acetylene is widely used as a fuel gas, other gases that are used are MAPP, propylene, natural gas, and propane. Tests have shown that propane and MAPP have a considerably shorter preheat time than acetylene.

*Flame Machining.* Although manual cutting is used extensively for salvage and repair work, machine flame cutting provides greater speed, accuracy, and economy. With a cutting torch mounted on a variable-speed electric motor, referred to as a radiograph, straight cuts can be made with the addition of a track and circular cuts can be made with a compass attachment.

Other methods of guiding the cutting torch are templates, photoelectric tracers, and numerical control. Hand-cutting and machine-cutting tolerances are compared in Table 6-10.

TABLE 6-10. *Flame-Cutting Tolerances*

| Method | Thickness (in.) | Tolerance (in.) | Cutting Speed |
|--------|-----------------|-----------------|---------------|
| Hand | 1/8 to 18 | ±1/32 | 10 to 14 ipm on 2″ thick |
| Machine | 1/8 to 18 | ±0.005 | 18 to 26 ipm on 1/8″ thick<br>12 to 15 ipm on 2″ thick |

*Oxygen-Assisted Gas-Laser Cutting.* A comparatively recent addition to the metal-cutting field has been the oxygen-assisted gas-laser cutting torch as shown in Fig. 6-45. The laser beam is directed downward by the focusing lens. High-pressure oxygen enters the chamber and goes out the jet at high speed. The 250 watts of power is focused onto the surface, which, with the oxygen, is enough to produce a cutting action on the steel. The advantages of the laser-cutting process are that it typically produces accurate narrow widths of cut (kerf), smooth edges, and a significantly reduced heat zone, thus minimizing distortion.

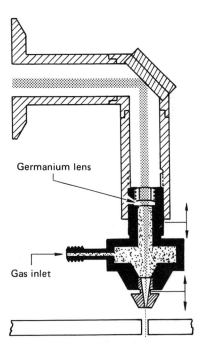

Germanium lens

Gas inlet

FIGURE 6-45. *An oxygen-assisted laser beam used for cutting steel. Courtesy Society of Manufacturing Engineers.*

## RESISTANCE WELDING

Resistance welding is based on the well-known principle that as a metal impedes the flow of electric current, heat is generated. The amount of heat generated is related to the magnitude of the electric current, the resistance of the current-conducting path, and the time the current is allowed to flow. Of course the metal makes a much better path for the current than the arc in arc welding, therefore the current must be higher. The interface between the two surfaces of the workpiece offers the greatest resistance to current flow in comparison to the balance of the circuit and is therefore the area of greatest heat. The heat generation is directly proportional to the square of the current times the resistance and is expressed as:

$$H = I^2 RT$$

*where:*    $H$ = heat generated in watt hours
$T$ = time in hours
$I$ = current in amperes
$R$ = resistance in ohms

The heat may be changed from watt hours to Btu's by multiplying the current by 3.412, since there are 3412 Btu's in one kwh.

The principle types of resistance welding are:

| | |
|---|---|
| Spot welding | Butt welding |
| Seam welding | Upset welding |
| Projection welding | Flash welding |
| High-frequency welding | |

## Spot Welding

Basic spot welding consists of clamping two or more pieces of metal between two copper electrodes, applying pressure, and then passing sufficient current through the metal to make the weld.

Pressure, weld time, and hold time are recognized as the fundamental variations in spot welding. For most spot welding, these variables must be kept within very close limits.

*Pressure.* Spot welds are made by cleaning the two pieces of metal to be lapped and placing them between the copper electrodes of the spot welder [Fig. 6-46($a$)], The pressure or squeeze time is used to bring the two workpieces together in intimate contact just prior to current flow [Fig. 6-46($b$)]. It may be considered the most important variable in resistance welding. The resistance at the joint interface is inversely proportional to the pressure ($R = 1/P$). The effect of high and low pressures on resistance are shown graphically in Fig. 6-47. Variations at low pressure have a great effect on resistance and vice versa. On welding equipment requiring high heat and pressure, the equipment is usually massive and air operated.

*Weld Time.* The weld time is the period when the current flows. It is usually timed in fractions of a second. For example, a spot weld can be made in two $\frac{1}{16}$ in. (1.6 mm) thick pieces of mild steel in 15 cycles or $\frac{1}{4}$ second, when using 60-cycle current.

Low-carbon steels offer little or no metallurgical problems when spot welded. Once the weld nugget is formed, the metal passes from austenite back to pearlite. If medium- or high-carbon steels were to be treated in the same manner, the rapid cooling of the spot would result in some embrittlement. This does not mean that medium- or high-carbon steels cannot be resistance welded, but other processes such as quench and postheat are needed to permit the proper heat treatment of the spot immediately after it is made.

*Hold Time.* Hold time is basically a cooling period. It is the interval from the end of the current flow until the electrodes part. Water-cooled electrodes serve to transfer the heat away from the weld rapidly.

Other variables that may be sequenced along with the steps already described are: low current for preheating; either high or low current as required for spot heat treatment; a forging action to help refine the grain of the weld. This is done by increasing the electrode pressure during the cool time. Some of these added functions are particularly useful in joining hardenable alloy steels that may crack if welded with one surge of current and allowed to cool.

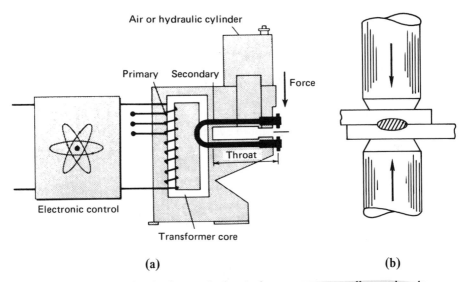

(a)                                                                 (b)

FIGURE 6-46. *The fundamental electrical components usually employed in resistance-welding machines (a). The two pieces of metal are placed between the two copper electrodes for spot welding (b).*

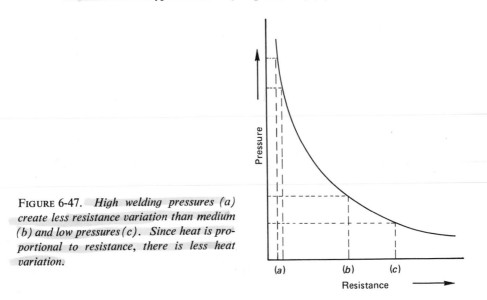

FIGURE 6-47. *High welding pressures (a) create less resistance variation than medium (b) and low pressures (c). Since heat is proportional to resistance, there is less heat variation.*

### Spot-Welding Power Supplies

There are two basic types of resistance-welding power supplies: *stored* and *direct energy*. The stored-energy machines draw current at a relatively constant rate and store it so it can be discharged instantaneously when needed to make a weld. The direct-energy machines discharge the power to make the weld just as they receive it.

The stored-energy machine has the advantage of low current input which makes for a more balanced load on the plant's three-phase electrical system. Also, variations in line voltage have relatively little effect on the efficiency of the machine. The most widely accepted method of energy storage for welders is by means of a bank of capacitors.

Direct-energy machines have the advantage of both lower initial cost and lower maintenance cost. The wave shape of the current is also more controllable, making it quite versatile for a greater variety of metals. An example of a wave form of the current together with other timed events is shown in Fig. 6-48.

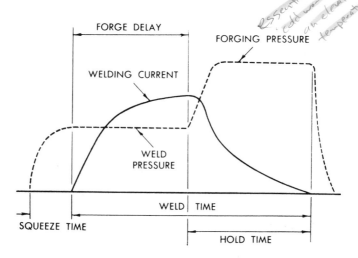

FIGURE 6-48. *The current-wave form in relation to other variables. Courtesy Kaiser Aluminum and Chemical Corp.*

*Slope Control.* Slope control refers to the rise and fall of the welding current, as shown in Fig. 6-49. The current may be made to build up gradually (up-slope) or to decrease gradually (down-slope or current decay).

Up-slope gives the electrodes a few impulses of time to sink into the metal and seat before they have to carry the full welding current. If this is not done in welding aluminum, overheating occurs at the electrode tip, resulting in unwanted metal on the tip. The down-slope or decay helps to reduce internal cracking of the weld nugget. Retarding the cooling action ensures a proper temperature during the forging action. An increase in electrode pressure as the weld solidifies produces the forging effect.

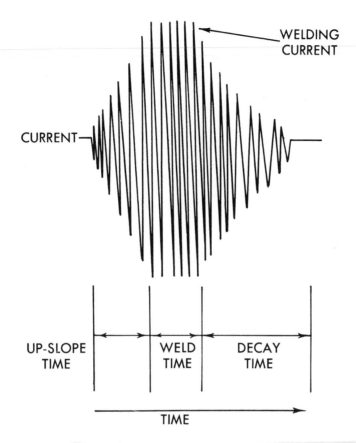

FIGURE 6-49. *Current up-slope and down-slope modification form for single-phase, direct-energy welding. Courtesy Kaiser Aluminum and Chemical Corp.*

**Spot-Weld Design Considerations.** For spot welds to obtain maximum strength, the following design factors should be considered: the amount of overlap, the spot spacing, and accessibility.

*Overlap.* The amount of overlap is determined by the weld size, which in turn is determined by the metal thickness. As an example, an acceptable weld size for a thickness range of 0.032 to 0.188 in. (0.812–4.775 mm) is roughly estimated at 0.10 in. (2.540 mm) plus two times the thickness of the thinnest member. The overlap should be equal to two times the weld size plus $\frac{1}{8}$ in. (3.2 mm), ($\frac{1}{8}$ in. is the tolerance for positioning the weld). If fixturing is used so that the spot will be positioned in the center of the overlap distance, the $\frac{1}{8}$ in. can be disregarded.

Welds placed too close to the edge will often squirt molten metal from the joint. Edge expulsion is called "spitting" and results in weak welds.

423

TABLE 6-11. *Spot-Welding Data for Mild Steel*

| Sheet Thickness | Tip Diameter* | Electrode Force | Weld Cycles | Amperes | Diameter of Fused Zone | Minimum Spot-Weld Spacing |
|---|---|---|---|---|---|---|
| 0.020 | $\frac{3}{8}$ | 300 | 6 | 6,500 | 0.13 | $\frac{3}{8}$ |
| 0.035 | $\frac{3}{8}$ | 500 | 8 | 9,500 | 0.17 | $\frac{1}{2}$ |
| 0.047 | $\frac{1}{2}$ | 650 | 10 | 10,500 | 0.19 | $\frac{3}{4}$ |
| 0.059 | $\frac{1}{2}$ | 800 | 14 | 12,000 | 0.25 | 1 |
| 0.074 | $\frac{5}{8}$ | 1,100 | 17 | 14,000 | 0.28 | $1\frac{1}{4}$ |
| 0.089 | $\frac{5}{8}$ | 1,300 | 20 | 15,000 | 0.30 | $1\frac{1}{2}$ |
| 0.104 | $\frac{5}{8}$ | 1,600 | 23 | 17,500 | 0.31 | $1\frac{5}{8}$ |
| 0.119 | $\frac{7}{8}$ | 1,800 | 26 | 19,000 | 0.32 | $1\frac{3}{4}$ |

* Outside diameter of the tip dressed down to a 20-deg. bevel. The resulting diameter of the cone ($d$) will be $2t + 0.1$ in.

*Spot Spacing.* The spots should not be so close that current is shunted to the previous weld, thus reducing the size of the weld being made. A general rule is to allow $16t$ between the welds, where $t$ is the thickness of the material. If distortion becomes a larger factor than strength, then $48t$ is used. If it becomes necessary to place spot welds where there is a liklihood of current shunting, the current must be increased to compensate. Shown in Table 6-11 is typical spot-welding data for mild steel.

*Spot Strength.* Spot strength varies directly with area. It is generally safe to assume that the strength will be equal to the weld area times the tensile strength of the metal in the annealed state.

**Testing of Spot Welds.** The appearance of a spot weld can be very deceiving. A weld may appear good even though, in essence, there is no weld.

Three common destructive tests for spot welds are the peel test, tensile test, and cross-section test, as shown in Fig. 6-50.

The peel test is probably the simplest to execute and yet give effective results. All that is required is a sample strip and a means of rolling the top edge back. If the weld is good, the weld nugget will pull a hole in either piece. This is true in materials up to 0.094 in. (2.387 mm) thick. Greater thicknesses may only pull out a slug of metal, leaving a crater in the other piece.

Portable tensile-testing machines are often used for spot welds. These machines are of 10,000 psi capacity and will pull single spot welds in mild steel up to 0.094 in. (2.387 mm) thick.

The cross-section test consists of cutting through the middle of the weld, polishing it, etching it in a suitable acid, and inspecting it under a microscope. The penetration of the weld nugget into the base sheet should be from 40 to 70 % of the sheet thickness. In addition to penetration, the metallurgical structure of the weld can be examined for grain size, microstructure, and any harmful effects such as carbide precipitation which may occur in stainless steel.

1250
lbs/spot

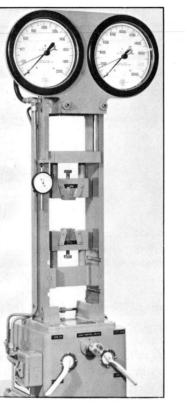

Portable 10,000 pound capacity tensile testing machine used both in laboratory and in production line for determining individual spot weld strength. Note how sample is unconfined in gripping jaws. At right are stages (a,b,c and d) in a destruction test of a spot weld sample. There is generally a "pull out" type of failure as shown. Rotation is caused by the two parts of the test piece being offset.

(a)

(b)

(c)

(d)

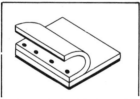

Peel test of a series of properly spaced spot welds. All welds should pull slugs of uniform size and shape.

Weld penetration

Cross-section test. To determine weld diameter and penetration, the spot weld is cut in half, polished and then etched in acid.

FIGURE 6-50. *Three common destructive tests for spot welds. Courtesy Acco, Wilson Instrument Division.*

### Seam Welding

Seam welding is a continuous type of spot welding. Instead of using pointed electrodes, the work is passed between copper wheels or rollers that act as electrodes. Thyratron and ignitron tubes (gas-filled electronic tubes) are used to make and break the circuit. The appearance of the completed weld is that of a series of overlapping spot welds which resemble stitches. Seam welding can be used to produce highly efficient water- and gas-tight joints. A variation of seam welding used to produce a series of intermittent spots is called *roll welding* (Fig. 6-51).

425

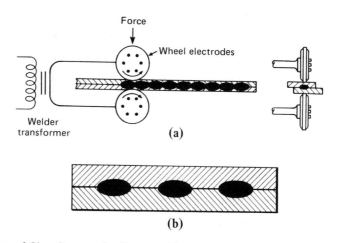

FIGURE 6-51. *Seam and roll spot welding. Seam welding can be done as shown, or one of the wheels can be replaced with a flat backing electrode that supports the work for the entire length of the seam.*

***Seam-Weld Testing.*** Seam welds can be tested as mentioned for spot welds but "pillow tests" can also be made. Two rectangular or square sheets are joined together by welding a seam all around the edges. A pipe connection is welded to the center of one sheet. Hydraulic pressure is applied through the pipe until the sheets expand and burst. The bursting pressure is recorded as is the length of the weld line. Failure should occur in the parent metal rather than in the weld.

### Projection Welding

Projection welding is another variation of spot welding. Small projections are raised on one side of the sheet or plate with a punch and die. The projections act to localize the heat of the welding circuit. During the welding process, the projections collapse, owing to heat and pressure, and the parts to be joined are brought in close contact (Fig. 6-52).

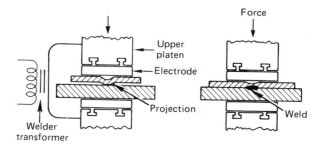

FIGURE 6-52. *An example of projection welding.*

One of the most common applications of projection welding is for attaching small fasteners, nuts, special bolts, studs, and similar parts to larger components. A wide variety of these parts are made with preformed projections.

*Advantages and Limitations.* Spot welding is an excellent way for joining lighter-gage materials, particularly if the work is of a repetitive nature. Welds can be made in a fraction of a second.

Production must be high if resistance welding is to be used since this type of equipment is rather costly. It is possible to buy a spot-welding machine for less than $1000. But to produce satisfactory welds consistently in a variety of materials, larger machines are needed with more sophisticated controls.

### High-Frequency Resistance Welding

High-frequency resistance welding of tubes and strip material is a newer adaptation of an old idea, that of causing current to flow across a metallic joint, heating the edges to the melting point and joining them together. What is new about the process is that a high frequency (450,000 Hz) is used instead of 60 Hz. The current does not follow a direct low-resistance path between the two contacts; rather it follows a long, low inductance path as shown in Fig. 6-53. This circuitous path of the current allows adjustment of the heating on either side of the weld (an asset when joining dissimilar metals). Also the shallow depth of the inductance path produces high-intensity heating in shallow zones, resulting in a minimum heat-affected zone. The

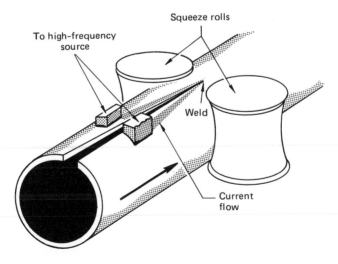

FIGURE 6-53. *A working section of a high-frequency welding unit. High-frequency current heats the metal edges to the welding temperature and at the same moment the squeeze rolls pinch the moving edges of the tubing together.* Courtesy *Metal Progress.*

localized high-intensity heating allows the whole edge of the material to reach welding temperature at the same rate as the strip progresses through the welding machine. Pressure applied at the root of the "V" brings the two edges together in a forge weld.

The oxide films on the metal surface act as a high-frequency conductor by virtue of their peculiar properties at high frequency.

Because the metals are pinched together (in tube welding) at the point of the "V", molten material, if present, is squeezed out (upset) and the weld is essentially forged.

*Advantages and Limitations.* The advantages of high-frequency resistance welding can be listed briefly as follows:

1. Joining can be done at high speed—up to 1000 fpm in light-gage materials up to 0.012 in. thick (0.56 mm).
2. Very thin sheet, ranging from 0.004 to 0.020 in. (0.10 to 0.51 mm) thick, can be lap welded. The contact pressure on the seam is so light that it creates no problem.
3. Metals up to 0.25 in. (6.35 mm) thick have been welded satisfactorily. Thicker metals sometimes require a device to preheat the edges to about 1000°F (537°C).
4. The power required is a fraction of that needed for standard resistance welding, the reason being that the effective resistance of the metal is so much higher at 450 kHz that, by the $I^2R$ law, the amperage required for a given amount of heat input is sharply reduced.
5. Most of the process variables that affect normal resistance welding, such as contact pressure, contact resistance, surface oxides, etc., do not apply because the joint faces are heated without actually touching each other.
6. The heat-affected zone is very small.
7. Dissimilar metals can be welded together. An unusual application is that of band-saw blades, where high-speed steel teeth are welded to a high-carbon-steel backing strip at 50 to 100 fpm.

The main disadvantage is that the process is limited to high-production operations.

## Resistance Butt Welds

*Upset.* The material to be welded, usually bar stock, is clamped so that the ends are in contact with each other. High-density current, usually from 2000 to 5000 amp per sq. in., is used to make the weld. The high resistance of the joint causes fusion to take place at the interface. Just enough pressure is applied to keep the joint from arcing. As the metal becomes plastic, the force is enough to make a large, symmetrical upset that eliminates oxidized metal from the joint [Fig. 6-54(a)]. The metal is not melted and no spatter results. For most applications, the upset must be machined off before use.

*Flash.* In flash butt welding, the ends of the stock are clamped with a slight separation before welding. As the current is turned on, it jumps the gap and creates a great deal of heat. Some of the metal burns away. The two pieces are moved toward each other in an accelerated motion and as they reach the proper temperature, they are forced together under high pressure and the current is cut off [Fig. 6-54(b)].

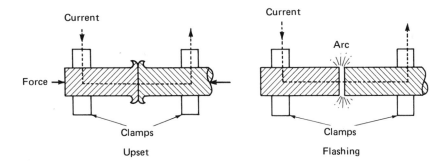

FIGURE 6-54. *Resistance butt welding: upset (a) and flash (b).*

## SPECIALIZED WELDING PROCESSES

Technological advances have produced many new welding processes since World War II. Some of these have come in a response to better methods of dealing with reactive materials. Some of the processes are new and others are modifications of older processes. These may be classified into three areas: fusion, resistance, and solid state. Significant newer specialized processes in the fusion area are electron-beam welding and laser welding.

### Electron-Beam Welding

Electron-beam welding is perhaps the most spectacular of the more recent welding developments to become commercially feasible. Electrons, under normal circumstances, are contained within the molecular structure of the parent material by inherent electrical forces. If the metal (usually titanium) is subjected to high temperatures, free electrons will "boil off" the surface of the materials. In many electron-beam welders, a wire filament is heated to thermionic-emission temperatures by an electrical current to produce electrons (Fig. 6-55). Another system uses an indirectly heated tungsten rod which is encircled by an auxiliary filament. In any case, free electrons are simultaneously accelerated and collimated into a beam by exposure to an electrical field. As the stream of electrons leaves the anode, it diverges due to the repulsion of individual electrons. At this point, they are exposed to a magnetic field in a focusing coil which refocuses the beam without affecting the velocity of the electrons. The electrons travel up to about 60 % of the speed of light but do not give off any heat until they strike the target. The intensity of the beam can easily be controlled by the focusing coil. Sharp focusing is desirable for most welding, but the out-of-focus broad circle can be extremely useful in scouring contaminants from the face of the work prior to welding.

429

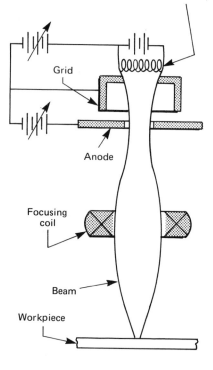

FIGURE 6-55. *A schematic view of the electron-beam welder. A wire filament is heated to thermionic emission temperatures by an electric current to produce electrons. The free electrons are collimated by exposure to an electrical field. As the stream of electrons leaves the anode, they are exposed to a magnetic field in a focusing coil for high concentration of energy.*

In generating free electrons, the emitters must be operated within a high-vacuum environment to minimize oxidation. A vacuum is also needed in the high-voltage gap, where the electrons are accelerated to prevent a gas discharge or arc. A vacuum or partial vacuum is desirable along the entire path of beam transmission to reduce the frequency of collisions between electrons and air molecules. The collisions cause the electrons to scatter, which increases the diameter of the beam and reduces its effectiveness. For high production, the necessity of a hard vacuum, 0.0001 Torr (as compared to atmospheric pressure of 760 Torr), presents problems. The nonproductive time required to pump down the vacuum can take anywhere from 3 to 30 min, depending on chamber size, pump size, and vacuum level required.

Production can be increased considerably by performing the welding operation at atmospheric pressure. In this type of arrangement, the beam-producing elements are surrounded by a hard vacuum produced by a diffusion pump [Fig. 6-56(*a*)]. A neutral gas is made to flow past the exit orifice to prevent metal vapor from being drawn up into the beam and to protect the fusion zone. The weld quality is not greatly reduced, but the distance through which the beam can penetrate is reduced somewhat. Helium used at the beam outlet increases both the speed and penetration of the weld [Fig. 6-56(*b*)].

The out-of-vacuum system frees the welding process from restrictions on production cycling caused by the time required to pump down the welding chamber. Beam penetration is restricted to about 1 in. (2.54 cm) (in larger systems), with a

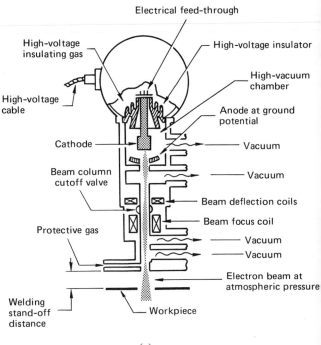

Electrical feed-through

High-voltage
insulating gas

High-voltage insulator

High-vacuum
chamber

High-voltage
cable

Anode at ground
potential

Cathode

Vacuum

Vacuum

Beam column
cutoff valve

Beam deflection coils

Beam focus coil

Protective gas

Vacuum

Vacuum

Electron beam at
atmospheric pressure

Welding
stand-off
distance

Workpiece

**(a)**

FIGURE 6-56. *The electron-beam welder can be operated at atmospheric pressure (a). In this case, the electron-beam-producing elements are maintained at a hard vacuum by a separate diffusion pump and a semi-vacuum is maintained along the beam transmission path to limit the spread of the beam caused by collisions between electrons and air molecules. The very end of the beam-exit nozzle is slightly pressurized by an effluent gas to prevent metal vapors from getting into the transfer column. With air, the practical distance between the exit nozzle and the workpiece is limited to about 1.2 in. Using helium, this distance can be increased to 1.8 in. The speed also can be increased as shown at (b).* Courtesy Machine Design.

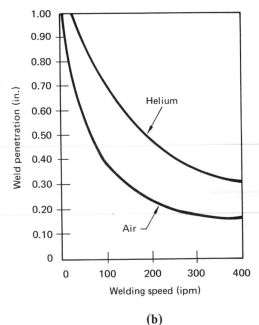

Weld penetration (in.)

Welding speed (ipm)

Helium

Air

**(b)**

431

correspondingly wider cross section of weld. The workpiece must be positioned approximately 0.5 to 0.75 in. (12.70 to 19.05 mm) from the beam-exit orifice.

When high-energy electrons impinge on a solid material and are suddenly decelerated, X rays are produced. When welding is done within the evacuated chamber, the X rays are of no consequence; but in the out-of-vacuum system the personnel must be protected. Because X rays travel in a straight line, a labyrinthine arrangement of lead shielding is usually effective.

*Advantages and Limitations.* Electron-beam welding confines heat to a narrow area, making it possible to have a depth-to-width penetration of about 25 : 1 (Fig. 6-57). Postwelding operations are negligible or unnecessary since there is no contamination of the weld. Welding speeds can be high, up to about 200 in. per min (508 cm/min) depending upon depth of penetration required and other factors. Electron-beam systems are available with power ratings ranging from 6 to 60 kW to handle a wide variety of work.

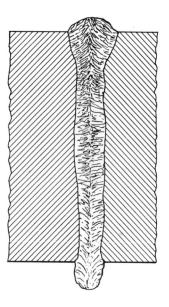

FIGURE 6-57.  *High depth to width welds are made with the electron beam.*

Numerical control has been combined with the electron beam to obtain the best speed and beam path. A lower-power scanning beam can be used to obtain detailed path description (Fig. 6-58). Once the path has been defined, the beam power is brought back to the welding level and made to traverse it at a constant speed, maintaining the point of impingement on the joint to close tolerances. Information concerning path variances can also be fed to the computer, which will cause predetermined corrective actions to be taken.

Edge and butt joints can be made in metals as thin as 0.001 in. (0.03 mm). Also, small, thin parts can be welded to heavy sections. Dissimilar metals are joined as effectively as similar metals.

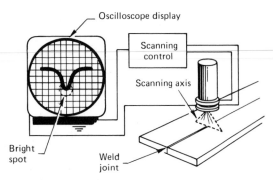

FIGURE 6-58. *The electron beam is made to traverse the weld area with a scanning beam to get the exact location of the joint. This information is then programmed to direct the beam for the actual welding. Courtesy Automation.*

Electron-beam welding is not practical where a wide gap filling is necessary. For butt welding parts thicker than 0.100 in. (.254 mm), the air gap should be not more than 0.005 in. (0.13 mm).

## Laser Welding

The term *laser* stands for "light amplification by stimulated emission of radiation." The idea of the laser goes back to Niels Bohr's theory that electrons are given off or absorbed as the energy level of the material changes. Electrons exist only in definite shells. The energy the electron possesses is determined by its distance from the nucleus. Shown in Fig. 6-59(a) is a schematic of an electron in shell $E_2$,

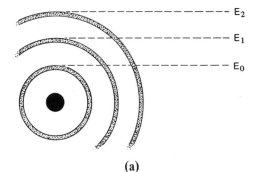

(a)

FIGURE 6-59. *When electrons are pumped or stimulated they go to a higher energy level (a). Energy is emitted when electrons move to a lower level and population inversion takes place as shown at (b). The stimulated emission produces an in-phase or coherent light beam.*

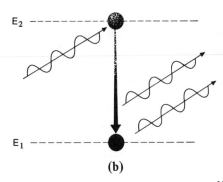

(b)

which has more energy than electrons in shells $E_1$ and $E_0$. Energy is given off as the electron falls back to a lower shell. Thus the electron can absorb or give off energy as it changes levels. A phenomenon that makes the laser possible is that in certain materials an electron can be in a semistable level [Fig. 6-59($b$)], so that when a photon of radiation of exactly the right frequency triggers the electron, the energy can cause the electron to fall to a lower level and in the process give off a photon of radiation which is exactly in phase with the photon that caused the triggering action. As shown schematically by the arrows, the triggered photon has twice as much radiant energy.

For laser action to occur, the majority of the electrons must be at the upper energy level. This condition is referred to as a population inversion and is produced through external excitation of the lasing medium, commonly referred to as a "pumping" action that may be achieved optically or electronically.

Essentially then, a laser consists of a reservoir of active atoms which can be excited to an upper energy level, a pumping source to excite the available active atoms, and a resonant cavity to ensure that the radiation will make many passes through the active laser medium, thereby obtaining the maximum propagation of stimulated emissions.

Lasers can be classified as solid-state, liquid, or gas types. Only solid-state and gas lasers (Fig. 6-60) are presently used in welding and metalworking.

A typical solid-state lasing medium consists of a single crystal rod of yttrium-aluminum-garnet (called yag), which contains ions of neodymium. Neodymium-doped yag (Nd : yag) has replaced most other materials, including the original ruby rod, for high-energy, crystal-laser applications because it allows high pulse rates, has relatively good efficiency, and can be operated with simple cooling systems.

The solid-state laser can be pumped by a flashlamp mounted in a reflecting cavity which also contains the laser rod. The setup as shown in Fig. 6-60($a$) utilizes a linear flashlamp parallel to the laser rod inside an elliptical gold-plated enclosure. Flash lamps are usually xenon- or krypton-filled.

At one end of the laser rod is a totally reflective mirror and at the other end is a partially transmitting mirror. The area between the mirrors forms a resonating cavity, causing a preferential buildup of energy until it bursts out of the transmitting end.

Gas lasers consists of an optically transparent tube filled with either a single gas or a gas mixture as the lasing medium. The principal gas lasers are He-Ne, Ar, and $CO_2$ lasers. The pumping source is some form of electrical discharge applied by electrodes or a special electron-beam arrangement. A typical $CO_2$ laser actually uses a mixture of three gases: carbon dioxide, helium, and nitrogen for high-power operation. In flowing-gas systems, this mixture is constantly pumped through the resonator to sustain the lasing action.

The cross-flow–design gas laser eliminates the gas tube, which can become quite cumbersome. The working gas is pumped at high speed at cross-flow to the direction of the laser beam. The system operates somewhat like a closed-cycle wind tunnel, with heated gas passing through a heat exchanger before it is returned to the cavity area. The discharge is stabilized by preionization of gas with a broad-area electron beam. This cross-flow/electron-beam preionization combination allows power levels of over 15 kW with a laser-cavity length of approximately 4 ft.

**Solid-state laser system**

**Axial-flow gas-laser system**

**Cross-flow gas-laser system**

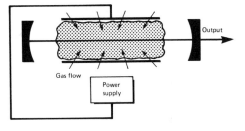

FIGURE 6-60. *A schematic comparison of three laser systems: solid state, axial-flow gas, and cross-flow gas. Courtesy* American Machinist.

For fabrication purposes, the total energy available represents the laser capacity for work. Energy is expressed in joules (J), which are equivalent to watts × seconds. A laser capable of delivering 25 J at a rate of one pulse per second is considered to be a 25-W laser. But if the 25 J are emitted in a single pulse of only $1 \times 10^{-3}$ sec, then the laser would achieve a peak power of 25,000 W.

Lasers are operated in either a continuous wave (CW) or a pulse mode. Solid-state lasers, such as Nd : yag, are usually operated in the pulsed mode because of flashlamp limitations, but continuous operation is also available. Gas lasers are usually operated in a CW mode and are capable of developing higher continuous average power than the solid-state lasers.

Historically, some of the first applications of lasers to welding involved the joining of very small wires. Now, however, solid-state lasers producing 50 J and more are common. Seam welds can be made by overlapping spots at the rate of 12 in. (30.48 cm) per min. High-power CW solid-state lasers are being used to produce full weld penetration up to 0.125 in. (3.175 cm) at speeds of 10 in. (25.4 cm) per min. Shallow welds are made at faster rates.

435

TABLE 6-12. *Laser vs. Electron-Beam Welding. Courtesy* Machine Design

| Characteristic | Laser | Electron Beam |
|---|---|---|
| Welding speed | Low | High |
| Material range | Extremely wide | Wide |
| Thickness range | Narrow | Wide |
| Joint-design versatility | Excellent | Excellent |
| Range of dissimilar metals | Very wide | Wide |
| Need for vacuum chamber | No | Yes |
| Gas coverage | No | No |
| Tooling costs | Low | Moderately high |
| Initial cost | Moderate | High |
| Operating costs | Low | Moderate |
| Ease of automation | Excellent | Excellent |
| Controllability | Very good | Good |
| Heat generation | Low | Moderate |
| Miniature welding | Limited | Excellent |

Laser welding is most often compared to electron-beam welding since both use a stream of energy to fuse the metals. The actual processes are quite different, as shown in Table 6-12.

***Advantages and Limitations.*** Laser welding has the advantage of controllability, which makes it ideal for tiny precision parts. Intense energy can be brought to bear on a tiny area. For this reason high-melting-point materials and refractories which are virtually unweldable by any other process can be fused together. Also, it is a noncontacting technique which makes it possible to weld two pieces of material together at the bottom of a small hole or some other inaccessible spot. Most lasers operate in the visible-light range and thus only affect what they can "see." This makes it possible to weld components inside a clear tube envelope, as in vacuum tubes.

The heat generated in laser welding is very concentrated, thus the physical and chemical properties of the materials are not significantly affected. A 10-in. (25.4-cm) seam weld feels cool to the touch immediately after welding.

The laser power source is only 30 % as expensive as that of EB (electron beam). As an example, a complete EB system may cost a minimum of $100,000, whereas a good 60-joule laser capable of welding materials up to 0.040 in. thick (1.016 mm) can be obtained for $20,000.

One of the drawbacks to the use of lasers in welding has been the comparatively short life span. The ruby and Nd : yag units are good for about 10,000 shots at maximum rating before the flash tube must be replaced. If operated well below the maximum, a flash tube may last for 100,000 shots.

The laser welding is basically a slow process since only a certain pulse rate is possible. At present, only thin materials are being welded. The depth of penetration is normally restricted to about $\frac{1}{8}$ in. (.317 cm).

Although the laser is an excellent means of welding dissimilar materials, there are some limitations. For example, it is not practical to weld titanium to stainless steel as the two are incompatible.

## Solid-State Welding Processes

Theoretically it is possible to place two perfectly clean metallic surfaces into such intimate contact that the cohesive forces between the atoms would be sufficient to hold them together. This condition is approached in "wringing in" two precision gage blocks as described in Chapter 9. As good as this is, it does not approach a perfect bond. Astronauts experienced some welding of parts that were in outer space due to the nonoxidizing atmosphere.

Despite difficulties, some progress has been made toward solid-state bonding. Welding processes that may be classified under this heading are: ultrasonic, explosive, diffusion, and friction processes.

### Ultrasonic Welding

Man's ability to hear sound extends from frequencies of 20 to about 14,000 cycles per second. It is quite easy to produce high-frequency electrical pulsations by several commercially available types of power supplies. A device is then needed to convert the electrical pulsations into mechanical vibrations. This device is called a transducer. There are two common types of ultrasonic transducers. One depends upon a piezoelectric effect in a quartz or ceramic crystal of particular geometry; a second type depends upon a magnetostrictive effect in a metal stack as used in welding. A coupler and horn are attached to the transducer to provide the physical displacements at the welding tip. The tip displacement is usually between 50 and 600 micro-inches at the excitation frequency of 60 kHz. The horn, which is tuned to vibrate at exactly 20,000 Hz, transmits the vibratory energy into the workpiece (Fig. 6-61). Depending upon the power supply, the mechanical output may range from 1700 to 3200 inch pounds per second at the tip of the horn. The horn is usually designed to meet the requirements of a specific application. Since the mass and shape of the horn determine the length at which it will oscillate for the required frequency of 20,000 Hz, no two horns will have the same length. Horns may very in length as much as 1.25 in. (26.03 mm). The tips are usually contoured to conform to the shape of the parts being assembled. Horns are usually constructed from a special titanium alloy that has an exceptionally high strength-to-weight ratio, or they may also be made from laminated nickel.

The sonic timer controls the duration of the ultasonic exposure and the duration of the pressure.

When used for automatic assembly, the sonic converter and horn are usually mounted on a pneumatically operated assembly stand which operates at regulated pressure to bring the horn in contact with the work surface.

Automatic parts handling to increase production includes a means of fixturing the part to prevent lateral or other movement during the ultrasonic exposure cycle. A variety of rotary indexing tables are used that include nesting devices for the use of

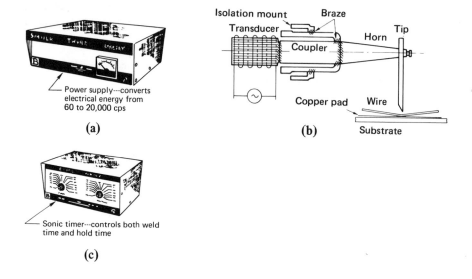

Power supply---converts
electrical energy from
60 to 20,000 cps

**(a)**

Isolation mount    Braze

Transducer               Horn    Tip

Coupler

Copper pad    Wire

**(b)**    Substrate

Sonic timer---controls both weld
time and hold time

**(c)**

FIGURE 6-61. *A power source (a) converts conventional 60-Hz electrical energy to 20,000-Hz electrical energy. A transducer assembly then changes the electrical energy to mechanical vibrations at the same frequency (b). The high-frequency mechanical vibrations are then transmitted into the workpiece through a resonant section called a horn. All functions are controlled by a timing mechanism (c) since successful results depend on the right combination of time, pressure, and energy output.*

single or multiple horns. Shown in Fig. 6-62 are examples of recommended joint designs and a sketch of an integrated closed-loop system. Parts are automatically oriented and fed from the vibratory feeder to the conveyor where they are welded. Several ultrasonic-welding stations permit welding of different components at different stages of assembly.

*Advantages and Limitations.* Ultrasonic power is especially useful for the assembly of plastic parts. The process eliminates the need for solvents, heat, or adhesives. In addition it may be used for metal inserting and staking of plastic parts, as shown in Fig. 6-63. A further assembly application is in the reactivation of adhesives. Thus components may have the contacting surfaces coated with the desired adhesive but permanent assembly may be delayed until a convenient time.

The heat produced by the vibratory energy is not enough to affect the metallurgical properties of the materials joined. Bonds have from 65 to over 100 % of parent-metal strength. The reduction of the joint thickness is usually less than 5%. The process is ideally suited for welding thin sheet and foil to itself or to any thickness. Several thin pieces can be joined simultaneously. Dissimilar metals may be joined, but aluminum and its alloys are the most weldable by this process. The ultrasonic

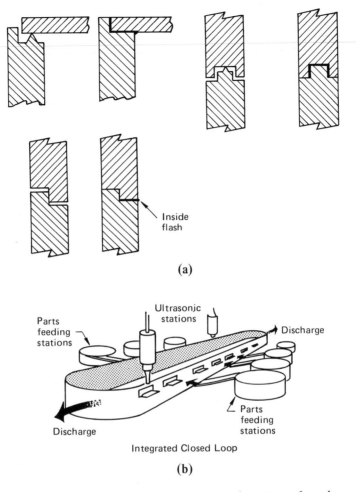

(a)

(b)

FIGURE 6-62. *Some recommended joint configurations for ultrasonic welding (a). An integrated, closed-loop system used to automate the assembly of parts by ultrasonic welding (b). Courtesy of* Automation.

process is especially good because the vibratory action breaks up and disperses thin layers of oxides that form, thus providing sound metal-to-metal welds. Normally, degreasing is all that is necessary. The time required to make a spot weld in aluminum sheet stock is 1 sec or less, in foil it is practically instantaneous.

The maximum thickness of aluminum presently weldable is approximately 0.080 in. (2.03 mm) for high-strength aluminum alloys, e.g., 7075-T6, and about 0.120 in. (3.05 mm) for softer aluminum alloys, e.g., 1100-O.

Equipment life is relatively short due to fatigue encountered by the horn. Some materials being welded tend to weld to the coupling anvil.

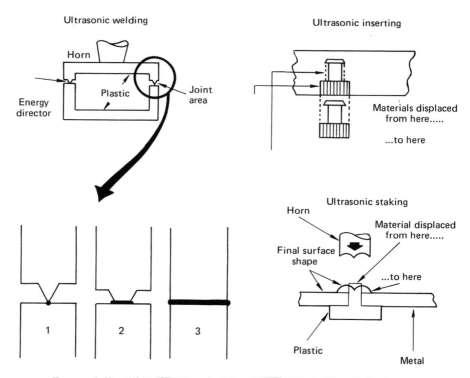

FIGURE 6-63. *Three basic methods used in the assembly of plastic parts with ultrasonic energy. Note that in inserting, the plastic melts momentarily, permitting the part to be forced into place. The plastic then flows around the insert and encapsulates it. Courtesy* Automation.

## Explosive Welding

Explosive bonding is the technique of using explosive charges to form a metallurgical bond between two pieces of metal at pressures of up to several million psi under conditions that produce a minimum of melt at the interface.

The basic technique involves placing two pieces of metal parallel to each other and separated by a small space. One of the metals is usually placed on a firm platform; the other will be impacted against the fixed piece by igniting a layered explosive charge (Fig. 6-64). As may be expected, this description is oversimplified. Usually the surfaces have oxides, nitrides, or absorbed gases. In explosive welding, the high pressure developed at the angled collision surface causes a portion of the surfaces, called a jet, to become fluid and be expelled. The severe deformation of the colliding surfaces and the resulting jet breaks up any surface film, forcing the surfaces into intimate contact. The jetting phenomenon, required for bonding, causes the collision point to become plastic and flow into the space between the two plates. By this mechanism, the necessary conditions for the formation of a direct metal-to-metal bond occur. Melted zones are formed, but at discrete intervals along the bond rather than as a continuous layer. This surface jetting contributes greatly to the strength of the weld, providing a mechanical lock in addition to the metallurgical bond (Fig 6-65).

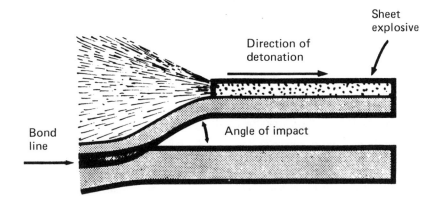

FIGURE 6-64. *The basic mechanism of explosive bonding is simple. An explosive charge is detonated, driving two or more plates against each other at high velocity. The resulting collision produces a strong, permanent bond.*

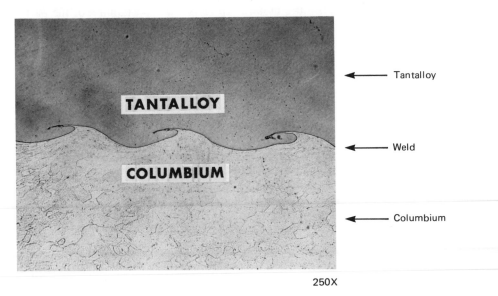

250X

Etchant: Lactic, HNO$_3$, HF

FIGURE 6-65. *Photomicrograph of tantalloy explosively welded to columbium C103 alloy. Note the "rippled" effect at the interface, providing a mechanical lock in addition to the metallurgical bond. Courtesy Aerojet Ordinance and Manufacturing Co.*

441

*Advantages and Limitations.* The big advantage of explosive bonding is that it is essentially a "cold-welding" process. The heat generated is considered insignificant and is rapidly dissipated. Explosive welding is especially useful in joining dissimilar materials and metals that are metallurgically incompatible. As an example, normal welding processes may cause two metals to melt but a brittle alloy forms at the fusion zone rendering it useless. Explosive welding eliminates this problem. Also, some materials are considered unweldable because the strain-hardening properties will be destroyed by the elevated temperatures. Except for small hardness and elongation changes, explosive bonding leaves the materials in the as-received condition.

The process is relatively inexpensive since very little equipment is required. Also it is very "portable."

One problem associated with explosive bonding it that it requires so many disciplines (chemical, metallurgical, mechanical, etc.). The process cannot be carried out in an ordinary factory due to the high noise level, so a remote location must be used unless adequate soundproofing can be found.

Theoretically there is no size limitation on the process; plates as large as $10 \times 30$ ft have been clad. However, there are difficulties when the parts or sections become very small. It is difficult to keep the effects of the explosion confined to the area desired. Despite this difficulty, most of the current interest is directed toward smaller-scale operations.

Flat plates are relatively easy to weld. However, as the configuration complexity increases, difficulties mount. In some cases positioning of the parts may require expensive jigs and fixtures. Also the charges must be shaped to fit the configuration.

### Diffusion Bonding

Diffusion bonding refers to the metallurgical joining of metal surfaces by the application of heat and pressure so as to cause co-mingling of atoms at the joint interface. The process presupposes that the surfaces to be joined are clean and free of contaminants.

Diffusion bonding differs from fusion welding in that fusion welding is dependent on melting of the metals at the joint interface whereas, in most instances, diffusion bonding is accomplished entirely in the solid state, about at one-half to two-thirds the melting point of the metal.

There are three major conditions that require careful control for successful diffusion bonding. These are pressure, temperature, and time (at temperature and pressure).

The temperature accelerates the co-mingling of atoms at the joint interface and provides metal softening, which aids in surface deformation and more intimate contact. The temperature also aids in breaking up surface oxides.

The time is controlled to be at a minimum. The only time allowed is that sufficient to assure that surfaces are in intimate contact and some atomic movement has occurred across the interface.

Other variables that influence the quality of the bond are atmosphere and surface finish. A good vacuum ($> 10^{-3}$ Torr) or oxygen-free inert-gas environments are best suited for diffusion bonding. Moisture-free argon or helium are commonly employed at less than $-70°F$, dew point.

It is essential to provide smooth conforming surfaces at the bond face. Soft materials conform readily and usually do not require other than normal, commercially obtained finishes. Hard materials such as refractory metals, tool steels, and super alloys require special attention. A surface finish no rougher than 16 microinches is recommended unless softer intermediate bonding films which can deform are used.

Surface cleanliness is mandatory. However, this does not imply that unduly restrictive handling is necessary. It merely means that good practice in surface preparation is essential. Vapor degreasing followed by pickling, rinsing, and thorough drying has been used effectively in preparing certain steels, copper, titanium, and other materials for diffusion bonding.

***Diffusion Bonding Methods.*** Common techniques used in diffusion bonding are: hot-press bonding, isostatic hot-gas pressure bonding, vacuum-furnace bonding, explosive bonding, and friction bonding.

*Hot-Press Bonding.* Hot-press bonding is a convenient method of diffusion bonding large, flat sheets or where the part configuration requires application of pressure in one axial direction. Usually a hydraulic press equipped with heated platens is used. The parts to be welded are sealed in a flexible metal envelope before inserting between the press platens. Bonding is accomplished in about 20 min at the required temperature and pressure.

The first aircraft to use diffusion-bonded titanium for primary load-carrying structures is the B-1. It uses a total of 66 diffusion-bonded parts in critical fracture areas. Shown in Fig. 6-66(*a*) is a joint made by diffusion bonding. The 18,000-ton diffusion-bonding press is shown at (*b*).

*Isostatic Hot-Gas Pressure Bonding.* Isostatic pressure refers to the uniform application of pressure to all surfaces of a body at once. This may be accomplished with gases or fluids. For example, the bonding may be accomplished by placing the part in an autoclave and subjecting it to 15,000 psi pressure of inert gas at 1000°F (538°C). Another method may employ 60,000 psi pressure at 3632°F (2000°C) with the pressurizing media being hydraulic fluid in combination with silica sand.

*Vacuum-Furnace Bonding.* Radiant-heated vacuum furnaces are used for some types of diffusion bonding. A vacuum of less than $10^{-4}$ Torr is adequate for bonding such metals as copper-aluminum, copper-copper, iron-copper, and titanium-titanium A pressure may be applied by a hydraulic ram, dead-weight loading, or clamping.

*Explosive Bonding.* This type of welding was discussed previously in this chapter.

### Friction Bonding

Friction bonding is often termed friction welding or inertia welding. The principle is that of rubbing surfaces to both clean them and provide heat. In practice, the process is limited in that one of the parts to be joined must be cylindrical or easily mounted in the chuck or holding fixture of the friction-welding machine. The equip-

(a)

(b)

# DIFFUSION BONDING PROCESS

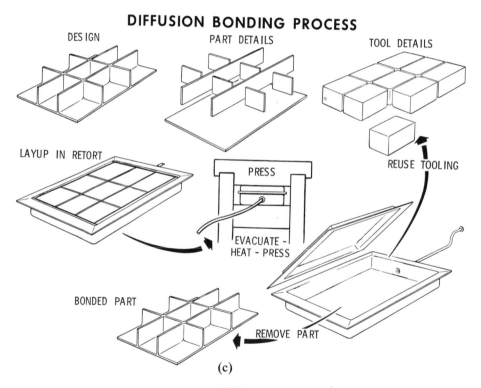

(c)

FIGURE 6-66.   *Shown at (a) is a macroetched section of a diffusion-bonded titanium T-joint.   The 18,000-ton diffusion-bonding press is shown at (b). At diffusion-bonding temperature, titanium acts like a viscous superplastic liquid.   By proper adjustment of time, temperature, and pressure, the metal is made to flow into a myriad of shapes within the dies.   A schematic of the entire process is shown at (c).   Courtesy Rockwell International.*

ment, which has been developed by several different companies, usually includes the following features:

1. A stationary holding fixture to hold one of the components for welding.
2. A rotating holding fixture which is coupled directly to a flywheel and driving source.
3. A constant-load source (hydraulic cylinder) to bring parts together for joining.

Rotating member is brought up to desired speed.

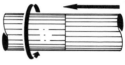

Nonrotating member is advanced to meet the rotating member and pressure is applied.

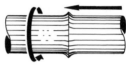

In heating phase, pressure and rotation are maintained for a specific period of time.

In forging phase, rotation is stopped and pressure is either maintained or increased for a specific period of time. Total Welding time is 2 to 30 sec.

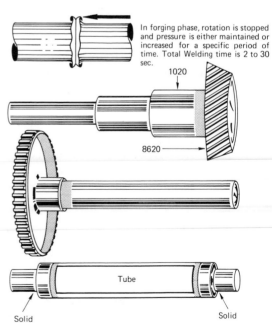

FIGURE 6-67. *Basic steps in friction welding and some typical applications.*

445

The basic steps are shown in Fig. 6-67 along with typical applications. The joint is generally accomplished without the occurrence of melting. The process has been found effective, repeatable, and efficient.

*Advantages and Limitations.* Diffusion bonding results in complete metallurgical-joint continuity without brittle phases or voids, even over comparatively large areas. The problems of dissimilar-metal fusion welds are eliminated. Diffusion bonding permits the utilization of materials based on their condition before bonding. The isostatic process is not limited by complex exterior or interior surfaces.

Dissimilar metals require careful tailoring of the process to prevent built-in bond stresses. Diffusion bonds may be subject to thermal stress degradation if not properly provided with an intermediate transition film. For example, a diffusion bond was made between an aluminum alloy and a 300-series stainless steel. The joint was developed for use at 800°F (427°C) in organic nuclear-power reactors. The joint was required to remain helium tight for 30 years. Failure of the joint would entail shutdown and major repairs. To prevent joint failure by the formation of iron-aluminum intermetallics, it was necessary to provide a tungsten diffusion-barrier film. By accelerated testing at 1000°F (538°C) the joint was shown to have a 30-year life.

Factors to consider in choosing an intermediate metal, where needed, are that it should diffuse well with the base metal, have, ideally, a relatively low yield strength (so that it has good plastic flow), and does not form intermetallic compounds with the base metal. Diffusion welds have high ductility which would be lost by the formation of intermetallics.

## WELDING DESIGN

Many factors enter into the proper design of weldments, such as the material selection, joint design, selection of the welding process, and allowable stresses.

### Material Selection

Materials were introduced in Chapter 2; however, in welding the emphasis is primarily on construction-type materials. The American Institute of Steel Construction (AISC) lists twelve steels that have been approved and classified by ASTM. The main types of these steels are shown in Table 6-13. Structural steels may be placed in three broad categories: carbon steels, high-strength low-alloy steels (HSLA), and heat-treated low-alloy steels.

*Carbon Steels.* Most of the structural steels used today are of the A36 class. These steels provide an excellent material where unit stresses are low and rigidity is a main consideration. The 36 refers to a minimum yield strength of 36ksi. Carbon steels have excellent weldability and require neither preheat nor postheat treatments.

*High-Strength Low-Alloy Steels (HSLA).* All the structural steels listed in Table 6-12, with the exception of A36, are HSLA steels. The addition of alloying elements enhances their hot-rolled strength.

TABLE 6-13.  *ASTM Structural Steel, Cast Steel and Steel Forging Designations*

| ASTM Designation | Description | Yield Strength in psi (MPa) | Tensile Strength in psi (MPa) | % Elongation in 8 in. (20.32 cm) | % Elongation in 2 in. (5.08 cm) |
|---|---|---|---|---|---|
| **Structural Steels** | | | | | |
| A36 | Structural steel. | 36,000 (248.2 MPa) | 58 to 80,000 (399.9 to 551.6 MPa) | 20 | 23 |
| A242 | High strength, low alloy. | 50,000 (344.7 MPa) | 70,000 (482.6 MPa) | 19 | 22 |
| A375 | Low alloy, hot-rolled steel sheet. | 47,000 (324.1 MPa) | 67,000 (462.0 MPa) | 19 | |
| A440 | High strength structural steel. | 50,000 (344.7 MPa) | 70,000 (482.6 MPa) | 18 | |
| A441 | Low alloy structural manganese vanadium. | 50,000 (344.7 MPa) | 70,000 (482.6 MPa) | 18 | |
| A588 | High yield strength, quenched and tempered alloy steel plate. | 50,000 (344.7 MPa) | 70,000 (482.6 MPa) | 18 | |
| A572 | High strength, low alloy columbium–vanadium steels of structural quality. | 42,000 to 65,000 (289.6 MPa to 448.2 MPa) | 60,000 to 85,000 (413.7 to 586.1 MPa) | 20 to 40 | 24 |
| **Cast Steels** | | | | | |
| A27 | Grade 65–35—Mild-to-medium strength carbon steel castings. | | | | |
| A148 | Grade 80–50—Steel castings for structural purpose. | | | | |
| **Carbon Steel Forgings** | | | | | |
| A235 | Class C1, F, and G—General industrial use. | | | | |
| A237 | Class A—General industrial use. | | | | |

***Heat-Treated Low-Alloy Steels.***  Steels in this classification (AISI 588 and 514) may be heated, quenched, and tempered to obtain maximum yield strengths of 80 to 110 ksi. These steels are weldable when the proper procedures are used and no additional heat treatment is required except moderate preheating and postheating.

### Welded-Joint Design

From the standpoint of geometry, metal plates may be joined together in five main types of joints, as shown in Fig. 6-68.  The two types of welds most often used are butt and fillet.

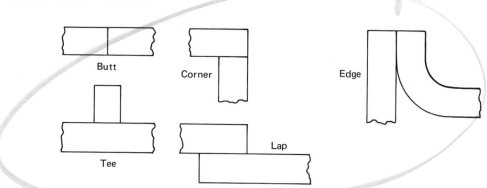

FIGURE 6-68. *The basic types of welded joints. There are many varia-tions of each of these. Variations of the butt joints are shown in Fig. 6-71.*

### Butt-Weld Joint Design

Butt welds, also known as groove welds, are made between abutting plates in the same plane and are generally described by the way the edges are prepared. In general, the least preparation required, the more economical. However, when plates are thick, the spacing required to obtain complete penetration for manual welding may cause the amount of weld metal to be excessive (Fig. 6-69). The type of edge preparation is largely dependent on how complete penetration can be achieved with the minimum amount of weld metal. Shown in Fig. 6-70 is a single-V butt joint, together with common terminology.

**Standard Butt-Weld-Edge Preparation and Spacing.** The square-edge butt joint re-quires no preparation other than providing a straight square edge and a means of retaining the proper spacing for the weld. This is usually done with tack welds at frequent intervals. Oftentimes a square-edge butt joint will be made with a backing strip. This strip may remain as part of the joint or it may be machined off. The backing strip is commonly used when all welding must be done from one side or when

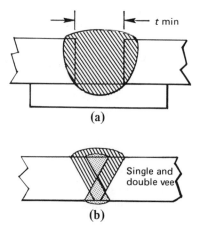

(a)

(b)

FIGURE 6-69. *The minimum preparation may lead to excessive weld deposit as in the case of the square butt joint in heavy plate(a). Two times as much weld deposit is required for a single-V butt joint as compared to a double-V butt joint(b).*

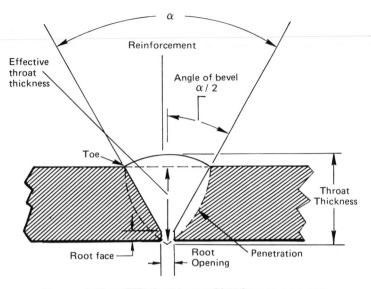

FIGURE 6-70. *A single-V butt weld with common terms.*

the root opening is excessive. Shown in Fig. 6-71 are standard butt-joint types and spacing for various thicknesses of metal.

A *backing bar* is similar to a backing strip; however, it is only intended to be temporary. It consists of a heavy copper bar with a groove that provides room for full penetration of the weld metal. It is also used in channel form with a flow of inert gas to protect the back side of the weld from the atmosphere. More recently a tape has been developed that acts as a backing strip for closed butt joints. It consists of a flexible, granular refractory layer mounted with a wider pressure-sensitive adhesive foil protected by a paper liner (Fig. 6-72).

***Butt Weld Depth-to-Width Ratio.*** Butt welds that are designed to maintain a balance between width and depth are generally less susceptible to weld cracks, as shown in Fig. 6-73. The critical width-to-depth ratio is almost entirely limited to the first pass. The second and subsequent passes are almost never subject to the same magnitude of stress. Thus, because of the importance of the width-to-depth factor, the joint preparation may be inherently crack-sensitive.

***Butt Weld Strength.*** Full-penetration butt welds are generally regarded as having the same load-carrying capacity as the base metal. Hence there is no need to calculate the strength of the weld. This assumes that the weld deposit corresponds to that of the base plate and that either a backing bar was used or a *sealing run*. A sealing run refers to backgouging the back side of the weld by grinding or with an arc-air torch and then depositing a weld bead.

| WELD TYPE | Sides Welded | Thickness (in.) | Gap (in.) | Root Face (in.) | Min Included Angle (deg) |
|---|---|---|---|---|---|
| 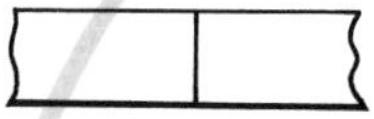 Closed Square Butt | One / Both | Up to 1/16 / Up to 1/8 / Up to 5/8 With DP electrodes | | | |
|  Open Square Butt | Both | Up to 3/16 | 1/16 | | |
| 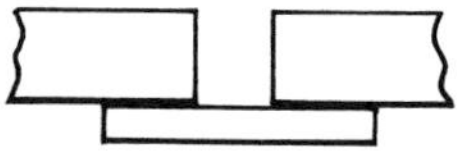 Square Butt with Backing Bar | One | Up to 3/16 / Up to 1/2 / Up to 1/2 With DP electrodes | 3/16 / 5/16 / 1/4 | | |
| 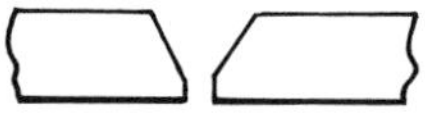 Single "V" Butt Weld | One or Both | Over 3/16 and up to 1 / If made with backing strip | 0 to 1/8 / 1/8 to 3/16 | 0 to 1/8 | 60 |
| 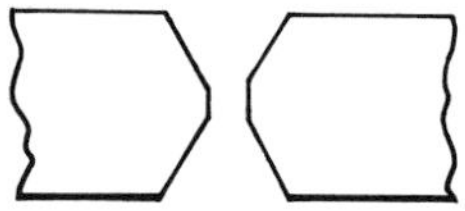 Double "V" Butt Weld | Both | Over 1/2 | 0 to 1/8 | 0 to 1/8 | 60 |
| 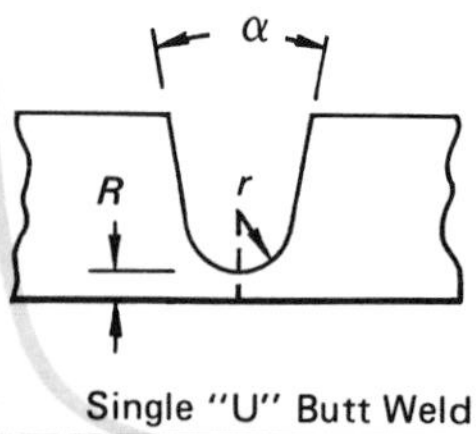 Single "U" Butt Weld | One or Both | Over 3/4 | 0 to 1/16 | 1/8 to 3/16 | 10 to 30 |

FIGURE 6-71.  *Standard types of welded butt joints.*

| WELD TYPE | Sides Welded | Thickness (in.) | Gap (in.) | Root Face (in.) | Min Included Angle (deg) |
|---|---|---|---|---|---|
| Single "J" Butt Weld | Both | Over 3/4 | 1/8 | 1/8 to 3/16 | 20 to 30 |
| Single Bevel Butt Weld | Both | Up to 1 Unlimited with backing strip | 0 to 1/8 | 0 to 1/8 | 45 to 50 |
| Double "U" Butt Joint | Both | Over 1-1/2 | | 1/8 to 3/16 | 10 to 30 |
| Double "J" Butt Weld | Both | Over 1-1/2 | 1/8 | 1/8 to 3/16 | 20 to 30 |
| Double Bevel Butt Weld | Both | Over 1/2 | 0 to 1/8 | 0 to 1/8 | 45 to 50 |

**(a)**  **(b)**

FIGURE 6-72. *A tape-type backing strip utilizes refractory granules on a carrier (a) and in use on an aqueduct (b). Courtesy 3M.*

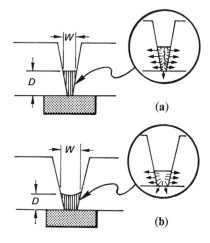

**(a)**

FIGURE 6-73. *The W/D ratio of weld preparation is related to crack sensitivity. The wider root opening (b) provides a better W/D ratio than (a) and is not crack sensitive. Courtesy The James F. Lincoln Arc Welding Foundation.*

**(b)**

### Fillet-Weld Joint Design

Fillet welds are those used to fill in a corner. They are the most common welds in structural work. Shown in Fig. 6-74 are fillet welds with a detailed view of a T-joint. The leg of a fillet weld is usually equal to the height of the base material (*h*). The nominal weld outline shown as dashed lines in the detail view makes a 45° triangle

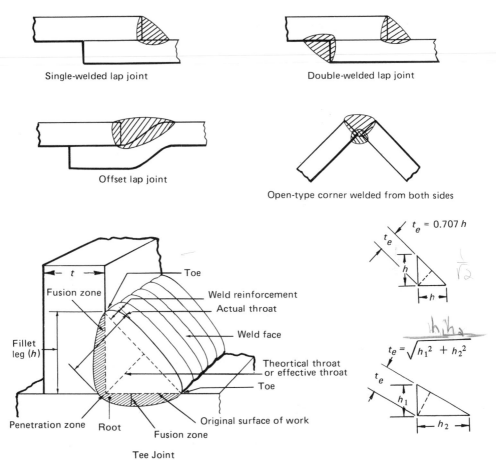

Single-welded lap joint

Double-welded lap joint

Offset lap joint

Open-type corner welded from both sides

Tee Joint

FIGURE 6-74. *Lap joints, open corner joints, and tee joints all use fillet-type welds. The fillet weld may have equal or unequal legs but the effective throat ($t_e$) is taken perpendicular to the weld face from the root of the joint, as shown by the dashed line.*

when each leg is equal to $h$. The effective throat ($t_e$) of the fillet is the distance from the root of the joint to a point perpendicular to the hypotenuse. For a 45° fillet weld, the throat is equal to 0.707 (sine of 45°) times the leg dimension.

Fillet welds are not always made with leg sizes equal to $h$. The *maximum* size permitted by AISC specification 1.17.6, for the design, fabrication, and erection of structural steel for buildings, states:

1. Along edges of material less than $\frac{1}{4}$ in. thick, the maximum size may be equal to the thickness of the material.
2. Along edges of material $\frac{1}{4}$ in. or more in thickness, the maximum vertical leg shall be $\frac{1}{16}$ in. less than the thickness of the material unless the weld is especially designated on drawings to be built out to obtain full throat thickness.

TABLE 6-14. *Minimum\* Fillet Weld Size* (*AWS Structural Welding Code D1.1–72*)

| Material Thickness of Thicker Part Joined in Inches (mm) | | Fillet Weld (*h*) in Inches (mm) | |
|---|---|---|---|
| to $\frac{1}{4}$ incl. | (6.35 mm) | 1/8 | (3.175 mm) |
| over $\frac{1}{4}$ | to $\frac{1}{2}$ (6.35–12.70 mm) | 3/16 | (4.7625 mm) |
| over $\frac{1}{2}$ | to $\frac{3}{4}$ (12.70–19.05 mm) | 1/4 | (6.35 mm) |
| over $\frac{3}{4}$ | to $1\frac{1}{2}$ (19.05–38.10 mm) | 5/16 | (7.9375 mm) |
| over $1\frac{1}{2}$ | to $2\frac{1}{4}$ (38.10–57.15 mm) | 3/8 | (9.525 mm) |
| over $2\frac{1}{4}$ | to 6 (57.15–152.40 mm) | 1/2 | (12.70 mm) |
| over 6 | >(152.40 mm) | 5/8 | (15.875 mm) |

\* Except that the weld size need not exceed the thickness of the thinner joined part.

The *minimum-size* fillet weld is governed by the amount of heat, and therefore the size of the weld, required to ensure fusion. The minimum-size fillet welds permitted by the AWS Structural Welding Code are shown in Table 6-14.

***Strength of Fillet Welds.*** The strength of a fillet weld varies with the direction of the applied load, depending upon whether it is parallel or transverse to the weld. In either case, the weld will fail in the shear plane that has the maximum shear stress. For loading parallel to the weld, the plane of rupture is at 45°, as shown in Fig. 6-75(*b*). Transverse loading in the plane of maximum shear occurs at 67.5° to the horizontal, as shown at (*a*). Because of this and the fact that the stress distribution for this type of loading is more uniform, the weld will carry about one-third more load in transverse shear than in parallel shear. Two 45° fillet welds with leg dimensions equal to $\frac{3}{4}t$ will develop the full strength of the plate for either type of loading, assuming that the weld metal is equal to that of the base plate and average penetration is obtained.

### Allowable Stresses

Recent changes in the AISC and AWS welded-construction codes have made obsolete the allowable stress of 13,600 psi for 45° fillet welds or partial-penetration bevel-groove welds. It is now 30% of the minimum specified electrode tensile strength, as shown in Table 6-15. The allowable unit force per lineal inch is also given and may be used to calculate the size fillet weld required for a given application according to the electrode used.

*Example:* Determine the allowable unit stress ($\sigma$) per inch for a 0.5-in. fillet weld made with an E70 electrode.

$$\sigma = .707t(0.30)(EXX)$$
$$= .707 \times 0.5 \times .30 \times 70$$
$$= 7.42 \text{ kips/lineal inch (see Table 6-14)}$$

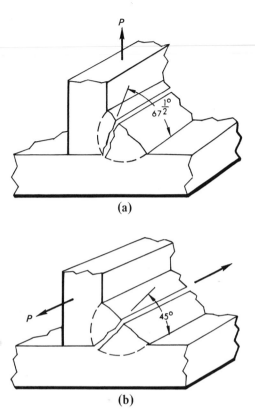

FIGURE 6-75. *The maximum shear stress occurs at an angle of 67.5° for transverse loading (a) and at 45° for parallel loading (b).*

The new AISC code (1.14.7) gives limited credit for penetration beyond the root of a fillet weld made by the submerged-arc process. Thus for welds $\frac{3}{8}$ in. or smaller, the $t_e$ is now equal to the leg size of the weld. For submerged-arc fillet welds, larger than $\frac{3}{8}$ in., the $t_e$ is obtained by adding 0.11 to $.707h$.

The strength of transverse fillet welds loaded in tension (transverse shear) can be found by:

$$\tau = \frac{P}{.707hl}$$

*where:*  $\tau$ = shear stress, psi
$P$ = load, lbs
$h$ = leg of weld, in.
$l$ = total length of weld, in.

## Welding Symbols

This brief presentation of welding symbols is for the purpose of showing how easily they can be used to convey the exact desired information. A more complete presentation of standard welding symbols is given in Appendix 6B and in the sixth edition of the AWS Welding Handbook.

TABLE 6-15. *AISC and AWS Allowable Loads for Various Sizes of Fillet Welds*

| | Strength Level of Weld Metal (EXX), ksi (in MPa) | | | | | |
|---|---|---|---|---|---|---|
| | 60 (413.7) | 70 (482.6) | 80 (551.6) | 90 (620.5) | 100 (689.5) | 110 (758.4) |
| | Allowable Shear Stress on Throat, ksi of Fillet Weld or Partial Penetration Groove Weld | | | | | |
| T = | 18.0 (124.1) | 21.0 (144.8) | 24.0 (165.5) | 27.0 (186.2) | 30.0 (206.8) | 33.0 (227.5) |
| | Allowable Force on Fillet Weld, Kips/Linear Inch | | | | | |
| f = | 12.73h (87.71) | 14.85h (102.32) | 16.97h (116.92) | 19.09h (131.53) | 21.21h (146.14) | 23.33h (160.74) |
| Leg Size (h) in | Allowable Force for Various Sizes of Fillet Welds, Kips/Linear Inch (in MPa/linear cm) | | | | | |
| (cm) | | | | | | |
| 1″ (25.4) | 12.73 (87.71) | 14.85 (102.32) | 16.97 (116.92) | 19.09 (131.53) | 21.21 (146.14) | 23.33 (160.74) |
| 7/8″ (22.22) | 11.14 (76.75) | 12.99 (89.50) | 14.85 (102.32) | 16.70 (115.06) | 18.57 (127.95) | 20.41 (140.62) |
| 3/4″ (19.05) | 9.55 (65.79) | 11.14 (76.75) | 12.73 (87.71) | 14.32 (98.66) | 15.92 (109.69) | 17.50 (120.58) |
| 5/8″ (15.87) | 7.96 (54.84) | 9.28 (63.94) | 10.61 (73.10) | 11.93 (82.20) | 13.27 (91.43) | 14.58 (100.46) |
| 1/2″ (12.70) | 6.37 (43.89) | 7.42 (51.12) | 8.48 (58.43) | 9.54 (65.73) | 10.61 (73.10) | 11.67 (80.41) |
| 7/16″ (11.11) | 5.57 (38.38) | 6.50 (44.79) | 7.42 (51.12) | 8.35 (57.53) | 9.28 (63.94) | 10.21 (70.35) |
| 3/8″ (9.52) | 4.77 (32.87) | 5.57 (38.38) | 6.36 (43.82) | 7.16 (49.33) | 7.95 (54.78) | 8.75 (60.29) |
| 5/16″ (7.93) | 3.98 (27.42) | 4.64 (31.97) | 5.30 (36.52) | 5.97 (41.13) | 6.63 (45.68) | 7.29 (50.23) |
| 1/4″ (6.35) | 3.18 (21.91) | 3.71 (25.56) | 4.24 (29.21) | 4.77 (32.87) | 5.30 (36.52) | 5.83 (40.17) |
| 3/16″ (4.76) | 2.39 (16.47) | 2.78 (19.15) | 3.18 (21.91) | 3.58 (24.67) | 3.98 (27.42) | 4.38 (30.18) |
| 1/8″ (3.17) | 1.59 (10.96) | 1.86 (12.82) | 2.12 (14.61) | 2.39 (16.47) | 2.65 (18.26) | 2.92 (20.12) |
| 1/16″ (1.58) | 0.795 (5.48) | 0.93 (6.41) | 1.06 (7.30) | 1.19 (8.20) | 1.33 (9.16) | 1.46 (10.06) |

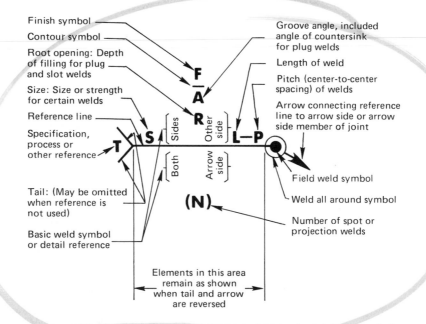

Finish symbol
Contour symbol
Root opening: Depth of filling for plug and slot welds
Size: Size or strength for certain welds
Reference line
Specification, process or other reference
Groove angle, included angle of countersink for plug welds
Length of weld
Pitch (center-to-center spacing) of welds
Arrow connecting reference line to arrow side or arrow side member of joint
Tail: (May be omitted when reference is not used)
Field weld symbol
Weld all around symbol
Basic weld symbol or detail reference
Number of spot or projection welds
Sides
Other side
Both
Arrow side
Elements in this area remain as shown when tail and arrow are reversed

FIGURE 6-76. *The location of the elements of a standard welding symbol.*

Shown in Fig. 6-76 is a welding symbol with the standard placement of information concerning the weld. In actual practice a symbol would never have this much information since there is usually not that much to say about one particular weld. Note, for example, that the tail may be omitted when the specification of the process used is omitted. A further breakdown of the symbols is shown in Fig. 6-77. Here, as one example, is a chain intermittent fillet welding symbol showing the desired leg size along with the length and increment spacing of the weld. Thus symbols allow the designer to give rather complete information to the welder without resorting to lengthy notes. Shown in Fig. 6-78 are some typical examples of the use of welding symbols in structural work.

## Welded Connections

The term *welded connections* refers to the design and method of joining standard structural members such as angles, channels, beams, and columns, some of which are shown in Fig. 6-79.

A simple direct method proposed by Blodgett for finding the properties of welded connections is to treat the welds as lines having neither leg nor throat dimensions. The forces, which can be taken from Table 6-16, are considered on a unit-length basis, thus eliminating the knotty problems of combined stresses. This is not to neglect stress distributions within a weld, which can be very complex, but the actual fillet welds tested and upon which the unit forces are based has these same conditions.

457

Back or Backing Weld Symbol

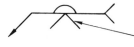

Any applicable single
groove weld symbol

Surfacing Weld Symbol Indicating Built-up Surface

Size (height of deposit).
Omission indicates no
specific height desired

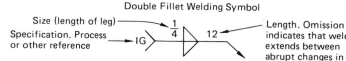

$\frac{1}{8}$

Orientation. Location
and all dimensions
other than size are
shown on the drawing

Double Fillet Welding Symbol

Size (length of leg)
Specification. Process
or other reference

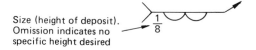

IG $\frac{1}{4}$ 12

Length. Omission
indicates that weld
extends between
abrupt changes in
direction or as
dimensioned

Chain Intermittent Fillet Welding Symbol

Size (length of leg)

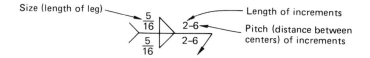

$\frac{5}{16}$  2-6
$\frac{5}{16}$  2-6

Length of increments
Pitch (distance between
centers) of increments

Staggered Intermittent Fillet Welding Symbol

Size (length of leg)

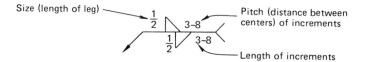

$\frac{1}{2}$  3-8
$\frac{1}{2}$  3-8

Pitch (distance between
centers) of increments

Length of increments

Single-Vee-Groove Welding Symbol

Size (depth of chamfering).
Omission indicates depth
of chamfering equal to
thickness of members

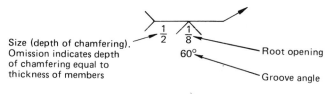

$\frac{1}{2}$  $\frac{1}{8}$
60°

Root opening

Groove angle

FIGURE 6-77. *Typical welding symbols.*

When a weld is treated as a line, the property of the welded connection is inserted into the standard design formula for that particular type of load and the force on the weld is found in terms of pounds per lineal inch. A sample problem will serve to illustrate the use of this method.

**Example:** Shown in Fig. 6-80 is a clip that has been welded to a supporting structure. Will this clip be able to sustain a load of 2 tons that at times has some impact force also?

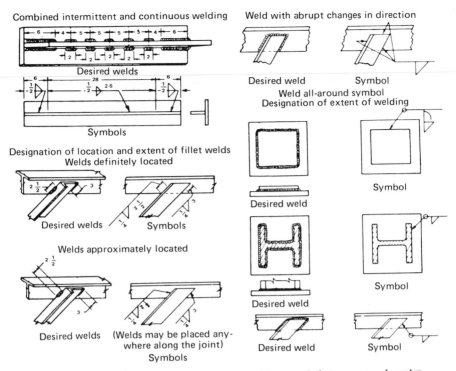

Combined intermittent and continuous welding

Desired welds

Symbols

Designation of location and extent of fillet welds
Welds definitely located

Desired welds         Symbols

Welds approximately located

Desired welds     (Welds may be placed any-
                    where along the joint)
                    Symbols

Weld with abrupt changes in direction

Desired weld          Symbol
Weld all-around symbol
Designation of extent of welding

Symbol

Desired weld

Symbol

Desired weld

Desired weld          Symbol

FIGURE 6-78. *An example of the use of welding symbols in structural work.*

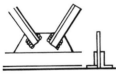

Roof Truss
with Gusset Plate

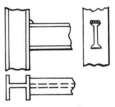

Column and Beam
Connection (Rigid)

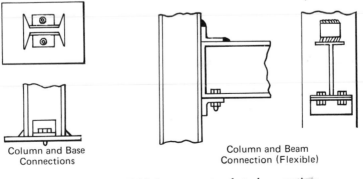

Column and Base
Connections

Column and Beam
Connection (Flexible)

FIGURE 6-79. *Welded-type structural-steel connections.*

TABLE 6-16.  *Stress Formulas for Weld Joints*

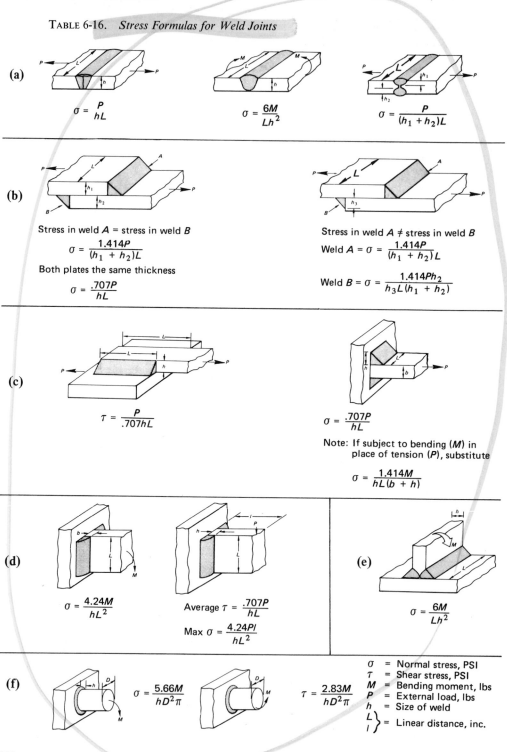

**(a)**

$$\sigma = \frac{P}{hL}$$

$$\sigma = \frac{6M}{Lh^2}$$

$$\sigma = \frac{P}{(h_1 + h_2)L}$$

**(b)**

Stress in weld $A$ = stress in weld $B$

$$\sigma = \frac{1.414P}{(h_1 + h_2)L}$$

Both plates the same thickness

$$\sigma = \frac{.707P}{hL}$$

Stress in weld $A \neq$ stress in weld $B$

Weld $A = \sigma = \dfrac{1.414P}{(h_1 + h_2)L}$

Weld $B = \sigma = \dfrac{1.414Ph_2}{h_3L(h_1 + h_2)}$

**(c)**

$$\tau = \frac{P}{.707hL}$$

$$\sigma = \frac{.707P}{hL}$$

Note: If subject to bending ($M$) in place of tension ($P$), substitute

$$\sigma = \frac{1.414M}{hL(b + h)}$$

**(d)**

$$\sigma = \frac{4.24M}{hL^2}$$

Average $\tau = \dfrac{.707P}{hL}$

Max $\sigma = \dfrac{4.24Pl}{hL^2}$

**(e)**

$$\sigma = \frac{6M}{Lh^2}$$

**(f)**

$$\sigma = \frac{5.66M}{hD^2\pi}$$

$$\tau = \frac{2.83M}{hD^2\pi}$$

$\sigma$ = Normal stress, PSI
$\tau$ = Shear stress, PSI
$M$ = Bending moment, lbs
$P$ = External load, lbs
$h$ = Size of weld
$\left.\begin{array}{c} L \\ l \end{array}\right\}$ = Linear distance, inc.

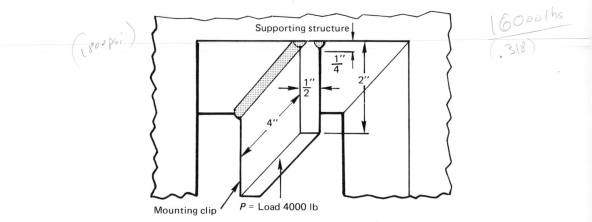

FIGURE 6-80. *A mounting clip is welded to a supporting structure.*

*Solution:* Choose a matching stress equation from Table 6-16. What is the maximum load the clip will hold? The maximum allowable stress for the weld is 3710 psi with an E70 electrode (Table 6-15).

$$\sigma = \frac{0.707P}{hl}$$

Solving for $P$,

$$P = \frac{\sigma hl}{0.707} = \frac{3710 \times \frac{1}{4} \times 8}{0.707} = \frac{7420}{0.707} = 10,459 \, lb$$

Applying a safety factor for occasional impact (Table 6-17), use 1.5.

$$P = 10,459 \div 1.5 = 6,696 \, lb$$

Therefore the welded clip will be able to withstand a load of 2 tons (4000 lb) plus occasional impact.

## Distortion

Distortion in weldments is the result of nonuniform heating and cooling. Expansion and contraction of the weld metal and adjacent base metal may be compared to heating a metal bar between vise jaws (Fig. 6-81). The restricted bar is not able to

TABLE 6-17. *Weld Safety Factors*

| Type of Weld | Factor |
|---|---|
| Reinforced butt | 1.2 |
| Toe of transverse fillet, occasional impact | 1.5 |
| End of parallel fillet | 2.7 |
| T-butt with sharp corners | 2.0 |

expand uniformly, therefore it is upset in the area of greatest heat. As it cools it will contract or shrink in all directions, with the result that the bar is now shorter and thicker than it was before heating.

A weld made on restricted members cannot shrink when cooled, therefore residual stresses are set up in the weld and in the heat affected zone (HAZ) adjacent to it (Fig. 6-82).

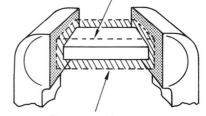

Unrestrained bar before heating and after cooling to room temperature

Unrestrained bar when uniformly heated to a specific temperature

**(a)**

Restrained bar at room temperature

Bar when uniformly heated while restrained

**(b)**

FIGURE 6-81. *A schematic presentation of how the heating and cooling cycle, as used in the welding of metals, causes distortion. Courtesy The James F. Lincoln Arc Welding Foundation.*

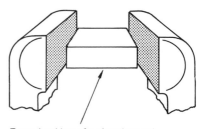

Restrained bar after heating and cooling to room temperature is shorter, thicker and wider

**(c)**

FIGURE 6-82. *Tensile stresses are set up between the weld metal and the cooler base metal as solidification and contraction take place.*

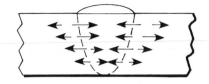

In an unrestricted weld, the members are free to expand and contract. Since the members move, the degree of stress in the weld and HAZ will not be as great as that of a restricted weld. Some of the locked-in stresses are relieved as the base metal shifts or distorts.

Distortion is evidenced in many different forms, but basically there are two main types: angular and longitudinal (Fig. 6-83). Transverse shrinkage is the main cause of angular distortion. It is also directly related to the volume of weld deposit, as shown in Fig. 6-84.

***Distortion Control.*** Shrinkage of hot metal cannot be prevented, but a knowledge of how it operates is useful in minimizing distortion. The following principles are used in minimizing distortion:

1. *Minimize shrinkage forces.* Use only the amount of weld metal required. Overwelding not only adds to the shrinkage forces but it is also uneconomical. Proper edge preparation and fit-up help keep weld deposit to the required minimum. Intermittent welds can also be used where strength requirements are not critical. However, in this regard, the AISC Code 1.1717 limits the minimum effective length of fillet welds to four times their nominal size. If this condition cannot be met, the size of the weld is considered to be one-quarter of its effective length.

Where transverse shrinkage forces cause distortion, as in the butt weld, the number of passes should be kept to a minimum. The shrinkage of each pass tends to be cumulative. The use of larger electrodes allows more weld deposit per pass with a smaller total volume of heat input into the base metal.

2. *Help shrinkage forces work in the desired direction.* Parts may be positioned or preset out of position before welding so that shrinkage forces will bring them into alignment, as illustrated by the T-joint and butt weld in Fig. 6-85.

3. *Balance shrinkage forces with other forces.* A common example of balancing shrinkage forces is the use of fixtures and clamps. The components are locked

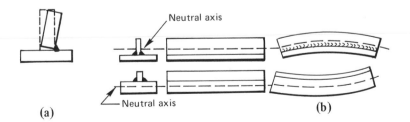

(a)

Neutral axis

Neutral axis

(b)

FIGURE 6-83. *Distortion types: angular (a), longitudinal (b).*

in the desired position and held until the weld is finished. This will of course cause shrinkage forces to build up until the yield point of the metal is reached. For typical welds on low-carbon steel plate this would be about 45,000 psi. You may visualize considerable distortion taking place as soon as the clamps are removed. This does not happen due to the small amount of unit strain ($\varepsilon$) compared to the amount of

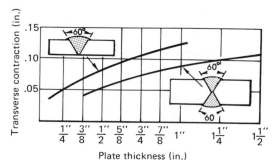

Plate thickness (in.)
Transverse contraction—single Vee vs. Double Vee

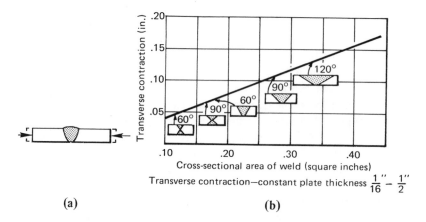

Cross-sectional area of weld (square inches)
Transverse contraction—constant plate thickness $\frac{1}{16}'' - \frac{1}{2}''$

(a)                                    (b)

FIGURE 6-84. *Transverse shrinkage (a) is directly proportional to the amount of weld deposit (b). Courtesy The James F. Lincoln Arc Welding Foundation.*

FIGURE 6-85. *Prepositioning (with tack welds) (a) allows weld shrinkage forces to bring parts into alignment (b).*

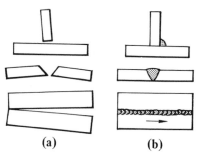

(a)                    (b)

movement that would occur if no restraint existed during the welding process. The strain for steel may be calculated as follows:

$$\varepsilon = \frac{\sigma}{E}$$

$$= \frac{45,000}{30 \times 10^6}$$

$$= 0.0015 \text{ in./in.}$$

where: $E$ = modulus of elasticity

Another common example of balancing shrinkage forces is a butt joint welded alternately on each side.

4. *Remove shrinkage forces after welding.* Peening is often used as a means of stress-relieving after welding. In theory the metal is made to stretch slightly under the force of each blow, thus relieving the stress. Care must be exercised in peening to avoid concealing a crack or work-hardening the metal.

In more specialized cases, such as alloy steels, preheating and postheating are used to minimize the residual stress.

5. *Place welds near the center of gravity.* It is not always possible to place welds on or near the center of gravity; however, even the amount of penetration as shown in Fig. 6-86 can change the amount of distortion. Shown at (*a*) is a T-weld made by manual welding and at (*b*) by the submerged arc.

*Metal Properties and Distortion.* Distortion is related to the coefficient of expansion, thermal conductivity, modulus of elasticity, and yield strength of the material welded. A high coefficient of expansion tends to increase the shrinkage of the weld metal and the adjacent metal, thus increasing distortion.

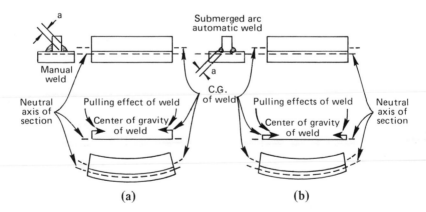

(a)                              (b)

FIGURE 6-86. *Deeper penetration (b) achieved with automatic submerged arc results in welds that are nearer the center of gravity with less distortion than the manual deposit (a). Courtesy The James F. Lincoln Arc Welding Foundation.*

465

TABLE 6-18. *Pounds of Electrodes and Steel Deposit per Lineal Foot of Weld*

## Horizontal Fillet Weld

| Size of Fillet L (in inches) | Pounds of Electrodes Required per Linear Foot of Weld* (approx.) | Steel Deposited per Linear Foot of Weld—Pounds |
|---|---|---|
| 1/8 | 0.048 | 0.027 |
| 3/16 | 0.113 | 0.063 |
| 1/4 | 0.189 | 0.106 |
| 5/16 | 0.296 | 0.166 |
| 3/8 | 0.427 | 0.239 |
| 1/2 | 0.760 | 0.425 |
| 5/8 | 1.185 | 0.663 |
| 3/4 | 1.705 | 0.955 |
| 1 | 3.030 | 1.698 |

* Includes scrap end and spatter loss.

## Square Groove Butt Joints

... welded one side

$R = 0.07$

$\frac{1}{2}t$    $t$

... welded two sides

If root of top weld is chipped or flame gouged and welded, add 0.07 lb. to steel deposited (equivalent to approx. 0.13 lb. of electrodes).

$R = 0.07$    $W$    $t$    $G$    $R = 0.07$

| Joint Dimensions (in inches) | | | Pounds of Electrodes Required per Linear Foot of Weld* (approx.) | | Steel Deposited per Linear Foot of Weld—Pounds | |
|---|---|---|---|---|---|---|
| T | B | G | Without reinforcement | With reinforcement** | Without reinforcement (lbs) | With reinforcement** (lbs) |
| 3/16 | 3/8 | 0 | — | 0.16 | — | 0.088 |
| | | 1/16 | 0.04 | 0.20 | 0.020 | 0.109 |
| 1/4 | 7/16 | 1/16 | 0.05 | 0.23 | 0.027 | 0.129 |
| | | 3/32 | 0.07 | 0.26 | 0.039 | 0.143 |
| 5/16 | 1/2 | 1/16 | 0.06 | 0.27 | 0.033 | 0.153 |
| | | 3/32 | 0.09 | 0.30 | 0.050 | 0.170 |
| 1/8 | 1/4 | 0 | — | 0.21 | — | 0.119 |
| | | 1/32 | 0.03 | 0.24 | 0.013 | 0.132 |
| 3/16 | 3/8 | 1/32 | 0.04 | 0.36 | 0.020 | 0.199 |
| | | 1/16 | 0.07 | 0.39 | 0.040 | 0.218 |
| 1/4 | 7/16 | 1/16 | 0.10 | 0.47 | 0.053 | 0.261 |
| | | 3/32 | 0.14 | 0.53 | 0.080 | 0.288 |

* Includes scrap end and splatter loss.
** $r$ = (height of reinforcement).

## Vee Groove Butt Joint

| Joint Dimensions (in inches) | | | Pounds of Electrodes Required per Linear Foot of Weld* (approx.) | | Steel Deposited per Linear Foot of Weld—Pounds | |
|---|---|---|---|---|---|---|
| T | B | G | Without reinforcement | With reinforcement** | Without reinforcement (lbs) | With reinforcement** (lbs) |
| 1/4 | 0.207 | 1/16 | 0.15 | 0.25 | 0.085 | 0.143 |
| 5/16 | 0.311 | 3/32 | 0.31 | 0.46 | 0.173 | 0.258 |
| 3/8 | 0.414 | 1/8 | 0.50 | 0.70 | 0.282 | 0.394 |
| 1/2 | 0.558 | 1/8 | 0.87 | 1.15 | 0.489 | 0.641 |
| 5/8 | 0.702 | 1/8 | 1.35 | 1.68 | 0.753 | 0.942 |
| 3/4 | 0.847 | 1/8 | 1.94 | 2.35 | 1.088 | 1.320 |
| 1 | 1.138 | 1/8 | 3.45 | 4.00 | 1.930 | 2.240 |

* Includes scrap end and spatter loss.
** $r$ = (height of reinforcement).

NOTE: Variations in joint configuration not shown may substitute appropriate figures in the formula $W_e = M_d/(1 - L)$. Where, $W_e$ = (weight of electrode required), $M_d$ = (weight of deposited metal), and $L$ = (total electrode losses). The weight of the steel deposited is based on the volume of the joint (area of the groove × length) which is converted to weight of steel by the factor 0.283 lb/cu in.

A metal with a low thermal conductivity such as stainless steel (Appendix 2A) retains the heat in the weld area. This results in a steep temperature gradient and greater distortion.

As a weld deposit cools and contracts, the adjacent metal "stretches" to satisfy the volume demand of the weld joint. Thus the higher the yield strength of the base metal, the greater the distortion. If the modulus of elasticity is high, the material is more likely to resist distortion.

### Economics of Welding

The economics of welding begins with the design and follows through to the final inspection of the finished product. It is at the design stage where the type of joint and the method of assembly are pretty well determined. Some feedback may come from production regarding modifications and desirable design changes. Thus the designer should be well acquainted with the wide variety of joining methods available and the comparative advantages and disadvantages of each for the applications at hand.

Economics is also very much in the hands of the production engineer. For example, the design may call for a welded assembly but the exact process may be left to production. The production engineer must weigh the cost of the joint preparation and decide whether assembly will be manual, semiautomatic, or fully automated. As an example, if welding is to be the method of assembly, a cost comparison of automatic versus manual welding can be estimated from standard data. The cost of manual welding is quite well established and may be based on the following: labor and overhead, $10/hr; cost per pound of deposited electrode, $6.67 based on an 8-hr day and $\frac{5}{32}$-in. electrodes. Of course these figures will vary as labor costs and material fluctuate.

The cost per pound of deposited metal by the semiautomatic method, such as in-hand manipulation of the GMA process or flux-cored wires, drops to about $1.90. If fully automatic welding can be used, the cost will be about $1.00/lb.

The higher initial investment cost for automatic equipment can result in greater savings only if the operating time is favorable. Usually this is a problem of production scheduling and the availability of versatile tooling to keep the setup time down and to minimize downtime during loading and unloading.

### Estimating Welding Costs

*Arc Welding.* Arc-welding costs can often be quickly estimated based on labor and overhead, welding consumables, and profit.

Table 6-18 can be used to determine the pounds of electrodes required per linear foot of weld or the weight of deposited metal per foot for various types and sizes of joints. Shown in Table 6-19 is a tabulation of the cost of deposited weld metal in dollars per pound, including electrode cost, labor, and overhead at $6.00/hr.

TABLE 6-19. *Approximate Cost of Deposited Weld Metal, $/lb (Includes Electrode Cost and $6.00/hr for Labor and Overhead)*

| Type | Size | Current | Operating Factor | | | | |
|------|------|---------|------|------|------|------|------|
| | | | 60% | 50% | 40% | 30% | 20% |
| E6010 | 1/8 | 100 dc(+) | 4.38 | 5.22 | 6.47 | 8.55 | 12.72 |
| | 5/32 | 140 dc(+) | 3.20 | 3.80 | 4.70 | 6.19 | 9.17 |
| | 3/16 | 180 dc(+) | 2.67 | 3.16 | 3.90 | 5.12 | 7.57 |
| E7018 | 1/8 | 130 dc(+) | 4.05 | 4.82 | 5.96 | 7.85 | 11.63 |
| | 5/32 | 180 dc(+) | 2.89 | 3.43 | 4.23 | 5.55 | 8.20 |
| | 3/16 | 260 dc(+) | 2.21 | 2.61 | 3.20 | 4.18 | 6.15 |
| | 1/4 | 325 dc(+) | 1.89 | 2.23 | 2.73 | 3.55 | 5.20 |
| E7028 | 5/32 | 230 ac | 1.94 | 2.28 | 2.78 | 4.63 | 5.32 |
| | 3/16 | 280 ac | 1.51 | 1.76 | 2.14 | 2.77 | 4.04 |
| | 7/32 | 330 ac | 1.22 | 1.42 | 1.71 | 2.20 | 3.18 |
| | 1/4 | 400 ac | 0.97 | 1.12 | 1.33 | 1.70 | 2.43 |
| E6012 | 3/16 | 245 ac | 2.44 | 2.88 | 3.55 | 4.66 | 6.87 |
| | 7/32 | 285 ac | 1.85 | 2.18 | 2.67 | 3.49 | 5.12 |
| | 5/16 | 405 ac | 1.50 | 1.76 | 2.19 | 2.81 | 4.07 |
| E7024 | 1/4 | 375 ac | 1.16 | 1.35 | 1.62 | 2.07 | 2.97 |
| Automatic submerged arc | Full | 1000 dc(+) | 0.64 | 0.70 | 0.80 | 0.96 | 1.29 |
| | Semi | 500 dc(+) | 1.00 | 1.15 | 1.57 | 1.71 | 2.41 |

*Estimating Example.* Four 20-ft long 0.5-in. horizontal, V-groove butt joints are required to make a square beam (Fig. 6-87). The electrodes used are 0.5-in. E7024. The operating factor (percentage of time welding is actually taking place) is 40%. What is the cost of the deposited metal?

*Solution:* Total cost of weld deposit = 20 × 4 × 0.489 = 39.2 lb (See Table 6-18). The cost of the deposited metal = 39.2 × $1.62 = $63.50 (see Table 6-19).

If fully automatic submerged arc can be used, the operating factor can be changed to 60% and the cost would then be 3.9 × 0.64 = $25.10.

*Oxyacetylene Welding.* Gas-welding costs may vary considerably due to the skill of the operator, the size of the tip used, and rod diameter used. The figures given in Table 6-20 are based on average working conditions with allowances made for lost time, etc.

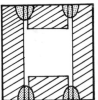

FIGURE 6-87. *A method of fabricating a square beam using 1-in. plate and ½-in. weld penetration. A root gap of ⅛ in. was provided.*

TABLE 6-20. I. *Average Cost Data for Manual Oxyacetylene Welding of Iron and Steel, and II. Approximate Weight of Weld Metal in 60° and 90° Single Vee Joints. Reprinted from the December, 1971 issue of Welding Design and Fabrication. Copyright 1971 by Industrial Publishing Co., Division of Pittway Corp.*

I.

| Thickness of Steel Inches | Joint Preparation No Spacing | Diameter of Rod Inches | Tip Drill Size | Oxygen—Cubic Feet per Hour | Oxygen—Cubic Feet per Linear Foot Welded | Acetylene—Cubic Feet per Hour | Acetylene—Cubic Feet per Linear Foot Welded | Pounds of Rods per Hour | Pounds of Rods per Foot | Speed Foot per Hour |
|---|---|---|---|---|---|---|---|---|---|---|
| 1/64 | Square Butt | 1/32 | 75 | 0.7 | 0.03 | 0.7 | 0.03 | | | 26.0–30.0 |
| 1/32 | Square Butt | 1/32 | 75–60 | 1.0 | 0.05– 0.04 | 1.0 | 0.05– 0.04 | | 0.013 | 22.0–25.0 |
| 1/16 | Square Butt | 1/16 | 60–56 | 2.4 | 0.13– 0.11 | 2.3 | 0.13– 0.11 | 0.23–0.27 | 0.030 | 18.0–21.0 |
| 3/32 | Square Butt | 3/32 | 60–54 | 5.1 | 0.36– 0.30 | 4.9 | 0.36– 0.29 | 0.42–0.51 | 0.053 | 14.0–17.0 |
| 1/8 | Square Butt | 1/8 | 56–53 | 8.8 | 0.80– 0.68 | 8.5 | 0.77– 0.65 | 0.58–0.69 | 0.150 | 11.0–13.0 |
| 3/16 | 90° Single V | 3/16 | 53–49 | 17.7 | 2.36– 2.08 | 17.0 | 2.27– 2.00 | 1.13–1.28 | 0.265 | 7.5–8.5 |
| 1/4 | 90° Single V | 3/16 | 49–44 | 27.0 | 4.50– 3.86 | 26.0 | 4.33– 3.72 | 1.59–1.86 | 0.414 | 6.0–7.0 |
| 5/16 | 90° Single V | 1/4 | 44–40 | 33.0 | 7.40– 6.05 | 32.0 | 7.11– 5.82 | 1.87–2.28 | 0.597 | 4.5–5.5 |
| 3/8 | 90° Single V | 1/4 | 43–36 | 45.7 | 11.42– 9.13 | 44.0 | 11.0 – 8.80 | 2.39–2.98 | 0.637 | 4.0–5.0 |
| 1/2 | 60° Single V | 1/4 | 40–36 | 58.2 | 11.65– 9.70 | 56.0 | 11.2 – 9.33 | 2.90–3.48 | 0.872 | 5.0–6.0 |
| 5/8 | 60° Single V | 5/16 | 36–32 | 73.8 | 21.10–16.42 | 71.0 | 20.30–15.79 | 3.06–3.92 | 1.307 | 3.5–4.5 |
| 3/4 | 60° Single V | 5/16 | 32–30 | 91.5 | 36.60–26.16 | 88.0 | 35.20–25.17 | 3.27–4.57 | | 2.5–3.5 |

## II.

| Thickness of Metal Inches | Weld Metal in 1″ Length of 60° Vee cu. in. | Weight of Weld Metal in 1-inch Length of 60° Vee Joint in Pounds | | | | | Weld Metal in 1″ Length of 90° Vee cu. in. | Weight of Weld Metal in 1-inch Length of 90° Vee Joint in Pounds | | | | |
|---|---|---|---|---|---|---|---|---|---|---|---|---|
| | | Steel | Armco Iron | Stainless Steel | Nickel | Page Bronze | | Steel | Armco Iron | Stainless Steel | Nickel | Page Bronze |
| 1/4 | 0.035 | 0.0098 | 0.0099 | 0.0101 | 0.0112 | 0.0105 | 0.062 | 0.0174 | 0.0176 | 0.0179 | 0.0198 | 0.0187 |
| 3/8 | 0.080 | 0.0224 | 0.0227 | 0.0232 | 0.0255 | 0.0240 | 0.140 | 0.0392 | 0.0397 | 0.0405 | 0.0446 | 0.0421 |
| 1/2 | 0.144 | 0.0403 | 0.0408 | 0.0417 | 0.0459 | 0.0432 | 0.250 | 0.0700 | 0.0709 | 0.0723 | 0.0796 | 0.0751 |
| 5/8 | 0.225 | 0.0630 | 0.0638 | 0.0651 | 0.0716 | 0.0676 | 0.390 | 0.1092 | 0.1105 | 0.1128 | 0.1241 | 0.1172 |
| 3/4 | 0.324 | 0.0907 | 0.0918 | 0.0937 | 0.1031 | 0.0973 | 0.562 | 0.1574 | 0.1593 | 0.1625 | 0.1789 | 0.1689 |
| 7/8 | 0.441 | 0.1235 | 0.1250 | 0.1275 | 0.1404 | 0.1325 | 0.765 | 0.2142 | 0.2168 | 0.2211 | 0.2435 | 0.2298 |
| 1 | 0.577 | 0.1616 | 0.1635 | 0.1668 | 0.1837 | 0.1734 | 1.000 | 0.2800 | 0.2833 | 0.2890 | 0.3182 | 0.3004 |
| 1 1/8 | 0.729 | 0.2041 | 0.2066 | 0.2107 | 0.2320 | 0.2190 | 1.265 | 0.3542 | 0.3584 | 0.3656 | 0.4026 | 0.3801 |
| 1 1/4 | 0.901 | 0.2523 | 0.2553 | 0.2604 | 0.2867 | 0.2707 | 1.562 | 0.4371 | 0.4425 | 0.4515 | 0.4971 | 0.4690 |
| 1 3/8 | 1.090 | 0.3052 | 0.3088 | 0.3151 | 0.3469 | 0.3275 | 1.890 | 0.5292 | 0.5355 | 0.5463 | 0.6014 | 0.5678 |
| 1 1/2 | 1.298 | 0.3634 | 0.3678 | 0.3752 | 0.4131 | 0.3899 | 2.250 | 0.6300 | 0.6375 | 0.6503 | 0.7160 | 0.6760 |
| 1 5/8 | 1.523 | 0.4265 | 0.4315 | 0.4402 | 0.4847 | 0.4576 | 2.640 | 0.7392 | 0.7480 | 0.7630 | 0.8401 | 0.7932 |
| 1 3/4 | 1.766 | 0.4945 | 0.5003 | 0.5094 | 0.5620 | 0.5306 | 3.062 | 0.8574 | 0.8675 | 0.8850 | 0.9744 | 0.9200 |
| 1 7/8 | 2.028 | 0.5679 | 0.5746 | 0.5861 | 0.6454 | 0.6094 | 3.515 | 0.9842 | 0.9958 | 1.0159 | 1.1185 | 1.0560 |
| 2 | 2.308 | 0.6462 | 0.6539 | 0.6673 | 0.7345 | 0.6934 | 4.000 | 1.1200 | 1.1332 | 1.1560 | 1.2728 | 1.2018 |
| 2 1/4 | 2.920 | 0.8176 | 0.8273 | 0.8439 | 0.9292 | 0.8773 | 5.062 | 1.4174 | 1.4341 | 1.4630 | 1.6108 | 1.5209 |
| 2 1/2 | 3.606 | 1.0097 | 1.0216 | 1.0422 | 1.1475 | 1.0834 | 6.250 | 1.7500 | 1.7707 | 1.8063 | 1.9888 | 1.8778 |
| 2 3/4 | 4.363 | 1.2217 | 1.2361 | 1.2610 | 1.3884 | 1.3109 | 7.562 | 2.1174 | 2.1423 | 2.1854 | 2.4063 | 2.2720 |
| 3 | 5.196 | 1.4549 | 1.4721 | 1.5017 | 1.6534 | 1.5611 | 9.000 | 2.5200 | 2.5497 | 2.6010 | 2.8638 | 2.7040 |

NOTE:   The data in the above table considers the metal veed from both sides of the joint; if the metal is veed from only one side of the joint, the volume and weight of the weld metal required are one-half of the above figures.

471

### Welding Positioners and Fixtures

Welding positioners and fixtures are used extensively in production welding to reduce costs and improve quality. By just tilting a fillet weld 10° in the downhand position, an increase in speed of over 60% can be realized, as shown in Fig. 6-88.

Weld positioners are also important in improving the operating factor. A high operating factor will be dependent on convenient positioning of the weld, as shown in Fig. 6-89, as well as having all needed supplies convenient and operating controls readily accessible.

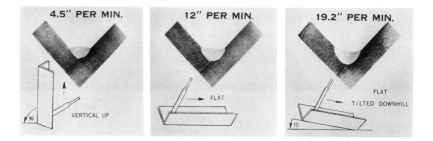

FIGURE 6-88. *Effect of position on speed of welding fillet welds in ⅜-in. plate or thinner. Courtesy The James F. Lincoln Arc Welding Foundation.*

FIGURE 6-89. *This 12.5-ton welding positioner is able to put all welds on the large rotor housing in a downhand position. Courtesy Harnischfeger Corp.*

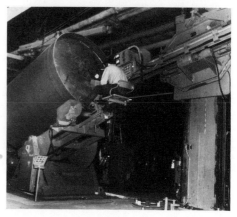

FIGURE 6-90. *A ram-type manipulator being used with power rolls.*

Power rolls and idlers are a convenient means of moving the work when making circumferential welds. Shown in Fig. 6-90 is a ram-type manipulator which provides a mount for the welding head that allows it to be raised, lowered, extended, retracted, or swung in an arc of 360°. The manipulator can be used with either rolls or positioners as mentioned or with headstocks and tailstocks, as shown in Fig. 6-91.

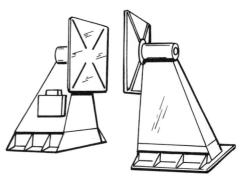

FIGURE 6-91. *A headstock and tailstock as used for positioning work for welding.*

## ADHESIVE BONDING

Adhesives were formerly thought of, especially by engineers, as a joining method to be used for noncritical applications. The aircraft industry led the way for the initial boost in the development of chemical fastening systems. Structural bonding of parts is now being extensively used for automotive, office, and industrial equipment.

An example of the use of adhesives in the auto industry is that of a custom-wheel specialist. A structural adhesive was used to join the die-cast aluminum hub and spoke center to the steel rims. The bond had to withstand both high and low extremes of temperature and a great amount of stress, especially shear stress. Two tests were used by the company to estimate the performance of the bond. A press was used to apply pressure evenly to the hub-and-spoke center in order to determine the pressure required to break the bond from the rim. The minimum pressure required was 42,000 lb. The bond withstood 70,000 lb. The second test was a rotary fatigue test. This test indicated the casting would break before the bond.

Structural adhesives have been used for quite a number of years to bond brake shoes to the brake drum. The bond develops strengths of 10,000 lb in shear compared to 2500 lb shear stress developed by the rivets formerly used.

### Classification of Structural Adhesives

There are many ways of classifying adhesives. A broad classification is that of *nonstructural* and *structural*. In the first category are household-type glues of vegetable or animal base. Typical materials involved are casein, rosin, shellac, and

473

asphalt. They are characterized by poor moisture resistance, but generally good resistance to heat and chemicals within their range of usage.

Structural adhesives are those that are capable of sustaining load-bearing joints for a long period of time. These may be classified into four main types: thermosetting, thermoplastic, unmodified elastomers, and adhesive alloys.

### Thermosetting and Thermoplastic Adhesives

Thermosetting adhesives are characterized by long chainlike molecules with many cross-links which contribute to a high modulus of elasticity and the ability to support considerable stress (Fig. 6-92). These adhesives characteristically soften with heat during the curing cycle and upon completion of the curing reaction (cross-linking) they remain relatively infusible and do not change when reheated. In this family are epoxies, phenolics, alkyds, melamines, and urea compounds. In contrast, the thermoplastics do not form cross-links when subjected to elevated temperatures. The long chain molecules merely become more active, allowing greater flexibility up to the point where they start to deteriorate.

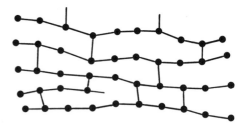

FIGURE 6-92. *A schematic representation of molecular cross-linking. As adjacent chains are anchored together, they become more rigid.*

*Epoxies.* Epoxy resins, a reaction product of acetone and phenol, are the most popular of the adhesives used today. Epoxies may be cured (cross-linked) by the addition of a catalyst, resulting in a room-temperature cure, or by subjection to an elevated temperature.

The two-part, room-temperature curing types are satisfactory for many applications, but higher strengths and better heat resistance can be obtained with the heat-cured type. Epoxies bond well to most surfaces, even with slight oil films; however, for maximum strength (up to 6000 psi in tensile shear) (420 kg/cm$^2$) surfaces should be cleaned and specially prepared. Although the tensile shear strength is good, the bond has low flexibility and poor impact strength. Little or no setting pressure is required but only contact pressure upon application.

*Modified Epoxies.* Since thermosetting-plastic adhesives exhibit "brittle" characteristics when cured, modifications have been incorporated that include thermoplastics and rubber. Some popular modifications are epoxy-nylon, epoxy-polysulfide, epoxy-phenolic, and epoxy-nitrile.

The modified epoxies are excellent for metal-to-metal bonding and bonding to concrete. One of their advantages is that they consist of 100 % solids, so there is no problem of solvent evaporation and shrinkage when joining impervious surfaces. Another advantage is their ability to *wet* metal, glass, and cementitious surfaces readily. Wetting refers to the ability of a liquid to spread over a solid surface evenly without voids. For example, water on an oily surface stays in droplets, but on a clean, dry surface it will spread out uniformly.

In addition to being more flexible, modified epoxies show an improvement in *peel strength*. Peel strength refers to the ability to withstand separation at the edge of the joint (Fig. 6-93).

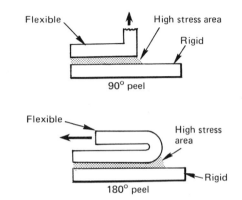

FIGURE 6-93. *Peel strength is measured in the ability of the adhesive to withstand separation at the edge of the joint.*

***Phenolics.*** Phenolics are formed by a reaction of phenol with aldehyde. Phenols have been used extensively in bonding plywood since the early part of this century. Now, with improvements in technology and increased availability of substituted phenols and more complex aldehydes, the physical characteristics of the phenol resins have been modified with thermoplastics and rubbers. These modifiers are nylon, neoprene, vinyl, or nitrile rubber, each of which adds to the flexibility. Phenolics without modifiers have very high strengths but are too brittle for most structural bonding. The neoprene phenolics, for example, are noted for their excellent peel strength.

Modified phenolics have high shear strengths, (up to 6000 psi or 420 kg/cm$^2$). Satisfactory bonding requires high pressures (up to 250 psi or 125 kg/cm$^2$) and high curing temperatures, which may go as high as 400°F (204°C).

Phenolic adhesives are available as liquids, powders, films, and reinforced tapes. Water and formaldehyde are produced during the curing time at approximately 300°F (149°C). The joint must be cured under a clamping pressure of about 100 psi (7 kg/cm$^2$) to prevent the joint from being forced apart.

***Polyurethane.*** Polyurethane is a synthetic rubber produced by reacting polyesters with diisocyanates. Isocyanates are very reactive and readily combine with compounds containing active hydrogen, such as —NH, —OH, and —COH. This makes it possible to use them either to lengthen the molecular chain or to introduce diisocyanate molecules as cross-linking groups.

475

Polyurethane adhesives are especially good for cryogenic applications where shear strengths of up to 6000 psi (420 kg/cm²) at −400°F (−240°C) may be required. Polyurethanes are difficult to apply and cure. They also show excessive creep at room temperature and age more rapidly than most other adhesives. Shown in Fig. 6-94(a) is a graphical comparison of the shear strengths versus temperature of the types of adhesives discussed. The general effect of cross-linking in shear strength versus temperature is shown in Fig. 6-94(b).

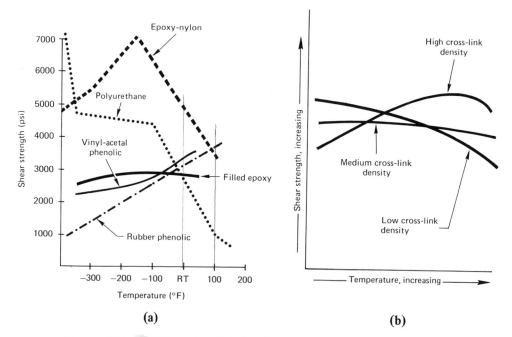

(a)                                           (b)

FIGURE 6-94. *A comparison of the shear strength vs. temperature of various adhesives (a). A comparison of the tensile shear strength of various types of thermosetting adhesives vs. temperature (b).* Courtesy Adhesives Age.

### Chemically Blocked Reactants

Chemically blocked reactants refer, in this case, to the adhesives that will not become activated by themselves but can be catalyzed by atmospheric moisture or the absence of oxygen. Examples of this type of adhesive are the cyanoacrylates, so called because they possess anaerobics and an organic cyanide group in the basic structure.

*Cyanoacrylates.* The prime advantage of these adhesives is their rapid cure, usually within seconds. Cure is affected by contact with moisture or alkali. Almost any surface has sufficient moisture to affect a cure. It has medium shear strength (up to 3000 psi shear) (210 kg/cm²). See Fig. 6-94 for a comparison with previously discussed adhesives.

Cyanoacrylates are costly, but the amount used is small since a very thin glue line is required to cure properly. The cure is too fast for many applications. Cyanoacrylates are used to a limited extent in surgery where their ability to cure almost instantaneously and give a "chemical" suture is extremely useful. Often the cyanoacrylates can be used to retain components in position while other operations such as welding are carried out. In this way expensive jigs or difficult clamping can be avoided.

***Anaerobic Adhesives.*** Anaerobic sealants are so called because they harden satisfactorily only in the absence of atmospheric oxygen. Hence they must be stored in the presence of air. When in the liquid form, they are capable of flowing into the most minute crevice and thus the whole of free space between assembled components is readily and completely filled with fluid. After the fluid has cured and formed a tough wedge, the assembly is sealed because there is virtually no shrinkage and therefore no passage for gases or liquids.

There are four common applications for anaerobic materials which involve compression loads, shear and compression forces, actual shear, and shear followed by destructive crushing, as illustrated in Fig. 6-95.

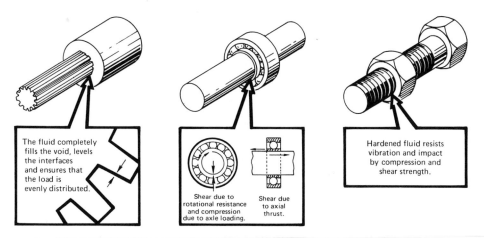

The fluid completely fills the void, levels the interfaces and ensures that the load is evenly distributed.

Shear due to rotational resistance and compression due to axle loading.

Shear due to axial thrust.

Hardened fluid resists vibration and impact by compression and shear strength.

FIGURE 6-95. *Four common applications for anaerobic materials involve compression loads, shear and compression forces, and actual shear followed by destructive crushing. Courtesy* Adhesives Age.

### Surface Preparation

Reliable adhesion depends upon the cleaning and preparation of the adherend surface. Unclean materials will not be receptive to adhesion regardless of the quality of the adhesive used. Cleaning processes may be classified as mechanical, passive chemical, and active chemical.

*Mechanical.* Common examples of mechanical cleaning are: sanding, abrasive scrubbing, wire brushing, grit blasting, grinding, and scraping.

*Passive Chemical.* Passive chemical treatment of a surface consists of removing soil from the surface by chemical means without chemically altering the parent material. This may be done by vapor degreasing, solvent washing, alkaline and detergent cleaning, or ultrasonic cleaning.

*Active Chemical.* An active chemical treatment alters the surface physically and chemically to increase its free-energy level and make it receptive to adhesion. This may be done with acids, caustic and sodium etchants, anodic films, or chemical films.

If high strengths are not required, the surface treatment may be either a mechanical or passive chemical treatment. The oxide film on aluminum, for example, may be removed by wiping with methyl ethyl ketone (EK) or trichloroethylene.

For some metals, particularly aluminum, an active chemical treatment is needed to give maximum strength. For aluminum, this consists of a vapor degrease with trichlorethylene, followed by a rinse, then a sulphuric acid chromic etch, and then a rinse and forced-air drying.

Generally, carbon steels offer maximum bonding strength by mechanical treatment, such as a dry-abrasive blast of 80-grit aluminum oxide followed by flushing with trichloroethylene and then a rinse. For detailed accounts of surface preparation, see references at the end of the chapter.

Because of the likelihood of contamination and oxide formation, it is desirable to use the prepared materials as soon after treatment as possible. If storage is necessary, the metal should be kept in airtight containers. Care must be taken so that surfaces that have been etched are not touched. Handlers must wear clean cotton gloves.


## Adhesive Application and Cure

*Application.* Adhesives can be applied in a variety of ways. Epoxies, for example, can be formulated for spraying, brushing, dipping, roll-coating, dusting with dry powder, extruding, and trowelling.

Tape-type adhesives have become quite popular because they eliminate the need for mixing and the application will have a known, uniform *glue-line thickness.* Glue-line thickness refers to the amount of adhesive that remains after the pressure is applied and cured. For example, to achieve an ultimate glue-line thickness of 1 to 3 mils, anywhere from 5 to 15 mils of 20 % solid, wet-type adhesive must be applied.

If the adhesive is applied in solution, time must be allowed for evaporation of the solvent before the two members can be brought together. Solvent reduces the viscosity of the adhesive, allowing it to penetrate the pores of the substrate. The amount of penetration into the substrate is not important as long as good wetting of the surface has occurred. In fact, too much penetration of the adhesive into the substrate may be detrimental since it may result in glue-line starvation.

Rough surfaces must have enough adhesive to fill in the small depressions, plus enough to achieve the desired glue-line thickness. The gap between the two surfaces should not exceed a few thousandths of an inch.

*Cure.* Adhesives may be cured by solvent release, cross-linking caused by heat and pressure, or a catalyst. Some adhesives, such as one-part epoxies, contain latent catalysts which are activated by heat and cross-linking results with only contact pressure. There are no volatile solvents to be driven off. Phenolics, on the other hand, do have volatile solvents and need high pressure during cure to ensure against solvent entrapment. The time required varies according to the heat, pressure, and catalyst used.

Some adhesives, known as *dry-film* adhesives, may be applied several days before assembly. The surfaces are then activated by one of two methods.

1. The surface is rubbed lightly with a solvent or a liquid adhesive of the same type.
2. The adhesive is heated, usually by infrared lamps, to the reactivation temperature.

*Holt-melt* adhesives are copolymers of polyethylene with polyvinyl acetate, polypropylene nylon, or polyester. Hot melts bond instantly. They set by contacting a surface, cooling, and solidifying. Hot-melt systems are often automatic, applying the adhesive from a pressurized tank through hoses to the applicator. They have limited toughness and heat resistance and are not used where high structural strengths are required.

### Adhesive-Joint Design

The strength of an adhesive joint is dependent on the joint geometry, material properties of the constituents (discussed previously), and the type of loading the joint is subjected to.

*Joint Geometry.* As with brazing and soldering, joint strength for adhesives is based on having an adequate area over which the stress may be distributed. As a consequence, the principal joint design is a lap type. When a lap joint is subjected to tensile shear, the ends have a tendency to deflect. Hence the ends of the lap experience the maximum stresses and the center of the bond experiences the minimum stress as shown in Fig. 6-96. When the ends of the joint are tapered as shown in Case 3 of Fig. 6-96, the tensile stress in the adherends and the shear stress in the adhesive are constant, thus giving a stress concentration of unity. Both Case 2 and Case 3 require relating the material properties of the two adherends to the taper ratios. However, Case 3 is almost impossible to manufacture since the thickness of the adherend must diminish to zero at the joint ends.

*Joint Loading.* Adhesive joints are stressed primarily in tension and shear, as shown in Fig. 6-97. The stresses produce shear, peel, and cleavage. Lap shear strengths are directly proportional to the width of the overlap, but the unit strength decreases with length. The optimum shear strength of the bonded joint is largely dependent upon the shear modulus of the adhesive and its optimum thickness. The thickness may vary from 0.002 in. (0.0051 mm) for high-modulus adhesives to 0.006 in. (0.0152 mm) for low-modulus material. Failure occurs when stresses overcome the cohesive forces of the joint.

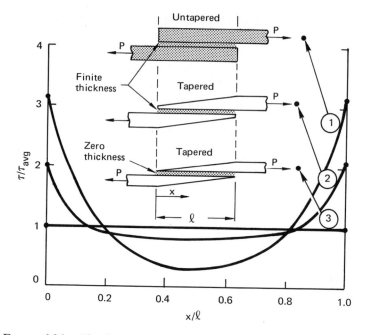

FIGURE 6-96. *The change in stress distribution of a lap joint with geometry of the joint.*

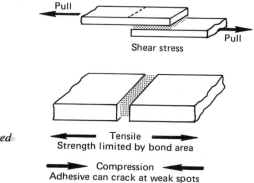

FIGURE 6-97. *Adhesive joints are stressed primarily in shear and tension.*

Adhesive lap joints made out of thin aluminum strips and subjected to peel tests have shown that failure need not be in the adhesive. Nylon-epoxy, for example, has sufficiently high peel strength to cause failure in the substrate rather than the adhesive bond. A standard ASTM peel-test specimen is shown in Fig. 6-98.

Adhesive joints may be tested for flexibility or bend qualities by wrapping the completed joint over a mandrel. In general, the joint should be designed to:

1. Reduce bending of the joint for a given load by stiffening the adherends or decreasing the thickness of the adherends, particularly toward the edges to allow the joint to have more flexibility.
2. Decrease the length of overlap required.

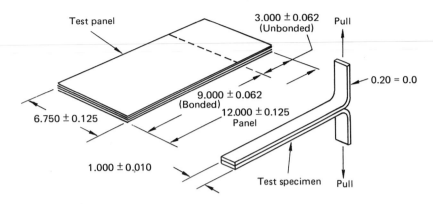

Test panel

3.000 ± 0.062
(Unbonded)

Pull

0.20 = 0.0

9.000 ± 0.062
(Bonded)

12.000 ± 0.125
Panel

6.750 ± 0.125

1.000 ± 0.010

Test specimen  Pull

ASTM D 1876 (dimensions, inches)

FIGURE 6-98.  *A standard ASTM peel-test specimen.*

## Adhesives and Temperature

Adhesives are constantly being subjected to increasingly hostile environments. Consequently, formulations have been made that maintain good strength at both cryogenic and elevated temperatures.  One approach toward withstanding elevated temperatures has been to add poor heat conductors to the adhesive, such as carbon fibers, cork composites, and cast silicon elastomers.  There are relatively few adhesives available that will perform well above 350°F (166°C).  (See Table 6-21.)  A high softening temperature is not the only requirement for a successful high-temperature adhesive.  The material must be able to stand up for a reasonable amount of time under attack by agents that oxidize and degrade the resin.

Most high-temperature adhesives, which consist of cross-linked networks of large molecules, have no melting points.  Their limiting factor is strength reduction due to oxidation.  Oxidation starts a progressive breaking (scission) of the molecular bonds in the long chains.  This scission results in loss of strength and elongation leading to failure.

Oxidation and degradation of an adhesive is affected by any catalytic reaction between the adhesive and the substrates.  As an example, under identical conditions, an adhesive used for bonding stainless steel may oxidize more when bonding aluminum but an adhesive with a different chemical makeup may show exactly the opposite effect (Fig. 6-99).

Newer polyamide-imide adhesives are able to retain considerable strength at temperatures as high as 650°F (343°C), even after 1000 hours continuous exposure in air.  This is a significant accomplishment when compared to solders that melt at 370–450°F (188–232°C) and organic chemicals that usually burn and catch fire at much lower temperatures.  Putting it another way, two pieces of steel could be adhesively bonded with this material and then immersed in a pot of molten lead (621°F). They could remain there for several hours and they would not lose their bond strength.

TABLE 6.21. *Typical Properties of Resinous High-Temperature Adhesives. Courtesy Machine Design*

| Adhesive | Cure Condition | | | Maximum Service Temperature (°F)* | | Lap-Shear Tensile Strength† (psi) | Environmental Recommendations |
|---|---|---|---|---|---|---|---|
| | Temperature (°F) | Time (min) | Clamping Pressure (psi) | Short Term | Long Term | | |
| Epoxy | 75–350 | 120–10 | 0 | 500 | 400 | 2500 | Resists moisture, brine, petroleum fuels, and oils; degrades slowly in air at maximum service temperature |
| Epoxy-phenolic | 350 | .. | 15–50 | 1200 | 500 | 2000 | Similar to epoxy, but degrades faster in air at maximum service temperature |
| Polyimide | 500 | 60 | 15–100 | 1000 | 600 | 2200 | Resists oxidation at >600°F better than other adhesives |
| Polybenzimidazole | 600 | 60 | .. | 1000 | 500 | 1900 | Oxidation resistance at high temperature almost equal to that of polyimide |
| Phenolic | 300 | 30 | 100 | .. | 600 | 2000 | Degrades rapidly at high temperatures |
| Cyanoacrylate | 75 | 2 | 0 | 325 | 285 | 1160 | Degraded by UV, moisture, and alkali materials |
| Polysulfone | 700 | 5 | 80 | .. | 300 | 2200 | Resists acids, alkalis; dissolved by polar organic solvents and aromatic hydrocarbons; degraded by moisture |
| Silicone | 300 | 5 | 0 | 700 | 500 | 38 | Resists most oils, acids, and bases; non-oxidizing; dissolved by solvents such as xylene and toluene |
| Fluoresilicone | 70 | 120 hr | 0 | 700 | 500 | .. | Particularly resistant to fuels and oils |

* Less than 2 min.
† At maximum long-term service temperature.

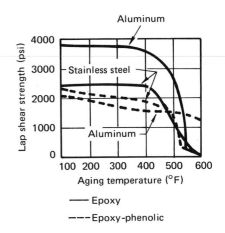

FIGURE 6-99. *A comparison of the strength of epoxy and epoxy-phenolic adhesive bonding aluminum to aluminum and stainless steel to stainless steel.* Courtesy Machine Design.

## Advantages and Limitations

There are many advantages that make adhesive bonding attractive in the metal-working industry. The fact that industry is now spending over 100 million dollars per year for adhesives attests to the benefits brought to manufacturing. Some of the main advantages in using adhesives for metal fabrication are:

1. The ability to join unlike materials.
2. The ability to join like or unlike materials without the stress concentrations usually associated with other joining methods such as spot welds, rivets, or bolts.
3. Adhesives do not conduct electricity, therefore electrolytic corrosion is not set up in joints between dissimilar metals.
4. Flexible adhesives can absorb shock and vibration, increasing the fatigue life. In many cases, adhesive-bonded metal joints have up to ten times more fatigue life than riveted joints.
5. Bonded joints present a smooth, often improved, appearance.
6. Dissimilar thicknesses can be joined, e.g., foils can be joined to heavy-gage plate.
7. Less joint cleaning is required after it is made.
8. Adhesives often permit extensive design simplifications.
9. Large areas can be bonded in a relatively short time.
10. Adhesives provide sealing action in addition to bonding.
11. The elimination of fastener holes allows lighter-gage material to be used. For example, a 0.020-in. bonded-aluminum joint has been used to replace a 0.051-in. riveted-aluminum joint and still maintains equal or better mechanical properties.

The limitations involved in adhesive joining are:

1. Adhesives are more subject to deterioration by environmental conditions than is metal bonding.
2. Adhesive joints are difficult to inspect once assembled.
3. Adhesives have less strength than some other joining methods. The poor resistance to peeling may require the usage of additional fasteners at stress points.

4. Adhesive properties tend to degrade with time.
5. Some rather elaborate jigs and fixtures may be needed to apply heat and pressure, depending upon the bonding cycle.
6. Adhesives are not suited to elevated temperatures, only a few are reliable over 600°F (316°C).
7. Most adhesives have a limited shelf life.

## MECHANICAL FASTENERS

Mechanical fasteners, or in everyday vernacular "nuts and bolts," have become a way of life that is quite taken for granted in all parts of the civilized world. However, just as in other branches of science and engineering, new concepts are continually being developed. The almost infinite variety of mechanical fasteners, over two million different kinds, has caused some manufacturers who have large assemblies to carefully re-evaluate the function of each fastener, with the view of better standardization. One manufacturer has calculated that $2150 is saved each time an unnecessary fastener is purged from the parts list.

This relatively brief presentation of fasteners is not intended to include all types, but rather to give a classification of the main types and a brief discussion of each with greater emphasis on threaded fasteners.

Fasteners may be divided into five main types:

1. Threaded
2. Rivets
3. Washers and retaining rings
4. Pin fasteners
5. Quick-operating fasteners

### Threaded Fasteners

One of the keys to meeting the demands for fasteners created by the Industrial Revolution was the development of better screw-threading technology. At first no great accuracy was required since most machines were assembled individually by carpenters. However, when a machine was torn down for repairs, all the nuts and bolts had to be labeled since they were not interchangeable. The one man most responsible for starting threaded screws on their way to precision was an Englishman by the name of Henry Maudslay. One of the first great tool engineers, he made many contributions to production machinery, but he devoted most of this career to the development of screw-cutting machines. As a result, the stage was set for the mass production of fasteners so accurate that they could be freely interchanged.

Screw-cutting lathes were the principal means of threading fasteners during the Industrial Revolution; but today most threads are rolled, a process that is faster and more efficient, and that produces a stronger, smoother surface.

FIGURE 6-100. *Threads may be rolled in high production by flat reciprocating dies. Courtesy Reed Rolled Thread Die Company, Division of Litton Industries.*

Thread-rolling dies re-form the surface of a screw blank into threads, rather than cut the material from it (Fig. 6-100). Production rates are high. For example, an $\frac{1}{8}$-in. machine can produce 175 screws per minute and a $\frac{3}{4}$-in. machine can produce 50 bolts per minute.

*Threaded-Fastener Materials.* Fasteners are available in almost any material. The key to the proper material selection is knowing what job the fastener is expected to do and then making the specifications accordingly. Common considerations are environment, weight, load characteristics, cost, reuse, and life expectancy.

The majority of fasteners are made from standard steels. The most often used SAE grades and identification of bolts are shown in Fig. 6-101.

Aluminum weighs about one-third as much as mild steel but can equal or exceed the tensile strength of mild steel. Therefore it is used where a high strength-to-weight ratio is required. It is also used for its corrosion resistance and appearance.

Brass in a cold-drawn state has a greater tensile strength than mild steel and has a higher resistance to corrosion. It will take a high lustrous finish and is non-magnetic.

Stainless-steel fasteners are used where problems of corrosion, temperature, and strength exist.

**SAE Grade 1**                **ASTM A 307**

Low-carbon steel suited for cold and hot heading. Used for large upsets such as square-head bolts, carriage bolts, etc.

**SAE Grade 2**

Low-carbon steel, bright finish. Primarily intended for cold-headed products. Cold working increases strength. Widely used for cold-headed hexagon bolts. Sometimes necessary to stress relieve fasteners with large upset heads.

**SAE Grade 5**                **ASTM A 449**

A quenched and tempered medium-carbon-steel bolt. Recommended where high preloading of the bolt is practical. This grade is considered the most economical on a highest-clamping-force per dollar-of-bolt-cost basis.

**SAE Grade 7**

A quenched-and-tempered medium-carbon alloy steel. Threads are cold rolled after heat treatment for improved fatigue strength.

FIGURE 6-101. *SAE and ASTM identification of the most-used steel bolt materials.*

**SAE Grade 8**     **ASTM A 354, Grade BD**

Quenched-and-tempered medium-carbon alloy steel which has a higher strength than Grade 7 material.

Plastic fasteners have excellent corrosion resistance. Salt water has no effect on nylons and mineral acids have no effect on polyvinyl chloride. Most plastics are good thermal and electrical insulators. Plastics are not recommended where high tensile or shear strengths are required or where there are high operating temperatures. Only Teflon and Kel-F can be used at temperatures over 250°F (121°C), and these are not recommended for use above 400°F (204°C). Tensile strengths of common plastic fastener materials are shown in Fig. 6-102(*a*). A comparison of service temperatures of various plastics with those of zinc and aluminum is shown at (*b*).

### Screw-Thread Designation

The standard method of designating a screw thread is by specifying, in sequence, the nominal size, number of threads per inch, thread series symbol, and thread class symbol. For example:

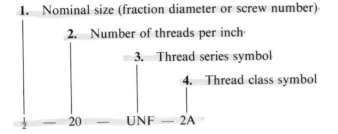

1. Nominal size (fraction diameter or screw number)
2. Number of threads per inch
3. Thread series symbol
4. Thread class symbol

$\frac{1}{2}$ — 20 — UNF — 2A

1. The thread diameter is the maximum diameter measured over the threads (Fig. 6-103). Below $\frac{1}{4}$ in. diameter the size is given by a number ranging from 10 down to 2.
2. There are two common series of threads for every diameter, National Coarse and National Fine. For example, the fine series for the $\frac{1}{2}$ bolt would be 20 threads per inch and the coarse series would be 13 threads per inch.
3. In 1946 a standard thread form was agreed upon by Great Britain, Canada, and the United States. This became known as the Unified Standard. The letter following the designation may be C or F, designating the coarse or fine series as UNC or UNF.
4. The class of fit is given by numbers and letters. The three general classes of fit are: Class 1A/1B, loose; Class 2A/2B, medium; and Class 3A/3B, tight. Class 1A/1B is seldom specified. Class 2A/2B is the most common. Class 3A/3B does not have an allowance between the nut and the bolt, so there is complete contact of the mating threads.

The basic metric thread profile is shown in Fig. 6-104. Many of the terms describing the thread profile found in the inch system are missing. The most significant omission is the term "pitch diameter." Instead, the major or outside diameter is considered to be the significant measurement. Thus the major diameter is the base line, and the linear advance of one revolution is the pitch ($P$). The depth of the thread is exactly $0.5P$. The flat crest is $0.125P$.

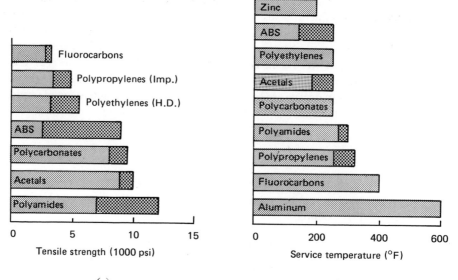

FIGURE 6-102. *A comparison of tensile strengths of various plastics used in fasteners (a). A service temperature comparison of plastics and metals (b). The shaded portion of the bar represents the range.* Courtesy Machine Design.

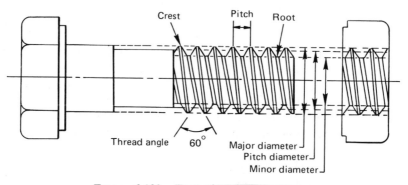

FIGURE 6-103. *Basic thread nomenclature.*

***Metric Threads.*** The metric system of thread sizes is similar to the Unified system, with slight variations. Diameter and pitch are used to designate the series as in the inch system, with the following modifications: for coarse threads, only prefix M and the diameter are necessary; but for fine threads, the pitch is shown as a suffix. For example, M16 is a coarse-thread designation representing a diameter of 16mm with a pitch of 2mm understood. A similar-diameter fine thread would be designated as M16 × 1.5 or 16mm diameter with a pitch of 1.5mm.

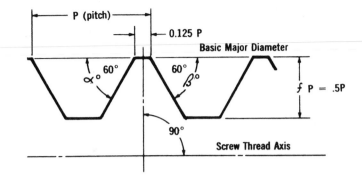

FIGURE 6-104. *The basic OMFS thread profile.*

For those familiar with the inch system only, there may be a couple of differences that can be confusing. In the inch system the threads per inch are referred to as pitch, but actually the pitch is the reciprocal of the number of threads per inch. Thus a thread referred to as 16 pitch has an actual pitch of $\frac{1}{16}$ in. or 0.0625 in. Also, small-size threads are identified by numbers and large sizes by fractions. For example, #4-40, #6-32, and #10-32 represent small-size threads. Larger sizes start at $\frac{1}{4}$-in. diameter. In the inch system, fine-pitch threads are thought of as having a larger number than coarse pitch.

In the metric system, diameters are given in millimeters, but the pitch designation is really the pitch. Consequently the coarse thread has a larger number. In order to become familiar with both thread series, Table 6-22 shows a comparison of standard sizes of each from #0 to 1 in, or from M1.4 to M27. This table can be used to become familiar with comparing metric-size screws to inch-size screws. For example, the closest metric-size screw to a $\frac{1}{4}$-20 in. would be M6, which would have an outside-thread diameter of 6 mm or 0.236 in., instead of 0.250.

### Design Considerations

The strength of mating threads depends on adequate engagement or overlap of the thread in the transverse direction. It would appear that the closer fit 3A/3B thread would be consistently stronger. This is not necessarily true. Tests have shown that the stressing of a bolt and the transfer of loads through the threads is an extremely complex problem in elasticity and plasticity of materials. There is ample evidence to show the desirability of having plenty of "breathing room" between mating threads to accommodate local yielding, thread bending, and adjustment due to elastic deformations throughout the length of the engagement.

In general the Class 2A/2B thread fit is used for low- and medium-strength materials (up to 150,000 psi or 1034 MPa). However, with materials of higher tensile strength and lower ductility, a closer fit (Class 3A/3B) is best. This is because brittle materials cannot "adjust" to distribute the load over the length of engagement. In addition, the need for rounded fillets at the roots of the thread become more acute as the plasticity of the material decreases.

TABLE 6-22. *A Diameter and Pitch Comparison of Inch and Metric Screw Threads*

| Dia. in Inches | INCH SYSTEM | | | | | Dec. Eq. (in) | METRIC SYSTEM | | | | |
|---|---|---|---|---|---|---|---|---|---|---|---|
| | Nominal Size | Dec. Size | Coarse TPI | Fine TPI | Nominal Size | | Coarse Pitch mm | TPI | Fine Pitch mm | TPI | Nominal Size |
| 1.050 | | | | | M27 | 1.063 | 3 | 8.5 | 2 | 12.5 | M27 × 2 |
| 1.000 | 1 | 1.000 | 8 | 12 | | | | | | | |
| .950 | | | | | M24 | .945 | 2 | 8.5 | 2 | 12.5 | M24 × 2 |
| .900 | | | | | | | | | | | |
| .850 | 7/8 | .875 | 9 | 14 | M22 | .866 | 2.5 | 10 | 1.5 | 17 | M22 × 1.5 |
| .750 | 3/4 | .750 | 10 | 16 | M20 | .787 | 2.5 | 10 | 1.5 | 17 | M20 × 1.5 |
| .700 | | | | | M18 | .709 | 2.5 | 10 | 1.5 | 17 | M18 × 1.5 |
| .600 | 5/8 | .625 | 11 | 18 | M16 | .630 | 2 | 12.5 | 1.5 | 17 | M16 × 1.5 |
| .550 | | | | | M14 | .551 | 2 | 12.5 | 1.5 | 17 | M14 × 1.5 |
| .500 | 1/2 | .500 | 13 | 20 | M12 | .471 | 1.75 | 14.5 | 1.25 | 20 | M12 × 1.25 |

| Size | Decimal | TPI | TPI |
|---|---|---|---|
| 7/16 | .437 | 14 | 20 |
| 3/8 | .375 | 16 | 24 |
| 5/16 | .312 | 18 | 24 |
| 1/4 | .250 | 20 | 28 |
| 10 | .190 | 24 | 32 |
| 8 | .164 | 32 | 36 |
| 6 | .138 | 32 | 40 |
| 5 | .125 | 40 | 44 |
| 4 | .112 | 40 | 48 |
| 3 | .099 | 48 | 50 |
| 2 | .086 | 56 | 64 |
| 1 | .073 | 64 | 72 |
| 0 | .060 |  | 80 |

Scale: .450  .400  .350  .300  .250  .200  .150  .100  .050

| Metric | Decimal | Pitch | | Pitch | | Designation |
|---|---|---|---|---|---|---|
| M10 | .393 | 1.5 | 17 | 1.25 | 20 | M10 × 1.25 |
| M8 | .315 | 1.25 | 20 | 1.0 | 25 | M8 × 1 |
| M6 | .236 | 1.0 | 25 | .75 | 34 | M6 × .75 |
| M5 | .196 | .8 | 32 | .5 | 51 | M5 × .5 |
| M4 | .157 | .7 | 36 | .5 | 51 | M4 × .5 |
| M3 | .118 | .5 | 51 | .35 | 74 | M3 × .35 |
| M2.5 | .098 | .45 | 56 | .35 | 74 | M2.5 × .35 |
| M2 | .079 | .4 | 64 | .25 | 101 | M2 × .25 |
| M1.6 | .063 | .35 | 74 | .2 | 127 | M1.6 × .2 |
| M1.4 | .055 | .3 | 85 | .2 | 127 | M1.4 × .2 |

**Fastener Loading.** The materials-joining engineer must consider the principal types of stresses that will be encountered by the fastener in service, tensile, shear, shock, or fatigue loads.

The formulas for tensile, shear, and shock loads are as follows:

Tensile yield strength,
$$S_y = \frac{P}{A}$$

Shear loads,
$$P_s = S_s A_n$$

Shock loads,
$$G = \frac{PN}{2W}$$

Ultimate shock loads:

Tensile,
$$G = \frac{S_t AN}{2W}$$

Shear,
$$G = \frac{S_s AN}{2W}$$

where:  $S_y$ = tensile yield strength in psi
  $P$ = total load in lb
  $A$ = net working area in.$^2$ (Normally the area at the root of the thread is used. In rolled threads the pitch diameter may be used.)
  $P_s$ = fastener shear load in lb
  $S_s$ = fastener shear stress in psi
  $n$ = number of shear planes
  $G$ = shock resistance (G-factor)
  $N$ = number of fasteners
  $W$ = the weight in lb of the part fastened in place (A sudden load is considered twice as severe as the same load applied statically, therefore the safety factor is 2.)

**Example.** Two bolts are used to join two $\frac{1}{2} \times 2$ in. metal straps as shown in Fig. 6-105. The load supported in shear by the bolts is 65,000 lb. Assume the bolt chosen is a 1010 carbon steel with a shear strength of 30,000 psi (Table 6-23). Calculate the exact size of the bolts required if the threaded section is entirely out of the shear zone. *Note:* If shear is in the threaded part of the bolt, Table 6-24 would be used.

$$A = \frac{P}{S_s} = \frac{65,000}{30,000} = 2.16 \text{ in.}^2 = \frac{\pi d^2}{4}$$

$$d = \sqrt{\frac{2.16 \times 4}{3.14}} = 1.66 \qquad \text{or two } \tfrac{7}{8} \text{ in. diameter bolts}$$

**Preload.** Preload is the amount of load put on the bolt by tightening the nut. A properly tightened nut is one that applies a tensile load to the bolt that is less than the

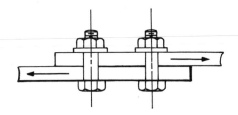

FIGURE 6-105. *Two bolts are used to hold the two steel straps for a shear load of 65,000 lb.*

elastic limit but greater than the service load, thus holding the part together. However, working loads are added to residual stress in the bolt shank. Thus to prevent fatigue failure, the sum of the two stresses must be held below the elastic limit of the bolt. That is why it is so important to apply only the correct tension when setting up or preloading a bolt.

Two types of torque wrenches are available. The one type indicates on a scale the amount of force being applied and the other employs a dial to show the torque. The amount of torque to be applied can be determined by the formula:

$$T = CDP$$

*where:*   $T =$ torque in in. lbs
$C =$ coefficient of sliding friction
$D =$ bolt diameter in in.
$P =$ total load in lb

The coefficient of friction will of course vary with the condition of the bearing surfaces, but in general cases may be taken as 2.

### Internal Threads

The process of cutting or forming threads on the surface of a hole is usually done by tapping. Common types of taps are taper, plug, and bottoming taps, as shown in Fig. 6-106. On a through hole, it is only necessary to use a taper tap. However, on a blind hole it is customary to start with a taper or plug tap and finish with a bottoming tap.

TABLE 6-23. *Ultimate Strength of Fastener Materials.*
*Courtesy* Machine Design

| Material | Tensile Strength, $s_t$ (psi) | Shear Strength, $s_s$ (psi) |
|---|---|---|
| 1010 Carbon steel | 50,000 | 30,000 |
| 303 Stainless steel | 90,000 | 65,000 |
| 316 Stainless steel | 90,000 | 60,000 |
| Brass (yellow) | 60,000 | 35,000 |
| Silicon bronze | 70,000 | 40,000 |
| 2024-T4 Aluminum | 68,000 | 41,000 |

TABLE 6-24. *Stress Area for Fasteners.* *Courtesy* Machine Design

| Bolt Size | Stress Area | Bolt Size | Stress Area |
|---|---|---|---|
| 2–56 | .0034 | $\frac{5}{16}$–24 | .0560 |
| 2–64 | .0037 | $\frac{3}{8}$–16 | .0747 |
| 3–48 | .0045 | $\frac{3}{8}$–24 | .0853 |
| 3–56 | .0049 | $\frac{7}{16}$–14 | .1028 |
| 4–40 | .0056 | $\frac{7}{16}$–20 | .1154 |
| 4–48 | .0062 | $\frac{1}{2}$–13 | .1376 |
| 5–40 | .0074 | $\frac{1}{2}$–20 | .1560 |
| 5–44 | .0078 | $\frac{9}{16}$–12 | .1769 |
| 6–32 | .0085 | $\frac{9}{16}$–18 | .1980 |
| 6–40 | .0090 | $\frac{5}{8}$–11 | .2201 |
| 8–32 | .0132 | $\frac{5}{8}$–18 | .2505 |
| 8–36 | .0140 | $\frac{3}{4}$–10 | .3266 |
| 10–24 | .0166 | $\frac{3}{4}$–16 | .3660 |
| 10–32 | .0190 | $\frac{7}{8}$–9 | .4518 |
| $\frac{1}{4}$–20 | .0303 | $\frac{7}{8}$–14 | .5006 |
| $\frac{1}{4}$–28 | .0349 | 1–8 | .5937 |
| $\frac{5}{16}$–18 | .0503 | 1–12 | .6520 |

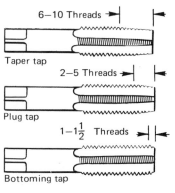

6–10 Threads

Taper tap

2–5 Threads

Plug tap

1–1$\frac{1}{2}$ Threads

Bottoming tap

FIGURE 6-106. *A hand-tap set consists of a taper, plug, and bottoming tap.* *Courtesy* *DoAll Co.*

The drill size to be used for tapped holes is given in Table 6-25. This table may also be used to select the proper drill size for the percentage of thread desired. A 75 % thread (one with the top 25 % of the thread removed) is recommended for commercial practice.

Shown in Fig. 6-107 are a number of examples of threaded fasteners in common use. They include screws, bolts, nuts, and threaded inserts.

Of more recent development are screws of the self-tapping type. In one case the threads are formed by the screw and in the other the threads are cut by the screw. Shown in Fig. 6-108 is a more recent development, that of a combined drilling and tapping screw. Tapping screws may be used in almost all materials including steel, cast iron, aluminum, zinc, brass, plastics, asbestos, and resin-impregnated plywood.

Many types of nuts have been developed in recent years, particularly of the self-locking type. Nuts with plastic inserts are referred to as *prevailing-torque nuts*. Spring-action types spin on easily but lock by the pressure of the arched base against the work.

Threaded inserts are especially good as a method of providing threads for soft materials such as plastics, wood, or soft metals. Some are made to cut their own thread into the soft material and then lock in place.

## Rivets

Permanent fastening is often done with rivets. Sizes are designated by body diameter and range from those used in bridges to those used in small toys and watches. Many different head styles are available to suit special conditions. Heading is usually done with a die held in a powered hammer and the rivet held up with a backup bar or anvil.

Blind rivets get their name because parts can be fastened together when only one side of the assembly is accessible. Most of them are made with a central shank that can be pushed or pulled to upset the head by some type of power tool (Fig. 6-109). Other types employ a small explosive charge to affect a head on the blind side.

## Pin Fasteners

Pin fasteners are often used in place of rivets or bolts, as shown in Fig. 6-110. The most-used types are groove, tapered, roll, and cotter, as shown.

## Retaining Rings

Retaining rings are inexpensive stampings on wire-formed items made to fit securely into grooves acting as a stop or artificial shoulder. They are usually applied as retainers or axial locators on shafts, as shown in the example in Fig. 6-111.

**TABLE 6-25.** *Tap Drill Sizes*

| Thread Size | Pitch Series | Tap Drill Size | Decimal Equivalent of Tap Drill | % of Thread (Approx.) |
|---|---|---|---|---|
| 0–80 | NF | 56 | .0465 | 83 |
| 1–64 | NC | $\frac{3}{64}$ | .0469 | 81 |
|  |  | 54 | .0550 | 89 |
| 1–72 | NF | 53 | .0595 | 67 |
| 2–56 | NC | 53 | .0595 | 75 |
|  |  | $\frac{1}{16}$ | .0625 | 58 |
|  |  | 51 | .0670 | 82 |
| 2–64 | NF | 50 | .0700 | 69 |
|  |  | 49 | .0730 | 56 |
| 3–48 | NC | 50 | .0700 | 79 |
|  |  | 49 | .0730 | 64 |
|  |  | 48 | .0760 | 85 |
| 3–56 | NF | $\frac{5}{64}$ | .0781 | 77 |
|  |  | 47 | .0785 | 76 |
|  |  | 46 | .0810 | 67 |
|  |  | 45 | .0820 | 63 |
| 4–40 | NC | 46 | .0810 | 78 |
|  |  | 45 | .0820 | 73 |
|  |  | 44 | .0860 | 56 |
| 4–48 | NF | 44 | .0860 | 80 |
|  |  | 43 | .0890 | 71 |
|  |  | 42 | .0935 | 57 |
|  |  | $\frac{3}{32}$ | .0938 | 56 |
| 5–40 | NC | 43 | .0890 | 85 |
|  |  | 42 | .0935 | 68 |
|  |  | $\frac{3}{32}$ | .0938 | 67 |
|  |  | 41 | .0960 | 59 |
|  |  | 40 | .0980 | 83 |
|  |  | 39 | .0995 | 79 |
|  |  | 38 | .1015 | 72 |
|  |  | 37 | .1040 | 65 |
| $\frac{1}{4}$–20 | UNC | 14 | .1820 | 73 |
|  |  | 13 | .1850 | 67 |
|  |  | 9 | .1960 | 83 |
|  |  | 8 | .1990 | 79 |
|  |  | 7 | .2010 | 75 |
|  |  | $\frac{13}{64}$ | .2031 | 72 |
|  |  | 6 | .2040 | 71 |
|  |  | 5 | .2055 | 69 |
| $\frac{1}{4}$–28 | UNF | 3 | .2130 | 80 |
|  |  | $\frac{7}{32}$ | .2188 | 67 |
| $\frac{5}{16}$–18 | UNC | F | .2570 | 77 |
|  |  | G | .2610 | 71 |
| $\frac{5}{16}$–24 | UNF | H | .2660 | 86 |
|  |  | I | .2720 | 75 |
|  |  | J | .2770 | 66 |
| $\frac{3}{8}$–16 | UNC | $\frac{5}{16}$ | .3125 | 77 |
|  |  | O | .3160 | 73 |
| $\frac{3}{8}$–24 | UNF | Q | .3320 | 79 |
|  |  | R | .3390 | 67 |
| $\frac{7}{16}$–14 | UNC | T | .3580 | 86 |
|  |  | $\frac{23}{64}$ | .3594 | 84 |
| $\frac{7}{16}$–20 | UNF | W | .3660 | 79 |
| $\frac{1}{2}$–13 | UNC | $\frac{25}{64}$ | .3906 | 72 |
|  |  | $\frac{27}{64}$ | .4219 | 78 |
| $\frac{1}{2}$–20 | UNF | $\frac{29}{64}$ | .4531 | 72 |
| $\frac{9}{16}$–12 | UNC | $\frac{15}{32}$ | .4688 | 87 |
|  |  | $\frac{31}{64}$ | .4844 | 72 |
| $\frac{9}{16}$–18 | UNF | $\frac{1}{2}$ | .5000 | 87 |
|  |  | 0.5062 | .5062 | 78 |
| $\frac{5}{8}$–11 | UNC | $\frac{17}{32}$ | .5312 | 79 |
| $\frac{5}{8}$–18 | UNF | $\frac{9}{16}$ | .5625 | 87 |

| Size | Series | Tap Drill | Decimal | % Thread | Size | Series | Tap Drill | Decimal | % Thread |
|---|---|---|---|---|---|---|---|---|---|
| 5–44 | NF | 38 | .1015 | 80 | 3/4–10 | UNC | 0.5687 | .5687 | 78 |
| | | 37 | .1040 | 71 | | | $\frac{41}{64}$ | .6406 | 84 |
| 6–32 | NC | 36 | .1065 | 63 | | | $\frac{21}{32}$ | .6562 | 72 |
| | | 37 | .1040 | 84 | 3/4–16 | UNF | $\frac{11}{16}$ | .6875 | 77 |
| | | 36 | .1065 | 78 | 7/8–9 | UNC | $\frac{49}{64}$ | .7656 | 76 |
| | | $\frac{7}{64}$ | .1094 | 70 | | | $\frac{51}{64}$ | .7969 | 84 |
| | | 35 | .1100 | 69 | 7/8–14 | UNF | 0.8024 | .8024 | 78 |
| 6–40 | NF | 34 | .1110 | 67 | | | $\frac{13}{16}$ | .8125 | 67 |
| | | 33 | .1130 | 62 | 1–8 | UNC | $\frac{55}{64}$ | .8594 | 87 |
| | | 34 | .1110 | 83 | | | $\frac{7}{8}$ | .8750 | 77 |
| | | 33 | .1130 | 77 | 1–12 | UNF | $\frac{29}{32}$ | .9062 | 87 |
| | | 32 | .1160 | 68 | | | $\frac{59}{64}$ | .9219 | 72 |
| 8–32 | NC | 29 | .1360 | 69 | $1\frac{1}{8}$–7 | UNC | $\frac{31}{32}$ | .9688 | 84 |
| | | 29 | .1360 | 78 | | | $\frac{63}{64}$ | .9844 | 76 |
| | | 28 | .1405 | 65 | $1\frac{1}{8}$–12 | UNF | $1\frac{1}{32}$ | 1.0312 | 87 |
| | | $\frac{9}{64}$ | .1406 | 65 | | | $1\frac{3}{64}$ | 1.0469 | 72 |
| 8–36 | NF | 27 | .1440 | 85 | $1\frac{1}{4}$–7 | UNC | $1\frac{3}{32}$ | 1.0938 | 84 |
| | | 26 | .1470 | 79 | $1\frac{1}{4}$–12 | UNF | $1\frac{5}{32}$ | 1.1562 | 87 |
| | | 25 | .1495 | 75 | | | $1\frac{11}{64}$ | 1.1719 | 72 |
| | | 24 | .1520 | 70 | $1\frac{3}{8}$–6 | UNC | $1\frac{3}{16}$ | 1.1875 | 87 |
| 10–24 | NC | 23 | .1540 | 66 | | | $1\frac{13}{64}$ | 1.2031 | 79 |
| | | $\frac{5}{32}$ | .1562 | 83 | $1\frac{3}{8}$–12 | UNF | $1\frac{7}{32}$ | 1.2188 | 72 |
| | | 22 | .1570 | 81 | $1\frac{1}{2}$–6 | UNC | $1\frac{9}{32}$ | 1.2812 | 87 |
| 10–32 | NF | 21 | .1590 | 76 | | | $1\frac{19}{64}$ | 1.2969 | 72 |
| | | 20 | .1610 | 71 | $1\frac{1}{2}$–12 | UNF | $1\frac{5}{16}$ | 1.3125 | 87 |
| 12–24 | NC | $\frac{11}{64}$ | .1719 | 82 | | | $1\frac{21}{64}$ | 1.3281 | 79 |
| | | 17 | .1730 | 79 | $1\frac{3}{4}$–5 | UNC | $1\frac{13}{32}$ | 1.4062 | 87 |
| | | 16 | .1770 | 72 | | | $1\frac{27}{64}$ | 1.4219 | 72 |
| 12–28 | NF | 15 | .1800 | 67 | 2–$4\frac{1}{2}$ | UNC | $1\frac{17}{32}$ | 1.5312 | 84 |
| | | 16 | .1770 | 84 | | | $1\frac{35}{64}$ | 1.5469 | 78 |
| | | 15 | .1800 | 78 | | | $1\frac{25}{32}$ | 1.7812 | 76 |

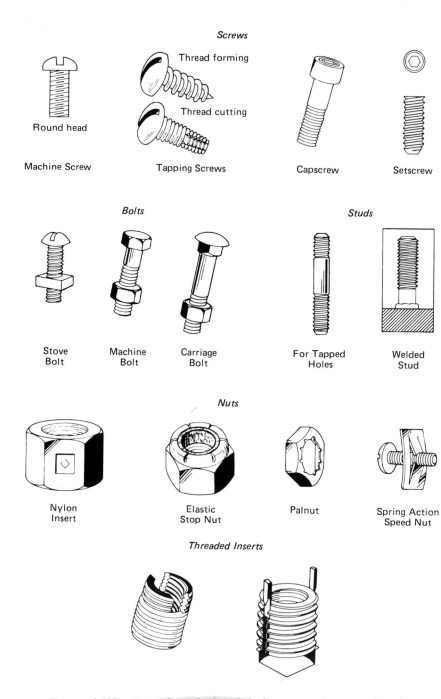

FIGURE 6-107. *Examples of threaded fasteners.* *Courtesy* American Machinist.

### Quick-Operating Fasteners

Most quick-operating or quick-release fasteners are specialized items made to operate on or against spring pressure. They are of many types, ingenious and varied, two examples of which are shown in Fig. 6-112. The most frequent application is on sheet-metal parts that must be periodically opened and securely closed.

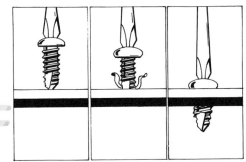

FIGURE 6-108. *Self-drilling, tapping screws are able to provide their own hole and threads in one operation.*

FIGURE 6-109. *A sketch to show the installation of a blind rivet.*

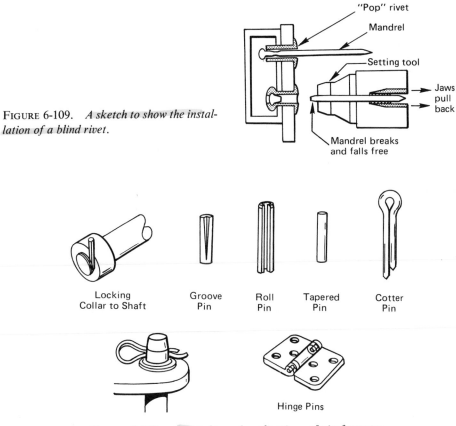

FIGURE 6-110. *Examples and applications of pin fasteners.*

FIGURE 6-111. *An example of one type of retaining ring used to hold a shaft in place.*

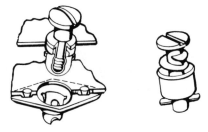

FIGURE 6-112. *Examples of quick-release fasteners.*

## PROBLEMS

**6-1.** Suppose an E6024 electrode was used for overhead or vertical welding. What difficulties might be experienced?

**6-2.** State the relationship between: (a) arc length and voltage, (b) increasing the wire speed and voltage (on an automatic machine), (c) the effect of (a) and (b) on current on a drooping volt-ampere machine.

**6-3.** What is the advantage of a machine that has a rising volt-ampere characteristic?

**6-4.** A 250-amp arc-welding machine is running continuously for 5 hours. If it is a 60%-duty-cycle machine, what is the total time it could have been used at capacity?

**6-5.** What is the significance of polarity? How is it used to advantage?

**6-6.** When is it particularly advantageous to have a superimposed high-frequency current on a regular ac welding current?

**6-7.** If you encountered a strong arc blow in attempting to weld the inside corner of a box, what would you do to try and overcome this condition?

**6-8.** How can a weldor do a reasonably good job of welding plates that have a poor fit up if he is using a flux-cored process?

**6-9.** Is there any advantage in using $CO_2$ along with the flux-cored electrode while welding mild steel?

**6-10.** Select the electrodes for the joints (sheet-metal, structural, and plate) shown in Fig. P6-1 and give a reason for your choice of each.

Fillet and groove welds in sheet metal (10 to 18 gage)

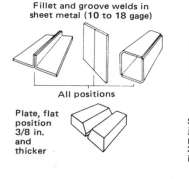

All positions

Plate, flat position 3/8 in. and thicker

Structural joints, all positions 3/16 to 5/8 in. plate

FIGURE P6-1

**6-11.** (a) What is the advantage of using helium over argon in some welding applications? (b) Why isn't helium used very much for a shielding gas for welding aluminum or magnesium? (c) What is the special requirement for $CO_2$ GMA welding?

**6-12.** What would be the difference in the amount of filler wire required for 10 ft. of weld made in $\frac{3}{4}$ in. thick plate with stick electrode and submerged arc? Assume conditions as given in Fig. 6-11.

**6-13.** (a) Make a sketch of the current characteristics of a drooping-arc-voltage machine with a long arc, normal arc, and short arc. Shade the area that shows the variation in current. (b) Why may a machine of this type be thought of as "constant current"?

**6-14.** When may the terms TIG and MIG be used correctly?

**6-15.** Contrast the use of short-circuiting arc with spray transfer as to general use.

**6-16.** What is the meaning of plasma arc?

**6-17.** What would be the main advantages of using high-current plasma welding for making a butt weld in $\frac{1}{2}$ in. thick aluminum plates?

**6-18.** (a) Compare the cost of cutting 10 ft. of 1 in. thick plate between that of plasma and oxyacetylene cutting under the conditions given in Fig. 6-27. (b) What is another main advantage of plasma cutting?

**6-19.** (a) What shielding gas would be used when welding stainless-steel tubing by the GMA process? (b) What GMA process and shielding gas would be used if corrosion resistance was very important?

**6-20.** Why is it that propane, which has a high heat in terms of Btu per cu ft, is considered a cool gas?

**6-21.** Compare the cost of fabricating a 3 ft. diameter, 2 ft. high mild-steel tank by shielded electrode and GMA with $CO_2$ shielding. The material is $\frac{1}{4}$ in. thick steel plate. The

plate is rolled up and butt welded and the cover plates are welded on each end. A 40 % operating factor is used for the shielded arc and 50 % for the GMA process. Assume the operating conditions given in Table 6-7.

**6-22.** Compare the amount of oxygen consumed while using 25 cu ft of acetylene with the amount required for 25 cu ft of MAPP.

**6-23.** A 2 in. thick steel plate was machine cut with an oxyacetylene torch (radiograph). In examining the cut surface you determine the drag to be about 0.050 in. Would this be classified as a high-quality cut? Explain.

**6-24.** What is wrong with a flame cut that is irregular and where the top edge shows some evidence of melting?

**6-25.** Two pieces of aluminum tubing are to be joined as a brazed lap joint. The larger tube, 2 in. in outside diameter, has a wall thickness of $\frac{1}{8}$ in. The nearest standard size $\frac{1}{8}$ in. thick tube wall that will fit inside is 1.750 in. in diameter. (a) What do you recommend be done to make a proper fit for brazing? (b) What size should the smaller tube be? (c) Both tubes are made out of 2024-T3 aluminum with a tensile strength of 64,000 psi, and a shear strength of 15,000 psi. How much overlap should be used? The shear strength of the brazing alloy is approximately 20,000 psi. The desired factor of safety is 3.

**6-26.** A stainless-steel pipe 2 in. in diameter is to be joined with a copper pipe. The inside diameter of the stainless steel pipe is 1.750 in. The copper pipe will have to be machined slightly on the outside diameter to obtain the desired size for a brazed joint. The coefficient of expansion for stainless steel is $8.5 \times 10^{-6}$/in./deg. F, and for copper, it is $9.8 \times 10^{-6}$. What size should the outside diameter of the copper pipe be? A nickel-alloy filler material will be used with a brazing temperature of 2100°F.

**6-27.** (a) What is meant by a eutectic solder? (b) What is the disadvantage of using a strictly eutectic solder?

**6-28.** What are two advantages of using MAPP and propane over acetylene as a fuel gas?

**6-29.** Make a sketch of how the spot welds should be laid out when joining two pieces of 16-gauge metal. Show overlap distance, spot size, and spot spacing. No fixturing is used and no distortion is anticipated.

**6-30.** What would the spot-welding requirements be for welding two sheets of 12-gauge (0.1046 in.) mild steel? State amperage, spot spacing, and spot size.

**6-31.** Estimate the weld time required to join two steel plates 3 in. thick and 10 ft wide if the electroslag process is used. The spacing between the plates is 1.5 in. Steel weighs approx. 0.100 lb/cu in.

**6-32.** (a) How are the electrons obtained for electron-beam welding? (b) What causes the electrons to have so much force?

**6-33.** (a) If two pieces of 2 in. thick aluminum plate are welded together by the EB process, what would the approximate width of the weld be? (b) What should the air gap be between these two plates before welding?

**6-34.** (a) In general terms compare the initial and operating cost of an electron-beam welder with that of a laser welder. (b) How do they compare for the thickness of the material they can be used to weld?

**6-35.** How can an EB welder be made to weld an irregular path automatically?

**6-36.** At the present time what are the main applications of ultrasonic welding?

**6-37.** Does any melting take place in the metal during explosive welding?

**6-38.** Why is explosive bonding especially good for joining dissimilar metals?

**6-39.** What are the conditions that make for good diffusion bonding?

**6-40.** How may dissimilar metals be successfully joined together by using the friction process?

**6-41.** What is meant by a structural adhesive?

**6-42.** What are the two types of cure used with epoxy resins?

**6-43.** How have phenolic resins been improved to increase their useful range in adhesives?

**6-44.** How did *anaerobic* adhesives obtain that name?

**6-45.** (a) Compare the shear strength of lap-joint samples made from strips of 16-gauge aluminum 2 in. wide at room temperature. The lap length is $\frac{1}{2}$ in. made with the following resins: rubber phenolic, polyurethane, and epoxy nylon. (b) Make the same comparison at 100°F.

**6-46.** An aluminum fuel-gas pipeline connection is shown in Fig. P6-2. The line will be buried in the ground and will be required to remain leakproof at 500 to 700 psi for

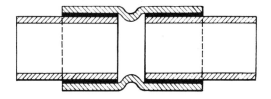

Swaged from each end toward center

FIGURE P6-2

30 years. State the following: (a) Recommended surface preparation. (b) Recommended adhesive. (c) Any special considerations that may affect the joint that are not mentioned in the problem. If a straight epoxy adhesive were used, would it exceed the basic strength of the material? (d) Plan a preliminary test program that gives practical assurance that the pipeline joints are going to withstand actual operating conditions as required.

**6-47.** A 50-gal stainless-steel tank is fabricated to hold a hot brine solution at 500 psi. The brine temperature will not exceed 250°F. The ends are 3 ft in diameter and are made with a $\frac{1}{2}$-in. lap joint. (a) What type of adhesive do you recommend? (b) Give reasons for your choice of adhesive. (c) What shear strength would be developed in the joint? Is this adequate?

**6-48.** Why is it better to taper the ends of the material when making an adhesive lap joint that is subject to tensile shear?

**6-49.** Compare the temperatures the most heat-resistant adhesives are able to withstand with those of solders.

**6-50.** (a) Use the metric thread comparison table and determine the size of a 10 mm diameter, 1.5 mm pitch thread in terms of inches. (b) What would be the nearest standard inch thread to this? See Table 6-22.

**6-51.** A weight of 2000 lb is to be attached to a mild-steel screw eye bolt. The bolt is threaded into a $\frac{3}{4}$ in. thick steel plate. (a) What size bolt would you recommend and why? Use a safety factor of 2. (b) To what depth should the hole be tapped (with full threads) if a 75 % thread is used? (c) What size drill should be used? (d) If hand tapping is done, what type of tap should be used? (e) Approximately how deep should the hole be, including clearance at the bottom? Assume the thread runout and the clearance each will equal about 2 threads.

**6-52.** Shown in Fig. P6-3 is a joint requiring two 4-in. fillet welds. Assume the tensile shear load is 50 ksi with occasional shock loads.

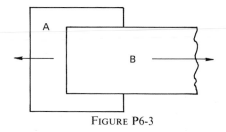

FIGURE P6-3

Are the welds adequate if they are made with E7010 electrodes?

**6-53.** (a) A butt weld is made in a HSLA steel. The material is 2 in. thick and 4 in. wide. What is the normal stress on the joint if it is placed in tension at 20 ksi? (b) If the butt weld of (a) has a $\frac{1}{2}$ in. thick plate, 3 in. wide, placed on the top and bottom of it and welded along both ends of the plate only, what will be the normal stress on these plates?

**6-54.** As shown in Fig. P6-4, plate B is to be welded to plate A. Upon completion, the weldment must resist 38 ksi tensile shear at the axis of the plates as indicated by the

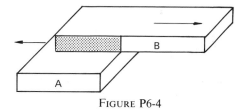

FIGURE P6-4

arrows. The following information is provided. (1) Plate B is low-carbon steel $\frac{3}{8}$ in. $\times$ 3 in. $\times$ 15 in. with a tensile strength of 45,000 psi. (2) Plate A has the same tensile strength and is $\frac{3}{8}$ in. $\times$ 4 in. $\times$ 4 in. (3) An E7016 electrode is prescribed.

How many inches of fillet weld are required on each side of plate B to guarantee the proper strength?

## BIBLIOGRAPHY

**Books**

AITKEN, D., (editor), *Engineers' Handbook of Adhesives*, Machinery Publishing, Brighton, England, 1972.

AWS, *Current Welding Processes*, United Engineering Center, New York, 1965.

AWS, *Welding Handbook*, Sixth Edition, United Engineering Center, New York, 1969.

DeFollis, N., *Adhesives for Metals: Theory and Technology*, Industrial Press, New York, 1970.

Giachino, J. W., W. Weeks and G. S. Johnson, *Welding Technology*, American Technical Society, Chicago, Ill., 1968.

Ingham, H. S., and A. P. Shepard, *Flame Spray Handbook*, Metco Inc., Long Island, New York, 1965.

Kaiser Aluminum & Chemical Sales Inc., *Welding Kaiser Aluminum*, Kaiser Center, Oakland, California, 1967.

Kisielewski, R., *Statistical Analysis of Fuel Gases for Flame Cutting*, Masters Thesis, University of Wisconsin, under the supervision of N. R. Braton, Mechanical Engineering Dept., June, 1972.

Lindberg, R. A., *Materials and Manufacturing Technology*, Allyn and Bacon, Boston, 1970.

SME., *Adhesives in Modern Manufacturing*, Society of Manufacturing Engineers, Dearborn, Michigan, 1970.

Sehgal, S. D. and R. A. Lindberg, *Materials—Their Nature, Properties and Fabrication*, Kailish Publishers, Singapore, 1973.

**Periodicals**

Alm, G. V., "Diffusion Bonding—Methods and Applications," *Adhesives Age*, in three parts: July, Aug., and Sept., 1970.

Bellware, M. D., "The Nickel Stainless Steels: Welding Processes and Procedures," *The Tool and Manufacturing Engineer*, Oct., 1965.

Dunlap. T. W. and E. G. Snype, "Argon and Other Shielding Gases," *Welding Design & Fabrication*, Dec., 1971.

Editor, "Locating Weld Flaws Ultrasonically," *Welding Engineer*, June, 1972.

Editor, "Plasma—MIG—A New Welding Process," *Welding Engineer*, June, 1972.

Editor, "Should Your Design be Joined With a Structural Adhesive?" *Product Engineering*, July, 1973.

Editor, "Submerged Arc Provides Uniform Welds, High Speeds," *Plant Engineering*, Nov., 1967.

Editor, "What's The Best Power Source For You?" *Welding Engineer*, April 1971.

Green, R. D., "Gases are Different," *Modern Machine Shop*, May, 1973.

Lavoie, F. J., "Explosive Welding," *Machine Design*, July, 1969.

MacDonald, N. C., "Standard Test Methods for Adhesives," *Adhesives Age*, Sept. 1972.

McClocklin, R. S., "Machine Wear—How to Eliminate It," *Manufacturing Engineering and Management*, Feb. 1973.

Norcross, J. E., "Flux-cored Electrode Wire for Automatic and Semiautomatic Welding," *ASTM Technical Paper No. SP65-92*.

Obeda, E. G., "Using Ultrasonics to Assemble Plastic Parts," *Automation*, Nov. 1967.

Snogren, R. C., "Selection of Sufrace Preparation Processes," Parts 1 and 2, *Adhesives Age*, July and Aug. 1969.

# 7

# *Automation and Numerical Control*

## Introduction

"Automation" is a relatively new word, coined in 1935 by D. S. Harder, then vice president in charge of manufacturing for the Ford Motor Company. The word, short for "automatization," implies doing something automatically or without human assistance. This may refer to a single operation or a whole continuous process.

In some of the foregoing chapters, certain operations were described as manual, semiautomatic, or automatic. These terms refer to hand operation, some human assistance, and without human assistance, respectively.

Automatic control systems are not a product of this generation. James Watt, as early as 1788, conceived of an automatic speed control for his steam engine in the form of a *fly-ball governor*. The two fly balls, acting by centrifugal force, would extend as the rpm of the engine increased. This, in turn, caused a motion which reduced the opening at the intake valve. With less steam intake, the speed reduced and the balls moved closer to the control shaft, thus reopening the intake valve.

Another example of early automation that is still common in many shops is the fully automatic screw machine. The inventor, Christopher Spencer, had worked on manually operated turret lathes. He realized that by attaching a cam shaft, strip cams, levers, and segment gears for actuating the turret, collet, and cross slide, it could be made to operate automatically. Thus the automatic screw machine was born. Surprisingly enough, Spencer did not realize the import of his invention but was content to let the machine produce screws and other similar items rather than to develop the machine itself.

About the same time as Spencer's automatic screw machine was built in the United States, the Swiss were working out similar ideas for an automatic screw machine. There seems to be some reluctance to give any one man credit for the invention of what has come to be known as the Swiss automatic screw machine. However, the contributions of a brilliant young engineer by the name of Bechler cannot be ignored. His concept of moving the material axially as it was being cut provided a means of generating an infinite variety of shapes with a minimum of tooling. The machine, entirely cam-operated, was fully automatic. One man could look after five or six of these machines, primarily to see that they did not run out of material and that the parts produced were within tolerence limits.

## SEQUENCING CONTROLS

An essential element of automation is being able to make a series of events take place at the desired time and in the desired sequence. These controls are either mechanical or electrical. Mechanical controls are usually cams, valves, or specially constructed mechanisms. Electrical controls are of a wide variety—such as limit switches, control relays, timer relays, and solenoids.

The automatic screw machine is an example of one of the first machines that was based on making a series of events occur as planned or essentially in sequence. Controls used to accomplish this are mechanical (cams), electrical, or combinations (electromechanical). The power required is usually pneumatic, hydraulic, electrical, or combinations of these.

*Sequencing Control of a Pneumatic Circuit.* It is intended in this brief presentation to show only some basic concepts of automatic controls. Because of their relative simplicity, compactness, versatility, low initial cost, and wide acceptance, some basic principles of pneumatic circuits will be presented.

Shown in Fig. 7-1 is a simple sequential, or automatic-cycling, pneumatic circuit. To start it the operator pushes button A which energizes solenoid A. This shifts the two-position valve to the right. One can visualize the action by mentally sliding the symbol of the two crossed arrows to the square on the right. In this position, the line pressure is able to flow into the left side of the cylinder, extending the piston until it closes the normally open limit switch, (1LS). Closing this switch energizes solenoid B which shifts the two-position valve back to its original position. Line pressure now flows into the right-hand end of the cylinder, causing the piston to shift to the left, completing one cycle. To make an automatic reciprocating circuit, a limit switch (2LS) can be added to the left side of the cylinder which would be used to activate solenoid A.

A variation of this circuit is one which incorporates speed control, as shown in Fig. 7-2. In this circuit the line pressure is shown as passing through the two-position value into the back end of the cylinder. As shown, the *flow-control valve* offers no restriction to the flow of air since the bypass to the right is open. The piston extends at full speed until the cam on the end of it strikes a secondary two-position

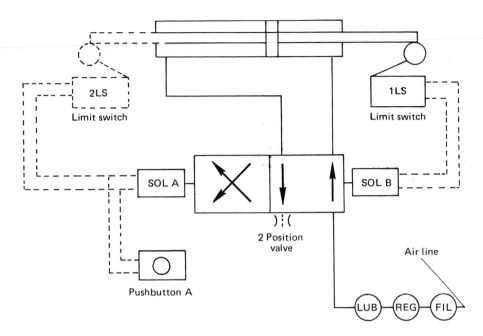

FIGURE 7-1.   *An automatic-cycling pneumatic circuit.*

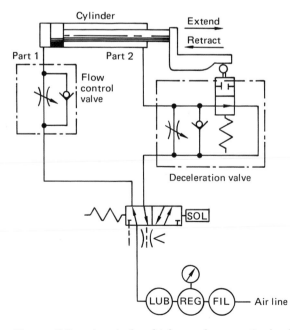

FIGURE 7-2.   *A typical multiple-speed pneumatic circuit.*

valve. The normally open air passage is now closed and it must go through the restricted orifice of the deceleration valve. As the piston extends to the right, a limit switch is activated (not shown) which causes the solenoid attached to the main valve spool to shift to the left. Line pressure can now go through the unrestricted passage to the right-hand end of the cylinder. The piston will not retract at full speed, however, since the air from the back side of the cylinder must pass through the restricting orifice of the flow-control valve.

These two circuits serve to show how the combination of pneumatic (or hydraulic) and electrical components can be combined to achieve relatively simple automation. Compressed air is usually used where economy of the initial investment, simplicity of control, and ease of maintenance are of primary consideration.

Although the two circuits shown were pneumatic, essentially the same symbols are used for hydraulics. Hydraulic circuits have the main advantage of being able to produce higher forces with smaller components. As an example, a 2000-lb force can be exerted by a 1.5 in. diameter hydraulic cylinder, whereas a 5 in. diameter pneumatic cylinder is barely adequate. This assumes the normal line pressure of 80 to 100 psi for pneumatics and 2000 psi for hydraulics. However, it is not uncommon to have hydraulic pressures operating in the range of 3000 to 5000 psi.

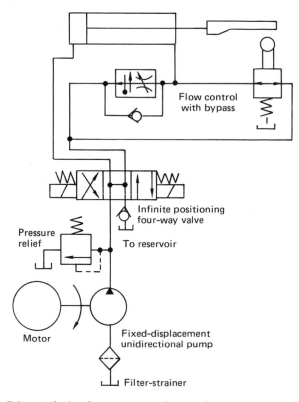

FIGURE 7-3. *A hydraulic circuit similar in function to the pneumatic circuit shown in Fig. 7-2.*

Shown in Fig. 7-3 is a hydraulic circuit that is very similar to the pneumatic circuit shown in Fig. 7-2. You will note a motor and pump are shown, as well as a return line to the reservoir rather than to exhaust. Hydraulic fluids have the advantage of being virtually incompressible, whereas pneumatic action may be somewhat spongy under heavy load conditions. Pneumatic valves, on the other hand, have a shorter response time and will, for a given size, shift from three to four times as fast as a solenoid-operated hydraulic valve.

In some cases it is advantageous to combine air and hydraulic circuits into one system. This is usually done where a smooth, even movement is required, as in machine-tool feeds. The hydraulic circuit is considered "captive." That is, it is built into the machine and air is supplied as shown in Fig. 7-4.

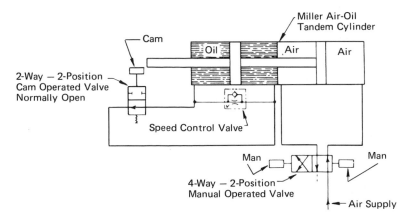

FIGURE 7-4. *A combination air-hydraulic circuit. Courtesy Miller Fluid Power Division, Flick-Reedy Corp.*

## AUTOMATIC POSITION CONTROL AND MEASUREMENT

In addition to sequencing, automation is concerned with accurate position control. This implies the measurement of a location relative to some reference point. As was shown in Fig. 7-1, a signal was sent from the limit switch to the two-position valve when the piston had advanced the required distance. Position control may be sensed mechanically, pneumatically, optically, electrically, etc. Of these methods, the most popular is electrical due to its flexibility, small size, and low power requirement. After limit switches, the next most-used electrical method of sensing position is with transducers.

*Transducers.* Transducers may be defined as any device which converts one form of energy into another. For example, thermocouples transform heat energy into electrical energy and a loud speaker converts electrical energy into mechanical motion.

# LENGTH

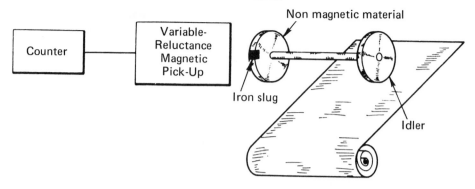

FIGURE 7-5. *The measurement of material length with a variable-reluctance transducer. Courtesy Baird Atomic, Inc.*

Transducers are used extensively in automation because they can be used to sense a physical quantity and convert it into an electrical value for measurement.

An example of the use of a transducer to aid in length measurement is shown in Fig. 7-5. Two wheels are firmly attached to a common axle. The idler wheel rides on the material to be measured and must have a high coefficient of surface friction. The sensing wheel is usually plastic and contains an iron slug imbedded in its periphery. As the material is moved, the idler rotates and causes the sensing wheel to rotate. Each time the iron slug passes a variable-reluctance magnetic pickup, a pulse is initiated and registered on the counter. Proper sizing of the idler and sensing wheels can make each pulse represent inches, feet, meters, etc. Thus, not only length can be measured but velocity as well. Another method often used to measure velocity of a shaft is by coupling it to a generator. The voltages produced can be calibrated directly in terms of rpm.

Potentiometers, or resistive transducers, are particularly good for measuring low-speed rotation. As the shaft of the potentiometer rotates, it varies the resistance in the circuit which may be translated into a given control signal. For example, the cam shown in Fig. 7-6 represents a mechanical signal that is picked up through linkage by the potentiometer, which in turn produces a variable electrical signal.

Velocity measurements can be made by the use of photocells, as shown in Fig. 7-7. Two photocells are set up with individual light sources a fixed distance apart. When the beams are interrupted by a passing object, each photocell emits a pulse. With the spacing of light beams known, the velocity of the moving object can be computed. The time interval can be measured with an electronic counter or shown on the screen by an oscilloscope.

The measurement of quantity is often an important part of automation. Parts may be counted automatically by capacitive, magnetic, or photoelectric transducers. Counting with capacitive and magnetic units requires that the items be metallic, such as tin cans. Photoelectric transducers can be used with either metallic

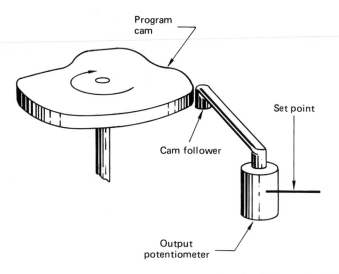

FIGURE 7-6. *The cam represents a program (mechanical type) which is picked up through linkage by the potentiometer, converting into a variable electrical signal.*

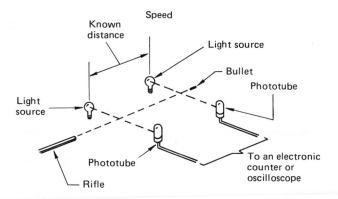

FIGURE 7-7. *Photoelectric transducers are positioned to measure the velocity of a bullet. Courtesy Machine Design.*

or nonmetallic materials. Shown in Fig. 7-8 is a photoelectric counting operation used in packaging. Containers on a conveyor belt are placed so that the products being discharged from a hopper interrupt a light beam to a photocell. The interrupted signal is sent to a preset counting unit. When the required number of products have been deposited in the container, the preset counter activates the device controlling the gate and permits the filled container to move along the conveyor. An empty container moves into position and the cycle is repeated.

## COUNTING

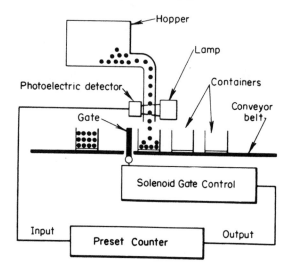

FIGURE 7-8.  *An automatic counting system for the control of the number of items fed into each container.  Courtesy of Baird Atomic, Inc.*

This brief presentation of transducers serves only to introduce them and show their importance in the area of automation.  For a more complete discussion refer to the bibliography at the end of the chapter.

## AUTOMATIC ASSEMBLY MACHINES

Automatic assembly of manufactured components has received a great deal of attention in the last decade.  Estimates show that 55 % of the total production man-hours in major industries is devoted to this area.  The major elements of a semi-automatic assembly line consist of workpiece orientation, part transfer, and part placement.

### Workpiece Orientation

Many devices have been developed to automatically orient and move manu-factured components.  Best known among those used for small-part orientation is the vibratory hopper (Fig. 7-9).  The bowl is made to vibrate in a twisting motion by means of leaf-type springs and electromagnets.  The motion causes the parts to climb the ramp on the inside of the outer edge of the bowl in a series of tiny hops.  Orienting devices such as *wipers*, *dish-outs*, *pockets*, and *hold-downs* are incorporated along the track (Fig. 7-10) to reject parts that have not assumed the proper position.  The track leading out of the bowl can be used to turn the parts to any desired position. Feeds of 10 to 15 fpm are considered normal.  This does not mean that all parts will

FIGURE 7-9.  *This four-track vibratory bowl is used to orient this gear and pinion assembly at the rate of 300 parts per minute.  Courtesy Moorfeed Corporation.*

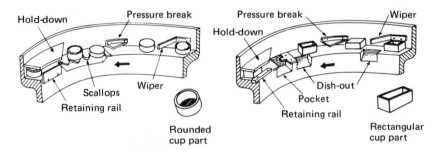

FIGURE 7-10.  *Examples of round and rectangular piece-part orientation on the track of a vibratory bowl.*

be oriented correctly at this rate.  A part takes several positions as it travels along. The desired feed rate is set at the number of correctly oriented parts required per minute.  More than one track can be used at one time, as was shown in Fig. 7-9.

## Basic Transportation of Workpieces

The basic method of automatically moving parts from one station to another may be classified as continuous motion, synchronous (or intermittent) motion, and nonsynchronous motion.  In order to select the most appropriate system for a particular application, three principal factors should be considered: (1) the kind and number of operations to be performed, (2) the size and weight of the components, and (3) the anticipated production rate.

***Continuous Motion.*** This method, by definition, is the transportation of piece parts and/or work-holding fixtures at a constant or variable speed without interruption. This type of system requires that the work stations move with the work-holding fixtures. Continuous-motion transfer is widely accepted in such operations as wave soldering, bottling, packaging, printing or marking, and surface coating. Continuous systems can operate at relatively high speeds since there are no stop-and-start, inertia, or acceleration problems.

Continuous motion systems, either rotary or linear, are the easiest to design, simplest to operate, and least expensive. However, unless the continuous motion is being used for materials handling only, or for simple assembly operations, work must be performed "on the fly." Although this is not difficult if the work is done manually, automatic work stations, which must match with the constant-motion transportation system, can be troublesome and expensive. In terms of fully automatic, high-speed assembly, metal removal, or many other processing applications, continuous motion is not acceptable.

***Synchronous or Intermittent Motion.*** Shown in Fig. 7-11 are four basic types of rotary motion used extensively in automation equipment. They all provide for a positive rotary motion and a dwell time. The cycle time can be made to vary by changing the speed on the drive input, but the relationship between index and dwell is synchronous or fixed. For example, in the Geneva mechanism shown there will be six stops or indexing motions for a complete revolution. The time ratio is based on 120° index and 240° dwell. Each indexing motion will require 20 degrees and each dwell time 40 degrees. Shown in Fig. 7-12 are two types of dial-index machines.

***Nonsynchronous Motion.*** In this system pallets are free to move at a predetermined rate but the overall machine cycle rate is dependent upon the speed with which the individual operations are performed. This system is sometimes referred to as *power-and-free* and is usually used in semiautomatic machine operations, that is, where some manual contact is involved. As an example of the nonsynchronous system, parts may be assembled at most stations with 2 sec of dwell time and a manual spring-insertion step at another station might require 6 or 8 sec. While the rest of the machine can operate at a faster cycle rate, the work-holding carrier must stop at the manual station for the slower nonsynchronous operation, as shown in Fig. 7-13(*a*). This type of operation may be speeded up by varying the number of manual stations. Stations may be added or subtracted with relative ease as shown schematically at (*b*).

***Advantages and Limitations.*** Continuous and synchronous type systems can be fully automated and usually cost less than linear systems. To hold close tolerances, rotary-type indexing-machine work stations can be equipped with floating fixtures.

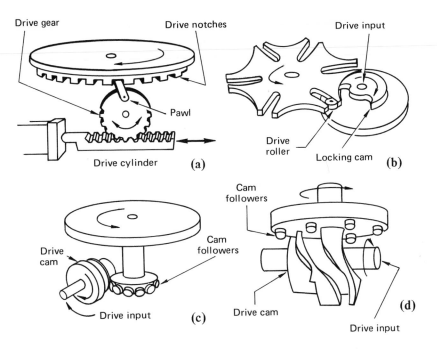

FIGURE 7-11. *Four basic types of rotary-drive systems. The ratchet and pawl system (a) is suitable for light-duty and low-speed applications. The power source is usually a pneumatic or hydraulic cylinder. The Geneva mechanism (b) is suitable for larger jobs. During the dwell time, the locking cam holds the driven member for positioning accuracy. Cam mechanisms are the most consistent and accurate indexing mechanisms with high torque capacity, smooth acceleration, and longer wear life. They can be furnished with 3 to 48 stops. The cam drive shown at (c) applies power to the cam followers which are mounted with their axes perpendicular to the table spindle. The cam followers shown at (d) are driven by a small-diameter drive cam in a compact unit. Courtesy* Automation.

Locating pins (Fig. 7-14) help move the part into exact location before the machining or assembly operation.

Linear transportation machines offer a wide latitude of design. Since the action is not confined to movements around a circle, large parts can be handled with greater flexibility for tooling. Two or more basic linear machines can be joined to provide greater capacity when needed.

Rotary tables have some limitations. They can be used only for relatively small parts. Indexing time must be geared to the longest operation. Trouble at one station stops all stations. Accessibility for maintenance and troubleshooting is quite limited. As the number of stations increase, dial size must increase, which adds to the starting and stopping problems. Shown in Fig. 7-15 is a dial-type assembly machine equipped with vibratory-bowl feeders.

The disadvantage of the linear machine is the initial cost. If close tolerances are necessary, parts must be mounted on floating fixtures which are clamped to pallets. The pallets must be pinned in place before machining or assembly operations can start. Some nonsynchronous systems are designed with individually energized pallets in which case operators are usually needed to pace the work.

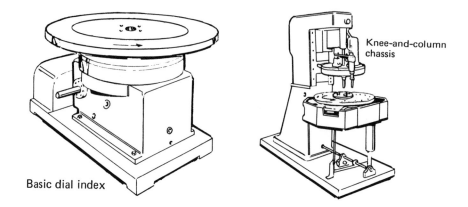

FIGURE 7-12. *Examples of the two basic types of dial or rotary indexing machines. Courtesy* Metalworking.

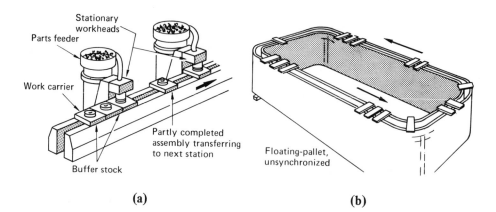

(a)                                  (b)

FIGURE 7-13. *In the nonsynchronous system, pallets are transferred by means of a continuously moving belt or chain. When pallets arrive at a work station, a clutch or latch mechanism operates to permit the pallet to remain stationary. Once the operation is performed, the clutch is released automatically or manually and the pallet moves to the next work station (a). Pallets may be added or taken off from the nonsynchronous carrier (b) with relative ease. Courtesy* Automation.

FIGURE 7-14. *In order to maintain accuracy of location (±0.001 in.), the holding fixture is built with a floating slide in which holes for the securing pins are provided. A set of securing pins is attached to the tool in such a way that they reach the holding fixture before the tool, thus securing the holding fixture in exact location.*

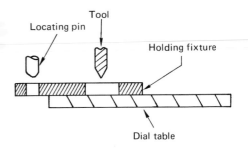

FIGURE 7-15. *A dial-type indexing table equipped with a vibratory-bowl feeder. In this illustration hose clamps are being assembled at the rate of 1800 per hour. Courtesy Hill-Rockford Co.*

## Part Placement

After each work station, parts must be transported to the next station and placed in an exact location for the next operation. It is important to have the placement done consistently, without damage to the part or loss of control, and at relatively high rates of speed. Placement mechanisms must be simple in design and ruggedly made to provide reasonably good service.

Basic types of parts-placement devices include feed tracks with escapement mechanisms, pick-and-place mechanisms, and various types of industrial robots.

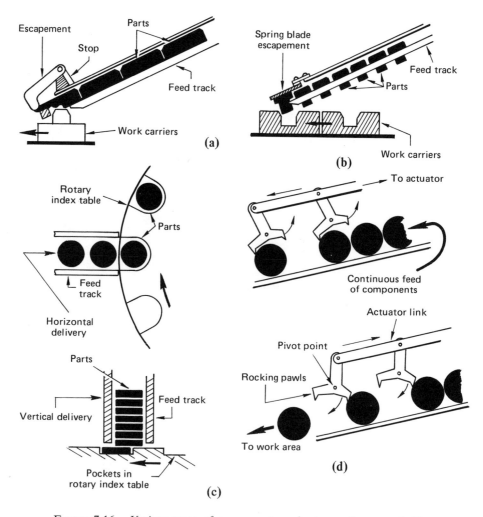

FIGURE 7-16. *Various types of escapement mechanisms. Parts are held in place by a rocker arm (a) or a spring (b) or are simply allowed to drop (c). The ratchet-type escapement (d) involves the use of a linear motion. Courtesy Automation.*

*Feed Tracks and Escapements.* Feed tracks are usually gravity-powered and are used to bring the part from the point of orientation or prior work station to the next operation.

As the part moves down the feed track to the end, it is released by an *escapement* mechanism. An escapement may be defined as a mechanical device that allows a predetermined number of parts to be released from a feeding system on demand, as shown in Fig. 7-16. Usually the escapement loads the part directly into the work-holding fixture. Direct loading is inexpensive and reasonably reliable (at slow speeds) and for some parts it is the best way to affect final placement. Direct placement may also have some disadvantages such as: (1) clearance problems between the escapement and the workhead or the fixture, (2) the force of gravity may not be sufficient for proper seating, and (3) loss of control of the part can cause misfeeds and/or jam-ups in the machine and possible damage to the part.

*Pick-and-Place Mechanisms.* Because of the possible problems cited for escapement mechanisms, parts are often positioned in the work-holding fixture by more elaborate methods, as shown in Fig. 7-17. Some of these are commercially available and others are made up for in-plant use.

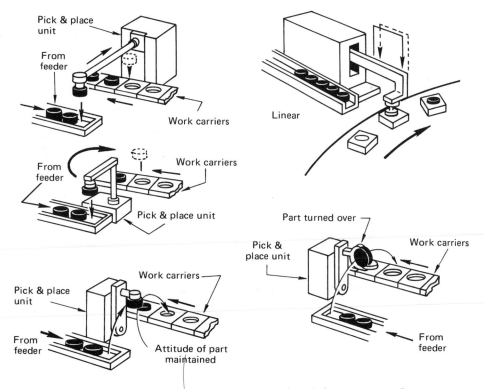

FIGURE 7-17. *Various types of automatic pick-and-place systems. Courtesy* Automation.

519

**(a)**

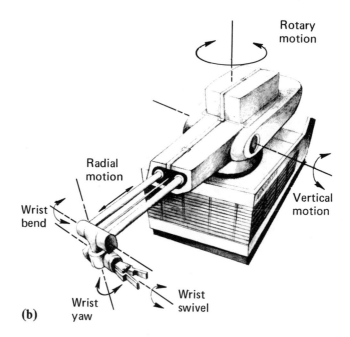

**(b)**

FIGURE 7-18. *Examples of two types of industrial robots: the Versatron, capable of positioning parts on a point-to-point program or by continuous path (a); and the Unimate, which also has extensive programming capabilities (b). Courtesy AMF Incorporated.*

*Industrial Robots.*  Typical industrial robots are the Unimate (made by Unimation, Inc., Danbury, Conn.) and Versatron, (made by American Machine and Foundry Co., York, Pa.) (Fig. 7-18).  Robots of this type have sophisticated memory systems, are easily programmable for routine or alternate actions, and have a wide variety of movements.  Shown in Fig. 7-19 are a few of the "arm" and "finger" movements that can be programmed.

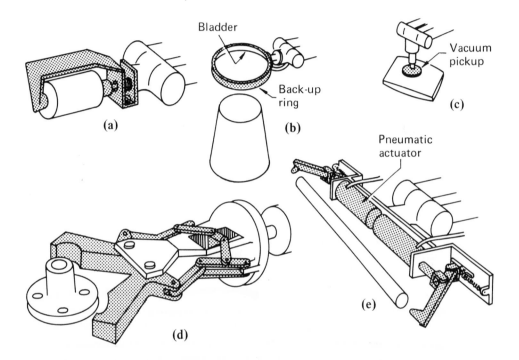

FIGURE 7-19.  *Some examples of various types of clamping devices developed for the Unimate robot.  The gripper can be lowered onto the part by a wrist action (a).  The expandable bladder (b) is useful for flexible and tapered walls.  Many parts can be handled by a simple vacuum pickup (c). If the part is not always in the same spot, a wide-opening hand device is used (d).  Long, slender parts may require special clamping techniques such as fingers that enter the tube (e).  Courtesy* Automation.

Some robots are built with an electromagnetic memory drum that can store up to 200 sequential commands.  If additional flexibility is required, the robot can be interfaced with a digital computer.  The robot shown in Fig. 7-18(*b*) is capable of lifting an object weighing up to 75 lb from a position 4 in. above the floor to one 94 in. above the floor.  The robot can then place the object within a 3550 cu ft working area to an accuracy of 0.050 in. in any dimension.

TABLE 7-1. *Application and Benefits of Industrial Robots. Courtesy* Machine and Tool BLUE BOOK

| Versatran Unit Applications | Benefit |
|---|---|
| 1. Load steel billet from conveyor to two-die hot forge press | 30 percent increase in productivity, no recruiting problem |
| 2. Unload die castings, inspect, trim | 50 percent increase in rate, 300 percent increase in capital equipment utilization |
| 3. Load, unload, operate coating process | 400 percent increase in work rate, elimination of worker hazard |
| 4. Automated shot-peening of metal parts | Reproducibility of peening patterns. Costs lower than NC equipment |
| 5. Process injection molded parts | Eliminated two-man recruiting problem on six-station operation, increased handling rate |
| 6. TIG welding of auto assembly | Uniformity in welding oval track, with increase in productivity |
| 7. Load/unload assembly machine | 100 percent increase in rate, handled two machines, eliminated worker hazards |

Some typical jobs handled by robots are shown in Table 7-1 along with the benefit gained over previous methods.

***Economic Considerations.*** The decision of whether or not a robot should be used should be based on a favorable economic investigation. Several factors should make up this investigation: (1) availability of other methods, including suitable special-purpose automation; (2) number of shifts; (3) duration of job; (4) labor cost; (5) installation cost; (6) cost of auxiliary equipment such as other parts orienters, conveyors, or inspection equipment; (7) cost of service and operation versus special-purpose automation; and, (8) availability of funds for investment in equipment. A basic equation for determining the payback period for a robot installation is:

$$P = \frac{I}{L - M}$$

*where:*  $P$ = payback period in years
$I$ = total investment for robot and accessories
$L$ = annual labor savings
$M$ = annual maintenance expense

Typical inputs may be: $I$ = \$25,000, $L$ = \$11,000, $M$ = \$2000 for one shift and \$3000 for two shifts. Thus, substituting in the equation for a one-shift operation:

$$P = \frac{\$25,000}{\$11,000 - \$2,000} = \frac{\$25,000}{\$9,000} = 2.7 \text{ years' payback period}$$

The growing use of robots in industry is signified by the first International Symposium on Industrial Robots held in 1975. Also, standards for the development of robots are now organized under the American National Standards Institute (ANSI). These standards emphasize safety factors and guidelines for both producers and users.

## Transfer-Line Automation

The specialized transfer-line machine probably comes nearer the average conception of automation than any other system in the metalworking industry. It is on these specially built machines that hundreds of operations are performed on rough-cast engine blocks and many other complicated high-volume production parts. Shown in Fig. 7-20 is a transfer line used to produce 1200 pistons per hour. The line is arranged in sections of identical machines which are fed by a conveyor. Parts feeding along the conveyor are automatically deflected into loading chutes for each machine section that is empty at that particular time. Thus the operation is not held up due to difficulty with any one machine nor is progress based on the longest cycle time. As the sophistication of the line increases, so must the complexity of safeguards that must be incorporated into it. Parts placement must be kept within close tolerances and circuits must be designed so that trouble lights will immediately light up upon the detection of part absence, parts out of position, parts out of dimension, etc. Several types of transducers are usually incorporated into the line that monitor the line on the basis of light, sound, temperature, proximity, and torque. The signals received may lead to one or more of the following: (1) machine stoppage to apply corrective action, (2) a process slowdown to permit corrective action, (3) input into the memory system for subsequent disposition, (4) an ink or paint marker to identify the part in error, or (5) an automatic corrective process.

The incentive behind automation has been the desire to make products faster and at lower cost. However, the greatest benefit has been in doing jobs beyond the power of human workers. Instrumentation, mechanical gaging, and mechanical perception have made possible closer tolerances than human hands can achieve. The added blessing of automation is that it has eliminated many of the tedious, monotonous jobs formerly done by hand.

On the other hand, automation does not replace intelligence. Instead it imposes a greater demand on it. The need for clearer and more thorough planning is much more evident in an automated system than for a manual one. Automation demands engineers and technicians of broad backgrounds to design, to build, and to maintain the present systems and the ever more sophisticated automation systems of the future.

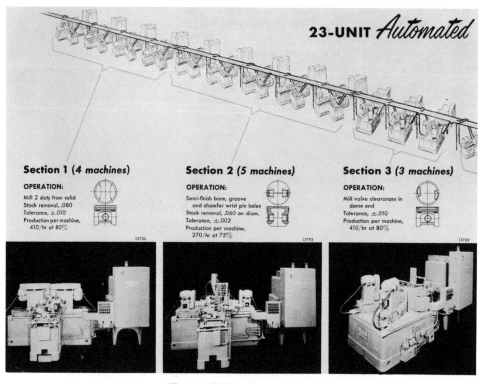

**23-UNIT** *Automated*

**Section 1 (4 machines)**

OPERATION:

Mill 2 slots from solid
Stock removal, .080
Tolerance, ±.010
Production per machine,
    410/hr at 80%

13786

**Section 2 (5 machines)**

OPERATION:

Semi-finish bore, groove
    and chamfer wrist pin holes
Stock removal, .060 on diam.
Tolerance, ±.002
Production per machine,
    270/hr at 75%

13795

**Section 3 (3 machines)**

OPERATION:

Mill valve clearances in
    dome end
Tolerance, ±.010
Production per machine,
    410/hr at 80%

13789

FIGURE 7-20. *An example of an automated transfer line used*

## NUMERICAL CONTROL

### Introduction

A relatively short time ago machines were operated by craftsmen who determined the many variables such as speeds, feeds, depth of cut, etc. Now much of this work is being assigned to computers and machines that are numerically controlled. The word numerical is defined as "the expression of something by numbers." Control is defined as "the exercise of directing, guiding, or restraining power over something." By combining the two definitions we find that numerical control consists of directing, guiding, or restraining power over something by the use of numbers. Since numbers are merely symbols and letters are used also, numerical control could well be called symbolic control.

Numerical control has become commonplace in the machine-tool world and has opened up new doors for more advanced concepts. Now complete preplanned and programmed design and production functions can be processed, stored, and finally executed with the aid of a computer. These are termed CAD, computer-aided design, and CAM, computer-aided manufacturing.

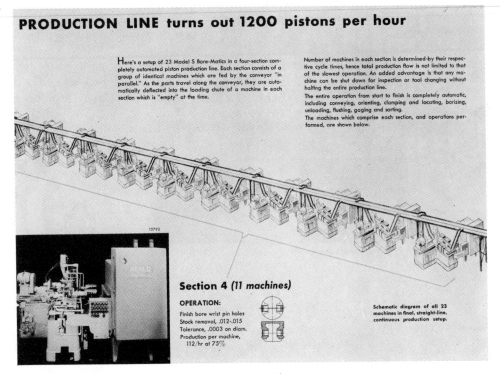

# PRODUCTION LINE turns out 1200 pistons per hour

Here's a setup of 23 Model S Bore-Matics in a four-section completely automated piston production line. Each section consists of a group of identical machines which are fed by the conveyor "in parallel." As the parts travel along the conveyor, they are automatically deflected into the loading chute of a machine in each section which is "empty" at the time.

Number of machines in each section is determined by their respective cycle times, hence total production flow is not limited to that of the slowest operation. An added advantage is that any machine can be shut down for inspection or tool changing without halting the entire production line.

The entire operation from start to finish is completely automatic, including conveying, orienting, clamping and locating, borizing, unloading, flushing, gaging and sorting.

The machines which comprise each section, and operations performed, are shown below.

15793

## Section 4 (11 machines)

**OPERATION:**

Finish bore wrist pin holes
Stock removal, .012-.015
Tolerance, .0003 on diam.
Production per machine,
112/hr at 75%

Schematic diagram of all 23 machines in final, straight-line, continuous production setup.

*for high-volume production. Courtesy Cincinnati Milacron.*

Numerical control (NC) is considered to be more than an improved control for machine tools that provides higher productivity. It is considered by many leading scientists to be the first significant step in what is referred to as the second industrial revolution. The first industrial revolution was ushered in by James Watt's steam engine. Human muscle power was replaced and extended into all types of industrial equipment. The second revolution is an extension of man's mind. NC, with the aid of the computer, takes a symbolic input and restructures it to an output that extends man's concepts and thinking into some creative, productive results.

The birth of NC may be traced back to the late 1940s. At that time a man by the name of John Parsons and his associates were manufacturing helicopter blades. They found the specifications, contours, and weights of the airfoil very critical and demanding. Struggling with the contour of the rotor blade, they came up with an incrementally moving dimensional inspection system. The gathered dimensional data was fed into an early model computer which contained the pure mathematical formula for the airfoil. The inspection data was compared to the basic formula. The idea worked so well that Mr. Parsons concluded that if a computer could tell what had happened after a part was made, why not use the same data to control a machine tool to make the part. Of course a concept is far from reality. However, in 1949, the U.S. Air Force, which recognized the value of the technique, was awarded a development contract. Later the contract was directed to the Massachusetts Institute of Technology and by 1952 the first numerically controlled machine tool was operational.

Numbers, letters, and symbols gathered together and logically organized to direct a machine tool for a specific task are called an *NC program*. The concept is revolutionary in that the control of a machine has moved away from operator skill

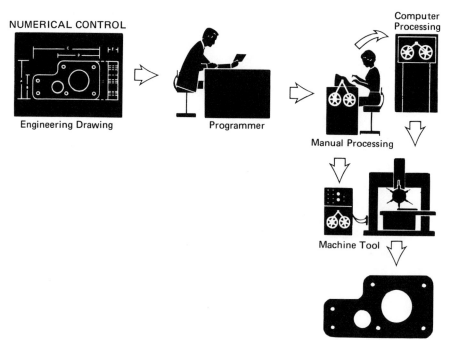

NUMERICAL CONTROL

Engineering Drawing

Programmer

Computer Processing

Manual Processing

Machine Tool

Finished Part

FIGURE 7-21.  *The operating efficiency of an NC machine is largely dependent on the programmer. He interprets the engineering drawing and writes the program that contains all the basic instructions for the machine control unit to follow. Some of the simpler programs are processed manually, the more complex require computer processing.*

and intuition to an entirely written program that can be interpreted by the machine control unit to perform all the required movements that produce a finished part (Fig. 7-21). Not only has NC replaced human skills and intuition, but it has made it possible to generate complex machine motions with a degree of precision that is far beyond any human capability. If a surface can be mathematically defined, NC can control the machining of it.

### NC Organization

As with anything to be manufactured, the concept must be formalized into a design. The part programmer studies the drawing and conceptually vizualizes all the machine motions that will be required to make it. He then proceeds to document them in a logical order on what is known as a programming manuscript. All steps must be logically conceived and clearly stated so that the machine control unit (MCU) can follow the coded commands. Most MCU's today have a basic electronic design, but a few hydraulic and mechanical units are still in use.

The programmed codes that the MCU can read may be one of four different media: perforated tape or "punched" tape, magnetic tape, tabulating cards, or signals directly from computer logic or some computer peripherals such as disc or drum storage.

Direct computer control (DNC) is the newest of these four methods and one which affords the help of a computer in developing a part program. Tape and card reading are slow and cumbersome by comparison. In addition, the storage of a program in a computer often affords an easier method of program modification. This topic will be discussed in more detail later.

*Tape Control.* The predominant method of communicating with the machine is through punched tape that is 1 in. wide and has eight channels running the length of the tape. The holes are punched in a predetermined pattern according to the number, letter, or symbol represented. The reader senses the pattern by one of several methods, the most common being that of a light beam passing through the tape's holes to complete an electric circuit with photoelectric cells. The MCU converts the input signals from the tape to output signals that can be utilized by the power elements actually running the machine tool.

*Other Control Methods.* Other control methods used on machine tools include dial controls in which the machine operator dials each axis dimension for the workpiece, pin-type controls in which pins are mounted in a mechanical program table with the machine table moving against the pins to determine axis location, and plugboard controls where machine cycles are determined by connector plugs inserted into a program board similar to that used by a telephone operator. Mechanical stops or similar techniques are used along with the plug-board to determine table or tool location.

### Closed- and Open-Loop Control Systems

*Closed-Loop Systems.* It is necessary in NC machining that the control unit know where the cutting tool is at all times. It must know, for example, if the table which was given a signal to move so many inches in the $+X$ direction actually did move that amount. Since the motion is accomplished by lead screws, the number of turns or fraction thereof must be translated to the controls. This is done by connecting an analog transducer, or "resolver," to the end of each lead screw. The resolver con-

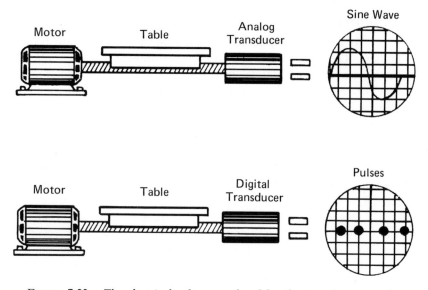

FIGURE 7-22. *The electrical voltage produced by the transducer may be analog (variable) or digital in nature. Both are in direct proportion to the rotation of the shaft input.*

verts the rotation of the lead screw into electrical feedback signals which the controller recognizes. These signals may be either analog (variable) or digital in nature, as shown in Fig. 7-22. The analog transducer produces a variable electrical output that is proportional to the rotation of the input shaft. The digital transducer produces individual, discreet electrical pulses.

As the punched tape is read by the control unit, it operates a series of switches for "on" or "off" conditions for digital control. In a closed-loop digital system, the feedback is also in digital form, making it easy to see if there is any difference between where the tool ought to be and where it is. If there is some discrepancy, an "error" signal is generated to make the electric motor turn the lead screw and move the table. When this is done, the feedback signal approaches the command signal and the error approaches zero. The system is shown schematically in Fig. 7-23.

*Open-Loop Systems.* The original NC machines were all of the closed-loop variety. They were used largely for carving complex aircraft and missile parts where a slip of the cutter often meant thousands of dollars down the drain. Also, in the late 1950s electric stepping motors were not yet reliable enough or powerful enough for applications emerging at the time. Thus, in the United States at least, NC came to mean a closed-loop system.

In Japan, during the 1960s, better stepping motors were becoming available and NC with the open-loop system began spreading to other countries. U.S. firms, however, continued to be strongly influenced by "tradition" or the precedent of closed-loop systems. Now, with the improved capacity, accuracy, and smaller size of stepping motors, the open loop is very common in the United States as well.

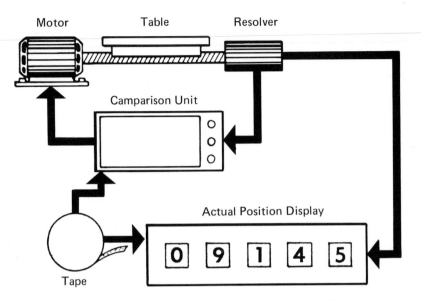

Motor    Table    Resolver

Camparison Unit

Actual Position Display

0 9 1 4 5

Tape

FIGURE 7-23. *The digital readout makes it easy to see if there is any difference between what the machine was programmed to do and what was actually accomplished. The feedback control should automatically correct for any discrepancy until the error signal is zero.*

***Stepping Motors.*** Stepping motors are designed so that each pulse of electric current exerts a given torque on the motor and makes it rotate a given amount. As an example, for a given-size motor each pulse may produce 1.8 degrees of rotation on the output drive. If coupled to a five-pitch lead screw, it will give an effective resolution of 0.001 in./step. For one inch of table travel, one thousand pulses are needed, and the positioning will be accurate to one thousandth. The errors are noncumulative, i.e., for a 1.8-degree motor, angular error (for a constant load) is 0.054 degree for one step or for 100,000 steps.

Shown in Fig. 7-24 is a sketch of three high-powered stepping motors as used on a milling machine. Motors of this type and application are about 4 in. in diameter and 7 in. long. Stepping motors are available with maximum torques of a fraction of an inch-ounce to 2000 inch-ounces, and maximum power output from 0.001 hp to at least 0.75 hp. Higher powers of 10 and 20 hp require hydraulically amplified units. These units are essentially hydraulic motors built integrally with electric steppers and with feedback to the steppers. Thus, the electric stepper is positioned by electric pulses as in the conventional open loop, but the stepper in turn positions a hydraulic valve which drives a hydraulic motor with more muscle than is available in an all electric motor.

In large NC systems, the pressing problem is not open or closed loop, but what is available in control logic. As the control system works its way up into the $20,000 or $30,000 bracket, an extra $2000 per axis for closed loop does not seem

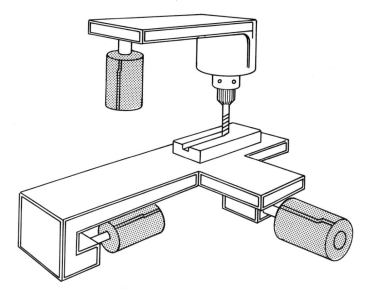

FIGURE 7-24. *High-powered stepping motors are used extensively as X and Y and sometimes Z drives on milling machines. The basic position information is fed to the control system by a tape reader and either point-to-point or contouring control is possible. Courtesy Automation.*

formidable. There is a tendency then to go to a complete-feedback system. The closed loop also has the advantage of being able to function with *adaptive control*. That is, the machine is sensitive to cutting conditions, and is able to make changes as they become necessary for continuous optimization.

Open-loop systems are most often used on large U.S. machines that have been *retrofitted*. This can be done much more easily with the open-loop than with the closed-loop system. Retrofitting refers in this case to equiping an ordinary machine with NC capability.

**The Coordinate System**

Although the cartesian coordinate system was conceived more than 300 years ago by a French mathematician, Descartes, no system has been found which surpasses it as a basis for NC control. Rectangular coordinates provide an excellent means of locating an NC tool with respect to the workpiece in either two or three dimensions. The three primary motions for an NC machine are given as X, Y, and Z. The primary X motion is normally parallel to the longest dimension of the machine table. The Y motion is normally parallel to the shortest dimension of the machine table and the Z motion is the movement that advances and retracts the spindle [Fig. 7-25(a)]. Letters A, B, and C designate angular and circular motions as shown at (b). Shown at (c) is the type of work that can be done on a five-axis, continuous-path control machine.

MACHINE ELEMENT MOVEMENTS
X Axis — Table side to side
Y Axis — Table front to rear
Z Axis — Spindle vertically into work

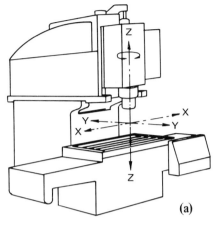

(a)

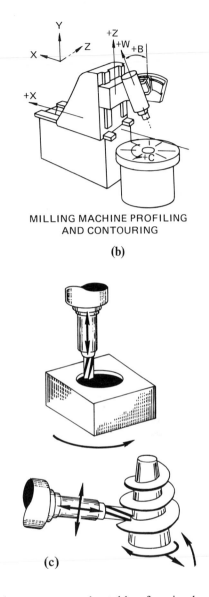

MILLING MACHINE PROFILING
AND CONTOURING

(b)

(c)

FIGURE 7-25. *A sketch showing the basic machine axes (a). Angular and circular motions about axis Z are designated by letters A, B, and C, as shown at (b). Shown at (c) is the type of work that can be done with a five-axis machine. An end mill is being used to generate a contour, or a circular form. It may also be used to make elaborate die cavities or airfoil contours. These operations are referred to as* profiling. *During profiling the cutter and workpiece move about several or all axes simultaneously.*

The diagram shown in Fig. 7-26 is used to represent the table of a simple drilling machine with the workpiece located in the upper right quadrant. Quite often the entire table will be designated as the upper right quadrant so that no minus dimensions are used. As shown in the diagram, a hole may be drilled at the +X3, +Y4 location or, if a slot is to be milled as represented by the broken line, the center point of the cutting tool is sent from the point of origin to +X3, +Y4. The slot width is determined by the cutting-tool diameter.

Considerable work is done on machines that have only two axes under NC control. In this case, the depth of holes and slots is determined by limit-switch set-

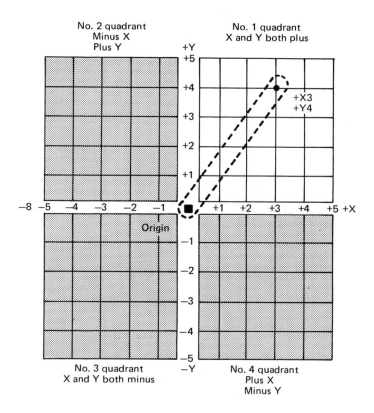

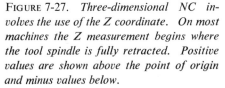

FIGURE 7-26. *The origin point may be the corner of the machine table or it may be relocated at a point on the workpiece itself. If a hole is to be drilled at the X3, Y4 location or a slot is cut as shown by the broken line, the MCU will be directed for three units of X motion and four units of Y motion. Although there is a tendency to seek setups that allow programming in the all-plus quadrant, programmers find no difficulty for any quadrant.*

FIGURE 7-27. *Three-dimensional NC involves the use of the Z coordinate. On most machines the Z measurement begins where the tool spindle is fully retracted. Positive values are shown above the point of origin and minus values below.*

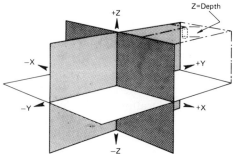

tings made by the operator where the part is placed on the machine. Advanced NC involves the Z coordinate. Positive values for all three axes are in the upper right quadrant, as shown in Fig. 7-27. The tool always points in the −Z direction. The coordinate system described assumes the tool movement with respect to the work. If the tool is stationary and the table moves the signs will be reversed, e.g., if the table on a vertical milling machine moves to the operator's right, it will be in the −X direction. In actual practice (bearing in mind the + and − signs just stated) one pretends the tool moves rather than the table.

### Numerical-Control Programming Systems

Numerical control programming can be divided into two main systems: *point-to-point* and *continuous path*. Although this demarcation was once very sharp, as far as machine-tool capability, it has become somewhat dulled due to a mixture of positioning and contouring capabilities.

### Point-To-Point Positioning (PTP)

PTP positioning is typified by drilling and punching machines. The drill or punch is directed, by X and Y coordinates, to a specific location on the workpiece. PTP positioning, as used for drilling, reaming, boring, punching, etc., requires a high degree of accuracy as to location, but places no restrictions on the path used in achieving it.

*Programming for PTP Positioning.* The programmer starts out with an engineering drawing visualizing all the necessary machining sequences needed to produce the part. These steps are written down in a manuscript from which a program tape can be developed. An important first step is that of establishing the reference point. Invariably a workpiece is loaded on the machine so that a flat surface is parallel to the table and a straight side will be parallel to machine axis X or Y. The point where the X, Y, and Z axes cross is called the *zero point*. The zero point may be either *fixed* or *floating*. The fixed-zero machine is usually made so that the zero location is at the lower lefthand corner of the X-Y coordinate. On some systems the fixed zero may be moved, by means of switches or dials, to anywhere within the full travel range of the machine. Ordinarily, most fixed-zero machines are made so that all machining will be done in quadrant 1 with only a small (0.050 in.) zero shift for adjustment after the part is loaded in place.

A floating zero allows the operator to place the zero at any convenient location without regard to the initial fixed zero point. The operator merely moves the table to the desired location by means of jog buttons. Jog buttons make it possible for the operator to "manually" move the table to the desired location at feed rates set on the dial. When the table is at the exact desired location, the operator depresses a button which establishes this point as the zero point. The part programmer usually determines where the zero should be and passes this information on to the operator. Shown in Fig. 7-28 is a comparison between a fixed and floating zero.

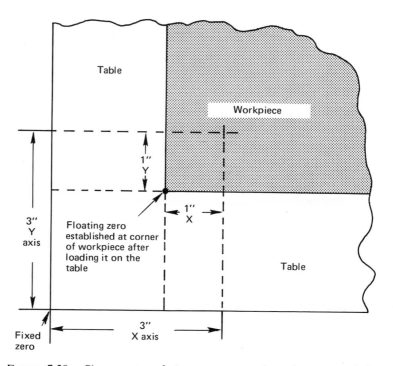

FIGURE 7-28. *Since most workpieces are mounted on the center of the table, the machine zero does not match the workpiece zero. In the example shown, the edges of the workpiece are 2 in. from the bottom and side of the table. The center of the hole to be drilled is 1 in. from the bottom and side of the workpiece. If the program is for a fixed zero, both X and Y axes will be programmed as 3 in. If the machine has a "floating zero" or "zero shift", the origin point will probably be set at the corner of the workpiece and X and Y dimensions will be programmed as 1 in.*

***Absolute and Incremental Dimensioning.*** Dimensions on an NC drawing may be either absolute or incremental in nature. Absolute dimensions always start from a fixed-zero reference point, as shown in Fig. 7-29(*a*). All incremental locations are given in terms of distance and direction from the immediately preceding point, as shown in Fig. 7-29(*b*). The direction of motion reverses itself when the magnitude of numbers decrease. Thus increasing numbers 2, 5, 7 causes motion in one direction, while decreasing numbers 7, 5, 2 will reverse the direction. This method eliminates the use of algebraic signs to indicate direction as long as the motions occur in the first quadrant.

## Continuous-Path Programming

Both PTP and continuous-path programming are based on rectangular coordinates. The main difference between the two systems is interpolation. Interpolation is a method of getting from one program point to the next so that the final

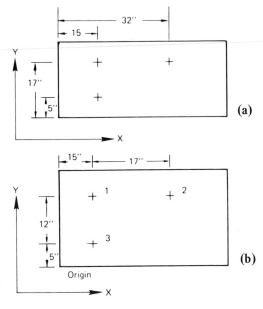

FIGURE 7-29. *Absolute measurements are given from the zero origin (a). Incremental locations are given in terms of distance from the immediately preceding point (b).*

Pt. 1 X = 15  Y = 17 (5 + 12)
Pt. 2 X = 17  Y = 17
Pt. 3 X = 15  Y = 5

workpiece is a satisfactory approximation of the programmed design. There are three interpolation methods of connecting defined coordinate points to generate angular or curved shapes: *linear*, *circular*, and *parabolic*. Of these, linear interpolation is the most used and is found on nearly all continuous-path machines.

*Linear Interpolation.* Shown in Fig. 7-30 is a slope-line motion as accomplished by PTP programming. The control will follow path *AB* because the ratio Y/X = 1, the feed rate in both axes is the same. At point *B*, the control interpolates to give an additional increment in the X direction, bringing the slope nearer that of the theoretical. The same interpolation would occur at point *D*, etc.

An arc can be approximated by a series of steps as shown in Fig. 7-31. The interpolation is similar to that of making a slope except that the lines are now tangential to an arc. The theoretical radius of the cutter, center line *R3*, is given as additional input data along with the tolerance. Because of the number of lines, the outline appears to be analog in nature. Numerical control, however, is not analog but digital. Therefore, to approximate an arc or a circle a sufficient number of points have to be described within the framework of rectangular coordinates. It can easily be seen that the length of the program tape is in direct proportion to the number of coordinate positions needed to define the arc. In some sculptoring operations the length of line may be only 0.0002 in. The standard procedure is to have a supply of tapes that have been programmed by the computer for various-size quadrants of a circle made for either clockwise or counterclockwise motion. These tapes are then copied into the regular part program when needed.

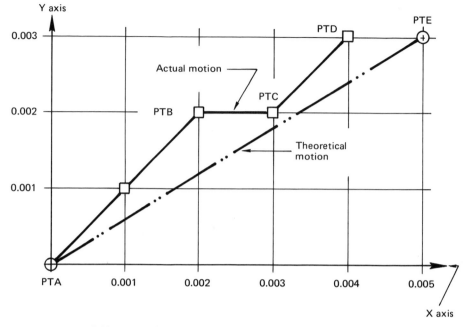

FIGURE 7-30. *Actual and theoretical tool motions using the incremental mode to move the tool from point A to point E.*

***Circular Interpolation.*** Circular interpolation may be done with a small computer module that is able to direct the cutting tool in a complete circle, or in arcs when given the coordinates of the center point, radius, direction of travel, and location of the arc end points.

***Parabolic Interpolation.*** Parabolic interpolation finds its greatest application in free-form designs such as mold work and automotive die sculpturing. Parabolic interpolation goes between three nonstraight-line positions in a movement that is either a complete parabola or a portion of one. Its advantage lies in the ability to closely approximate curved sections with as few as 2 % as many points as with linear interpolation. Parabolic interpolation, however, requires a more specialized approach.

***The Cutter Path.*** Assuming that a continuous-path program is to be made manually, the programmer will have to consider feed rates and the tool path. If a cutter, for example, is 1 in. in diameter and is used for profiling, it will have to be offset as shown in Fig. 7-32. Note that the offset must always be made perpendicular to the surface being machined, regardless of the direction that surface faces. Determining offsets not parallel to a machine axis will usually involve trigonometric calculations.

Some MCU's have cutter-compensation capabilities to take into consideration the variable diameter of cutting tools or tools that were programmed with one diameter in mind and have been changed or have been resharpened to a smaller diameter.

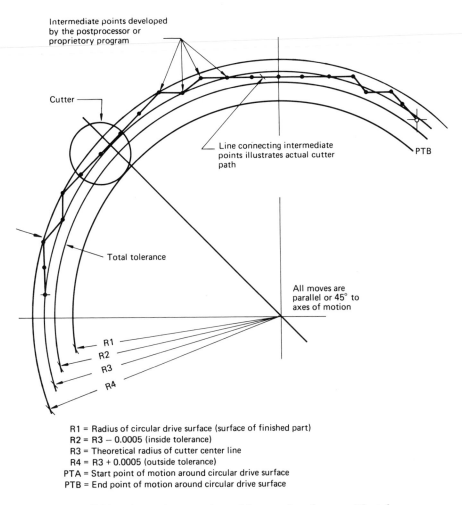

R1 = Radius of circular drive surface (surface of finished part)
R2 = R3 − 0.0005 (inside tolerance)
R3 = Theoretical radius of cutter center line
R4 = R3 + 0.0005 (outside tolerance)
PTA = Start point of motion around circular drive surface
PTB = End point of motion around circular drive surface

FIGURE 7-31. *An arc is approximated by a series of steps. The tolerance zone is represented by R4 and R3.*

***Programming by Scanning and Digitizing.*** Programming may be done directly from a drawing, model, pattern, or template by digitizing or *scanning.* A simple two-dimensional part that needs nothing more than positioning operations such as sheet-metal punching or hole drilling can be digitized (given X-Y coordinate locations) from a fully dimensioned drawing. The drawing is placed on the table (Fig. 7-33) and an optical reticle or other suitable viewing device connected to an arm or arms is placed over the location points. Transducers are made to identify the location and, upon a signal, transfer it to either a tape punch or other suitable programming equipment. Most digitizing equipment also has auxiliary input to allow the establishment of reference points or the addition of data such as position refinement.

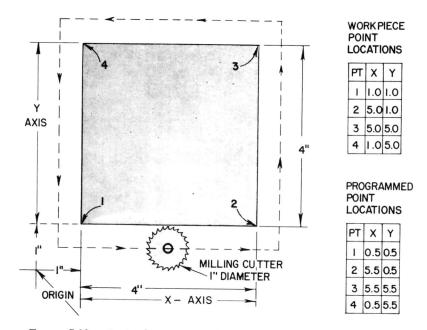

WORK PIECE
POINT
LOCATIONS

| PT | X | Y |
|---|---|---|
| I | 1.0 | 1.0 |
| 2 | 5.0 | 1.0 |
| 3 | 5.0 | 5.0 |
| 4 | 1.0 | 5.0 |

PROGRAMMED
POINT
LOCATIONS

| PT | X | Y |
|---|---|---|
| I | 0.5 | 0.5 |
| 2 | 5.5 | 0.5 |
| 3 | 5.5 | 5.5 |
| 4 | 0.5 | 5.5 |

FIGURE 7-32. *A simple program showing the difference between the workpiece location points and the cutter location points.*

A scanner enables an operator to program complex free-form shapes by manually moving a tracing finger over the contour of a model or premachined part (Fig. 7-34). Overall accuracy of the scanning system from model to tape is ±0.005 in.

Many digitizing and scanning units have the capability to "edit" or revise the basic data gathered. For example, the drawing or model may not be as accurate as required for the finished product. It may be possible to obtain the basic data and then refine it with an overriding corrective input. The design of a surface, for example, may call for an arc but the model may not be near the required tolerance. After the rough data has been scanned, the computer can be called on to smooth the data to a perfect arc.

## Standard Tape Control

The Electronic Industries Association (EIA) has developed standards to define the physical characteristics of punched tape and tape coding for the industry. The American Standards Code for Information Interchange (ASCII) also developed a perforated tape. The two standard tapes are very much alike in that they are both 1 in. wide and have provision for eight rows of holes plus the sprocket holes. The main difference is in the parity check. The EIA codes require an odd number of holes on every row across the tape and the ASCII has an even number of holes as a parity check. Most machine control units for NC have EIA tape readers but some are made to read either one.

It is assumed that the tape preparation is made on a machine such as the Flexowriter shown in Fig. 7-35. The manipulation of the keyboard actuates a tape-perforating mechanism attached to the side of the typewriter and the type also prints on paper. By means of a tape reader, the original tape may be read to reproduce additional tapes and/or reproduce printouts in English for all the data represented by the codes on tape.

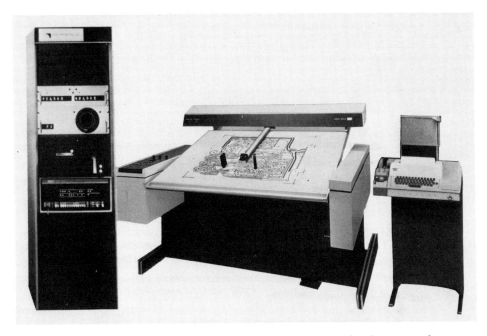

FIGURE 7-33. *Coordinate points are picked up from the drawing and digitized. The unit in which the digital readouts are located also contains buffer storage, scale factors, and programming aids. The tape is punched by a unit at the right of the digitizer. Courtesy Data Technology, Inc.*

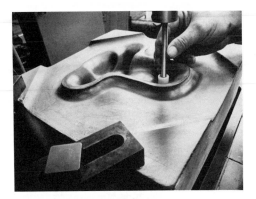

FIGURE 7-34. *Scanning provides simplified programming of a nonmathematical configuration. Data obtained through tracer movements is converted into punched tape by a minicomputer. Courtesy Cincinnati Milacron.*

FIGURE 7-35. *A Flexowriter as used for the preparation of punched tape for NC. Courtesy Singer Business Machines.*

**Tape Format.** Eight rows of holes can be punched on the tape plus the small sprocket holes. Each row of holes lengthwise of the tape is referred to as a channel. Each of the eight channels are numbered 1 through 8.

It is apparent from the amount of space on the tape that a shorthand method of representing both numeric and alphabetic information is necessary. For this purpose, the *binary-coded decimal* (BCD) system is used.

**Binary-Coded Decimal System.** Binary is based on two digits, 0 and 1. Channels 1, 2, 3, and 4, or each hole across the width of the tape, have respectively the binary values of 1, 2, 4, and 8, which are the exponential equivalents of $2^0$, $2^1$, $2^2$, $2^3$. Any hole punched in channel 1 will have the value of $2^0$, or numerical 1. Channel 2 will be $2^1$ or 2, channel 3 will be $2^2$ or 4, and channel 4 will be $2^3$ or 8. Zero normally would be indicated by an absence of a punch, but since there is greater reliability in reading a punch for 0, channel 6 is assigned to represent 0. These five channels can be used to represent any digit (single number) of numerical data. For example, 3 is represented by holes in channels 1 and 2, and 7 is represented by holes in channels 1, 2, and 4. Channel 5 is used for *parity check*. The parity system is to make sure there is an odd number of holes in each row (EIA standard). The parity check reduces the possibility of error, as in the case of holes being punched but the paper remaining in place. The tape punch automatically adds the punch to make an odd number of holes. The control equipment has circuitry which will detect an even number of holes anytime during data reading, stopping the reader and lighting a "read error" light when detection occurs.

The last two channels are assigned respectively an X code and an EB code. The X code is used in combination with alphabet codes to designate miscellaneous functions. The EOB or simply EB is the *end-of-block* channel. It signifies the end of all pertinent data for one particular action. This data group could be compared to a sentence. It also provides for easy visual examination of the tape. Shown in Fig. 7-36 is a section of EIA-standard BCD-punched tape.

Letter codes are punched in channels 6 and 7 in combination with other channels. Alphabetic punches are equal to numeric positions in the alphabet. For example, A is 1, B is 2, etc.   In addition, letters A through I are identified by having both channels 6 and 7 punched.   Thus A is 1, 6, and 7.

The next group of nine letters, J through R, are always punched in channel 7 plus the required numbers.   Thus J is 1 and 7, with 5 for parity check.

The last group has only eight letters, S through Z.   In this case, each letter is numbered from 2 through 9 and channel 6 is the common code punch for this group. An example of a tape command is shown in Fig. 7-37.   The command is to " move on

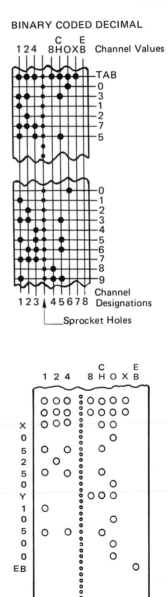

FIGURE 7-36.  *A section of EIA standard BCD punched tape.*

FIGURE 7-37.  *An example of tape command for X, 5.25 in. and Y, 10.5 in.*

FIGURE 7-38. *The NC machine-control unit. Courtesy Kearny and Trecker Corp.*

X-axis 5.25 in. and on Y-axis 10.5 in." The manuscript, which is typed while the tape is being made, would appear as X05250, Y10500. This is known as a block of data. It contains all the instructions required for one command. Other pieces of information that should be included are feed rates, spindle speeds, etc. It is assumed that the decimal is between the second and third digits.

The tape is placed in the MCU so that it can be fed through the *reader* (Fig. 7-38). The reader senses the presence or absence of holes in the tape. These sensors may be mechanical fingers, light beams, or air. In most applications, it is done with light beams acting on photoelectric cells. The circuits contacted provide the comparison unit with a command voltage or number of pulses which represent the desired position, or may provide a signal for an auxiliary function.

### Manual Part Programming

Manual part programming involves the detailed step-by-step listing of operations that the NC machine will perform. The steps are written in a definite order and form in what is known as the *manuscript*. The one who writes the instructions is the *part programmer*.

As described previously, the control functions are coded in a sequential order within a block as specified by EIA standards. There is, however, some variation in the format used since individual machines vary. In general there are three basic control-tape formats:

1. Word-address format
2. Tab sequential format
3. Fixed-block format

***Word-Address Format.*** In the word-address format, each *word* in a block is preceded by an *address* letter character. An NC word can stand by itself and have a definite meaning. Thus the X coordinate or Y coordinate are words that identify a device or mode of operations. The address character immediately precedes the word and serves to identify it.

Coordinate words are identified by X, Y, or Z axes. Noncoordinate words consist of a letter and two digits as shown in the following examples:

1. n—sequence number, e.g., n05, n10, n11, etc. The sequence number is optional but it helps locate errors easily, especially in large programs.
2. g—preparatory function, e.g., g00 for rapid move, g01 for incremental move, g60 for tape-code feed rate, g80 for quill up, and g88 for quill down. g words are given in Appendix 7A.
3. f—feed rate. The feed rate is generally expressed as a three-digit coded number in which the second and third digits express the feed rate in in. per min rounded off to a two-digit accuracy. The first digit has a value of 3 greater than the number of digits to the left of the decimal point. If there are no digits to the left of the decimal point, the number of zeros immediately to the right of the decimal point is subtracted from 3 to provide the value of the first digit. Examples of feed rates and the corresponding code are shown in Table 7-2.
4. s—speed, The speed (rpm) is similar to that of feed. It is represented by 3 plus the number of digits to the left of the decimal. Thus 1800 is coded 718 (3 + 4 digits = 7) and 400 is coded as 640 (3 + 3 digits = 6).
5. t—tool code. Calls out the code number of the tool that will automatically be selected by the tool changer. Machines that have a single tool position, upon reading the tool function code, will immediately display the code number on the

TABLE 7-2. *Examples of Feed Rate Coding*

| Feed Rate (ipm) | Code |
|---|---|
| 50 | f 550 |
| 20.2 | f 520 |
| 0.250 | f 325 |
| 0.0125 | f 212 |
| 0.0012 | f 112 |

Example: 50, 3 + number of digits left of the decimal
point: $50 = (2) + 3 = 5, f = 550$.

visual read out. The motion for that block of data will not be executed until the operator has changed the tool and has depressed the cycle-start button to restart the reader.

6. m—miscellaneous functions. Miscellaneous functions are auxiliary functions that do not relate to dimensional movements of the machine, as "on" or "off." Examples of this command are "coolant off" or "on" and "program start" or "stop." The code for miscellaneous functions usually consist of two digits although special characters such as + and − are also used. Some miscellaneous functions are shown in Appendix 7B.

***Word Order.*** Commands are given to the MCU through the tape in blocks of data which are composed of a word or a group of words considered a unit and arranged in a definite sequence. The same function word should not be repeated within any one block. The sequence used is as follows:

1. End-of-block code symbolized by an asterisk.
2. Sequence-word number.
3. Preparatory-function word.
4. Coordinate words in the order of X, Y, Z, A, B, C.
5. Feed-function word.
6. Spindle speed.
7. Tool function.
8. Miscellaneous function.

> *Example.* The part shown in Fig. 7-39 has a sequence of drilling operations as numbered. The feed rate and spindle speeds are selected and coded as f211 and s717 for the ⅛-in. holes and f185 and s625 for the ⅜-in. holes.

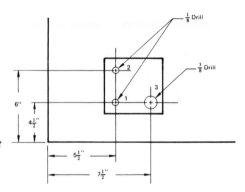

FIGURE 7-39. *Examples of a simple part for manual NC programming.*

> With the word-address format, each word in a block is preceded by an address letter character like n, g, x, y, m, etc., as specified by the standards. The block length is variable depending on whether the information for a particular word is changed from the previous block. The coded information for programming is shown in Fig. 7-40.

MANUSCRIPT

Part No.                                    Date
Part Name                                   Prepared by
Machine                                     Checked by

| Sequence No. | Preparation Statement | x | y | f | s | m |
|---|---|---|---|---|---|---|
| n001 | g81 | x5500 | y4500 | f211 | s717 | m07* |
| n002 | | | y6000 | | | m06* |
| n003 | | x7500 | y4500 | f185 | s625 | |
| | | | | | | |
| | | | | | | |
| | | | | | | |
| | | | | | | |

FIGURE 7-40. *The word-address manuscript for the coordinates shown in Fig. 7-39.*

Each block is preceded and ended by EOB or a carriage return character (symbolized by asterisks). The g-, x-, f-, or s- words in the second block are omitted because there is no change from the first block.

This control tape will produce a typed listing on a Flexowriter as follows:

n001g81x5500y4500f212s717m07
n002y6000m06
n003x7500y4500f185s625

The asterisk causes the carriage to return and is not shown in the type out. It is obvious from the above that the blocks are not all the same length.

***The Tab-Sequential Format.*** The word "tab" is short for tabulation. The tab key on a tape punch perfoms two operations at the same time: it spaces the column on the typeout and it punches five holes (23456) on the tape.

Word-address and tab-sequential programming are very much alike. The only difference is that the tab symbol replaces all letter addresses as used in the word-address format except the tab character is omitted for sequence numbers. Thus the first line as shown in the manuscript of 7-40 would be as follows:

001 Tab 80 Tab 5500 Tab 4500 Tab 211 Tab 717 Tab 07*

Where there are no changes in the tab, the character only is inserted.

***Fixed-Block Format.*** In the fixed-block format all blocks are the same length. The control system will interpret each block of information and divide it into words of specific length in a fixed order. As a result, there is no need for the word-address or tab characters to separate the words. However, each word has to be filled up even though there is no change in information from the previous block. Thus the first two lines of the manuscript shown in Fig. 7-40 would be as follows:

```
001   81   5500   4500   211   717   07   *
002   81   5500   6000   211   717   06   *
```

## Program Language

Today there are dozens of languages that a programmer can use. The largest and most powerful however is referred to as Automatic Programmed Tool or APT. At the outset of NC, the aerospace complex was the most advanced in the use of computers. Forseeing the gigantic task of developing a suitable language, the members of the Aerospace Industries Association (AIA) decided to pool their efforts, resources, and talents in the development of APT. In 1961 it was turned over to the Illinois Institute of Technology Research. Its membership was later broadened to other industries who supported APT by contributing $15,000 a year. The program is constantly being refined. Once a specific version of APT is two years old, it may be released to nonmember organizations.

At first APT was so powerful that only the more advanced computers could handle this language. Now however, after much development work, it is adaptable to smaller computers. This is done by storing the basic APT processor language not in the computer logic, but rather in a peripheral data-storage bank. With this system a minicomputer can take out of the data bank portions of the APT that are necessary for processing, make the necessary calculations, return the data, and insert new data into the computer logic. In computer parlance, this is known as the *overlay* system. The cost savings of this approach is considerable and is finding ever-growing acceptance.

The APT general-processor program has a language of about 300 "words." The words are all English-like and have specific meanings. None have more than six letters. Included are such terms as fedrat, cutter, from, point, plane, golft, gorght, circle, past, etc. Some are descriptions of geometric shapes or patterns and others are instructions regarding tool movements as explained in more detail later.

The part programmer must know the precise meaning and capability of each word and how it can be used before he can expect to write a part-program manuscript.

When a programmer writes an APT part-program manuscript, the first computer output will not be the program tape for use on the machine tool. The first output will be what is known as the "CL," which is the abbreviation for "center line" or "cutter line." This merely tells where the center line of a cutter, with a given configuration, will have to be located in a rectangular coordinate system to machine a part to the desired geometry as was shown in Fig. 7-32. An additional step, *post-processing*, is required to adapt the CL output to the particular machine-tool/control-unit combination which will be used to machine the workpiece. The tape that will be used is a product of the post processor.

***The Postprocessor.*** Perhaps no other term in NC is more misunderstood than that of the postprocessor. It is often thought of as a separate piece of hardware or a postmachining analysis.

The postprocessor is an individual program written to take into account the particular variables of the machine-tool/control-unit combination on which the part will actually be machined. Each machine has its own characteristics, such as indexing table, tool changers, swiveling head, etc. For example, there is no point in having tool-change instructions in a tape program if there is no automatic tool changer on the machine. Most machine-tool and control-unit builders have developed programs and postprocessors for their particular equipment.

***Other Processing Languages.*** Another well-known processing language for NC is ADAPT (Air Material Command Developed). This is a subset of APT. It is essentially limited to two-axes contouring. AUTOSPOT (Automatic System for Positioning of Tools) is a system for computer-aided NC programming. It is used for positioning and straight-cut systems and does not have contouring capabilities.

There are also general processing languages which are not quite so comprehensive as APT that have found their niche because they require less computer capability, less training, and fewer financial resources. They are usually designed for a specific type of machine tool or situation such as drills, lathes, or milling machines.

### The Apt Language and Computer Programming

As mentioned previously, APT is a three-dimensional language used to direct a tool in a well-defined path. There is far less arithmetic required in preparing a manuscript for a computer program than for a manual part since the computer performs the bulk of the mathematical calculations. However, the part programmer must be precise in the use of all words and notations as called for in the APT language.

The two primary surfaces that control a tool motion are called the *drive* surface and the *part* surface, as shown in Fig. 7-41. The tool continues until it encounters a third surface called a check surface. The distinction between the part and drive

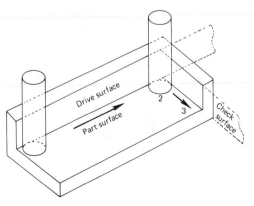

FIGURE 7-41. *The path of the cutter is guided by the* part *surface and the* drive *surface. The third surface is termed the* check *surface.*

surface is this: the surface that changes with each tool motion is the drive surface and the surface that serves as a continuing control surface in a series of tool movements is the part surface. In the illustration, as the tool moves to position 2, the check surface becomes the drive surface but the part surface remains the same.

*Elementary APT Grammar.* As with other languages, there are some rules that pertain to all occasions and others that can be followed for most occasions. The following are some of the basic APT rules.

*Statement,* A statement is like a sentence in English. It should contain enough information to stand by itself. As an example, the statement:

P6 = PØINT/XSMALL,INT ØF,L9,C3

defines a point labeled P6 which is at the intersection of line 9 and a circle having the smaller X coordinate as shown in Fig. 7-42.

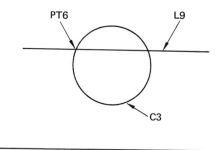

FIGURE 7-42. *Both symbols L9 and C3 must be defined before stating that PT6 is at the intersection of the circle (C3) and the line (L9).*

*Words.* The APT vocabulary consists of about 300 words. Some rules that govern APT words are given as follows:

1. It should not contain more than six characters (excluding blanks).
2. Words are always written in capital letters.
3. Blank spaces between the letters do not affect the word, e.g., GØTØ or GØ TØ mean the same.

There are three types of words: major, minor, and modal. Major words are those to the left of the slash and represent the geometry being defined, e.g., CIRCLE, PLANE, PØINT, etc. Minor words, to the right of the slash, describe the major word in detail, e..g, CENTER, RADIUS, INTØF, PARLEL, etc.

Modal words are those that have a continuing effect, that is, once entered in the program, they are effective from that point on to the end of the program or until a legal countermanding statement is made, e.g., MIST denotes the condition of coolant until another word like FLØØD or ØFF countermands it.

*Symbols.* Symbols are substitutes for geometric definitions and numerical values. They are also used for statement identification. Once defined, the symbols can be used to identify the defined variables. The programmer can assign any combination of alphanumeric characters as a symbol but the following rules must be observed:

1. Cannot be an APT word.
2. Not more than six characters.
3. One must be an alphabetical character.
4. No special characters can be used, e.g., + - * /, $ ( ).

Whenever a symbol is used it should be followed by an equals sign, e.g., P1 = PØINT/X,Y.

*Punctuation.* The special punctuation used in APT includes:

Slash (/) divides a statement into two sections, major and minor, e.g., FRØM/PTA, GØFWD L1,PAST,L2.

Comma (,) is used to separate individual terms in a statement such as APT words, symbols, or numerical data, e.g., PT1 = PØINT/1,2,0., LINE1 = LINE/ PØINT/1,2,0), (PØINT/2,2,0)

Parentheses ( ) are used to enclose a nested definition, to enclose function arguments in completing a statement, and to enclose subscripted values. An example of enclosing a nested definition by the use of parentheses is illustrated as follows:

PT1 = PØINT/1,2,0
PT2 = PØINT/2,2,0
LIN1 = LINE/PT1, PT2
LIN1 = LINE/(PØINT/1,2,0),(PØINT/2,2,0)

Equal sign (=) to the left of the equal sign is the accepted substitute for the geometric definition or numerical value.

Dollar sign ($ and $$) is placed at the end of a line and indicates that the statement continues on the following line. The double dollar sign tells the computer that the space from that point to the end of the line contains no more information for the computer processing. This allows the programmer to make notations at any logical and convenient location.

## Geometric-Definition Formats

Formats have been developed to define unique geometry without ambiguity as shown by the following examples for point, line, circle, and plane.

PØINT/$x,y,z$ or PØINT/$x,y$
LINE/$x,y,z,x',y',z'$ or LINE/$x,y,x',y'$
CIRCLE/XLARGE, *line*, XLARGE, *line*, RADIUS
PLANE/ *point*,PARLEL,*plane*

(See Fig. 7-43 for some examples of point definitions).

## Tool Definition

"Cutter" is used to describe a tool. For a flat-bottom tool there is just one parameter in the minor section, e.g., CUTTER/ *diameter*. For flat-bottom tools with edges rounded the format becomes CUTTER/ *diameter, corner radius.*

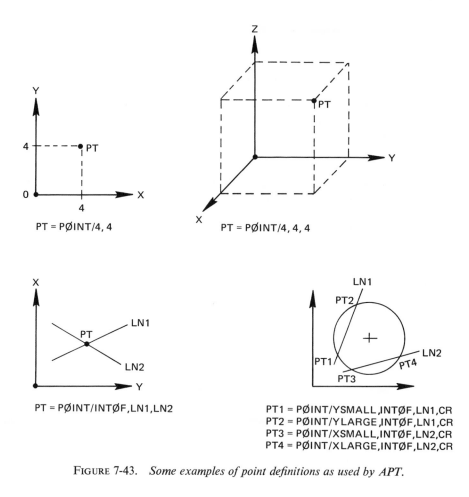

FIGURE 7-43. *Some examples of point definitions as used by APT.*

### Motion Statements

Regardless of whether the controlled motion is PTP or continuous path, the initial location of the tool must be specified before the plan of motion can be interpreted by the APT system, e.g., FRØM/SET PT— — — FRØM/ *x,y,z*. It should be noted that there is normally only one FRØM statement in a program and it comes before the motion statement.

*Point-to-point Motions.* PTP motions have no controlling surfaces involved. The tool is commanded to go directly to a particular point in space. There are only two motion commands, GØTØ or GØDLTA, which are used to activate these discrete positioning needs, e.g., GØTØ/2,2,0 — — — GØTØ/PT1, etc.

The commands, GØDLTA/2,2,0 and GØTØ/2,2,0 have entirely different meanings, as shown in Fig. 7-44.

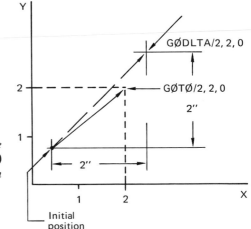

FIGURE 7-44. *The difference between the motion command of* **GØDLTA/2,2,0** *and* **GØTØ/2,2,0.** **GØDTLA** *is an incremental command which adds the values given to the initial position.*

**Continuous-Path Motions.** Unlike PTP, the continuous-path motion is always under the control of a pair of intersecting surfaces. To bring the tool from the initial point into contact with the controlling surfaces, the format is:

GØ/modifier, DS, modifier, PS, modifier, CS. Where DS, PS, and CS are the symbols of a previously defined drive surface, part surface, and check surface. The modifiers can be TØ, ØN, PAST, or TANTØ.

The modifiers are used to specify the exact relationship of the tool to the surface, as shown in Fig. 7-45.

The six APT words to direct the tool along the pair of surfaces (DS and PS) for contouring are GØLFT, GØRGT, GØFWD, GØBACK, GØUP, GØDØWN. The

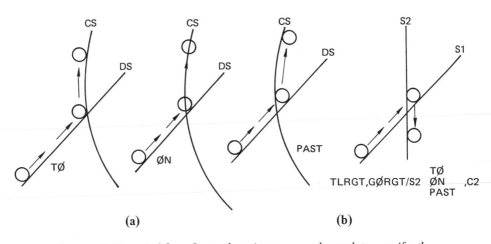

FIGURE 7-45. *Modifiers for tool-motion commands used to specify the exact relationship of the tool with respect to drive surface (DS) and check surface (CS)(a). The tool path is defined as shown at (b), C2 is the next check surface.*

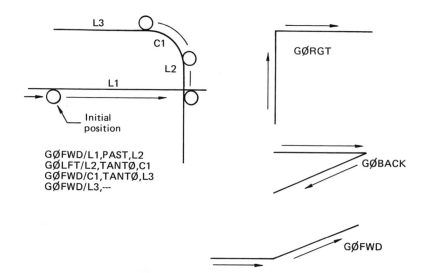

GØFWD/L1,PAST,L2
GØLFT/L2,TANTØ,C1
GØFWD/C1,TANTØ,L3
GØFWD/L3,---

FIGURE 7-46. *As motion statements continue, it is normal practice to make the check surface of the previous statement the drive surface for the next statement.*

direction for any statement is referred to as the tool direction and is given before the required motion. The motion statement has the general format:

GØ——/drive surface, modifier, check surface (as illustrated in Fig. 7-46).

*Examples of APT part program.* The program for the part shown in Fig. 7-47 is as follows:

```
        PARTNØ APT TESTI
        CLPRNT
$$          THE FØLLØWING ARE GEØMETRIC DEFINITIØNS
PL1     =  PLANE/(PØINT/0,4,0), (PØINT/4,0,0)
L1      =  LINE/(PØINT/0,0,0), (PØINT/3,0,0)
L5      =  LINE/(PØINT/0,0,0), (PØINT/0,3,0)
L2      =  LINE/PARLEL, L5, XLARGE, 3
L3      =  LINE/(PØINT/3,1,0), ATANGL, 135
L4      =  LINE/(PØINT/0,1,0), ATANGL, 45
C1      =  CIRCLE/XLARGE, L4, XSMALL, L3, RADIUS, 1
C2      =  CIRCLE/CENTER, 1, 5,1,0, RADIUS, .5
SETPT   =  PØINT/—1, —1, .5 $$ WELL ØFF THE PLATE
           CUTTER/.5 $$ USE END MILL
           FEDRAT/1 $$ 1" PER MIN
$$ THE FØLLØWING ARE MØTIØN STATEMENTS
           FRØM/SETPT
           GØ/TØ, L5, TØ, PL1, TØ, L1
           GØRGT/L1, PAST, L2
           GØLFT/L2, PAST, L3
```

```
GØLFT/L3, TANTØ, C1
GØFWD/C1, TANTØ, L4
GØFWD/L4, PAST, L5
GØLFT/L5, PAST, L1
GØDLTA/0,0,15 $$ UPWARD MØTION ØF TØØL
GØTØ/(PØINT/1.5,1, .3)  $$ CENTER ØF C2 AND .05" ABØVE PLATE
GØDLTA/0,0,—.3 $$ DRILLING ØPERATIØN
GØDLTA/0,0,+.3
GØTØ/SETPT
FINI
```

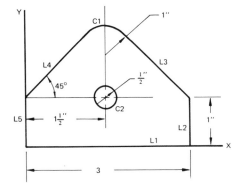

FIGURE 7-47. *An example part for APT programming. The part is 0.25 in. thick and the bottom surface of the plate is in the XY plane.*

You will note that the first part of the program defines the planes (in this case only one), the lines, the circle centers, the radius, and the set point. The motion statements are relatively easy to follow once these definitions have been given. For ease of reference, the common commands and definitions as used by APT are given in Table 7-3.

## Three-Dimensional Designing With NC

Three-dimensional designing with NC in mind can be considerably different than that of the conventional methods. The regular engineering drawing can be bypassed and in its place can be just a sketch, as shown in Fig. 7-48. On this sketch all the various geometric entities are identified. Once this data is identified for the computer, it will be able to generate the finished drawing as well as the mathematical models that can be used in making an analysis of the part.

The programmer plans the relationships between the various geometric entities. If a plane is tangent to a cone it must be defined as such to ensure tangency, or if a point is at the intersection of a circle and a line it must be so defined. Some common geometric shapes are shown and defined in Fig. 7-49. For example, to ensure appropriate tangency, the correct APT statement for the circles in respect to the two intersecting lines is

CIRCL1 = CIRCLE/XLARGE,LINE2,YSMALL,LINE1,RADIUS,R

TABLE 7-3.   *Common APT Commands and Definitions*

| Command | Definition |
|---|---|
| TLØN | Tool or cutter is positioned on surface |
| TLRGT | Tool or cutter is positioned to right of surface |
| TLLFT | Tool or cutter is positioned to left of surface |
| FRØM | Reference position |
| GØ | Positions tool relative to three surfaces |
| GØFWD | Tool is to move forward at next surface |
| GØRGT | Tool is to move right at next surface |
| GØLFT | Tool is to move left at next surface |
| GØDLTA | Tool is to move in the X, Y, Z directions incrementally |
| GØBACK | Reverse direction of motion along a surface |
| GØUP | Go in the positive Y-direction along a specified surface |
| GØ DØWN | Go in negative X-direction along a specified surface |
| GØTØ | Tool is to move to a specified location |
| INDIRV | Tool surface search direction vector |
| INDIRP | Tool surface search direction point |
| TØ | Tool to be positioned to surface |
| ØN | Tool to be positioned on surface |
| PAST | Tool to be positioned past surface |

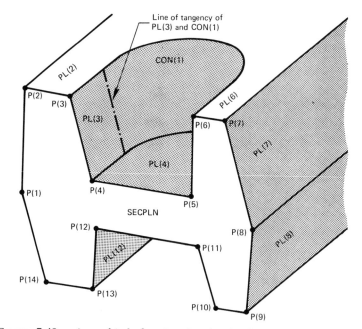

FIGURE 7-48.   *A new kind of engineering drawing is emerging.  It consists of a sketch with all the key points, planes, and geometric surfaces identified by number.  From this an NC program can be written which in turn can provide the drawings and math models for analysis.  Courtesy* Machine Design.

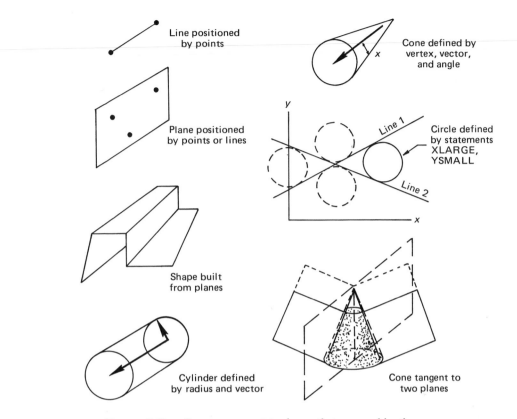

FIGURE 7-49.  *Common geometric shapes that are used by the programmer when designing for three-dimensional NC.  The surfaces are quite easily defined in APT language.*

The modifiers XLARGE and YSMALL distinguish among the four possible positions of the circles placed tangent to the two lines.  XLARGE,LINE2 indicates that the circle is to be on the high side of line 2 and YSMALL,LINE1 indicates it will be on the low side of line 1.

From a practical standpoint, programs must be limited to simple routines such as those for planes, cones, and cylinders.  These routines are usually adequate for many industrial parts, including forging dies and sheet-metal forming dies.  For intricate dies, however, it is often easier to build a model and trace it with the aid of a scanner or use the sculptured surface version of APT.

### Direct Numerical Control

As mentioned previously, the computer is one of four media used to control the NC machine.  This method is popularly known as direct numerical control (DNC) since it bypasses the common tape or card input.

To be useful in NC programming, the computer must first be equipped with basic intelligence, or what is often referred to as "computer language" or "processor language." This may be explained by a simple illustration. If you are asked to find the circumference of a circle, you would multiply $\pi$ times the diameter. The computer is ignorant of this fact. It is only when this knowledge is incorporated into a processor program that the computer will be able to calculate the coordinate positions a cutter must follow to machine a circle. If this example is extended to all circumstances surrounding a machine-tool operation, it is possible to understand why it sometimes takes years to develop processor languages. Each bit of information is stored in the logic space of the computer. Thus if 15,000 words are needed for the language, a computer with 16,000-word storage capacity would have only a 1000-word capacity for part programming, which in all probability would not be enough.

DNC is defined by EIA as "a system connecting *a set* of *numerically controlled machines* to a common memory for part-program or machine-program storage with provision for on-demand distribution of data." The key words are "set (meaning more than one) of numerically controlled machines."

DNC offers several operating advantages. First of all, it bypasses the tape reader, which is the most likely trouble spot of the MCU. Secondly, the computer is much more accessible for revision and editing or for quick and easy interaction between the programmer and the machine tool. As an example, the programmer observes the machine as it goes through the various steps of the process. He notices one or two wasted motions. This often comes about through the lack of a specific instruction, in which case the tool retracts all the way to the "home" position. The programmer can then examine the readout and find the unwanted motion, "return to home." The correction is made on a teletypewriter mounted beside the control unit. The instruction is given to delete the stated number. This can all be done while the program is in action so that no time is lost removing tapes, repunching, rewinding, reloading, and researching to find the pickup point. Thus the modified program is ready to go on the next workpiece. With the tape-control system, it would probably not be corrected until the next shift, next day, or next week.

One such improvement may result in a savings of only 5 % of the machining time for a particular workpiece, which is not significant in itself. However, the constant day-to-day improvement of programs for a whole metal-working plant can be very significant.

DNC does put a premium on programmers and supervisors who have a very clear idea of what is to be accomplished and the optimum method of programming it. The initial cost of DNC is high and therefore to make it economically feasible it must be supported by those who understand the whole concept of machining and NC thoroughly.

## Minicomputer Machine Control

One of the most important developments in NC control for machine tools has been the use of the minicomputer. As the name implies, the minicomputer is a scaled-down version of the conventional computer, not only in capability but also in cost.  The main limitation of the minicomputer is its relatively small memory system or logic-processing capability.  The basic unit of logic is the "word" which is composed of "bits."  A "bit" is the charged or discharged binary notation usually accomplished by means of magnetic film or ferrite cores, as shown in Fig. 7-50.  A minicomputer may have a 4000- or 8000-word memory capability, whereas a "medium"-sized computer may have up to approximately 64,000 words and a larger computer will jump to about 128,000 to over a million units.  The term "4000- or 8000-word computer "is seldom if ever used.  The terminology that has grown up with the computer is "4K," "8K," "16K," etc., which is a quick way of stating "kilo" or thousand.

FIGURE 7-50.  *A typical computer word is composed of nine binary bits which are usually ferrite cores as shown here.  Each bit is capable of existing in either plus or minus electrical polarity.  The pattern of binary notations within the word determines the symbolic value of the word whether it be numeric, alphabetic, or a symbol such as a period, comma, slash, etc.  Each bit can change its polarity as often as 20,000 times per second.*

***Hardwired Controllers vs. Minicomputers.***  NC machines that do not utilize a computer are referred to as *hardwired controllers*.  With the advent of the minicomputer, some firms have found it more economical to bypass the tape control.  This has been true when a minicomputer can be used to operate at a ratio of 2.5 : 1 or greater.  That is, by the addition of some peripherals, a single minicomputer can operate two or more machine tools whereas a conventional hardwired controller would be needed for each machine (Fig. 7-51).  The implication is that the cost of three hardwire controllers exceeds the cost of a single minicomputer.

A more recent addition to minicomputer technology has been a microprogrammable computer.  The most expensive part of the minicomputer, the core memory, is eliminated.  It is based on a "read only" memory.  This is an inexpensive type of memory which accepts and transmits data but which cannot be altered or erased through operator error.  The significance of the microprogrammable minicomputer is that its price performance ratio is superior to any other programmable control.

The traditional tape-operated contouring controller (hardwired) contains the electronic interpolation logic for contours.  When the minicomputer is used, the interpolation becomes part of the software, alterable by programming or reprogramming as the need for modification arises.  The hardwired controller, in much reduced form, is retained in a minicomputer control system.  The printed-circuit boards

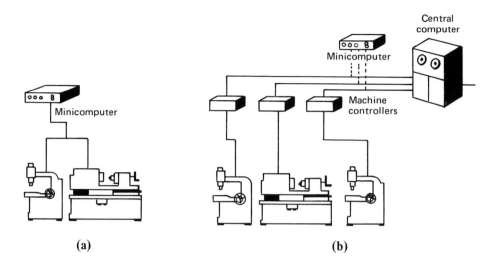

FIGURE 7-51. *The production of two or three machines can be controlled directly by a minicomputer, as shown at (a). In larger, more elaborate systems programmable machine controllers are used for each machine and these may in turn be interfaced with a minicomputer or, in extremely large systems, a central computer.*

associated with the interpolation function are all transferred to the software. This largely eliminates the concern over control obsolescence.

It now appears that in the near future the minicomputer will undergo vast development and the prefix " mini " will apply only to its physical size. Super-high-density, low-cost memories are being made to replace the expensive core memory now used.

The computer is now being used to advantage even in short programs that were formerly done manually. Of course in some simple cases the computer is only of minor help and in others it is almost a necessity even for positioning control, as shown in Fig. 7-52.

The question often arises as to the skill levels required to do part programming on a computer. For relatively simple programs done on a small computer with a simple processor language, a person with a knowledge of shop fundamentals may do very well with a minimum of training. On the other hand, a five-axis APT program for a complex part requires a knowledge of calculus and several weeks of intensive training in APT programming.

## Time Sharing

Time sharing is exactly what the term implies, sharing the time of a computer. The user neither owns nor leases the computer, but buys time as he needs it on a per-second bases. Many firms offer this service, along with consulting, on a fee

FIGURE 7-52. *In the point-to-point random pattern shown at (a), the computer is of only minor value since no common value exists between the different positions. If the random pattern is repeated one or more times, a computer can be of assistance. With a rational pattern of any kind, such as that shown at (b), the computer can be a real time saver. The location of the first position, the type of pattern, and the distance between each position is stated and the computer calculates the rest.*

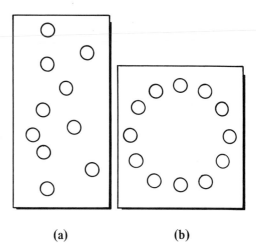

(a)  (b)

basis. Access to the computer is available either by direct telephone or teletype or both. Thus, a part-program manuscript is transmitted by wire to a computer which will immediately generate the information and transmit it directly to a tape-punching unit in the user's plant. Just as in DNC, one version of time sharing allows the programmer to correct errors as they occur. If, for example, the coordinate data sent to the computer is not within range of the machine slides, the programmer is so advised immediately. He then types the correction and continues to program from this point. This is known as *conversational* programming.

In another variation, known as *remote-batch* programming, the entire program manuscript is entered in the user's teletypewriter terminal. In the unlikely event that the remote computer, in comparing the manuscript with the stored machine-program file, finds the manuscript error-free, it immediately postprocesses it and makes a ready-to-use punched tape or card deck. If errors are detected in the manuscript, the remote computer generates a diagnostic listing through the user's terminal. Using this list and other diagnostic listings, the programmer continues to cycle corrections through the terminal. The computer will not begin postprocessing to create a machine tape until all corrections have been made.

### New Program Languages

New programming languages are constantly being developed. The goal is to be able to talk to a computer, give it a description of the part to be machined and the machining operations required, and let it do the rest. A programming language termed Part-Edge Programming (PEP), developed at the University of Wisconsin, has made considerable progress in this direction.

*PEP.* The PEP language was developed to utilize a 4K minicomputer in direct conversational mode. To write a program, the computer asks the question and the programmer answers it. This process is continued until the whole program is completed. For example, the computer asks, "What is the cutter radius?" Or, in the actual program, R will appear. The operator may designate 0.25 for example:

$\underline{R}$(IN.) = 0.25

*Note:* The programmer's response is always underlined.

The purpose of this section is not to describe the whole program language but merely to present enough to make the reader aware of developments that may become commonplace in tomorrow's NC world.

**Part Specification Commands.** In addition to the radius of the cutter, there are four other commands used for cutter-path definitions. They are S, L, C, and Z. It should be pointed out that with the R command is an automatic cutter compensation for the cutter path. Thus if a roughing and a finishing cut is desired, a slightly smaller cutter radius can be designated for the first cut followed by the actual size of the cutter for the second cut.

*Feedrate specifications: S.* The feedrate is given in in. per min for the X-Y plane and remains in effect until a new S command is given. This command has no effect on the Z-axis, which remains fixed at 1 ipm, hence it is termed a 2.5-axis part-programming language. The printout is $\underline{S}$(IMP) = $\underline{f}$.

*Motion specifications: Z.* This command specifies motion (total distance) along the Z axis at 1.0 ipm. The printout is $\underline{Z}$(IN.) = $\underline{dz}$. The command may be positive or negative: + = spindle upward, − = spindle downward.

*Linear interpolation specification: L.* Straight-line motions in the X-Y planes are specified by the command L. Data for the L command will be obtained directly from the part print.

The modifiers TØ/PAST may be required, depending upon the sequence of commands. In this case, PEP will respond first with a question for the TØ/PAST modifier before requesting data for the line dimensions. The printout will be:

$\underline{L}$
TØ/PAST? (T or $\underline{P}$)
DX = $\underline{dx}$; DX = $\underline{dy}$

**PEP Program Examples.** The dimensions of the part (shaded portion in Fig. 7-53) are specified and the dimensions for the cutter path are calculated by PEP. Note that after the command for a circle is given, $\underline{C}$, the computer asks for the intersection point and whether it is clockwise or counterclockwise. It also asks if the cutter should be on the inside or the outside. Knowing these few simple facts, the cut can be made around the specified radius and proceed with the straight line again.

Shown in Fig. 7-54 is a simple part program using the modifiers TØ/PAST. The cutter is initially touching the beginning of the first line. PEP automatically modifies the dimensions of the cutter path according to the dimensions of the lines and the TØ/PAST modifier so that on completion of the first L command, the cutter is *past* the second line and touching it. The dimensions of the cutter path along the second line are also modified by PEP so that the cutter ends up at the end of the second line as shown. A pause occurs at the corner between the two straight-line specifications on the order of 0.2 sec to enable the machine-tool axis clamps to turn on or off before initiating motion.

These examples may leave some unanswered questions in the reader's mind, but they serve to illustrate the relative simplicity of this program language. Refer to the bibliography for further reading on this subject. The PEP language may also be used with interactive graphics as described in the next section.

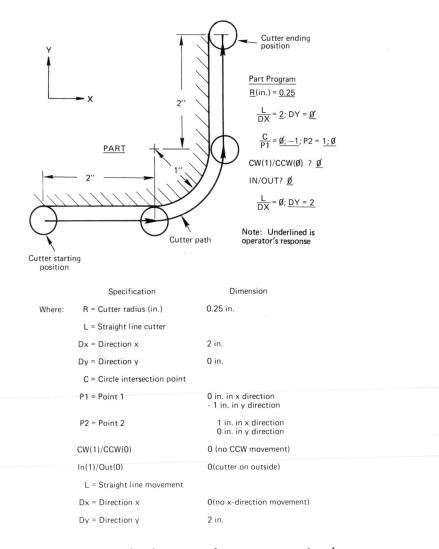

Where:

| Specification | Dimension |
|---|---|
| R = Cutter radius (in.) | 0.25 in. |
| L = Straight line cutter | |
| Dx = Direction x | 2 in. |
| Dy = Direction y | 0 in. |
| C = Circle intersection point | |
| P1 = Point 1 | 0 in. in x direction<br>- 1 in. in y direction |
| P2 = Point 2 | 1 in. in x direction<br>0 in. in y direction |
| CW(1)/CCW(0) | 0 (no CCW movement) |
| In(1)/Out(0) | 0 (cutter on outside) |
| L = Straight line movement | |
| Dx = Direction x | 0 (no x-direction movement) |
| Dy = Direction y | 2 in. |

FIGURE 7-53. *An example of a part and part program using the conversational Part Edge Program language.*

561

### Interactive Graphics

Interactive graphics involves the use of a computer and a cathode-ray tube (CRT) display as a drafting design aid. The user may "draw" a pattern on the CRT by moving a cursor on the screen and using a keyboard to call up specific geometrical shapes stored in the computer memory. The designer can build new shapes, move,

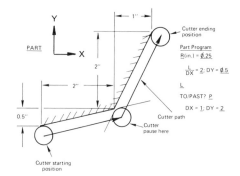

FIGURE 7-54. *An example of a part and part program using PEP, with the modifiers* **TØ/PAST**.

FIGURE 7-55. *The* light pencil *is used to pinpoint some area of the design either for enlargement or for dimensional change. The pencil functions through a conductive screen of thin, transparent tin oxide placed on the glass plate in front of the display tube. When the pencil touches the screen, it detects a voltage that is proportional to the distance from the top and left of the screen.*

expand, or otherwise alter the design and issue various instructions to the computer, plotter, or other peripherals.

Results are displayed instantaneously and may be amended as desired, using one or another of the input devices to delete and modify the data. One such input device, a *light pencil*, is shown in Fig. 7-55. The information may then be stored on computer tape for subsequent generation of scale drawings, numerical control tapes, etc.

A subroutine may be used to generate dynamic sectional properties of the geometric constructions. Thus a section may be looked at and at the same time the properties may be displayed with calculations. Properties displayed may be axes of maximum and minimum stress; centroid, shear-flow, and bending moment; weight per linear foot; surface area; volume; moment of mass; and radius of gyration. Construction parameters may be modified, in which case the effects of changes are displayed immediately.

Tool paths can be called for on the CRT screen. These paths are based on the tolerances given, scallop height, and bonding surfaces. Scallop height refers to the height of the ridges left between adjacent paths made by a ball-nosed end mill (Fig.

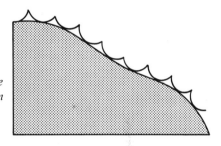

FIGURE 7-56. *The scallop height is the ridge left by a ball-nosed end mill when making adjacent paths.*

FIGURE 7-57. *The tool path is displayed on the CRT screen. In this case, a probable error is noted and further checked by calling for other views. Courtesy Tektronix, Inc.*

7-56). The designer defines the cutter to be used and the surfaces which the tool will move tangent to, upon, or past. The system then automatically generates the cutter paths, travelling in the direction indicated by the user (Fig. 7-57). If an error is noted in the display, the tape block in which the error was noted can be rerun and corrected.

## ADVANTAGES AND LIMITATIONS OF NC

The benefits of NC are many and are briefly summarized here as it affects various phases of manufacturing, tooling, engineering, and production control.

Effects on manufacturing:

1. Families of parts can be programmed with a minimum of time and effort.
2. Improved quality and closer tolerances.
3. Reduction in costs due to less variation in cutting speeds and feeds.
4. Fewer scrap parts.
5. Mirror-image parts can be made easily.
6. Reduced setup time.
7. More actual running time, 80 to 85 % vs. 20 to 40 %.
8. Power consumption is leveled out due to more continuous running.
9. Reduction of direct labor.
10. Ability to handle variables in raw material more easily.
11. Skills built into tape programs are retained through personnel changes.
12. NC is used extensively for the fabrication of electronic components where it is not uncommon to go to a large number of parts because they are often quite complex. A circuit board, for example, may require as many as 5000 drilled holes spaced at small fractions of an inch. Either a drilling machine with thousands of spindles or a transfer line with thousands of stations is clearly impractical. Thus NC provides the best method for handling these parts whether the production quantities are large or small. To avoid confusion on this point, NC machines are generally considered best for medium- and short-run production on metal parts but may be considered for long-run production on electronic parts.

Tooling used is reduced by:

1. Using an "offset adjustment" to compensate for tool wear or sharpening.
2. Less time needed for tool changing.
3. Being able to use more standard tooling and less specials.
4. Less cutting-tool storage.
5. Fewer and simpler jigs and fixtures.

Engineering cost is reduced by:

1. Less tool-engineering time, tool-engineering records, tool drawings, process sheets, etc. (printing costs).
2. Better control of weight-to-strength ratios in the finished products.

3. Ease of part-design change. Changes can be made via the visual display or, in the case of DNC, minor changes can be made by conversational programming.
4. Savings in setting and maintaining standards.
5. Savings in supervision.
6. Better product due to the flexibility of the system.
7. Improved estimating accuracy.

Effects on production control:

1. Scheduling and routing can be done more efficiently since estimated times are more accurate and many operations can be done on one machine such as the machining center shown in Fig. 7-58.
2. Less inventory is needed. Parts can be produced on relatively short notice.

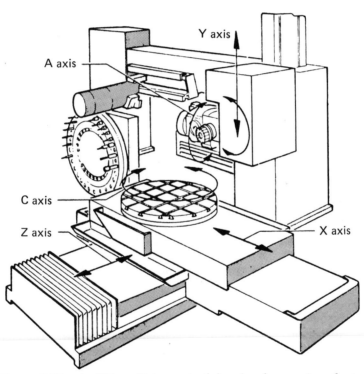

FIGURE 7-58. *An NC machining center (a) and a close-up view of a tool changer (b). Courtesy Sundstrand Machine Tool Division.*

Limitations:

1. Despite all the advantages listed, NC is not generally considered feasible for high production. Large-volume parts are usually more economically produced on specially built automated machines. The ideal lot size is usually in the range of 50 to 1000 pieces. Of course this can vary since special machines are often

costly even though they are made up of many standard components. A simple break-even formula can be used to provide an answer expressed in terms of production volume as follows:

$$P_A + \frac{T_A}{n} = P_B + \frac{T_B}{n}$$

Where $P_A$ and $P_B$, respectively, are the operating costs per piece for alternatives A and B; $T_A$ and $T_B$ are tooling and setup costs for each alternative; and $n$ is the production quantity.

*Example:*

|  | Cycle Time (min) | Operating Cost Per Piece | Tooling and Setup Cost |
|---|---|---|---|
| Special machine | 6 | $1.50 | $30,000 |
| One NC machine center | 24 | 4.00 | 1,000 |
| Three NC machining centers | 8 | 2.20 | 2,500 |

$$\$1.50 + \frac{\$30,000}{n} = 4.00 + \frac{1000}{n}$$

$$n\left[1.5 + \frac{30,000}{n} = 4.0 + \frac{1000}{n}\right]$$

$$30,000 - 1000 = 4.0n - 1.5n$$

$$29,000 = 2.5n$$

$$n = 11,600$$

Therefore, if there were more than 11,600 parts a special machine may be considered. However, if three NC machining centers are available, similar to that shown in Fig. 7-58, the picture will change considerably. The NC machines may then be considered competitive to the special machine up to 39,500 parts. In this case, it is assumed that one person will supervise all three NC machines.

2. The initial cost of NC equipment is considerably higher than for conventional machines.
3. NC is more complex than the conventional machine. The uninitiated manufacturer suddenly finds himself or herself immersed in a world of electronics programming.
4. Many jobs are quite complex and a higher degree of training is required for programming, operation, and maintenance of NC machines.


## COMPUTER-AIDED MANUFACTURING (CAM)

About 200 years ago, John W. Wilkinson developed the first of the present line of mechanically powered tools. With his water-powered boring mill, he was able to help James Watt bore the cylinders of the first steam engine. At that time the wonder of manufacturing was the ability to multiply man's muscle. Today the wonder of modern manufacturing is the control behind the machinery or what may be termed the "nervous system" of manufacturing.

CAM is a term used to denote the use of the computer in all manufacturing and related operations such as inventory control, purchasing, production, testing, packaging, and shipping. DNC becomes an integral part of the CAM system by controlling production machines. A typical production system, in very basic terms, decides *what* is to be manufactured, *when* it is to be manufactured, and *how* it is to be manufactured.

***Determining What and When.*** What is to be manufactured is based on forecasts of end-product sales and end-product backlog. Forecasts also include product requirement for service and repair. The data for the computer is obtained from bills of material which contain component and subassembly requirements used to build each end product. These are broken down into raw material, individual parts, subassemblies, assemblies, and packaging. Complementary computerized data is also obtained that has the required manufacturing routings operations and procurement lead times. All these are given to the computer at various intervals to help determine the future production calendar.

The computor combines product specifications with existing inventories and work in process to determine what and when to build. Planned component orders emerge from the system as an immediate response. Answers are instantly generated to problems such as a sudden increase in requirements or a "negative availability" of a given component in some future period.

***Determining How.*** The how of computer-aided manufacturing is based on computer-aided design and DNC of machine tools, including programmable controllers (PC, discussed later) and materials handling.

Industry is now at the point where it is possible to digitize the information from a designer's model and go directly to an NC program which will machine a die that will make the desired component. As another example, if the parts are castings that require more than one operation, they may be scheduled for an NC machining center. Since this machine is very versatile, in many cases all the machining operations will be accomplished without having to route the part from one machine to another.

The advantage of being able to make dies by NC is the relative speed with which they can be cut and finished compared to conventional methods. There is also the advantage of excellent match of male and female surfaces since both are obtained from the same computer-derived data.

A time savings results at die assembly and tryout due to the precise mathematical data given the NC machines from the computer.

The problem of moving materials out of storage to the various work stations for machining operations and then along the route is also taken to the computer. One fully automated system now in use is that of high-rise storage controlled by a *dedicated* minicomputer. ("Dedicated" in this sense refers to the assignment of the minicomputer to one particular task or group of related tasks.) The minicomputer controls the conveyor system and stacker cranes used to store and retrieve a pallet of material. The system also maintains a continuous inventory record on items stored in the facility. The material can be delivered to the work centers on an automatic replenishment basis exactly when it is needed. The timeliness of the delivery is based on an accurate feedback of the real status on the shop floor. Here DNC forms an important link between computerized numerical control (CNC) and planning, as shown by the diagram (Fig. 7-59). CNC is a dedicated stored-program computer with a read-write memory. ("Read" is the retrieval of information from memory and "write" is the act of storing information in memory.) Thus the CNC with its software logic is capable of handling a wide variety of tasks including continuous feedback from the machines operating under DNC. The changing status of the tool and the workpiece is kept under continuous control or, as it is often referred to, under "adaptive control."

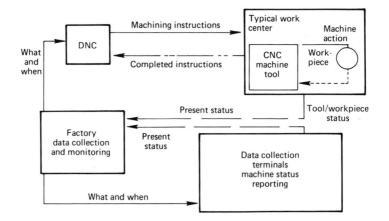

FIGURE 7-59. *As an example of a modern machining facility, each machine is under DNC and computerized numerical control (CNC). The DNC transmits the next machining instruction when it receives feedback of the previous instruction completion. Overall planning of what to do and when to do it requires feedback of the present status of the overall manufacturing effort.*

***Programmable Controllers (PC).*** An important part of automatic control of manufacturing operations is now done by programmable controllers. So important is this area that it is viewed as having an even brighter future than NC and may overlap and replace NC's in a number of areas.

Programmable controllers are essentially sequential-logic devices whose output is dependent on a number of independent inputs. For example, a relay coil will function depending on the status of several normally open or normally closed series/parallel contacts, as shown in Fig. 7-60.

Functions ordinarily handled by the PC are timing, counting, and arithmetic capabilities and memory storage. The memory is programmed so that the controlled machine follows a command sequence.

*Timing* may range from a simple, pneumatic, time-delay relay with a fixed time lapse to more elaborate continuous delays for both pick-up, drop-out, etc.

*Counting* is available on most PC's, typically with three-decimal-digit capacity. The counter accumulates the number of input-contact closures. When the accumu-

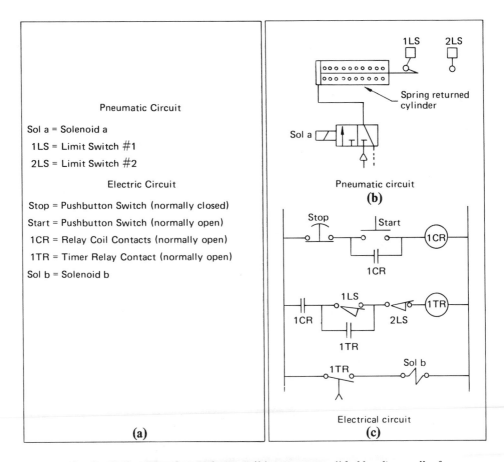

FIGURE 7-60. *The electrical circuit (b) represents a "ladder diagram" of the components and the sequence of operation required to operate the pneumatic circuit (a). As the normally open pushbutton switch is closed, the second normally open relay contact is closed. An untimed hold will keep the contact closed long enough to advance the piston rod to 2LS. As switch 2LS is actuated, it will interrupt the circuit coil of 1TR causing solenoid (a) to de-energize and retract the piston. Upon full retraction, switch 1LS will be actuated again. After a time delay, the piston rod will again extend.*

lated count equals or surpasses a preset count, a logic output is provided. A separate input then resets the counter.

*Arithmetic capabilities*, which are usually optional, include addition, subtraction, and comparison. One of the input quantities may be binary-coded decimal signals from external selector switches or electronic measuring instruments. Another input may be the result of some previous arithmetic operation or perhaps an accumulated time count. The arithmetic-operation result can be on an output display, another arithmetic function as an imput variable, a timing or counting function as a preset quantity, etc.

***PC Memory Types.*** The most widely used memory for programmable controllers is the Read-Write Memory (RWM). It is made up of small doughnut-shaped ferromagnetic cores with wires running through, as was shown in Fig. 7-50. Reading the memory is destructive, that is, it destroys the charge pattern. If the information is to be retained, it must be rewritten. No electric power is needed to maintain the information stored and therefore it is not lost during power failure. Such ability is termed *nonvolatile*.

The use of Read-Only Memory (ROM) has increased exponentially in recent years. These devices permit stored digital data to be read out of the memory, but will not accept new data, i.e., they cannot be altered by a *write* command. While

FIGURE 7-61. *Shown is the heart of the Electrically Alterable Read-Only Memory (EAROM), a plated-wire memory plane. The unit can be preprogrammed electronically through a write interface of the same type as used to program read-write memory. Courtesy Memory Systems Inc.*

this may at first appear to be a handicap, it may also be a big advantage. It is such an advantage that ROM's could change the way industry has been conditioned to think about control-system design.

If in any situation in which repetitive sequence commands are read out of a control unit to operate a machine or a process and the operator unintentionally alters the controllers contents, he can endanger or destroy the control sequence. Here the ROM offers protection against the accidental modification of the control sequence.

The Electrically Alterable Read-Only Memory (EAROM) now offers the advantages of both the RWM and ROM. EAROM is based on a plated-wire technology for the memory plane (Fig. 7-61). It is nonvolatile and it permits repetitive readout of stored data without destruction of the memory contents. Unlike conventional ROM, EAROM can be programmed in a variety of ways. Like the RWM, it can be programmed through a conventional *write* interface; then, with programming completed, the *write* circuitry is switched off or even physically removed. The nondestructive readout can then be executed at high speeds indefinitely with full memory protection.

***Programming the PC.*** One of the main advantages of a programmable controller over a computer is the ease of programming. No complex programming instructions written in special languages (software) are needed. Programs are easily entered into the controller memory with a variety of program entry devices (Fig. 7-62). Generally

FIGURE 7-62. *A programmable-controller entry device with a CRT display (a). This type allows programs to be entered in relay-logic format (b). Courtesy Industrial Solid State Controls, Inc.*

portable for onsite programming, many of these devices also serve as program and machine-operation monitors.

Some programming devices have displays in the form of ladder and logic diagrams (Fig. 7-63). A ladder diagram of high complexity can be stored in the

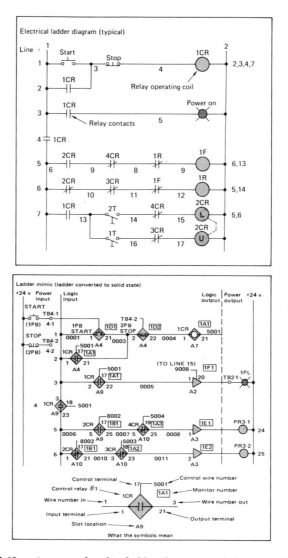

FIGURE 7-63. *An example of a ladder diagram and the equivalent as a ladder mimic. The mimic ladder substitutes solid-state devices for the electromechanical ones and puts them in a smaller panel. The advantage is that it is not necessary for the programmer to know computer logic. Many controllers are now made that use solid-state logic schematics. Courtesy* Product Engineering.

memory of the controller. All that is necessary to completely change the functions is to insert a new program, which may be in the form of a tape. This is termed *sequential memory* [Fig. 7-64(*a*)]. In some designs, the operator can electronically alter the device arrangements by pushing appropriate buttons on the front of the

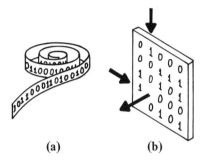

FIGURE 7-64. *The sequential memory (a) must be taken in the order it comes from storage. The random-access memories allow picking out the nuggets desired (b).*

(a)              (b)

FIGURE 7-65. *Examples of programmable controllers. Courtesy Square D Company.*

panel, termed *random-access memory* (*b*). Some even display the ladder diagram on a cathode-ray tube when needed for monitoring. The operator does not change any of the wiring, this is done only by specialists when malfunctions occur.

Since programming by use of a panel is slow and tedious, an interface is provided between the panel and any peripheral devices such as paper punches and readers, magnetic-tape devices, etc. Using the interface, a program which is known to be good can be dumped from the PC memory onto paper or magnetic tape. The program can then be loaded quickly whenever it is needed. Examples of programmable controllers are shown in Fig. 7-65.

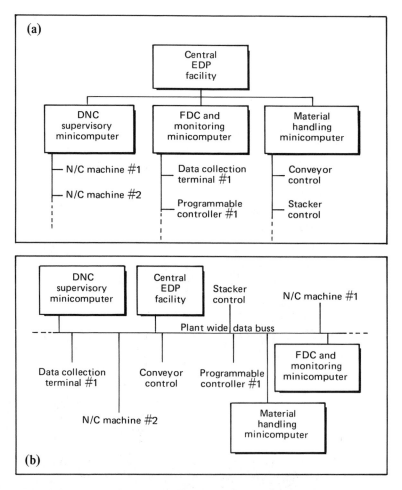

FIGURE 7-66. *A schematic comparison of centralized and decentralized computer control. Centralized control (a) is characterized by physical and logical relationships. Decentralized control (b) is built so that any element can communicate directly with any other element via a data bus. Courtesy* Automation.

## CAM and Future Developments

In general, it would appear that CAM is a centralized system where a large computer is used for all important functions. However, the advantages of decentralized control are becoming apparent. A significant trend is increasingly toward the concept of accomplishing a task by the lowest-level computer (or PC) capable of handling the operation, thus matching the best cost effectiveness and quickest response.

Shown in Fig. 7-66 is a comparison of a centralized and a decentralized CAM system. In the centralized system (*a*) is the traditional concept. The decentralized concept (*b*) is characterized by a communication system that operates on a plant-wide data bus. The data busses are able to transfer digital data at high rates, 5000 to 10,000 times faster than a telephone circuit. This makes it possible for any one unit to contact any other unit, eliminating the need for centralization. Also, computer technology has progressed to the point where individual computer support is both practical and preferable for many manufacturing problems.

## PROBLEMS

**7-1.** What are the essential components needed to make a simple automatic pneumatic circuit?

**7-2.** What function does a flow-control valve serve in a pneumatic or hydraulic circuit?

**7-3.** What are some advantages of a pneumatic circuit over a hydraulic circuit?

**7-4.** Could a turbine generator in a power station be classified as a transducer? Why or why not?

**7-5.** What would the actual index and dwell time be for a five-station Geneva drive driven at 1 rpm if the time ratio is 108° index and 252° dwell?

**7-6.** What are some disadvantages of rotary-table indexing machines?

**7-7.** Why is the rotary table often used for automation despite its disadvantages?

**7-8.** What type of automation system is best for operations that vary in length?

**7-9.** What are some elements in the vibratory hopper that make it possible to orient parts automatically?

**7-10.** Company ABC decided to buy three robots at a total cost of $70,000. Each robot will be used on two shifts. Preliminary studies show that each robot can do the work of an unskilled worker who is paid $4.00/hr (including fringe benefits). The plant runs 40 hrs/wk per shift but shuts down two weeks for vacation and eight holidays. How long would it take to recover the investment? Yearly maintenance = $200./machine.

**7-11.** What would be the coordinates of a point in quadrant 3 that is 2.125 units from the $x$ axis and 3.250 units from the $y$ axis?

**7-12.** A point having a plus sign for the $x$ coordinate could be in either of what two quadrants?

**7-13.** The coordinates of a point are $x = +4.250$, $y = 3.500$. A second point lies 6 units directly to the right of this point and down in the $-y$ direction a distance of 4.750 units. What are the coordinates for the second point?

**7-14.** Suppose the $xy$ plane is 3.000 in. above the work table. How many inches above the work table would a point be that has coordinates $x = -10.000$, $y = +7.750$; $z = -2.125$?

**7-15.** (a) Considering the *xy* plane only, point *A* is located at coordinates $x = +8.000$; $y = +4.000$. Point *B* is located at coordinates $x = -2.000$; $y = -4.000$. What are the incremental *x* and *y* distances when moving from point *A* to point *B*? (b) Show points *A* and *B* on coordinate graph paper, label each axis.

**7-16.** How would the following X-Y coordinate numbers appear on tape? (a) $x = +6.250$, (b) $y = -12.500$, (c) $x = -.005$, (d) $y = +.010$.

**7-17.** What would the feed-rate code be for the following: (a) 10 ipm, (b) 125 ipm, (c) 1.2 ipm, (d) .010 ipm.

**7-18.** What would the rpm code be for the following: (a) 650, (b) 2000, (c) 225, (d) 60.

**7-19.** Refer to Fig. P7-1 and identify the coordinate points. (Note + signs need not be shown.)

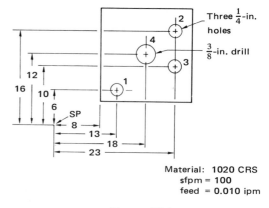

Material: 1020 CRS
sfpm = 100
feed = 0.010 ipm

FIGURE P7-2

**7-21.** Write a part program for cutting the form shown in Fig. P7-3 with a 0.25-in. diameter end mill. The material is Styrafoam and may be cut at a feed rate of 20 in./min. Use the manuscript form as shown in Fig. 7-40. Assume cutter offset is not available. The rpm used is 2000.

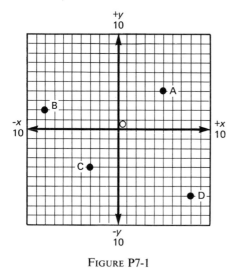

FIGURE P7-1

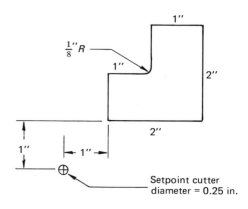

FIGURE P7-3

**7-20.** Write a part program in absolute form for drilling the four holes shown in Fig. P7-2. Use the manuscript form shown in Fig. 7-40.

**7-22.** A bolt circle is to be drilled as shown in Fig. P7-4. The set point is assumed to be in the center of the 8 in. diameter circle. Twelve equally spaced holes are to be drilled. List *x* and *y* coordinates for each of the hole.

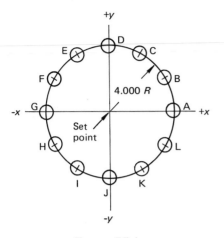

FIGURE P7-4

**7-23.** What would the distance be for a 0.5 in. diameter cutter to travel from point *A* to point *B* as shown in Fig. P7-5?

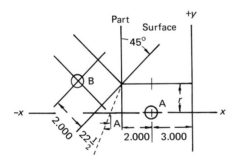

FIGURE P7-5

**7-24.** Tell which is the *drive* surface and which is the *check* surface when the cutter is at the positions shown in Fig. P7-6. Ex. Position A Drive Surf = 1, Check Surf = 2.

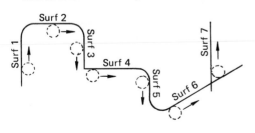

FIGURE P7-6

**7-25.** Write the APT *motion* statements for the sketch shown in Fig. P7-7. Start at the set-point and return to it. Assume the part geometry has already been defined.

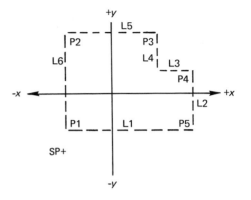

FIGURE P7-7

**7-26.** Use the APT language to guide a cutter around the surfaces shown in sketch of Fig. P7-8. The cutter will start from the set-point and return to it. The part surface (PSURF) lies below the bottom surface of the part.

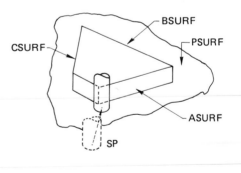

FIGURE P7-8

**7-27.** Describe the points shown in Fig. P7-9. Example:

P1 = PØINT/8,3,0.

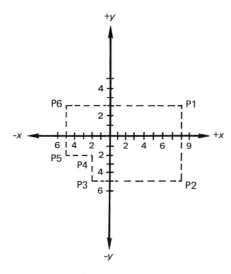

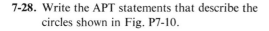

FIGURE P7-9

**7-28.** Write the APT statements that describe the circles shown in Fig. P7-10.

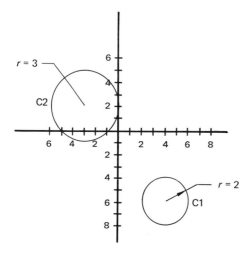

FIGURE P7-10

**7-29.** Write the statements that describe the circular arcs shown in Fig. P7-11. The radius of each of the corner arcs is 0.5 in. *Note:* The arcs are described as full circles in the geometry statements. The motion statements will determine the length of the arcs. This calculation is performed by the computer.

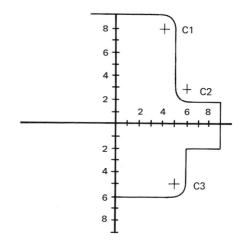

FIGURE P7-11

**7-30.** Write the complete APT program for the tool path as shown in Fig. P7-12. The programmer has the following information:

| | |
|---|---|
| Part no. | Sample program No. 1 |
| Machine AB | Coolant on |
| Tolerance .005 | Feedrate 5 ipm |
| Cutter .250 | |

The cutter is to set 0.125 in. below the bottom surface of the part which is the plane or setting for the z dimension.

The following format is given as a guide:

PARTNØ
MACHIN
INTØL
ØUTTØL
CUTTER
CLPRNT
FEDRAT
SP =
P1 =
P2 =
P3 =
P4 =
P5 =
C1 =
L1 =
L2 =
L3 =
PLN1 =

578

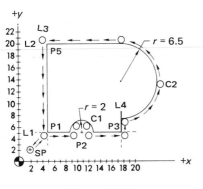

FIGURE P7-12

FRØM
RAPID
GØTØ
GØTØ
GØRGT/L1, PAST, C1
GØLFT
GØLFT/L4
GØFWD/
GØFWD/L2
GØLFT
RAPID
GØTØ
CØØLNT
FINI

**7-31.** Manually program the part shown in Fig. P7-13. Assume all radii are the radius of the tool. A spindle speed of 600 rpm is used and a feed of 0.010 ipr when cutting. A 2 in./min. feed is used as rapid for positioning the tool. Use the form as shown below.

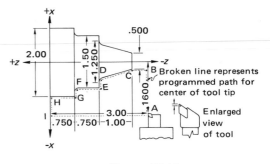

FIGURE P7-13

**7-32.** Write the complete APT program for the part shown in Fig. P7-14. The inside and outside tolerances are given as .001 and

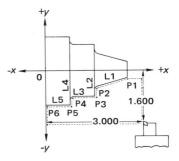

FIGURE P7-14

.0005 in. respectively. The cutter has a $\frac{1}{16}$-in. radius. The feed rate is 1 ipm. The spindle speed is based on 300 sfpm and coolant is used. *Note:* Dimensions are given in Fig. P7-13.

MANUSCRIPT

Part No._____                     Date:_____
Part Name_____                    Prepared by:_____
Machine No._____                  Checked by:_____

| seq No | g word | x word | z word | f word | s word | m word | EOB |
|--------|--------|--------|--------|--------|--------|--------|-----|
|        |        |        |        |        |        |        |     |

**7-33.** Sketch a tape as shown in Fig. 7-36 or use ordinary graph paper to show how it would be punched for positions 1 and 2 of the part shown in Fig. P7-15. Start from set point or origin. Show direction of tape. The first hole will be End of Block. Label at the side what each hole stands for.

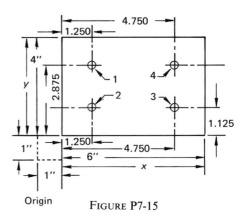

Origin     FIGURE P7-15

# BIBLIOGRAPHY

## Books

CHILDS, J. J., *Numerical Control Part Programming*, Industrial Press Inc., New York, 1973.

CHILDS, J. J., *Principles of Numerical Control*, Industrial Press Inc., New York, 1969.

HARRISON, H. and J. BOLLINGER, *Introduction to Automatic Controls*, International Textbook Co., Scranton, Pennsylvania, 1969.

HENKE, R. W., *Introduction to Fluid Power Circuits*, Addison-Wesley Publishing Company, Reading, Massachusetts, 1970.

NEWTON, D. G., *Fluid Power for Technicians*, Prentice-Hall, Inc., Englewood Cliffs, New Jersey, 1971.

PATTON, W. J., *Numerical Control*, Reston Publishing Company Inc., Reston, Virginia, 1972.

## Periodicals

Editor, "By-The-Numbers Digital Revolution Has Transformed Design Engineering," *Product Engineering*, Jan. 1974.

"Electric Controls," *Machine Design*, Reference Issue, 1973–1974.

"Fluid Power," *Machine Design*, Reference Issue, 1973–1974.

GAY, D. S., "Ways to Place and Transport Parts," *Automation*, June 1973.

GETTELMAN, K. M. (editor), "NC Guidebook," *Modern Machine Shop*, 1974.

JOHNSON, D., "Programming for Process Control," *Automation*, Oct. 1973.

KAPLAN, I. L., "Design of Pneumatic-Power Systems," Machine Design, July 7, 1960.

LINDBOM, T. H. and E. YEO, "Clamping Devices for Industrial Robots," *Automation*, Dec. 1970.

PECK, E. S., "EAROM: Next-Generation NC Controller," *Manufacturing Engineering and Management*, Nov. 1970.

PENN, R., "NC or Special?," *American Machinist*, Dec. 24, 1973.

POO, A. N. and J. G. BOLLINGER, *Control System and Part Edge Programming For Computer Numerical Control*, Proceedings of the International Conference on Production Engineering, Tokyo, 1974.

RUITER, H. H. JR. and R. G. MURPHY, "Transducers and What They Measure," *Machine Design*, Nov. 23, 1961.

# 8

## Nontraditional Machining Methods

Conventional machining methods always produce some stress in the metal being cut. Newer methods have been developed that are essentially stress-free. Even very hard metals or very thin metals can be cut without distortion or stress. Four common methods of stress-free machining are: photochemical machining (PCM): electric-discharge machining (EDM); electrochemical machining (ECM); and electrochemical grinding (ECG).

### PHOTOCHEMICAL MACHINING

Although PCM is not exactly new, it probably rates as one of the newer metal-machining methods. It is basically an etching technique that has been developed to compete on a production-line basis. The steps involved are: preparation of the part drawing, coating the metal with a photo resist, baking the resist, image printing, etching, and resist removal.

The drawing, as shown in Fig. 8-1(*a*), is very carefully and accurately made since the finished part can be no better than the artwork from which it originated. Drawings of very small parts are sometimes made 100 times actual size to improve accuracy and are then photographically reduced on film to the required size.

Before the photo resist is placed on the metal, it is thoroughly cleaned, usually with the aid of solvents. When the resist is completely dry, it is baked at about 180°F (82°C) for several minutes. The master negative is then placed against the resist (*b*). A powerful light source is used to expose the resist through the negative. Development of the resist leaves a chemical resistant image on the metal in the shape of the part.

The metal is then spray etched (*c*) to dissolve all unprotected areas and leave only the finished part, which is washed in a solvent and flushed with water to remove the resist.

For greater accuracy, particularly with thicker metals, two master negatives are made. One is a mirror image of the first and is exposed on the back side of the metal so the etchant can act on both sides of the metal at the same time.

(a)

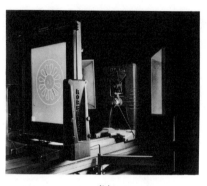

(b)

(c)

FIGURE 8-1. *Steps in the process of photochemical machining. Preparation of the drawing (a). The image is transferred to the metal (b). Spray etching the metal to leave only the finished part (c). Courtesy Chemcut Corporation.*

### Advantages

1. Eliminates die costs.
2. Excellent for thin parts (Fig. 8-2).
3. Provides for quicker delivery.
4. Produces stress-free parts.
5. Produces burr-free parts.
6. Virtually any metal can be used.
7. Has perfect repeatability.
8. Makes short runs feasible.
9. Makes design changes easy.
10. Both large and small parts can be made.
11. Produces close-tolerance parts.

FIGURE 8-2. *Representative parts that have been photochemically machined. Parts may range in thickness from 0.0005-in. (0.0013-mm) foils to 0.125-in. (3.17-mm) thicknesses. Courtesy Chemcut Corporation.*

### Disadvantages

1. If close accuracy is required, etching all the way through is limited to metal thicknesses of about 0.0625 in.
2. Etchant vapors are very corrosive and must be isolated from other plant equipment.

TABLE 8-1. *Removal Rates and Tolerances for Chemical Milling. Courtesy ASTM*

| Material | Depth of Etch per min. (in.) | Total Depth of Cut (in.) | Milling Tolerance (in.) |
|---|---|---|---|
| Aluminum | .001 | up to .020 | $\pm.001$ |
| | | .021 to .060 | $\pm.002$ |
| Magnesium | .0013 | up to .020 | $\pm.001$ |
| | | .021 to .060 | $\pm.002$ |
| | | greater than .060 | $\pm.003$ |
| Stainless steel | .005 | up to .020 | $\pm.001$ |
| and heat-resistant | | .021 to .060 | $\pm.002$ |
| alloys | | greater than .060 | $\pm.003$ |
| Titanium | .005 | up to .020 | $\pm.002$ |
| alloys | | .021 to .060 | $\pm.0035$ |
| | | greater than .060 | $\pm.005$ |

3. Production yields are relatively low. Part size, thickness of metal, and quantity required determine the breakeven point. The metal-removal rates and tolerances are shown in Table 8-1.
4. Because of the inherent undercut, there is a minimum-size limit for slots and holes and other piercings. For example, the smallest hole that could be pierced in 0.010-in. brass while still providing a near vertical wall would be 0.007 in. in diameter. The progressive undercutting action is shown schematically in Fig. 8-3. An etch factor of 3 : 1 means that for every 0.003 in. of etched depth, 0.001 in. of undercut will occur. The following limitations on hole size are characteristic: $0.7t$ for copper alloys $1.0t$ for steel alloys, and $1.4t$ for aluminum alloys.

## ELECTRIC-DISCHARGE MACHINING

At some time or other you may have observed an arc caused by an accidental short circuit and noticed the pitting that occurred on the surface of the shorted material. This is the principle that operates in EDM machining. In EDM the spark is pulsed at frequencies ranging from a few hundred to several hundred thousand kilohertz and the workpiece and electrode are separated by a dielectric, as shown schematically in Fig. 8-4. The dielectric is usually a low-viscosity hydrocarbon or mineral oil. As the tool and the workpiece are brought together, the gap between them is maintained by a servo mechanism comparing actual gap voltage to a preset reference voltage.

The gap is raised until the dielectric barrier is ruptured, which usually occurs as a distance of 0.001 in. (0.03 mm) or less and at about 70 volts. Technically, the dielectric is ionized to form a column or path between the workpiece and the tool so that a surge of current takes place as the spark is produced, as shown schematically in Fig. 8-5.

The spark discharge is contained in a microscopically small area, but the force and temperatures resulting are impressive. Temperatures around 10,000°C and pressures many thousand times greater than atmospheric are created, all in less than one microsecond for each spark. Thus a minute part of the workpiece is vaporized. The metal is expelled as the column of ionized dielectric vapor collapses. The tiny particles are cooled into spheres and are swept away from the machining gap by the flow of dielectric fluid.

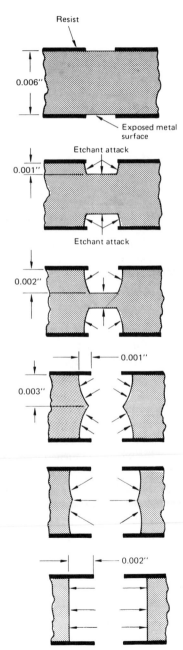

FIGURE 8-3. *The undercutting action of chemical machining explains why there is a lower limit on the size of small holes and slots that can be made.*

585

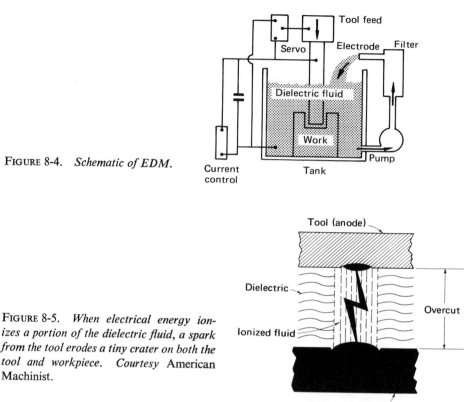

FIGURE 8-4.  *Schematic of EDM.*

FIGURE 8-5.  *When electrical energy ion-izes a portion of the dielectric fluid, a spark from the tool erodes a tiny crater on both the tool and workpiece.*  *Courtesy* American Machinist.

The wear of the electrode is relative to the heat and ion flow that attack it. Therefore, the higher the melting temperature of the electrode, the less it wears. For this reason, high-density carbon is one of the best overall electrode materials. However, the fine carbon dust produced while machining is quite objectionable and must be drawn off with an adequate vacuum system.

Shown in Fig. 8-6 is an example of a blanking die produced by EDM.  In this case the electrode is placed on an arbor and is started from the back side of the die. Since the hole produced is slightly tapered, it provides a positive taper for the die opening when it is turned right side up.  The clearance between the punch and the die can be regulated but may be as small as 0.001 in. (0.03 mm).  The cost of producing a graphite electrode is very small compared to having a broach made to produce the same hole.  In this case, the machining time for three holes on a $\frac{5}{8}$ in. (15.87 cm) thick die block was 47 min.  Although the actual machining time is relatively slow, other operations that are often done by hand, such as providing positive taper and controlling the surface finish and clearance, are taken care of all at the same time.

*Advantages.*  EDM has many advantages that stem from three basic facts:

1. The hardness of the workpiece is not a factor.  As long as the material can conduct current, it can be machined.

FIGURE 8-6. *A blanking die produced by EDM. Shown at the right is the graphite tool used to machine the serrated holes. Courtesy South Bend Lathe Company.*

2. Any shape that can be produced in a tool can be reproduced in the workpiece. Complicated tooling may be made up in segments and fastened together with an adhesive such as Eastman 910.
3. The absence of almost all mechanical force makes it possible to machine the most fragile components without distortion. A 0.002 in. (0.05 cm) diameter hole, for example, can be produced in a small, delicate part using a very fine wire as a tool.

*Disadvantages:*

1. Tool wear requires stepped tooling or redressing of tools for deep holes.
2. EDM leaves a recast layer at the surface of the cut. This may be an advantage or disadvantage. Where undesirable, it may be removed by a light finishing cut or by polishing. Depending upon the current density used, the recast layer may be from 0.0002 to 0.005 in. (0.005 to 0.13 mm) in thickness.
3. EDM is slow when compared to conventional methods or even to ECM. Whenever possible, the cavities are roughed out prior to heat treatment and then finished by EDM after heat treatment.

*Fine-Wire Numerically Controlled EDM.* While EDM has long been used to sink holes and cavities in hardened material, including carbide, it has the drawback of requiring an electrode of the same shape as the hole or cavity to be machined. Manufacturing the electrode can often be a complex problem.

A newer concept is to use a 0.015 in. (0.38 mm) diameter traveling wire. This process may be compared to a band saw. A small hole is first machined as in the regular EDM process. A fine wire is then threaded through the hole to act as the cutting blade. As in bandsawing, the workpiece is manipulated to obtain the desired shape. In the traveling-wire system, the work table is manipulated by numerical control. The taut traveling wire is unwound from one spool, passes through the workpiece, and then goes into another windup spool at the rate of 15 ipm (38.1 cm/min), as shown in Fig. 8-7. The purpose of having a traveling wire is to constantly bring an unworn portion of it into the cut; otherwise the wire would soon erode and break. The wire is quite inexpensive and is therefore used once and discarded.

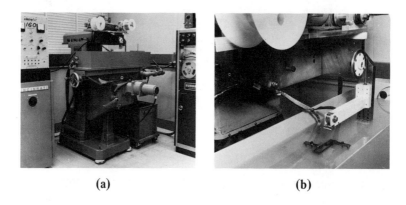

(a)                                                   (b)

FIGURE 8-7. *The thin traveling wire is used to cut a very fine line for intricate die openings. The workpiece configuration is obtained by moving the table with NC programming and stepping motors.* Courtesy Modern Machine Shop.

If the program-tape feed rate is set too fast for the actual cutting action, the voltage-sensing unit incorporated in the numerical controller will stop the tape reader, thus preventing shorting of the wire electrode. The reading rate is then adjusted to produce the proper machining gap.

The cut is made at 1 to 2 amp, which produces a very fine surface finish that is microscopically pocketed. This has proven beneficial in actual die operation since it tends to retain the lubricant used in stamping.

By utilizing the computer, several advantages are obtained. The same input information can be utilized to generate tapes with offsets for punch and die clearance, stripper clearance, backup-plate clearance, and shedder clearance. Thus one die may have four different tapes that will mate perfectly since they are all generated from the same data. It has been found that tools machined by this process resulted in a savings of 40 to 60 % over conventional EDM.

## ELECTROCHEMICAL MACHINING

Electrochemical machining is a machining process that may be described as the reverse of electroplating. High-density direct current is passed through an electrolyte solution that fills the gap between the workpiece (anode) and the shaped tool (cathode). The electrolyte must be considered, in effect, a part of the tool (Fig. 8-8). The electrochemical reaction deplates the metal of the workpiece as shown in the more detailed schematic (b).

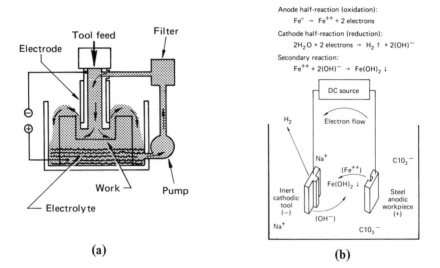

**(a)**

**(b)**

FIGURE 8-8. *A schematic of the ECM process (a). A detail of the chemical process (b).*

ECM is considered by many to be a new metal-cutting process. Actually, both the fundamentals and the concepts of its application to the metal-working field have been known since 1920. At that time a Russian, Waldimer Gusseff, first demonstrated how the fundamental laws of electrolysis could be successfully used to machine a desired shape in a metal part. The basic laws of electrolysis were discovered by Michael Faraday in 1933–34. Briefly, this law states the amount of material deposited or dissolved is directly proportional to the current density multiplied by the time.

***The Process.*** Although ECM is compared to reverse electroplating, it is much more. The fact that it can be used to produce specific shapes with precise dimensional control qualifies it as a modern machining process.

Basically, the tool is brought very close to the workpiece (anode). The distance ranges from less than 0.001 in. (0.03 mm) up to 0.010 in. (0.25 mm). The low-voltage, high-density current passes between the cathode and anode by means of the conductive electrolyte solution. A variety of electrolytes are available, the most common being a solution of caustic soda. Other electrolytes are sodium nitrate and sodium chlorate, as was shown in Fig. 8-8(*b*). The electrolyte is pumped through the gap at pressures often as high as 300 psi (204 kg/m²). The solution is maintained at a temperature of 100 to 120°F (38–49°C). The current used may vary widely. Some machines use up to 20,000 amp per sq in. However, the more usual range is from 500 to 5000 amp capacity. As the current passes from the workpiece to the tool, metallic particles (ions) on the surface of the workpiece are sloughed off into the electrolyte.

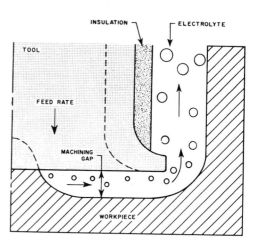

FIGURE 8-9. *The ECM process showing the machining gap.*

The machining gap (Fig. 8-9) is maintained by a servo control mechanism. Any contact of the tool with the workpiece will result in arcing, with damage to both members. On the other hand, if the gap becomes large (0.003 in; 0.08 mm), more electrolyte will flow, offering increased resistance for the current with a subsequent reduction in stock removal. In the application of Ohm's Law, $E = IR$, or as transposed, $I = E/R$. We can see that if the resistance (gap between the electrode and work) is small, a higher current density will result. However, when the gap gets too small, hydrogen bubbles tend to crowd together and increase the resistance across the gap, thus reducing the current flow. Hence the maximum penetration rate is based on a number of variables but is primarily a function of a precisely controlled machining gap.

***Surface Finish and Accuracy.*** Surface finish is dependent on the alloy being machined and the current density. Areas finished on the frontal surface of the cathode where the current densities are high generally have fine finishes. Finishes of 20 microinches are not uncommon, and on high-grade materials such as nickel-chromium alloys it may be as little as 2–5 microinches. Bearing races have been finished in production quantities by ECM to within 6 microinches.

The overcut per side of the cathode is about 0.005 in. (0.13 mm) and repeatability is good to within 0.0015 in. (0.04 mm).

*Advantages.* The advantages of ECM are similar to those given for EDM with the following additions:

1. The tool does not wear. Once the tool is developed, it can be used indefinitely.
2. There are no thermal or mechanical stresses on the workpiece.
3. Faster stock removal and better surface finish can be obtained. As a rule of thumb for estimating the metal removal rate, ECM is considered to have the ability to remove 0.10 cu in. (1.64 cm$^3$) of metal per 1000 amps used or 1 cu in./min (16.4 cm$^3$/min) for 10,000 amp as compared to up to 10 cu in./hr (164 cm$^3$/hr) for EDM. The surface finish range is generally from 4 to 50 microinches.

*Disadvantages*

1. The basic cost of the equipment is several times that of EDM. A representative ECM machine and control unit is shown in Fig. 8-10.
2. Rigid fixturing is required to withstand the high electrolyte flow rates.
3. The tool is more difficult to make than for EDM since it must be insulated to maintain the correct conductive paths to the workpiece. The tool may also serve as a device for introducing the electrolyte into the machining gap and must then have a center hole. However, this is not true if the tool is connected to the negative terminal and becomes the cathode.
4. The most common electrolyte, sodium chloride, is corrosive to the equipment, tooling, and workpiece.

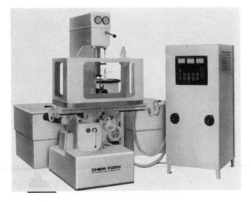

FIGURE 8-10. *A typical ECM machine showing electrolyte work enclosure and control unit. Machines normally range in size from 500 to 5000 amps in steps of 500 amps. Courtesy Chemform, Division of KMS Industries, Inc.*

A comparison of the process characteristics of EDM and ECM are given in Table 8-2. Shown in Fig. 8-11(*a*) is a forging die being machined by EDM. At (*b*) is a steel alloy part cut by ECM in less than 5 min.

## ELECTROCHEMICAL GRINDING

Electrochemical grinding is the removal of metal by a combination of electrochemical decomposition and the action of diamond abrasive particles contained in the grinding wheel. The electric current flows from the negatively charged abrasive

TABLE 8-2. *A Comparison of EDM and ECM Process Characteristics. Courtesy Chemform, Division of KMS Industries, Inc.*

| Data | EDM | ECM |
|---|---|---|
| Typical applications | Machining of: <br> 1) Dies (stamping, cold heading, forging, injection molding) <br> 2) Carbide forming tools <br><br> 3) Tungsten parts <br> 4) Burr-free parts <br><br> 5) Odd-shaped holes and cavities <br><br> 6) Small diameter deep holes <br><br> 7) High strength and high hardness materials <br> 8) Narrow slots (0.002″–0.012″ width) <br><br> 9) Honeycomb cores and assemblies and other fragile parts | Machining of: <br> 1) High temperature alloy forgings <br> 2) Turbine wheels with integral blades <br> 3) Jet engine blade airfoils <br> 4) Jet engine blade cooling holes <br> 5) Deburring of all kinds of parts <br> 6) Odd-shaped holes and cavities <br> 7) Small deep holes <br><br> 8) Honeycomb cores and assemblies and other fragile parts <br> 9) High strength high hardness materials |
| Tolerances | Practical: $\pm 0.0005″$ <br> Possible: $\pm 0.0001″$ | Practical: $\pm .002″$ <br> Possible: $\pm 0.0005″$ |
| Surface | Finish is affected by rate of removal <br> 0.010 in.³/hr–30 rms <br> 0.5 in.³/hr–200 rms <br> 3.0 in.³/hr–400 rms <br> Heat affected zone 0.0001″–0.005″ | 4–50 rms can be attained. No heat affected surface is produced. No burrs. Must guard against selective etching in remote areas exposed to electrolyte |
| Practical removal rates | Approximately 0.00025 in.³/amp./min. or up to 10 in.³/hr | 0.1 in.³/min./1000 amps is often used for rough approx. or 1 in.³/min. for 10,000 amp. |
| Machining characteristics | Process produces taper and overcut (finished cavity minus original electrode size) in workpiece as well as corner radii <br> Typical values: <br> Taper 0.001–0.005 in./in./side <br> Overcut 0.002–0.005 in./side <br> Min. corner radius 0.001–0.20 in. <br> Electrode wear ratio: <br> Metallic electrodes 3/1 <br> Carbon 5–70/1 <br> Corner wear ratio: 2/1 | Process does not introduce machining stresses. Virtually no electrode (tool) wear occurs. Process may produce overcut, taper, and corner radii. Tools subject to damage by arcing if process malfunctions |

FIGURE 8-11.  *An example of an EDM machine being used to cut a connecting-rod forging die cavity in a solid die block.  The electrode (top part) is machined from graphite.  Fast cutting speeds and feeds can be used and still provide a good finish.  The EDM process removes between 4 and 5 in.³/hr at 150–200 amps.  The tolerance is ±0.002to 0.003 in.( ±0.05–0.08 mm).  Courtesy Cincinnati Milacron.*

wheel to the positively charged workpiece through an electrolyte (saline solution), as shown schematically in Fig. 8-12.  The resulting electrochemical oxidation produces a soft oxide film on the workpiece surface.  The grinding wheel then "wipes away" this oxide film.  Since the oxides are very much softer than the parent metal, there is very little grinding-wheel wear.

The tool looks like an ordinary grinding wheel, and it is, except that the bonding material is mixed with a conductor such as pure copper or carbon.  The abrasive particles which are nonconductive tend to maintain a gap between the metal-bonding agent and the work.  Applying Faraday's Law, the metal-removal rate is almost directly proportional to current density.

*Form Grinding.*   The wheels may be dressed to facilitate form grinding with the aid of a single-point diamond dressing tool, a metal-bonded diamond dressing wheel, or a crushed-formed vitrified grinding wheel in a plunge grinding-wheel operation.

*Power Supply.*   The power used for electrochemical grinding varies in size up to 1000 amps and normally has a dc voltage output which ranges from 4 to 15 v.

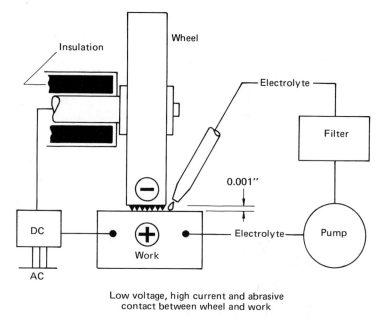

Low voltage, high current and abrasive
contact between wheel and work

FIGURE 8-12.  *Schematic of electrochemical grinding.*

***Types of Wheels.***  The type of wheel commonly used to grind carbide cutting tools is shown in Fig. 8-13(*a*).  The workpiece is usually plunged into the face of the wheel, allowing for maximum metal removal since the entire surface to the ground is in constant contact with the wheel.  Shown at (*b*) is the type of wheel most commonly used for larger pieces of work or for form ginding, as shown at (*c*).

### Advantages

1. Higher, about 80 % faster, metal-romoval rates are obtained with ECG than conventional grinding of hard materials such as carbides, germanium, cobalt, cast alloys, etc.  On hardened alloy steels, the metal removal rate is about 0.100 cu in./min/1000 amps ($1.64\,cm^3$)/min/1000 amps).  On carbides, it ranges down to 0.060 cu in./min/1000 amps ($1.0\,cm^3$/min/1000 amps).
2. Wheel wear is drastically reduced.  Depending upon feed rate, size of the wheel, and other process variables, the wheel life can be increased by a factor of 10 on most ECG applications.
3. No heat is generated, so there is no danger of burning or heat distortion.
4. Only a small amount of mechanical force is applied to the workpiece, so the possibility of burrs or distortion is also eliminated.

### Disadvantages

1. As with ECM, the salt solution is corrosive.  The machine should be washed down periodically to remove salt deposits.
2. The electrolyte should be changed about once a week.

3. Initial cost of the equipment is high when equipped with large power supplies.
4. Copper-resin–bonded diamond wheels may be dressed to simple forms only. Intricate forms with multiple radius tangents and very deep forms require single-layer, plated-metal, bonded diamond wheels.

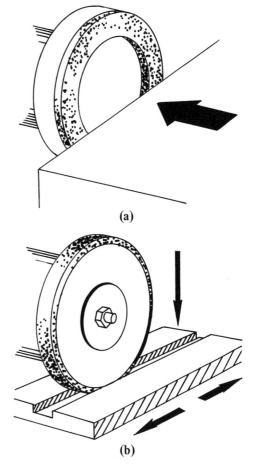

(a)

(b)

FIGURE 8-13. *Electrochemical-grinding wheels. Face-grinding wheels (a) are used for plunge cutting carbide tools and may also be used in tool grinding of carbide milling cutters. Larger workpieces may be ground by straight and formed wheels as shown at (b) and (c) respectively.*

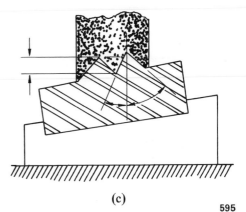

(c)

### ABRASIVE JET MACHINING (AJM)

Abrasive jet machining may be simply described as a fine stream of abrasive particles propelled through a nozzle by a high-pressure gas for the purpose of cutting a wide variety of materials. The gas used to obtain the high pressure may be carbon dioxide, nitrogen, or filtered compressed air. The nozzle through which the abrasive particles are propelled is usually made of tungsten carbide or synthetic sapphire. A typical AJM unit is shown in Fig. 8-14(*a*) and a sketch of the process at (*b*).

The jet can be used to make fine (0.005 in.; 0.13 mm) line cuts or a wide spray to "frost" large areas. The width of the cut is dependent on how far the abrasive must travel before it impinges on the workpiece. As the jet leaves the nozzle opening, which may be either round or rectangular, it begins to diverge at about $\frac{1}{16}$ in. (1.58 cm)

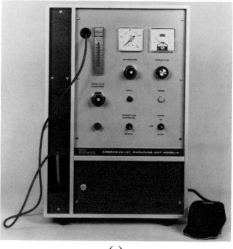

**(a)**

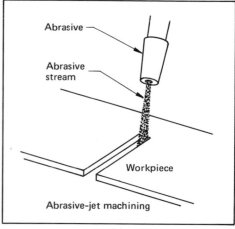

**(b)**

FIGURE 8-14. *A typical abrasive-jet machining unit. Courtesy S. S. White Industrial Products.*

from the opening. When the nozzle is kept relatively close to the work, the cut will be straight; at greater distances, the cut will become triangular in cross section. In materials that are too thick for a "straight" cut or where the nozzle distance is too great, a straight side is possible simply by angling the jet. Straight cuts have been made in steel to a depth of 0.060 in. (1.52 mm), and in glass to depths up to 0.250 in. (6.35 mm). Currently the narrowest cut possible is 0.005 in. (0.13 mm), made with a 0.003 × 0.060-in. (0.08 × 1.52-mm) rectangular nozzle opening held at a distance of $\frac{1}{32}$ in. (0.793 cm) from the work.

*Cutting Variables.* The cutting rate can be altered by several variables such as flow rate, distance of nozzle to the work, gas pressure, and type of abrasive used. The maximum flow of abrasive is about 10 gpm (37.8 lpm). Gas pressure is usually 75 psi (5.07 kg/cm$^2$), but it may vary from 1 to 125 psi (0.70 to 8.6 kg/cm$^2$).

The abrasive powders used range from about 10 to 50 microns. The hardness may vary from a soft sodium bicarbonate to a hard silicon carbide. Other commonly used abrasives are aluminum oxide, dolomite, and ground nutshells.

*Applications.* Etching and surface preparation are common applications of AJM. The process has been successfully used in cleaning corrosion and other contaminants from such diverse parts as electronic components to priceless leather artifacts. A common use is in trimming resisters of hybrid power-amplifier circuits for a specified value within a tolerance of ±0.5 %. The entire operation is automatic. The time for trimming each resistor is 0.5 sec.

### Advantages

1. The abrasive jet can be used to cut any material. Even diamonds have been cut, using diamond dust as the abrasive.
2. Because of the low velocity of the abrasive particles, there is virtually no impact and very thin brittle materials can be cut without danger of breaking. This is shown by the example of cutting an eggshell (Fig. 8-15).
3. Virtually no heat is generated in the workpiece.
4. The process is safe. The operator can pass his hand under the jet without injury.

### Disadvantages

1. Because of the very small stream of abrasive particles, the material removal rate is low. On plate glass, for example, the removal rate is about 0.001 cu in./min (0.03 mm$^3$/min), or the equivalent of a "slot" 0.020 × 0.010 × 5 in (0.57 × 0.25 × 127 mm).
2. The abrasive powders cannot be reused since the points and edges get worn down. However, the cost of most abrasives is relatively low.
3. Because of its nature, AJM usually requires some type of dust-collecting system. Most machines are used with an enclosure to which a vacuum hose has been attached.

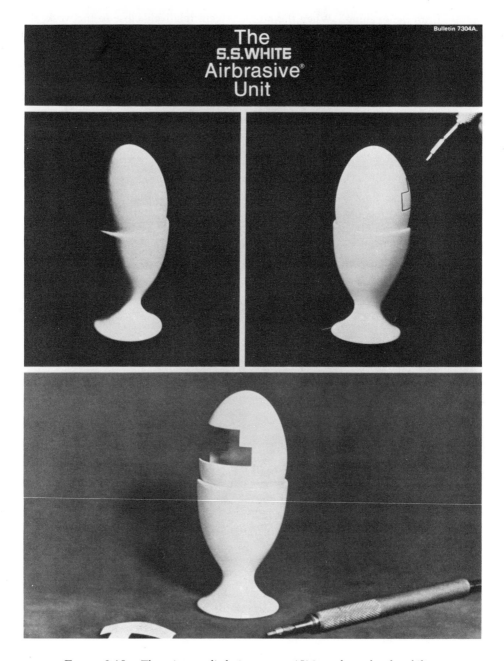

FIGURE 8-15. *There is very little impact to AJM as shown by the ability to cut an eggshell. Courtesy S. S. White Industrial Products.*

## LASER-BEAM MACHINING

The principles of the laser were discussed in Chapter 5 in connection with its use in welding. It may also be used in special applications for machining. The two types of lasers used in machining are the optically pumped solid-state laser and the continuous-wave (CW) $CO_2$-$N_2$ laser.

*Optically Pumped Solid-State Laser.* This laser, as shown schematically in Fig. 6-58, has a high energy and peak power output. The high-intensity beam is delivered for a short interval, which causes most materials to vaporize. By focusing the beam through an optical system, the energy output is delivered into a very small area, typically from 0.0005 to 0.050 in. (0.013 to 1.27 mm) in diameter.

*Advantages.* The advantages of laser drilling are as follows:
1. Small holes can be drilled in thin materials (typically 2–5 mils) at a rate of 50 holes/sec. The minimum hole size is given by the formula:

$$D = \frac{1.22 f \lambda}{d}$$

*where:*   $f$ = focal length of the lens or mirror used,
   $\lambda$ = the laser wavelength,
   $d$ is the diameter of the aperture.

2. Holes can easily be drilled on curved or angular surfaces.
3. Holes may be drilled inside a vacuum or through a transparent window.
4. The holes are burr-free.
5. The holes are drilled automatically and may be programmed by the addition of automatic or numerically controlled equipment.
6. The drilling time is measured in microseconds. The heat is extremely localized.
7. Holes can be drilled in hard-to-machine materials such as alumina-ceramic, tungsten, tool steels, tantalum, niobium, and diamonds.

*$CO_2$-$N_2$ Gas Laser.* In 1964 the first continuous-wave (CW) laser was announced. It had a power of several milliwatts. Since that time the power of the CW laser has grown astonomically to almost 100 kilowatts. The $CO_2$ laser lends itself to rapid pulsing with pulse rates of 1200 pps. Energies of 0.5 joules have been reported from a 5-ft laser tube.

Pulse powers of this magnitude, when focussed with a lens or mirror to an area of 1 mm$^2$, yield intensities of $10^5$ to $10^7$ watts/cm$^2$. This compares with energy intensities produced by a plasma torch of $10^4$ watts/cm$^2$ and electron-beam torches of $10^9$ to $10^{11}$ watts/cm$^2$.

*Advantages.* The advantages of a laser-beam cutting are very similar to those mentioned for laser drilling.

*Disadvantages*
1. The laser beam can be dangerous if it is not used carefully. The visible laser beam produced by the ruby can penetrate the cornea and do retinal damage. The $CO_2$ laser, on the other hand, is in the infrared region and cannot be seen.

Special glasses are necessary for the ruby laser but regular safety glasses will protect against the $CO_2$ laser.

2. Production laser machines are costly, ranging from about \$18,000 to \$100,000 for $CO_2$ units. Ruby lasers are somewhat below this range.

3. The laser depends on thermal properties of the material being cut. When exposed to the laser beam, some materials char and bubble, as in the case of fibreglass-reinforced structures. In general, a material's boiling and vaporization points determine how suitable it is for laser machining. The higher the vaporization point, the more difficult it is to laser machine. Also, the closer the material's boiling and vaporization points, the harder it is to machine.

*Applications.* In recent tests of a $CWCO_2$ laser with an oxygen-gas jet assist (Fig. 6.43) 0.5 cm (0.195 in.) thick stainless steel sheet was cut at the rate of 75 cm/min (30 ipm). Plywood of about the same thickness was cut at the rate of 520 cm/min (195 ipm) with an argon-gas jet assist. The same laser system can be easily changed from cutting metals to cutting flamable materials by substituting an inert gas for a reactive gas in the jet-assist apparatus.

At the present time, relatively little of this new technology has been exploited It is expected that, as the technology expands, there will be $CO_2$ lasers "on-line" as a production tool in a variety of industries.

## PROBLEMS

**8-1.** A hardened alloy-steel block is to be ground by the electrochemical process. The area to be machined is $1 \times 2$ in. and it is to be cut to a depth of 0.060 in. (a) How long will it take to machine this surface assuming 1000 amps will be used? (b) Assume a 300-amp machine will be used.

**8-2.** A carbide tool with a surface area of $0.5 \times 1$ in. is to be cut to a depth of 0.0625 in. Assume 900 amps will be used to ECG this surface. Power supplies are available in capacities of 300, 600, 1000, and 1500 amp. (a) What power supply will be needed? (b) What is the volume of metal to be removed? (c) Assuming a 0.010 cu in./min/100 amp metal-removal rate, what will the feed rate be? (d) What is the time required to make the total cut? (e) Determine the feed rate from the proportion: time for cut/60 = depth of cut/feed rate.

**8-3.** (a) What is the minimum hole size in centimeters that can be made with a laser beam under the following conditions:

Focal length of the lens = 4 cm
Wavelength of the laser light = $10^{-3}$ cm
Diameter of the aperture = 1.5 cm

(b) What is the minimum hole size in inches?

**8-4.** What is the minimum hole size that can be photochemically machined in $\frac{1}{32}$ in. thick copper, $\frac{1}{16}$ in. thick aluminum, and 0.010 in. thick brass?

**8-5.** If 2 in. diameter brass circles are to be photochemically machined from $\frac{1}{16}$ in. thick material, how much larger should the drawing be made? Assume an enlargement of $5 : 1$.

**8-6.** A material that has an etch factor of $3 : 1$ is to have several slots cut into it. If the material is $\frac{1}{16}$ in. thick and the slots are to be $\frac{1}{16}$ in. wide, what width should they be drawn? Assume a 10 : 1 enlargement.

8-7. Shown in Fig. P8-1 is a stamping that can be made in various ways. (a) What tolerance could be expected if the part were photochemically machined? (b) What size would the electrode punches be for the

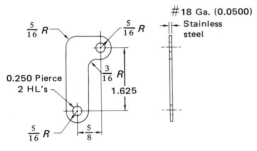

holes if the die cavity were to be made by EDM? (c) How long would it take to etch out one piece or a sheet of parts by photochemical machining using the double-sided etching method? (d) If just ten of the parts shown in Fig. P8-1 were to be made, how would you propose they be made?

8-8. Assume a forging die must have 15 cu in. of metal removed and have a 100 microinch finish or better. (a) What methods could be used to produce the die if it is to be machined in the hardened condition? (b) How long would the machining take?

8-9. A 1 × 1 in. square is to be removed from 0.25 in. thick plate glass. Approximately how long would it take if AJM were used? Assume the cut will be 0.020 in. wide.

## BIBLIOGRAPHY

### Books

Chemform, *ECM, ECD, ECG Simplified*, Chemform, Division of KMS Industries, Inc., Pompano Beach, Fl., 1970.

SPRINGBORN, R. K., (editor), *Nontraditional Machining Processes*, ASTM, Dearborn, Mich., 1967.

### Periodicals

DULEY, W. W., and J. N. GONSALVES, "Industrial Applications of Carbon Dioxide Lasers," *Canadian Research & Development*, Jan.–Feb. 1972.

Editor, "Photochemical Machining Offers Precise Method for Forming Metal," *Product Engineering*, Feb. 1974.

GETTLEMAN, K., "Thin Wire Act: NC and EDM Turn Out The Dies," *Modern Machine Shop*, March 1972.

LAVOIE, F. J., "Abrasive Jet Machining", *Machine Design*, Sept. 6, 1973.

LAVOIE, F. J., "The Blue-Collar Laser," *Machine Design*, Feb. 19, 1970.

# 9

# *Measurement—Quality Control— Product Liability*

## MEASUREMENT

### Introduction

The progress of measurement has played a large part in man's scientific advancement. Early attempts at standardization of length measurements were based on the human body. The width of a finger was termed a digit, and the cubit was the length of the forearm from the end of the elbow to the tip of the longest finger. These measurements were in use at the time of the construction of the Khufu pyramid (4750 BC) and are mentioned in connection with the building of Noah's Ark.

Careful checking of the Khufu pyramid at Gizeh reveals each side to be 756 ft long and the height to be 481 ft. There is a mean error in the length of the sides of only 0.6 in. and an error of angle from a perfect square of only 12 seconds.

*Metric Measurement.* From these simple beginnings, man progressed gradually toward standardization of basic measurements. A very significant advance was made at the Convention of the Meter held in Paris in 1875. At that time, the meter was defined as one ten-millionth of the distance from the north pole to the equator, measured on the meridian quadrant that passes through Paris. The metric system was worked out with related factors of 10 so that conversions could be handled by simple shifting of decimal points.

In 1960, 36 countries, including the United States, participated in the 11th General Conference of Weights and Measures, meeting in Paris. At that time, the modernized International System of Units was adopted. It may be abbreviated SI from its French name, *Systéme International d'Unités*.

TABLE 9-1. *International System of Units*

| Physical Quantity | Unit | Symbol |
|---|---|---|
| *Base Units* | | |
| length | meter | m |
| mass | kilogram | kg |
| time | second | s |
| electric current | ampere | A |
| thermodynamic temperature | kelvin | K |
| luminous intensity | candela | cd |
| amount of substance | mole | mol |
| *Derived Units* | | |
| area | square meter | $m^2$ |
| volume | cubic meter | $m^3$ |
| speed | meter/second | m/s |
| acceleration | meter/second squared | $m/s^2$ |
| density | kilogram/cubic meter | $kg/m^3$ |

*Combination Derived Units*

| Physical Quantity | Name of Unit | Symbol | Derived Units | Combination Base Units |
|---|---|---|---|---|
| force | newton | N | | $m \cdot kg \cdot s^{-2}$ |
| pressure & stress | pascal | Pa | $N/m^2$ | $m^{-1} \cdot kg \cdot s^{-2}$ |
| energy | joule | J | $N \cdot m$ | $m^2 \cdot kg \cdot s^{-2}$ |
| power | watt | W | $J/s$ | $m^2 \cdot kg \cdot s^{-3}$ |
| conductance | siemens | S | $A/V$ | $m^{-2} \cdot kg^{-1} \cdot s^3 \cdot A^2$ |

*Supplementary Units*

| Physical Quantity | Unit | Symbol |
|---|---|---|
| plane angle | radian | rad |
| solid angle | steradian | sr |

The International Organization for Standardization (ISO) is responsible for international development of industrial and commerical standards in almost every field of technology. It is supported by 55 nations, including the United States.

The SI system is built up from three kinds of units: base units, derived units, and supplementary units, as shown in Table 9-1.

An essential feature of SI is the systematic use of prefixes to designate decimal multiples and decimal fractions of base units. The most commonly used SI prefixes are shown in Table 9-2.

Tables of conversion factors, permitting conversion from English to metric units, are widely available and an abbreviated example is shown in Table 9-3 and in Appendix 9A. Some useful approximations are: a meter is about 10 % longer than a yard; a liter has 5 % more volume than a quart; to get kilograms, subtract one-tenth of the number of pounds and divide by 2; to calculate kilometers (or km/h),

TABLE 9-2. *Commonly Used SI Prefixes*

| Prefix | Factor | Symbol |
|--------|--------|--------|
| tera | $10^{12}$ | T |
| giga | $10^9$ | G |
| mega | $10^6$ | M |
| kilo | $10^3$ | k |
| hecto | $10^2$ | h |
| deka | $10^1$ | da |
| deci | $10^{-1}$ | d |
| centi | $10^{-2}$ | c |
| milli | $10^{-3}$ | m |
| micro | $10^{-6}$ | $\mu$ |
| nano | $10^{-9}$ | n |
| pico | $10^{-12}$ | p |
| femto | $10^{-15}$ | f |

add half and then one-tenth of the miles (mph); a Btu is about one kilojoule; and water freezes at 0°C and boils at 100°C.

Not standard at the present time, but widely used, is the system of using a comma to separate a decimal fraction rather than a decimal point. Thus $2\frac{3}{4}$ is written 2,75. Also, a blank space is used rather than a comma to group every three digits of a number. Thus a million is written as 1 000 000.

TABLE 9-3. *Basic Conversion Factors*

| To Convert from | To | Multiply by |
|-----------------|-----|-------------|
| Inches | Millimeters | 25.4 |
| Feet | Meters | 0.3048 |
| Yards | Meters | 0.9144 |
| Miles | Kilometers | 1.609 |
| Square inches | Square centimeters | 6.4516 |
| Square feet | Square meters | 0.0929 |
| Square yards | Square meters | 0.836 |
| Acres | Hectares | 0.405 |
| Cubic inches | Milliliters | 16.387 |
| Cubic feet | Cubic meters | 0.0283 |
| Cubic yards | Cubic meters | 0.765 |
| Quarts | Liters | 0.946 |
| Gallons | Liters | 3.784 |
| Ounces | Grams | 28.35 |
| Pounds (mass) | Kilograms | 0.454 |
| Pounds (force) | Newtons | 4.448 |
| Pounds per square inch (psi) | Kilopascals | 6.895 |
| Horsepower | Kilowatts | 0.746 |
| Btu | Kilojoule | 1.055 |

### The Development of Useable Standards

The use of the earth's meridian (or some specified fraction of it) was fine in theory, but it was difficult to translate into everyday needs. Even with the definition of the meter as the distance between two engraved lines on a bar of a platinum-iridium alloy at zero degrees centigrade (now Celsius), it was impractical for manufacturers to use.

*Gage Blocks.* Toward the close of the 19th Century, Carl Johansson of Sweden assumed the task of providing measurement standards that would be available to industry. They were to duplicate the international standards as nearly as possible. To accomplish this, he painstakingly made a set of steel blocks that were accurate to within a few millionths of an inch. This unheard of accuracy, and how it was achieved, remained a secret with Johansson for many years. Since these steel blocks were so accurate, they could be used to check manufactuers' gages and other measuring instruments. The blocks were made up in sets and are known as "gage blocks." A typical set consists of 81 blocks with sizes ranging from 0.05 in. to 4.0 in., as shown in Fig. 9-1.

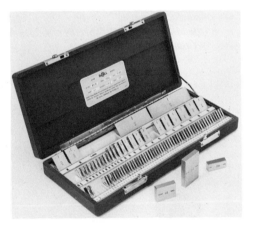

FIGURE 9-1. *A standard 81-block gage set. Courtesy DoAll Company.*

*Gage Block Use.* Gage blocks maintain accuracy even though several have to be used in combination to produce a specified dimension. By carefully sliding the gaging surface of one block over the gaging surface of another, two blocks can be "wrung" together so tightly they are difficult to separate (Fig. 9-2).

Gage blocks may be used in direct measurement, as in checking the size of a keyway or the distance between milling cutters (Fig. 9-3). They may also be used for indirect measurement, as in checking a snap gage or setting a dial indicator, as shown at Fig. 9-4.

*Gage Block Calibration.* We have seen how a useable standard was developed for industry, but how could the accuracy of these blocks be checked consistently within millionths of an inch?

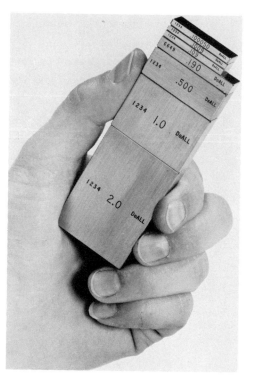

FIGURE 9-2. *Gage blocks that have gaging surfaces wrung together form effective length rods for specific measurements. Courtesy DoAll Company.*

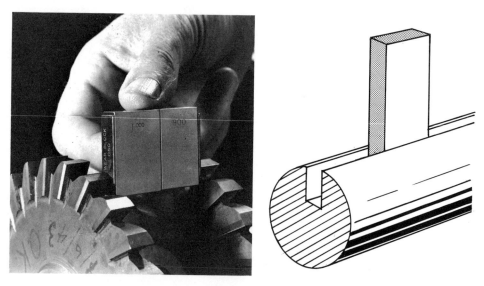

FIGURE 9-3. *Gage blocks may be used for direct measurement as in checking a keyway or measuring the distance between milling cutters. In the latter case, carbide-wear gage blocks are wrung in for protection of the gaging surfaces. Courtesy DoAll Company.*

(a)

FIGURE 9-4. *The dial indicator is an example of an indirect-measuring tool. The workpiece height is compared to that of a gage block (a). A gage block or a combination of gage blocks can be wrung in to check a snap-gage setting (b).*

(b)

As early as 1827 a French physicist, Jacques Babinet, suggested that the wavelength of light had possibilities as a standard of length. His proposal could not be realized until much later when the concept of interferometry had been developed. In 1892–93 Professor Michelson of the University of Chicago perfected methods of optical measurement and defined the international meter in terms of a number of light-wave lengths. The concept, which is relatively simple, is that when the crests of two light waves coincide, they reinforce each other; when a crest coincides with a trough, the two waves cancel. In the first case the interference is constructive and in the second case it is destructive, hence the alternate light and dark bands.

A simple interferometer can be made by placing one optically flat glass plate on top of another. Shown in the sketch [Fig. 9-5(a)] are two ceramic-glass gage blocks wrung in at the edge. Shown in Fig. 9-5(b) is a more detailed view of how the fringe lines are formed. The light of a single color, or wavelength, is directed to the upper surface. When this light reaches the lower surface of the upper flat, it is divided into two portions. One portion is reflected upward by the lower surface and the other portion continues downward to be reflected from the upper surface of the lower flat. The two series of reflected rays recombine in the upper flat, where they interfere to form alternate light and dark bands as seen from above. Such interference will occur at every point where the distance from the flat to the work is in odd multiples of one-half the wavelength of the light used. By counting the number of fringes between the

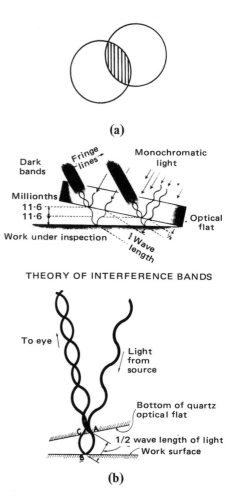

**(a)**

THEORY OF INTERFERENCE BANDS

FIGURE 9-5. *Interference bands show between the surfaces of two ceramic-glass gage blocks even under fluorescent lighting (a). The straightness of the lines indicate the flatness and the degree of wring. The theory of interference bands is shown at (b).*

**(b)**

edge of the upper plate and its point of contact with the object to be measured, the object's length can be determined. It will be equal to the wavelength of the light used multiplied by one-half the number of dark fringes to the point of contact. The most-used light source in metrology laboratories is helium, which has a wavelenth of 23.2 microinches. The krypton-86 isotope is the new basic international standard of length. The meter is defined as being exactly 1,650,763.73 wavelengths of this source, measured in a vacuum. The krypton lamp is generally used only in standardizing laboratories due to the cooling requirement, and because the lamp emits so many different wavelengths (spectral lines) that a fairly elaborate monochromator is required to separate them. It now appears that the helium-neon laser is the best light source for the measurement. It is capable of producing light that is far more monochromatic and more intense than other light sources, and therefore extremely sharp fringe lines are produced. The American National Standards Institute has related one laser line, the neon line, at a wavelength of 6328 angstrom units, to the krypton standard with an accuracy of one part in 100 million, which is close to the limit of accuracy inherent in the krypton wavelength.

Conceivably the laser could extend the limit of interferometric measurement to hundreds or even thousands of kilometers. Such measurements, however, present the difficulty of counting several hundred million interference fringes, a task that could be accomplished only by automatic means.

The flatness of the gage block can be checked by noting the pattern of the fringe lines. If they are straight and parallel, the surface is flat within one-half wavelength of the light used. If the surfaces are not flat, the fringe lines will not be straight but will show the amount of deviation by their contour as shown in the three sketches of Fig. 9-6.

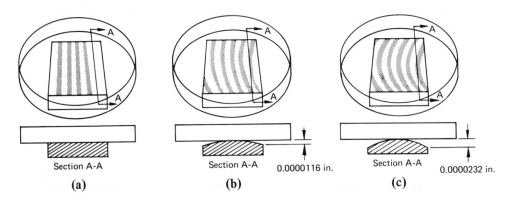

FIGURE 9-6. *A perfectly flat surface (a) produces uniform straight inter-ference bands when viewed through the optical flat. A slightly convex part (b) produces curved bands. The amount of curvature is one band or 0.0000116 in. Shown at (c) is a curvature of two full bands or 0.0000232 in.*

## Precision Measurement Principles

The science of measurement has progressed until "millionths of an inch" is becoming a commonplace term. It is difficult to grasp the true meaning of so fine a measurement. To calibrate a gage block or to make any measurement to within 1 microinch requires considerable precaution and effort since temperature, humidity, ambient air conditions and other factors must be at specified standards. Thus controlled environment laboratories, or metrology labs as they are often referred to, are necessary to insure precise measurements.

The temperature of standard metrology laboratories is maintained at the international reference level of 68°F (20°C) within ±0.5°F. Due to differences in linear coefficients of expansion of various materials, even small deviations in tempera-ture can cause significant dimensional differences.

Foreign particles cannot be tolerated in the atmosphere of a standards labora-tory due to the adverse effects on the accuracy of measurement and dependable oper-ation of instruments. Clean air is accomplished by careful filtering and by having a higher air pressure in the room than outside of it. Personnel enter through a double air lock.

Humidity is maintained at 45 to 50 % to eliminate the possibility of rust which forms quickly on highly polished surfaces in the presence of moisture.

*Precision and Accuracy.* Precision and accuracy, two key concepts in the science of measurement and gaging, are frequently misused. *Precision* may be defined as the repeatability of a measuring procedure. A carpenter measures a doorway $37\frac{1}{4}$ in. wide. He hands his tape to a friend who promptly measures the doorway $37\frac{1}{2}$ in. wide. Thus this measuring procedure is precise within $\frac{1}{4}$ in.

*Accuracy* is the traceability of a measurment to an ultimate standard, or theoretically true value. No two measurements are alike, so every measurement has some uncertainty in it. The degree of uncertainty, or deviation from the ultimate standard, is the accuracy level of the measuring device. Thus the expected accuracy of an ordinary micrometer is very close to 0.001 in.

A level of accuracy and precision is inherent in every measuring tool. Many of the hand-operated tools in the lower echelons of accuracy depend on operator skill and care. Instruments of high accuracy, however, are protected against operator-initiated errors.

Basically, all measuring devices can be broadly classified into two groups: direct and indirect. Direct-measuring instruments obtain the measurement without the aid of other equipment, as in the case of a micrometer. Indirect-measuring tools require a standard of reference, as in the example of a dial indicator shown in Fig. 9-4.

Measurement can be further divided into five main areas: linear (length, diameter), geometric forms (roundness, flatness), geometric interrelationships (parallelism, squareness, and concentricity), angular dimensions, and surface texture.

## Linear Measurement

### Direct

Linear measurements may be taken directly with rules (scales), micrometers, vernier scales, and measuring machines.

*Steel Rule.* The steel rule or scale is the most common measuring device. Graduations commonly used are fractions of an inch down to $\frac{1}{16}$ in. Millimeter scales are also available.

*Vernier Scales.* Vernier scales improve the accuracy of the scale by an order of magnitude. The vernier scales are used on calipers and height gages shown in Fig. 9-7. The reading on the insert scale is as follows.

2.000 inches (the zero on the vernier scale has passed 1 in.)
+ 0.300 hundred-thousandths on the scale
+ 0.050 two twenty-five thousandths lines
+ 0.018 thousandths, where the vernier line coincides with a line on the scale.
Total = 2.368 in.

*Micrometers.* Micrometers are universal direct-measuring tools. Using the mechanical reduction of a fine-pitched screw thread, the micrometer reduces great rotational motion to slight linear motion. One full rotation on the micrometer thimble or barrel (Fig. 9-8) produces only 0.025 in. of linear travel.

The basic micrometer principle leads to many special-purpose measuring tools such as the thread and depth micrometer shown in Fig. 9-9.

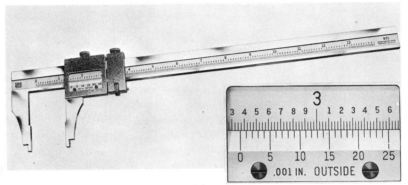

**(a)**

FIGURE 9-7. *The vernier scale is used for direct measurement in the vernier caliper (a) and the vernier height gage (b). Courtesy Brown and Sharpe Manufacturing Co.*

**(b)**

*Measuring Machines.* Measuring machines are highly sophisticated direct-reading tools (Fig. 9-10). While most hand-held tools frequently depend on operator skill, or feel, for their reliability, measuring machines are less subject to human error. Measuring machines are usually accurate to within ±0.00015 in. on all three axes.

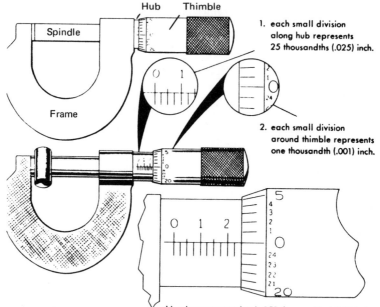

1. each small division along hub represents 25 thousandths (.025) inch.

2. each small division around thimble represents one thousandth (.001) inch.

this micrometer setting is 250 thousandths (.250) inch.

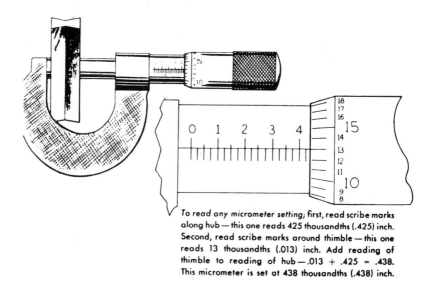

To read any micrometer setting; first, read scribe marks along hub — this one reads 425 thousandths (.425) inch. Second, read scribe marks around thimble — this one reads 13 thousandths (.013) inch. Add reading of thimble to reading of hub — .013 + .425 = .438. This micrometer is set at 438 thousandths (.438) inch.

FIGURE 9-8. *The micrometer with an enlarged view of a reading.*

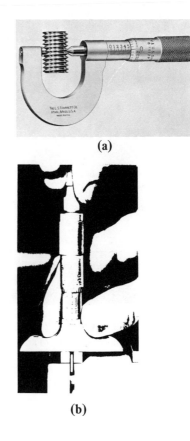

**(a)**

**(b)**

FIGURE 9-9. *A thread micrometer used to measure the pitch diameter of a thread (a). A depth micrometer used to measure the depth of a milled slot (b). Courtesy L. S. Starrett Co.*

In operation, a suitable probe tip is first selected and inserted into the stylus. To measure center distances between several holes, for example, a tapered plug probe would be used, as shown in Fig. 9-11. The probe is inserted into the first hole and the machine is set to a zero reference. As the probe is moved from one hole to another, movements indicating each position on $x$, $y$, or $z$ coordinates are read out on the digital display.

A typical measuring machine operates on the moiré fringe-pattern principle. A ruled grating extends the entire length of travel. An index grating with the same line pattern is mounted above it at a slight angle. The superimposed pattern is the familiar moiré fringe [Fig. 9-12(*a*)]. As the measuring head travels from point to point, the changing fringe pattern is converted to electrical impulses through photocells (*b*).

Measuring machines can quickly and easily do jobs that would be long and tedious with less sophisticated equipment. Moreover, the measuring machine is usually more accurate and reliable. The main disadvantage is the high initial cost.

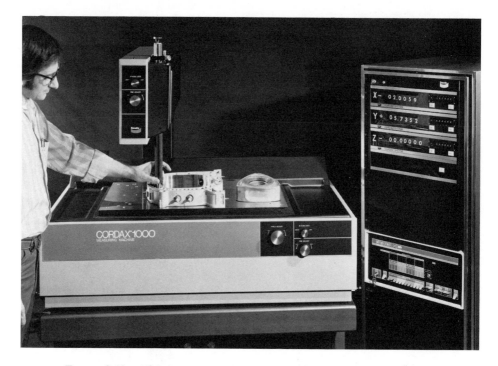

FIGURE 9-10. *The automatic measuring machiné or Cordax\* checks location points in three axes to an accuracy of ±0.00015 in. The probe can be made to move on command from the computer core memory or from an operator-held programmable control box. Courtesy Bendix Automation and Measurement Division.*
*\*Trade name registered by Sheffield Company.*

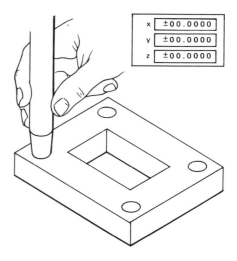

FIGURE 9-11. *A tapered-plug probe is used when measuring the center distances of hole locations.*

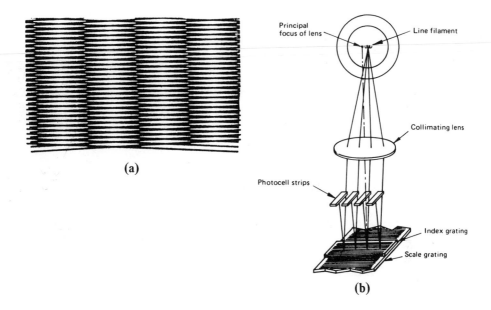

**(a)**

**(b)**

FIGURE 9-12. *A typical measuring machine operates on the moiré fringe pattern, which occurs when one grated plate is placed above another at a slight angle (a). As the measuring head travels from point to point, the changing fringe pattern is converted to electrical pulses (b). Courtesy* Machine Design.

## *Indirect*

Indirect-measuring tools are used to transfer or compare a measurement with a known standard. These *comparators*, as they are referred to, are of four main types: mechanical (dial indicator), electronic, optical, and air.

***Dial Indicators.*** The dial indicator has a stylus probe that is connected through a precision gear train to the dial pointer [Fig. 9-13(a)]. Small linear deflections of the stylus are multiplied into large pointer movements. An example of a comparison measurement is shown at (b). The part surface to be measured is compared to a gage block or a combination of gage blocks that have been wrung in. The gage is set at "0" over the blocks and the workpiece is measured as any deviation form "0".

Comparison measurement frequently requires a good frame of reference. As was shown in Fig. 9-13, the indicator base, the part, and the standard all rest on a precision surface plate for a uniform plane of reference. Any variation in the "flatness" of the surface plate will cause the measurement to be inaccurate. However, granite surface plates usually have guaranteed overall accuracies to within $\pm 0.0002$ in. in the larger sizes, $\pm 0.000025$ in. in the smaller sizes.

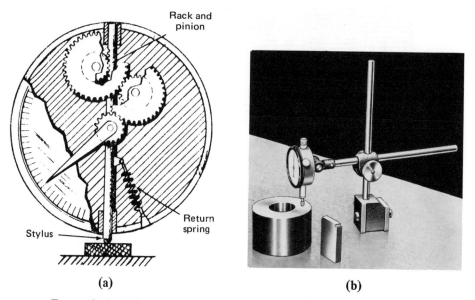

**(a)**                 **(b)**

FIGURE 9-13. *The typical mechanism of a dial indicator (a). A dial indicator is used to make a comparative measurement between the work-piece and the gage block from the same reference plane (b). Courtesy L. S. Starrett Co.*

***Electronic Comparators.*** Electronic comparators provide a high degree of precision measurement. The measuring stylus is coupled through a pressure-sensitive transducer, variable resistor, or capacitor to an electronic circuit. Small changes in stylus position are converted to electronic signals that register deflection on a calibrated meter (Fig. 9-14). The maximum gage-tip force is about 60 g, with the minimum about 5 g. Various scales can be selected, either inch or metric, with the finest graduation usually 0.000010 in.

FIGURE 9-14. *The electronic indicator is among the most accurate of comparators. A small (0.000010-in.) deflection on the stylus can be made to read as a large deflection on the meter. Courtesy Bendix Automation and Measurement Division*

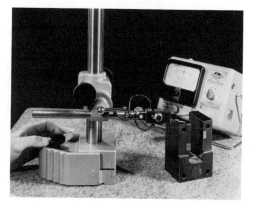

FIGURE 9-15. *The optical comparator magnifies the surface silhouette and is especially useful for checking intricate and complicated forms. Direct measurements may be seen on the digital readout. Courtesy Bendix Automation and Measurement Division.*

**Optical Comparators.** Optical comparators improve accuracy by magnification. When an optical comparator, such as that shown in Fig. 9-15, magnifies an image by 100X, an observed error of 0.10 in. (0.25 mm) is only 0.001 in. (0.03 mm) full scale.

Optical comparators are best suited for silhouette projections, although some models are equipped for surface projection. These machines are ideal for measuring complex outside dimensions as well as tiny intricate grooves, radii, or steps. Frequently templates matching the correct shape are placed on the projection screen as production parts are compared for contours and many dimensions at one time. Direct measurements may also be made with an accuracy of 0.0001 in. (0.003 mm).

**Air Gages or Comparators.** Air gages utilize the effect of minute dimensional changes produced by metered air. The air gage consists principally of plugs or rings that are calibrated by checking them with a master or standard. A gentle stream of compressed air escapes between the surface of the standard and the close fitting

617

AIRPLUG

**(b)**

**(a)**

FIGURE 9-16. *An air gage (a).  Variations in size cause variations in air pressure (b).  Courtesy Bendix Automation and Measurement Division.*

surfaces of the gage.  The differential in air pressure produces a reading which can be adjusted to zero.  As more or less air escapes when the workpiece is checked, the float in a glass tube will be correspondingly higher or lower.  Shown in Fig. 9-16 is an air gage (*a*) and an example of how the air plug is used to check for waviness (*b*).  Air gages are comparatively easy to use for both inside and outside measurements.  They are particularly well suited to checking long, small diameters that are hard to measure by any other method.

### Geometric Forms (Roundness, Flatness, and Angles)

*Roundness Measurement.* A large number of precision components that must be produced are of a circular nature, such as ball and roller bearing parts, pistons, gear blanks, gyroscope components, hydraulic-valve parts, crankshafts, etc.  In the past the relative ease of manufacturing circular parts meant that they could generally be produced to tolerances of thousandths and even tenths with little or no regard to their geometric form. This is no longer true. In working to tolerances of tenths of thousandths or better, parts are often found to be far from a true circular form.

There are several types of out-of-roundness that consist of lobing or bumps about the circumferences. An even number of lobes equally spaced about the circumference results in a variation in the diameter of the part. An odd number of lobes equally spaced around a circumference can result in an effective constant diameter. A third category of out-of-roundness consists of random lobing or a combination of even and odd lobing conditions. In all of these, the radius of the part will vary from the true center.

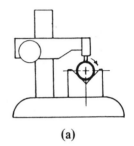

**(a)**

FIGURE 9-17. *The V-block and bench-center arrangements have only limited use in being able to check roundness geometry accurately. In order to measure the out-of-roundness with a dial indicator and V-block, it is necessary to have the correct angle for the number of lobes present. Thus a three-lobed out-of-round condition requires a 60° included V-block and a five-lobed condition requires 108°. The proper V-block angle is found by the formula 180 + (360/n), where n = the number of lobes.*

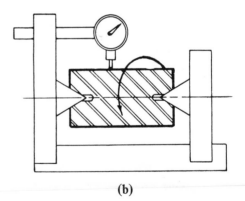

**(b)**

Formerly, V-block and bench centers were used to check the roundness geometry of parts, as shown in Fig. 9-17. The V-block is no longer considered accurate since erroneous readings may not be readily detected on parts having an odd number of lobes. The system of bench centers is applicable to only those parts made with center holes. If the center holes on the parts are in good condition, this method can be fairly accurate. However, errors in the center hole and roundness errors may cancel each other.

The preferred method of measuring roundness today is with an electronic trace-type roundness gage, as shown in Fig. 9-18(*a*). A polar chart is produced (*b*)

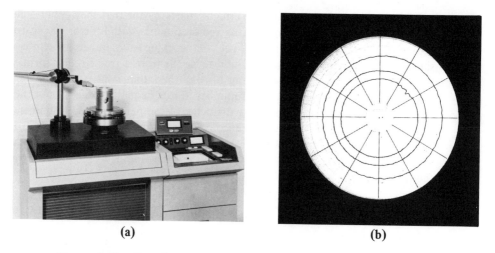

(a)                                        (b)

FIGURE 9-18.   *Roundness measurements are made by centering the work piece on the rotary table shown at the left.   As the part is rotated the gage-head probe follows the surface.   The polar recorder, at the right, shows the trace in amplified form on a polar chart.   Courtesy Bendix Automation and Measurement Division.*

that can be used for close examination and comparison.   The trace represents a profile in which the radial variation is greatly magnified in order that subtleties of geometrical error can be closely examined.

***Flatness.***   The accuracy of small, flat surfaces (up to 10 in. in diameter) may be checked with optical flats, as discussed in the calibration of gage blocks.

For larger surfaces, a quick rough check can be made by sweeping the surface several times with a straight edge, or more accurately with a specially made knife edge.   The light, shining from one side, will leak through the low spots of the surface plate.

Accurate checking of large, flat surfaces such as a planer or milling machine table is done by what is known as *optical tooling*.   A line of sight is used somewhat as a surveyor uses a transit or theodolite.

FIGURE 9-19.   *The autocollimator is used to calibrate a surface plate.   Courtesy Rahn Granite Surface Plate Co.*

The optical tooling used to check alignment and flat surfaces in machine-tool work is the autocollimator and mirrors, as shown in Fig. 9-19. The fundamental feature of the autocollimator is the perfectly straight optical axis of its small telescope. This straight line, augmented by an integral light source and illuminated reticle, will measure small changes of angle in a distant mirror as an image shift when the image is reflected back in the telescope, as shown in Fig. 9-20. This instrument is the basis of the most common method of calibrating surface plates and large machine-tool tables.

The autocollimator rests at one corner of the plate, its parallel beam of light projecting an image horizontally to a fixed mirror at a second corner. The beam is then reflected to a second mirror located at successive positions in a straight line across the surface plate, either perpendicular or diagonal to the original direction of the beam. The reflected beam from the second mirror retraces its original path according to the contour of the plate. As the image returns, it is observed to be either coincident with its origin or displaced by a measurable value. This measurement, twice the angular change of the second mirror, is caused by a change in slope from its previous position on the plate's contour. A sensitivity of 0.1 arc sec is typical and equivalent to elevations on the plate contour of 0.5 microinches per inch. Special features are avilable to make automatic photoelectric readings and recordings of the data, or a continuous recording of the trace.

Laser beams can now be used to calibrate surface plates in a fraction of the time required by earlier methods. The laser is also used to calibrate multiaxis machines for positioning accuracy, alignment, and squareness of axes, with a printed or plotted record of each measurement.

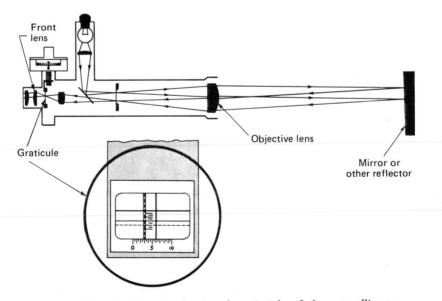

FIGURE 9-20. *A schematic showing the principle of the autocollimator. Courtesy Rank Precision Industries, Inc.*

### Geometric Interrelationships

*Parallelism.* Parallelism may be defined as a condition where a surface line or axis is equidistant at all points from a prescribed datum plane or axis.

Two examples of geometric tolerancing for parallelism are shown in Fig. 9-21. These parts may be checked by what is known as functional gages. As an example, a gage to check the parallelism of the two holes is shown in Fig. 9-22. There are other gage designs that could be used, as well as instrumentation to check the parallelism of the two holes. For example, if the part is not too large, the holes

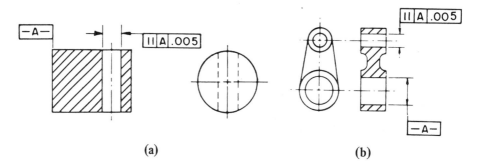

(a)                                                        (b)

FIGURE 9-21.   *Examples of geometric tolerancing for parallel axes. The notation in the box calls for the hole to be parallel to datum plane A within 0.005 in. (a).   The small hole must be parallel to the large hole (datum plane A) within 0.005 in. (b).*

FIGURE 9-22.   *An example of a functional gage used to check the parallelism of the part shown in Fig. 9-21(b).   The tolerances for the dowel pins and perpendicularity to datum plane A will have to be closer than that of the part.   This gage would also check the center distance of the two holes.   One pin is longer to ease assembly.*

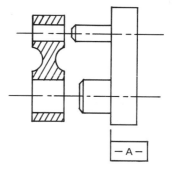

could be fitted with dowel pins and checked for parallelism on the optical comparator. If the part is large, the dowel pins could be checked for parallelism with the use of a surface plate and a dial indicator or electronic height gage.

*Squareness or Perpendicularity.* Perpendicularity refers to the condition of surfaces or axes which are exactly 90° from a given datum plane or axis [Fig. 9-23(a)].  Shown at (b) is another example of geometric tolerancing, using the abbreviated notation for

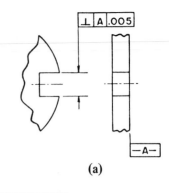

**(a)**

INTERPRETATION

FIGURE 9-23. *An example of a hole axis not being square with the reference or datum plane. The abbreviated notation as used in geometric tolerancing calls for the keyway to be perpendicular to datum plane A within 0.005 in.*

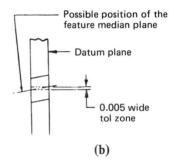

**(b)**

perpendicularity. The keyway is to be perpendicular to datum plane *A* within 0.005 in. Again, a functional-type gage can be made by the use of a square pin in a plate on which the part would be placed, as shown in Fig. 9-24. The allowance between the gage and the part would have to be less than $+0.005$ in.

***Concentricity.*** Concentricity may be defined as a condition in which two or more features (cylinders, cones, spheres, etc.) in any combination have a common axis.

Concentricity and wall thickness can be checked to an accuracy of millionths of an inch using differential techniques, as shown in Fig. 9-25. The two signals given

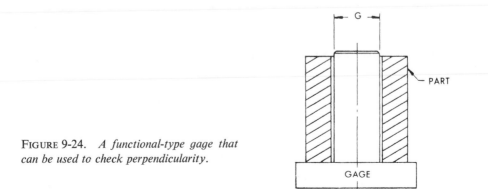

FIGURE 9-24. *A functional-type gage that can be used to check perpendicularity.*

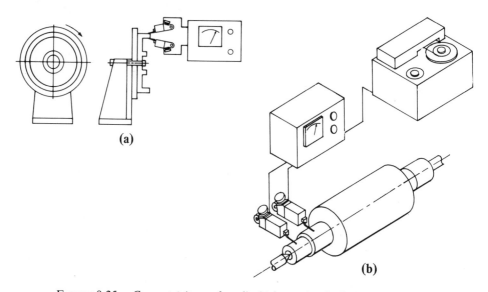

(a)

(b)

FIGURE 9-25. *Concentricity and wall thickness is checked at the same time (a). Roundness and concentricity may be checked right in the production machine to a radial accuracy of 10 millionths of an inch (b). Courtesy Society of Manufacturing Engineers.*

by the gage heads which are set opposite each other (*a*) can be combined by means of special circuitry to permit input data refinement. For example, when a source of error affects both heads equally, the error can be cancelled out. Errors of work support, gage support, temperature, and wear may all be cancelled out. High-precision concentric measurements are thus obtainable without high-precision fixturing. Routinely matched heads can cancel an error up to 100 times the desired gaging accuracy. Roundness and concentricity are checked at the same time, as shown at (*b*). The system is especially useful in checking bearing surfaces concurrently with their manufacture. Adjustments and corrections can be made without removing the part from the machine.

***Geometric Relationships and Machine Tools.*** Perhaps nowhere is there a greater need for geometric relationship accuracy than on machine tools. In the case of drilling, boring, milling, and other similar machines, there are two or three orthogonally related axes. Linear, angular, and straightness interferometers are capable of evaluating five of the six degrees of freedom of a rigid body, roll being the exception (Fig. 9-26). For machines where the workpiece rotates, as in lathes, parallelism between the tool and spindle axes is of fundamental importance. The laser straightness interferometer is able to make both perpendicularity and parallelism measurements, as shown in Fig. 9-27. This is done by comparing two consecutive straightness calibrations of adjacent axes and introducing either a 90° or 180° rotation into the straightness reflector (shown on the lathe cross slide) between traverses.

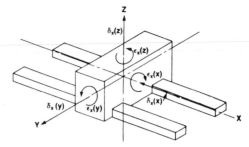

FIGURE 9-26. *The six degrees of freedom of a machine-tool table.*

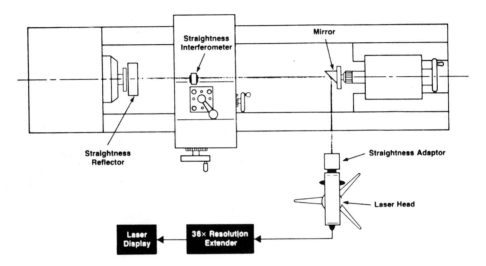

FIGURE 9-27. *A plan view showing how straightness and parallelism are calibrated on the lathe with the use of a laser and straightness interferometer.*

For perpendicularity, a separate right-angle reference is required, which causes the extended axis of the straightness reflector to be viewed with a 90° deflection. As an alternative, a precision indexing table can be used, which physically rotates the reflector assembly after the first axis has been measured.

### Angular Measurement

***Direct Angular Measurement.*** Direct angular measurements are commonly made with a protractor [Fig. 9-28(*a*)]. For more accurate measurements, the vernier bevel protractor is used, which reads in degrees and minutes (*b*). Direct angular measurements may also be made in degrees and minutes on the optical comparator (*c*), or with the autocollimator and a rotating table (*d*).

***Indirect Angular Measurement.*** Indirect angular measurement is made with the aid of sine bars and sine plates. A sine bar, as shown in Fig. 9-29(*a*), is a bar that has a

625

roll or plug on each end. The center distance between plugs on this particular bar is 10 in. The center distances are made to vary in increments of 5 in., ranging from 5 in. for the smallest sine bars to 20 in. on sine plates and bench centers made with sine-bar bases.

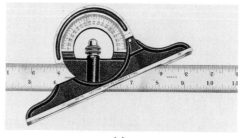

**(a)**

FIGURE 9-28. *Direct angular measurements can be made with the protractor (a). Angular measurements may be made in degrees and minutes with the universal bevel protractor (b), the optical comparator (c), and the autocollimator when the part is mounted on a precision rotary table (d). Courtesy L. S. Starrett Co.*

**(b)**

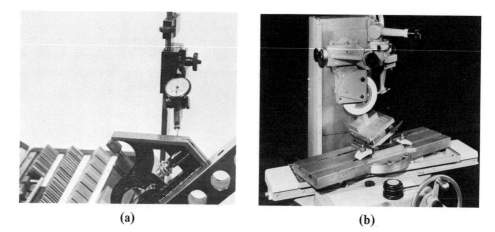

**(a)**          **(b)**

FIGURE 9-29. *A sine bar being used to check the 45° angle of a square head (a). A sine-bar table being used on a surface grinder to grind a compound angle (b). Courtesy DoAll Co.*

The sine bar in Fig. 9-29(a) is being used to check the 45° angle on a squaring head. The procedure used is as follows:

1. The sine of 45° = 0.70711.
2. Select gage blocks to make up this dimension and wring their gaging surfaces together.
3. With the use of a dial indicator, stand, and surface plate, check to see if the 45° angle is parallel to the base reference plate. Any change in the dial reading would indicate an error in the angle. The amount of error can be determined by the required length of the opposite side. Shown at (b) is a sine-bar table being used on a surface grinder.

### Measuring Microscopes

Measuring microscopes are used for direct linear and angular measurement of small parts. The contour of the part as observed through the microscope represents a magnified replica. Although not actually a shadow because it is not projected on a surface, it is commonly called a shadow image or, more correctly, a silhouette. Contours which are partially obstructed from viewing can be measured by special accessories.

The toolmakers' microscope, as shown in Fig. 9-30, is equipped with micrometer screws which fill the double function of moving the table laterally and horizontally and providing measurements to accuracies of 0.0001 in.

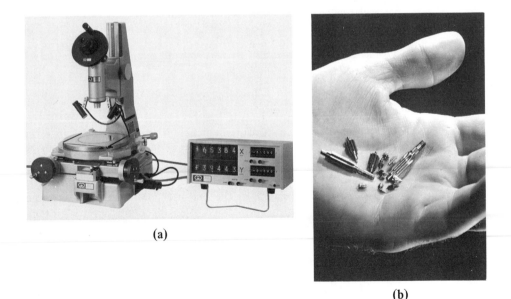

(a)

(b)

FIGURE 9-30. *The toolmaker's microscope and typical parts that can be accurately measured. Courtesy Gaertner Scientific Corp.*

The most common method of measuring is by the use of graticule hair lines. The edge of the part to be measured is lined up with the hair line. At that point, a zero setting can be made on the micrometer dial. The part is then moved until the other edge is lined up with the hair line and the distance noted on the dial. A protractor hair line is used for angular measurements. Parts may be viewed in magnifications of 10 X to 100 X. Typical applications are shown in Fig. 9-31.

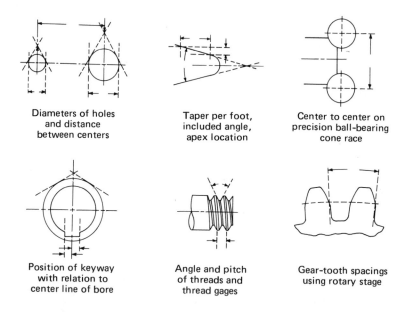

Diameters of holes
and distance
between centers

Taper per foot,
included angle,
apex location

Center to center on
precision ball-bearing
cone race

Position of keyway
with relation to
center line of bore

Angle and pitch
of threads and
thread gages

Gear-tooth spacings
using rotary stage

FIGURE 9-31. *Typical measurements made on a toolmakers' microscope. Courtesy Gaertner Scientific Corp.*

## Surface-Texture Measurement

The science of surface measurement has progressed rapidly in recent times. Not too many years ago an appraisal of surface finish was given by rubbing the thumbnail across the surface or by a purely visual examination.

Surface finish is usually specified by the design engineer, and the degree of finish should be carefully related to the end-use component. It is important to remember that the finest finish attainable by a skilled operator is not necessarily the best finish for a particular component. Also, costs increase rapidly as extra operations are added, as shown in Fig. 9-32.

Formerly the designer merely placed a symbol "$\sqrt{}$" on all surfaces to be machined, but in time a much more meaningful standard symbol evolved, as shown in Fig. 9-33. You will note that the symbol has considerable information surrounding it, including roughness, roughness width, roughness width cutoff, waviness height, waviness width, and lay. All of these conditions are now covered by the term *surface texture.*

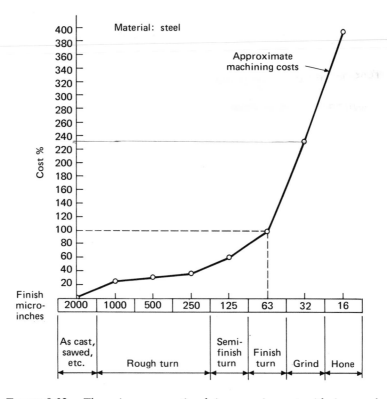

FIGURE 9-32. *There is a proportional increase in cost with improved surface finish up to about 63 microinches, after that there is a rapid rise.*

***Roughness.*** Roughness can best be defined as the marks left by the action of the production tool such as a lathe tool or grinding wheel. The common method of measuring roughness today is by the use of a shrap-pointed stylus. The vertical movement of the stylus as it passes over the surface is translated into a meter reading that is proportional to the surface roughness (Fig. 9-34). The stylus-tip radius must be as small as practicable (0.0005-in. radius is standard) to include the effect of fine irregularities. The average surface roughness as measured in microinches was formerly termed Root Mean Square (RMS) but is now given in terms of the arithmetical average (AA) in the U.S., center line average (CLA) in Britain, and roughness average (Ra) in Europe. For historic reasons, CLA and AA are generally given in microinches (0.000001 in.) and Ra in micrometers (0.001 $\mu$m). The determination of the surface roughness in AA is shown in Fig. 9-35. AA roughness is defined as the arithmetic average deviation from the center line; the center line in turn is defined by the ANSI B46.1 standard as "the line parallel to the general direction of the profile within the limits of the roughness width cutoff, such that the sums of the areas contained between it and those parts which lie on either side of it are equal." Cutoff is discussed later but may be briefly defined as the sampling length. A com-

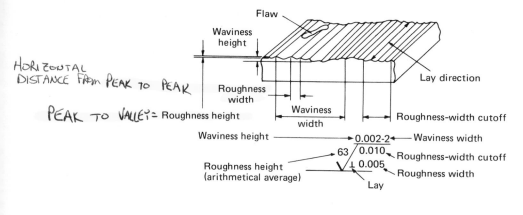

HORIZONTAL
DISTANCE FROM PEAK TO PEAK

PEAK TO VALLEY = Roughness height

Symbols indicating direction of lay

=   Parallel to the boundary line of the nominal surface indicated by the symbol

⊥   Perpendicular to the boundary line of the nominal surface indicated by the symbol

✕   Angular in both directions to the boundary line of the nominal surface indicated by the symbol

M   Multidirectional

C   Approximately circular relative to the center of the nominal surface indicated by the symbol

R   Approximately radial relative to the center of the nominal surface indicated by the symbol

FIGURE 9-33. *Standard surface-texture symbol and illustration of terms. Courtesy American Standards Assoc.*

FIGURE 9-34. *The Profilometer employs a diamond-stylus tracer to move over the surface to check the roughness in millionths of an inch. The vertical movement of the diamond stylus is translated into a meter reading as shown at the right. Courtesy Bendix Automation and Measurement Division.*

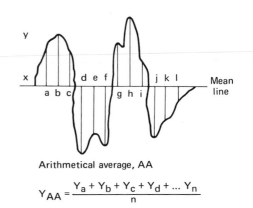

FIGURE 9-35. *Standard measurement of average surface roughness.*

Arithmetical average, AA

$$Y_{AA} = \frac{Y_a + Y_b + Y_c + Y_d + \dots Y_n}{n}$$

parison surface-roughness measurement is shown in Fig. 9-36. Shown at (*a*) is the conventional stylus trace with skid as used to measure the AA. At (*b*) is also a stylus trace, but it measures the whole surface roughness (peak to valley) against an optical flat datum. Peak to valley readings are usually referred to as the total roughness or Rt.

*RMS.* RMS is the average root mean square of deviations taken from a mean line through the surface texture. This was formerly an American standard, declared obsolete in 1955, but it may still be encountered occasionally. Its numerical value is about 11 % higher than AA or Ra.

*Waviness.* Waviness is that part of the surface texture that underlies the roughness. It is the height of wave form remaining after roughness has been filtered out at a selected wavelength cutoff value (Fig. 9-37). All types of machine vibration and imbalance can cause waviness. The waviness width is the distance over which the height is measured. A 0.003-2 specification means no waves over 0.003 in. high in any 2-in. length.

*Lay.* Lay is the direction of the predominant surface pattern, ordinarily determined by the machining process.

*Cutoff.* The cutoff or sampling length is a facility that is built into most surface-measuring instruments. It refers to the greatest spacing of surface irregularities to be included in the measurement of average roughness height. If the irregularities are finely spaced, the cutoff width can be relatively short, and vice versa. Standard roughness width values in both British and American standards are 0.003 in., 0.010 in., 0.030 in., 0.100 in., and 1.0 in. In metric standards they are 0.025 mm, 0.075 mm, 0.75 mm, 2.5 mm, and 7.5 mm.

In deciding which cutoff value to use, it is best to take three or more sample measurements at different points on the surface without any meter cutoff value. This will give an accurate representation of the surface because all irregularities, coarse and fine, will be included. Then, given the horizontal magnification on the graphs

I. Conventional surface roughness measurement with skid (shoe) guided stylus traverse.

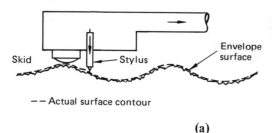

Skid    Stylus

Envelope surface

— — Actual surface contour

The stylus is sensing variations of the actual surface contour relative to the envelope surface.

(Used for average readings)

**(a)**

II. Positively referenced roughness measurement with independent datum for the stylus traverse.

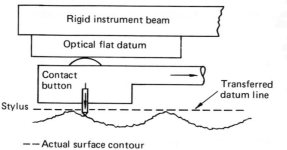

Rigid instrument beam

Optical flat datum

Contact button

Stylus

Transferred datum line

— — Actual surface contour

The stylus is sensing the true variations of the actual surface relative to the ideal contour represented by the datum element.

(Used for "peak-to-valley" recordings)

**(b)**

FIGURE 9-36. *Conventional average surface roughness, AA (a), compared to peak-to-valley roughness, Rt (b). Courtesy Society of Manufacturing Engineers.*

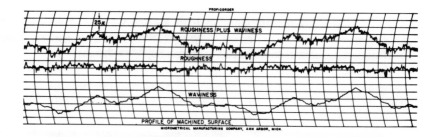

FIGURE 9-37. *The recorded profile of a machined surface, showing both roughness and waviness. Courtesy Bendix Automation and Measurement Division.*

TABLE 9-4.  *ISO Standard Traverse Lengths—Correlated with Cutoff*

| Meter Cutoff | | Traverse Length | | | |
|---|---|---|---|---|---|
| | | Minimum | | Maximum | |
| mm | in. | mm | in. | mm | in. |
| 0.075 | 0.003 | 0.40 | .015 | 2.00 | .080 |
| 0.25 | 0.010 | 1.25 | .050 | 5.00 | .200 |
| 0.75 | 0.030 | 1.50 | .055 | 8.00 | .350 |
| 2.50 | 0.100 | 5.00 | .200 | 15.00 | .600 |
| 7.50 | 0.300 | 16.00 | .650 | 40.00 | 1.500 |

produced, an estimate of the size of the spacings can be made and a meter cutoff value can be applied that will take these into account.

In measuring a surface that has little or no waviness (secondary texture), a stable measurement will be reached relatively quickly and a short cutoff value can be applied. If initial sampling reveals considerable waviness, a greater cutoff value may be needed because it may be desirable to include secondary texture in addition to primary values.

***Traverse Length.***  A standard traverse length has been designated by ISO as R-468, which is tied in with the meter cutoff used as shown in Table 9-4. The standard traverse length may vary depending upon conditions. For example, various patterns (lays) left by the grinding operation can be seen by the naked eye. These patterns will have a certain width due to the crossfeed of the wheel. It would not be a representative reading if less than the particular pattern were measured.

It is essential for the designer to realize what is being specified by the surface-texture symbol and what is not. Flaws are not specified. These are faults due to poor structure of the material or those imposed by handling. A good example of flaws is the pitted surface found on cast iron or hard coated aluminum. Although the roughness reading may be within specifications, the flaws may cause it to be rejected.

All elements of the surface specification are significant, but to overspecify can be as wasteful as to underspecify. The designer must look at the job and decide how the part is to function and how it will be processed before specifying the surface texture.

## Optical Methods of Assessing Surface-Texture Characteristics

Although stylus methods of surface measurement are used most, they are somewhat limited in that they provide a single profile and scratches may result because of the stylus' contact on highly finished surfaces. Optical methods that leave no mark on the surface include light-sectioning techniques, multiple-beam interference microscopes, and scanning electron microscopes.

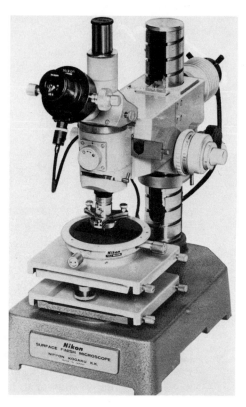

FIGURE 9-38. *The surface-finish microscope can be used in three modes: light section, multiple-beam interference, and metallographic. Courtesy Nikon.*

***Light Section.*** The surface-finish microscope shown in Fig. 9-38 can be used in three modes: light-section profile, multiple-beam interference, and metallographic. Shown in Fig. 9-39(*a*) is a sanded aluminum surface and at (*b*) is a light-section profile of the surface. A filar micrometer attachment permits a measurement of the surface roughness.

***Multiple-Beam Interferometry.*** This surface measurement is made with the aid of an adhesive film which is pressed against the surface and then removed. The process provides a highly reflective replica of the surface. Shown in Fig. 9-40 is a photomicrograph of a minute scratch on the surface of a gage block. The photograph was taken of the film replica using a multiple-beam interference mode. In this mode the fringes (dark bands) are produced when the workpiece replica (tape) is properly positioned with respect to the mirror and the objective lens of the surface-finish microscope. As shown, the gage-block surface is theoretically perfect except for the scratch. The distance between the fringes is one-half the wave length of the light used, which in the case of helium would be 11.6 millionths of an inch.

***Scanning Electron Microscope (SEM).*** The SEM makes possible resolutions at least ten times as great as the best optical microscope. In manufacturing applications, the SEM makes surface measurements that are capable of being specifically correlated with other material properties of engineering importance. Although the process is as yet experimental, a complete system for electronic measurement of surface topography is envisioned, as shown in Fig. 9-41. The system consists of a data-

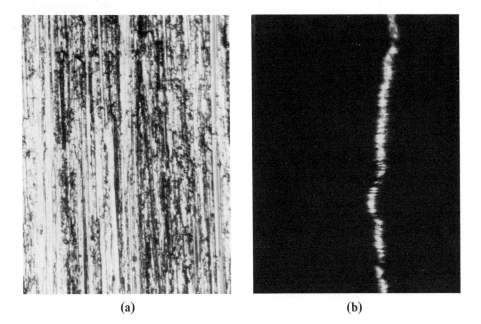

(a)                                                      (b)

FIGURE 9-39. *(a) The metallographic mode of examining surface texture by reflected light. This photograph is of a sanded aluminum surface with magnification of 130×. (b) A light-section photomicrograph taken of the aluminum sanded surface shown at (a).*

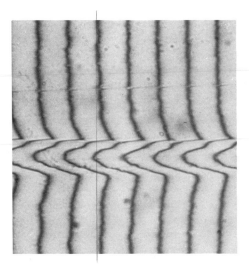

FIGURE 9-40. *A photomicrograph of a minute scratch on the surface of a gage block. The photomicrograph is taken by the multiple-beam interferometry mode using a pressed-film replica.*

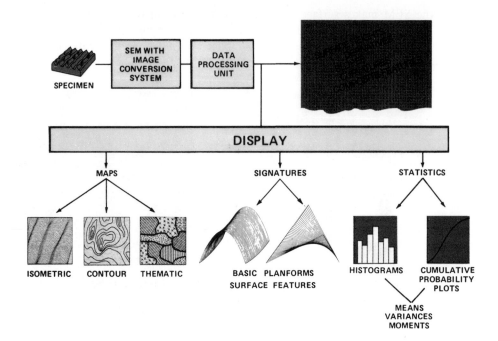

FIGURE 9-41. *A systems approach to surface texture using the scanning electron microscope (SEM) to obtain the information which can then be given to a computer to produce statistical data. Courtesy H. T. McAdams, Calspan Corp., and* Modern Machine Shop.

acquisition module based on scanning electron microscopy, a conversion module for transforming the SEM imagery to height measurements, a computer for data processing, and a display unit. The display can be of various forms, such as contour maps, or it can generate various statistical measures as required by the application at hand.

Although the SEM described is still a laboratory instrument, when combined with statistical sampling techniques a more thorough assessment of surface finishes for critical applications will be made available.

Design and production engineers must be aware of how each process affects the surface texture. Shown in Fig. 9-42 is a guide to surface-finish capabilities of various processes given in both micro and metric units.

### Gaging

The tools and instruments discussed thus far are largely used for individual part checking, with the exception of the comparators. Comparators such as the dial indicator, electronic type, and air type may be considered *adjustable-type gages* and those to be discussed now are *fixed-type gages*.

ROUGHNESS HEIGHT RATING MICROMETRES, μm (MICROINCHES, μin) AA

| PROCESS | Metric Micro | 50 (2000) | 25 (1000) | 12.5 (500) | 6.3 (250) | 3.2 (125) | 1.6 (63) | 0.80 (32) | 0.40 (16) | 0.20 (8) | 0.10 (4) | 0.05 (2) | 0.025 (1) | 0.012 (0.5) |
|---|---|---|---|---|---|---|---|---|---|---|---|---|---|---|
| Flame Cutting | | | | | | | | | | | | | | |
| Snagging | | | | | | | | | | | | | | |
| Sawing | | | | | | | | | | | | | | |
| Planing, Shaping | | | | | | | | | | | | | | |
| Drilling | | | | | | | | | | | | | | |
| Chemical Milling | | | | | | | | | | | | | | |
| Elect. Discharge Mach | | | | | | | | | | | | | | |
| Milling | | | | | | | | | | | | | | |
| Broaching | | | | | | | | | | | | | | |
| Reaming | | | | | | | | | | | | | | |
| Electron Beam | | | | | | | | | | | | | | |
| Laser | | | | | | | | | | | | | | |
| Electro-Chemical | | | | | | | | | | | | | | |
| Boring, Turning | | | | | | | | | | | | | | |
| Barrel Finishing | | | | | | | | | | | | | | |
| Electrolytic Grinding | | | | | | | | | | | | | | |
| Roller Burnishing | | | | | | | | | | | | | | |
| Grinding | | | | | | | | | | | | | | |
| Honing | | | | | | | | | | | | | | |
| Electro-Polish | | | | | | | | | | | | | | |
| Polishing | | | | | | | | | | | | | | |
| Lapping | | | | | | | | | | | | | | |
| Superfinishing | | | | | | | | | | | | | | |
| Sand Casting | | | | | | | | | | | | | | |
| Hot Rolling | | | | | | | | | | | | | | |
| Forging | | | | | | | | | | | | | | |
| Perm Mold Casting | | | | | | | | | | | | | | |
| Investment Casting | | | | | | | | | | | | | | |
| Extruding | | | | | | | | | | | | | | |
| Cold Rolling, Drawing | | | | | | | | | | | | | | |
| Die Casting | | | | | | | | | | | | | | |

The ranges shown above are typical of the processes listed.

Higher or lower values may be obtained under special conditions.

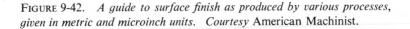

KEY — Average Application

Less Frequent Application

FIGURE 9-42. *A guide to surface finish as produced by various processes, given in metric and microinch units.* Courtesy American Machinist.

Gages in general are used for checking various part dimensions on a quantity basis. They are used as much to *prevent* scrap parts from being made as to check dimensions after the parts are made. In fact, some gages are made to be used right on the machine as the parts are being made. This is known as *in-process* gaging.

***The Rule of Ten-To-One.*** A general rule in gaging is that a gage must be ten times more accurate than the dimension being checked. For example, a part that must be checked to ±0.005 in. must have a gage that is accurate to ±0.0005 in., as shown in Fig. 9-43.

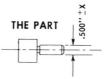

THE PART

**METHODS OF GAGING**

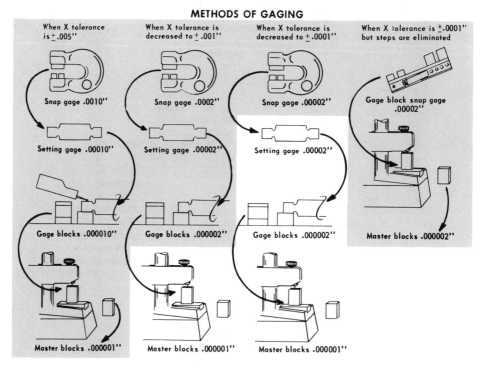

FIGURE 9-43. *Shaded areas show steps that comply with the ten-to-one rule of gaging. Steps outside are of doubtful reliability. Courtesy DoAll Company.*

***Fixed Gages.*** Fixed gages are those gages made for one dimension and are not easily altered. They may also be classified as *progressive* or *go/not-go*. These gages may be divided into groups according to the purpose for which they are used.

***Plug Gages.*** Plug gages are used to check internal diameters. They may be double-ended with a go dimension on one end and a not-go on the other. Plug gages may also be tapered or threaded, as shown in Fig. 9-44.

### Automatic Gaging

Automatic gaging provides a means of checking each part as it is being made or immediately afterward. The checking of parts as they are being made is termed *in-process gaging* and checking after they are made is termed *postprocess gaging*. It

A go—not-go progressive plug gage

A go—not-go double-ended-thread plug gage

A not-go ring gage

A go ring gage

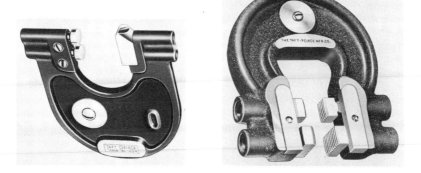

FIGURE 9-44. *Plug, ring, and snap gages. The snap gage with two anvils is a go–not-go snap gage. Courtesy Taft Peirce Manufacturing Co.*

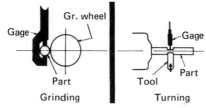

Grinding    Turning

"In-process" control or gaging during machining provides the machine or cutting tool with signals to:

* Automatically stop or change the tooling from a roughing to a finishing cut.
* Automatically retract the cutting tool when the part is to finished size.
* Automatically indicate and adjust the tool when the size trend is toward either limit of tolerance.
* Automatically stop the machine after the tools become worn and parts no longer can be machined within limits.

"In-process" control prevents faulty parts from being made by initiating signals for tool correction or replacement before part size is out of control.

FIGURE 9-45. *The " in-process" control gage shown continuously measures the size of the part. The control system produces a feedback signal to stop the machine when the nominal size has been achieved. In grinding, wheel wear is automatically compensated for. Courtesy Bendix Automation and Measurement Division.*

is done primarily by air or electronic means. It may be further classified as contact and noncontact automatic gaging.

***Contact In-Process Automatic Gaging.*** Shown in Fig. 9-45 is a contact-type automatic gage used on a cylindrical grinder. The size of the part is under continuous control of the gage; when the right size is obtained the grinding wheel retracts. The control circuit can be made to adjust the tool automatically when the size approaches the limit of the tolerance.

***Noncontact In-Process Gaging.*** The eddy-current principle is used for one type of noncontact in-process gage, as shown in Fig. 9-46. Ordinarily the eddy current spreads out too much to act as a probe, but both penetration and spread can be controlled by rapidly alternating the shaped inductive field in the air gap. Thus the device acts as a highly sensitive proximity gage. Depth control, or measurement, is accomplished by energy dissipation, which is shown by a Wheatstone bridge circuit. The normal gap range is 0.100 in. ±0.050 in. Scale readings on the meter range from 0.003 in. full scale to the smallest increment of 0.000020 in. Shown in Fig. 9-47 are two schematic presentations of this type of noncontact in-process gaging.

***Contact Postprocess Gaging.*** Postprocess gaging is applicable to a greater variety of machines than in-process gaging. It also has the advantage of being able to gage

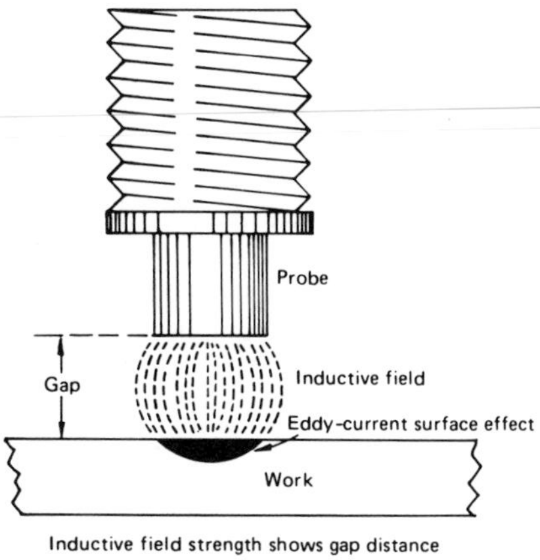

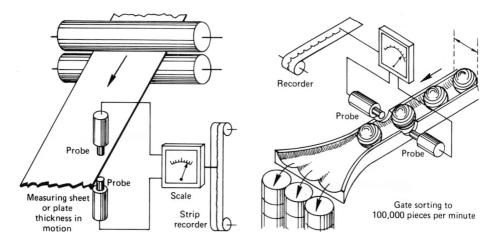

FIGURE 9-46. *A schematic of the eddy-current principle as used in noncontact gaging. Courtesy* American Machinist.

FIGURE 9-47. *Two examples of noncontact gaging using the eddy-current principle. Courtesy* American Machinist.

the parts under more stable conditions, away from any effects of heat or vibration. Shown in Fig. 9-48 is a schematic of postprocess gaging that includes sorting of parts according to size and automatic feedback to the machine controls. When the work exceeds marginal tolerance limits, the feedback will cause automatic adjustments to be made. In the case of tool failure, the machine will be stopped. An audible or visual signal is incorporated into the machine shutdown.

## Selective Assembly

Consumers are now in the habit of expecting long life on automobile and appliance parts. For example, a motor compressor unit for a home refrigerator is expected to last 15 years without trouble, or the total life for an automobile engine is

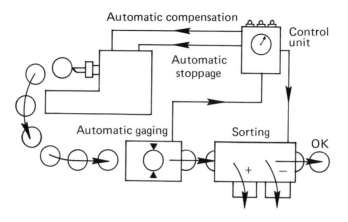

FIGURE 9-48. *A schematic diagram to show the principle of post-process contact gaging and sorting. Courtesy Federal Products Corporation.*

expected to be 100,000 miles. Production of parts of this type demands close tolerances, tolerances that are not compatible with today's prices. The alternative for the manufacturer is to have a system of selective assembly. To do this, the mass-production industries had to scrap the old idea of inspection in which parts were classified as "good" or "bad" and adopt a newer concept of selective assembly which allows much wider tolerances on the individual parts but maintains very tight tolerances on assembled units.

To accomplish selective assembly, two fundamental concepts are involved. The first involves the normal distribution curve which occurs in most production lots. This concept merely means that in any sizable production lot most of the components will cluster close to the specified dimensions if the process is basically capable of manufacturing to the stated tolerances. However, there will always be a few parts that are towards the extremes of the tolerance, usually $\pm 2\sigma$ limits, as indicated in the normal distribution curve in Fig. 9-49(*a*). Plus or minus one standard deviation ($1\sigma$) under the normal gaussian curve includes 68 % of all values that fall between the minimum and maximum values, and $\pm 2\sigma$ limits includes 95 % of all values, as shown in Fig. 9-49(*b*).

The second consideration in selective assembly, as in the case of shafts and sleeves, is that the relationship must be very close, possibly within 0.0001 in. (0.003 mm). However, the shaft or sleeve may have a relatively wide tolerance such as 0.001 in. (0.03 mm) or more. Thus, if in the production of shafts and sleeves the tolerance was held within 0.001 in. (0.03 mm), an automatic gage could segregate all components into ten different groups with 0.0001 in. (0.003 mm) limit for selective assembly of the individual units.

Of course, there is always the probability that the two component parts would not have a symmetrical or normal distribution. The shafts may all be on the high side and the sleeves on the low side. The net result would be a lot of parts that could

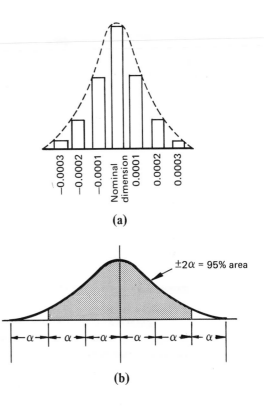

FIGURE 9-49. *(a) A histogram showing normal distribution of acceptable shaft diameters made to a tolerance of 0.001 in. (b) A normal curve with the shaded area representing $\pm 2\sigma$ limits, or 95% of the area.*

not be matched. The most recently developed answer to the problem is to combine automatic gaging with minicomputer control. This gaging procedure is being used in the automotive industry in grouping wrist pins into categories of 0.0001 in. diameter. As the bore for the wrist pin is measured, the proper pin is automatically selected and inserted at rates up to 1800 per hour.

## INSPECTION, QUALITY CONTROL, RELIABILITY, AND PRODUCTS LIABILITY

The terms inspection, quality control, and reliability may seem at first to be somewhat synonymous. They are, however, quite different in application. Inspection is concerned with *how well* the physical specifications for a product are being met. Quality control is not as concerned with the physical aspects of the individual part as it is with the prevention of bad parts by use of statistical techniques. Reliability is concerned with the probability of satisfactory operation of the product over a specified time interval in a specified environment. Products liability, a comparatively new term, is concerned with consumer protection.

### Inspection

Inspection has two equally important tasks. One of these is to detect errors in workmanship and defects in materials so that the end product will meet specifications; the other is to help minimize future errors. It is the latter that occasionally leads the inspection function to go beyond workmanship and materials and into design specifications. For example, inspection may reveal tolerances that are impractical or materials that are inferior for the intended use.

***Inspection Requirements.*** The inspection department must have a broad knowledge of the products being manufactured and their end use. The inspector must be familiar with the product design specifications and know the relationship of each component and how it functions in respect to the complete assembly. He or she must set up a sequence of inspections designating at what time in the manufacturing operations inspection takes place. He or she will also determine the type of equipment needed to carry out the inspection.

***Inspection Procedure.*** The manufacturing process is analyzed to establish the best sequence and methods of inspection. Common inspection steps are first-piece, sampling, and final inspection. Inspection procedures may involve various sampling methods.

*First-Piece Inspection.* Usually right after a machine has been set up for a new product or a change in dimension the operator will call the inspector to check the first piece before the production run is started. The inspector must have all the specifications, which are usually obtained from the part print. After the first piece has been inspected, critical dimensions are selected thereafter and used as the basis of inspection.

Specifications may often be abbreviated on a print by a reference to a given standard. For example, the reference might be for a standard 3/4-10 NC-3 thread. The inspector knows this refers to the American National Coarse Thread series, 10 pitch, and a class three fit as specified in the American National Standards *Screw Thread Handbook*. The size in this standard specifies an 0.0750 maximum diameter, 0.7410 minimum diameter, and a tolerance of 0.009 in. Many other standards are established by engineering societies such as ASTM, AISC, SAE, AWS, etc., and by plant usage. Once standards are set up and accepted, they not only make for interchangeability, but prevent much duplication of effort.

*Sampling Plans.* Between first-piece and final inspection, various sampling plans may be set up. A roving inspector may tour the shop inspecting parts at random or he may have certain set intervals and a specified sample size.

*Batch Sampling.* Batch sampling is generally done on smaller parts that are removed from the manufacturing floor. Each batch is chosen and sampled according to statistical methods.

*Final Inspection.* Inspection of parts and assemblies just before they are prepared for packaging is called final inspection. At this stage much of the inspection may be visual, checking for flaws, defects, and missing or damaged parts. Products such as engines or motors are usually given an operational test. Since all of them are tested, it is considered 100 % inspection.

## Quality Control

As stated previously, quality control is concerned with the prevention of bad parts. To achieve the desired product standards, agreed upon by engineering and management, quality control sets up sampling plans which assure the process is under control and, at the same time, inspection costs are being kept at an economical level.

Very little was done in the application of statistics to quality control until the beginning of World War II when, faced with the job of buying enormous quantities of war material, the U.S. Government adopted the use of statistical sampling plans. In general, contracts with suppliers indicated that material would be subject to sampling inspection. Failure of several lots to meet specifications as judged by those samples would result in disqualification of that particular manufacturer as a supplier. Individual manufacturers responded by attempting to train people to understand and apply sampling plans within their own industry and the government aided the effort with appropriate courses.

When industries found that their own inspection departments, using these sampling plans, were rejecting high percentages of their own parts, they recognized the desirability of installing "in-process" or "performance" control systems. By this method, they were able to detect and eliminate defective material as soon as possible after it was produced. In addition, the new system improved the possibility of finding and correcting the point in the process that caused the production of defective items. Thus a major tool developed for in-process control, *capability studies* and *control charts*.

*Process Capability.* Process capability studies are a quality control technique used for determining how much variation there is in the process and what the average variation is. For example, a lathe is used to turn out parts that have a 1-in. critical diameter with a tolerance of $\pm 0.005$ in. The result of a sampling plan used to inspect the parts is shown in Fig. 9-49. The curve gives the following information:

1. The variation pattern is consistent; that is, the process is in a state of stastical control.
2. The total variation of the lathe is 0.012 in., which is greater than the tolerance (0.010 in.).
3. The lathe is being operated on the high side, 0.003 in. above the nominal dimension.
4. Under current operating conditions, approximately 16 % of the product is beyond the upper specification limit and will probably require reworking (determined by computing the area of the normal curve in the out-of-specification segment).

Possibilities for action are:

1. Reset the tool to the nominal dimension. This will reduce the out-of-tolerance product from 16 % to 4 %. However, the *process capability* is still not acceptable.
2. Recondition the lathe to bring the total variation within the 0.010 in. tolerance.

3. Change the specifications to coincide with the process capability.
4. Transfer the job to more capable equipment.
5. Leave as is and screen defective parts.

Suppose the first possibility is chosen from the five foregoing alternatives The key factors in determining process capability are the natural tolerance ($\mu_{na}$) and the standard deviation.

The equation for natural tolerance is:

$$\mu_{na} = \sqrt{(\text{tol. of part 1})^2 + (\text{tol. of part 2})^2}$$

$$= \sqrt{0.006^2 + 0.008^2} = 0.010 \text{ in.}$$

The standard deviation may be defined as one-third of the natural tolerance:

$$\sigma = \mu_{na}/3$$

$$= 0.010/3 = 0.0033 \text{ in.}$$

*Note:* The standard deviation may be somewhat oversimplified since it will be dependent on the sample size. If the sample size is between five and ten, the example given is satisfactory.

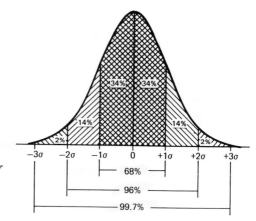

FIGURE 9-50. *The approximate areas for the typical normal curve.*

After determining the value of the standard deviation $\sigma$ we know that $\pm 3\sigma$ will include approximately 99.7 % of the product (Fig. 9-50) and is considered to be the total variation or the process capability. The only precaution necessary is to assure that the process is normally distributed and that it is in a state of statistical control. This can be determined visually from the histogram comparison to the normal curve and from plotting control charts.

In practical mass production, grinding tolerances may be held to 0.001 in. (0.03 mm). However, in hydraulic valves the fit tolerance between the sleeve and the shaft may be 0.0001 in. (0.003 mm). This requires selective assembly, that is, shafts from the regular production lot must be gaged and classified to the nearest "tenth" dimension. Likewise the sleeves. Such classifying work is totally impractical with manual gaging equipment but is entirely feasible on automatic equipment. Automatic gaging equipment is available that can gage and match 1200 assemblies per hour with 100 % reliability.

In production, the grinders are set at the nominal dimension and the bulk of the production will cluster at the midpoint in the normal distribution. However, the production may shift or "skew" to the right or left, as shown in Fig. 9-51.

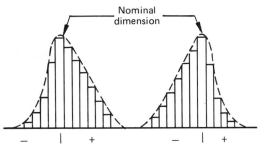

FIGURE 9-51. *Histograms of actual production lots may be skewed either to the right or left. It is important when matching close-tolerance parts by selective assembly that they have similar histograms.*

As mentioned previously the lots of shafts and sleeves brought to the automatic gaging machine may be skewed in opposite directions and there will be many parts that will not match. Thus a histogram of each production run is necessary for effective assembly tolerances. In automotive plants where parts are subcontracted, it has become a practice to include histograms of tolerances so that the assemblies can be more effectively matched.

***Control Charts.*** Control charts are basically a graphic means of presenting in-process data on a particular machine operation. The control chart as first introduced by Western Electric in the 1920s presented a new and powerful concept in that a certain amount of variation in a process, as distinct from the product, must be considered "normal" or expected.

***$\overline{X}$-and-R Charts.*** One of the most common of the quality control charts is the average-and-range chart or $\overline{X}$ ($X$-bar) and $R$ chart. For this chart, samples are taken at regular intervals. The size of the sample is usually four or five items. The average value of the critical dimension is plotted for the given time interval on the chart, as shown in Fig. 9-52.

The centerline of the chart is obtained from the grand average or $\overline{\overline{X}}$. This is usually shown on the chart as a red dashed line. The ranges are also averaged and a line drawn as shown at $\overline{R}$. The upper and lower control limits can then be calculated by multiplying the average range ($\overline{R}$) by a constant (Table 9-5) which is based on the number of observations in the sample ($n$).

Control limits can be established after a reasonable number of samples have been taken, say after 20 or 25 samples. Too few samples will not give a true picture.

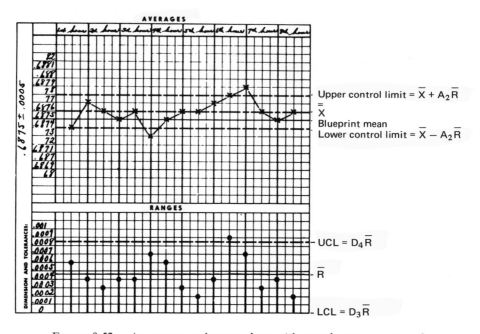

FIGURE 9-52. *An average-and-range chart with sample averages posted each half hour. The range of each sample is shown on the lower half of the chart. Courtesy Federal Products Corporation.*

The $\overline{X}$ chart indicates the level at which a machine is operating and the $\overline{R}$ chart indicates the amount of variation that exists in the machine output. Usually the operator is responsible for controlling the tool setting. Therefore the chart shows his ability to establish and maintain the process average.

The biggest advantage of in-process control is that a malfunctioning machine or an improper tool setting is discovered almost immediately. This reduces scrap and repairs.

TABLE 9-5. *Control Chart Constants for Small Samples. Courtesy ASTM*

| Number of Observations in Sample, $n$ | Chart for Averages | | | Factor for Central Line | Chart for Ranges | | | |
| | Factors for Control Limits | | | | Factors for Control Limits | | | |
| | $A_1$ | $A_2$ | $A_3$ | $d_2$ | $D_1$ | $D_2$ | $D_3$ | $D_4$ |
|---|---|---|---|---|---|---|---|---|
| 2 | 2.121 | 3.759 | 1.880 | 1.128 | 0 | 3.686 | 0 | 3.268 |
| 3 | 1.732 | 2.394 | 1.023 | 1.693 | 0 | 4.358 | 0 | 2.574 |
| 4 | 1.500 | 1.880 | 0.729 | 2.059 | 0 | 4.698 | 0 | 2.282 |
| 5 | 1.342 | 1.596 | 0.577 | 2.326 | 0 | 4.918 | 0 | 2.114 |

### Zero Defects

The term "zero defects" has become popular or unpopular depending upon one's point of view. In one sense it would appear to be asking for the impossible or for the acceptance of unduly rigorous surveillance of the manufacturing process. The latter would appear to include prohibitive costs.

Some plants have achieved this seemingly impossible goal by using what is known as *target-area control*. The approach is quite simple. One plant that adopted this plan has shipped over 30,000,000 small parts with zero defects.

As an example of the use of the plan, assume a part has a critical 0.5-in. diameter with a ±0.002 in. tolerance on a 0.5 in. long body. The *target area* is defined as the middle half of the tolerance or, in this case, 0.499 to 0.501, as shown in Fig. 9-53.

FIGURE 9-53. *Based on normal statistical laws 86% of the parts will fall within the middle half of the limits shown. In this case it is rounded out to nine out of ten parts. With this percentage in the middle half of the tolerance, the entire production will be within the allowable tolerance. Courtesy* Modern Machine Shop.

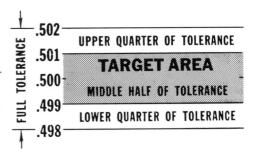

Two simple rules are set up to achieve the target-area tolerance: (1) Set up: at set-up time, five consecutive samples are taken. If they are found to be within the target area, the run is started. (2) Running: After the run has started, (say for 10 min), two samples are taken in a row. If the first piece is within the target area, both parts are placed with the good parts and the operation continues. If the first piece is within the tolerance but outside the target, the second piece is measured. If two pieces in a row are inside the tolerance but outside the target, step one must be repeated.

Some parts are allowed outside of the target, about one out of ten. Statistically the chances of picking two parts in a row outside the target area are only one in 196. Thus the operator knows with two pieces in a row outside the target area the odds are strongly against returning to it. He notes the size of the pieces; if they are both high he adjusts downward, and vice versa. If one is high and one is low, then there is too much variation or "spread" in the process and the process capability of the equipment should be examined.

### Adaptive and Computer Quality Control

Quality control has, in many cases, been taken over by automatic gaging and feedback, or what is known as *adaptive control*. As an example, a cylindrical grinder may have in-process gaging which compensates for wheel dressing and wheel wear.

Also, with the advent of the minicomputer, which is capable of monitoring two or three machines, part quality can be under continuous surveillance.

### Reliability

In more recent years, reliability has become a common term or, simply stated, both the public and industry have become "reliability-conscious." The main reason has been the customer's incessant demand for a more reliable product.

Reliability may be defined as the probability of a device giving adequate performance for a given length of time under a specified operating environment. Thus reliability is a property determined by many factors: design, production, operational demand, operator use, maintenance, personnel skills, etc.

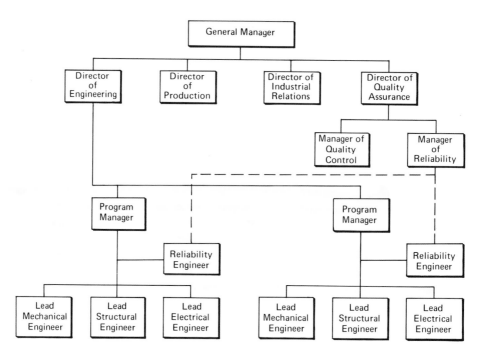

FIGURE 9-54. *The organization of a company showing the position of the reliability engineer as reporting to the manager of reliability, who in turn reports to the general manager.*

Manufacturers often provide a warranty on their products. Before this can be done, the reliability engineer must determine the probability of having to supply additional equipment or make repairs under the terms of the warranty. Thus some of the personnel in a reliability group should be engineers with statistical backgrounds.

Some conflict may develop between reliability programs and those people responsible for delivery and costs. A separation of responsibilities, where each group reports directly to the general manager, is shown in Fig. 9-54.

A reliability program is involved in eight areas:

1. Design criteria. Once preliminary designs are completed, estimates are made on the basis of the design criteria and these are compared to the original reliability goals. From this comparison it can be determined if further design work is needed to improve reliability.
2. Compenent part specifications. Close control of component parts is necessary to see that they meet specifications as stated on the drawing.
3. Design review. A design review is needed to appraise the product at critical stages of its development. This allows design and reliability engineers to evaluate and suggest improvements while the design is still fluid.
4. Reliability computations. A sound reliability program can be effective only with adequate statistical analysis. The prediction study should, among other things, isolate potential sources of trouble and indicate the magnitudes of risks of failure by source. The reliability engineer should point out the assumptions under which the analysis was conducted and the consequences of design changes.
5. Qualification testing. Reliability qualification testing of components and sub-assemblies should be carried out only when specifically required or when there is no other way to determine component reliability. Reliability can often be determined from analysis based on data accumulated from other tests.
6. Acceptance criteria. Quality control must establish a method of acceptance that does not fall below established standards.
7. Failure and deficiency reporting. This is a system for collecting, recording, summarizing, revising, and analyzing equipment-performance data. The procedure allows reliability to improve prediction accuracy.
8. Reliability check. Reliability should be predicted and measured during and at the end of each of the following design stages: conceptual design, preprototype design, preprototype demonstration, full-scale production, product improvement.

*Reliability Advancement.* The advancements made by reliability in recent years are often taken for granted. For example, for a machine with 1000 parts the probability of each part functioning for 24 hours may be 99.99 %. However, the probability of the entire system working for 24 hours would be reduced to 91 %. Yet a modern jet engine has thousands of parts with each part exhibiting several modes of failure. Each mode of failure has a certain probability of occurrence. This probability of failure has been reduced by continuing development programs. As an example, failure frequencies in the range of one every 10,000 hours to one every 100,000 hours now require an investigation and reliability design reviews. Thus reliability programs have made beneficial advances for both industry and the consumer. The engineer has given more rigorous treatment to such areas as reliability specifications, design reviews, analysis of failure and manufacturing problems, surveillance of production and environmental testing, vendor surveys, material developments, field testing, maintenance records, etc.

## Products Liability

As stated previously, product liability is concerned with consumer safety. An event that happened in 1911 first brought this subject into sharp focus.

A Mr. MacPherson was driving his automobile in upstate New York at a modest speed of 15 mph (24 kmph) when a wheel fell off. The car flipped over and MacPherson ended up in the hospital. He sued Buick for negligence. The case of *MacPherson v. Buick Motor Company* became a landmark case and is still widely quoted as a reference in current legal actions.

In its defense, Buick raised what was then a plausible argument: it had not sold MacPherson anything directly, therefore there was no direct contractual agreement between the company and the car owner. The judge felt otherwise. In the now famous opinion rendered in the New York Court of Appeals, he extended manufacturers' liability to third parties for any product "reasonably certain to place life and limb in danger when negligently made." His decision did leave one vital requirement, the plaintiff must prove that the manufacturer was negligent. This principle has been severely eroded in recent years.

Because proof of negligence is often highly elusive (it is presumed that only manufacturers know all the technical facts), and because products are becoming increasingly sophisticated, more and more courts are finding that plaintiffs need not prove negligence in manufacture. They need only prove that the product had a defect when it left the manufacturer's hands, whether or not that defect was known or could have been prevented.

In 1960 the New Jersey Supreme Court, in a case of *Henderson v. Bloomfield Motors Company*, allowed an injured customer to recover against the manufacturer as well as the dealer. This doctrine is being gradually adopted by courts across the country.

The government's National Commission on Product Safety has encouraged the accreditation of private testing laboratories, control of imported products, safety research and development of test methods and product analysis, and trade regulation rules for those who certify or endorse the safety of consumer products.

In 1972 the Consumer Product Safety Act was signed into law. This was the first real attempt by the government to enforce safety standards by preventive measures. The act has, in effect, forced manufacturers to evaluate their entire product safety organization and all management functions relating to getting a product on the market.

*Product Liability Loss Prevention.* From the foregoing discussion of products liability, it would appear that the odds are heavily against a manufacturer unless he takes the utmost precaution to produce safe products. How can this safety be assured without incurring undue cost?

Investigations have revealed that no one group within a manufacturing organization can assume the total responsibility for product liability loss prevention. The errors of omission and/or commission can be traced in many instances to routine or minor decisions made in nearly all areas of a company's operation and at all levels of responsibility. In product liability claims, the magnitude of the error bears little relationship to the magnitude of the potential loss resulting from it. Sometimes the cause is not the result of outright error, but of innocent, well-intentioned, production-line decisions.

These observations indicate that the best approach to the problem is total involvement by all company personnel. A special organization responsible for product liability is not only costly but produces a false sense of assurance to others. However, to be successful a liability loss program does require an organizational mechanism to provide coordination of the effort, follow-up audits, and a line of accountability to top management. A product liability committee can accomplish these ends.

A product liability loss prevention program can be organized into seven main steps:

1. Education. Emphasis in education is placed on the importance of the program.
2. New-product safety review. A special concerted effort is needed to minimize risk associated with new products. Plans for development and end use must be carefully reviewed.
3. Establishment of risk criteria. Some products pose greater risks than others. Products that are rated as "significant risk" must be subjected to more stringent production controls.
4. Control of warranties, advertisements, and technical information. A review by upper management of all warranties, publications, and advertising is needed before release.
5. Complaint and claims procedure. Complaint documents serve the accounting function and are a source of information for quality control. They also serve as a communications link between the manufacturing operation and the market. An early alert by a disgruntled customer often serves to resolve an adverse occurrence before it becomes a dispute.
6. Records retention schedule. The collection of design, production, and sales records is the first critial step in resolving an adverse claim or preparing for a product liability lawsuit. Records cannot be kept indefinitely but decisions as to what should be kept can be made on the expected life of the product.
7. Products liability audit. As mentioned, the key to developing an effective product liability loss prevention program is to obtain the total involvement of each department and not let the task fall on one or two groups closely related to the problem.

In short, a manufacturer's defenses must be built into his product by the effort of all concerned.

# PROBLEMS

**9-1.** (a) If you travel at 90 km/hr, what would it be in miles/hr? (b) At 65 mph what would the kilometer speed be?

**9-2.** Shown in Fig. P9-1 are three flat surfaces as seen through optical flats. How much out of flat are the surfaces?

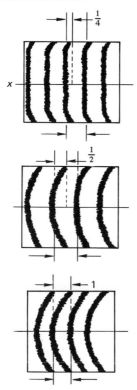

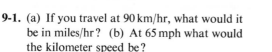

FIGURE P9-1

**9-3.** How could the accuracy of a micrometer be checked to within 0.001-in. at four or five points over its 1-in. length of travel?

**9-4.** A sine-bar fixture is used to check the included angle of a plug gage as shown in Fig. P9-2. (a) Determine the angle from the information given. (b) What instruments can be used to check the sine-bar plugs to an accuracy of 0.001 in.? The height over the plugs is 1.745 in. and 3.276 in.

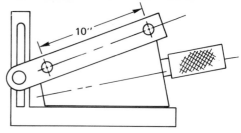

FIGURE P9-2

**9-5.** Gage blocks are used to calibrate a dial indicator in steps of 0.010 in. over a travel of 0.1000 in. Blocks available are as follows:

| Series, in. | No. of Blocks | Range, in. |
|---|---|---|
| .010 | 1 | .010 |
| .0001 | 9 | .1101 thru 1009 |
| .001 | 11 | 0.020 thru .030 |
| .010 | 9 | .110 thru .190 |
| .100 | 3 | 1.000, 2.000, 4.000 |

Select the blocks and state the procedure to use.

**9-6.** (a) A tapered gib is shown in Fig. 9P-3. State three methods that could be used to measure the angle of the gib in degrees and minutes. (b) What is the taper of the gib block in in./ft?

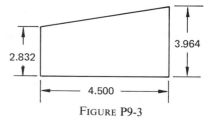

FIGURE P9-3

**9-7.** Using a 5-in. sine bar, what is the length of the side *c*, of a right triangle, or hypoteneuse of a setup for measuring tapers?

**9-8.** What is the percent increase in cost to go from a 63$\mu$ finish to a 32$\mu$ finish?

**9-9.** If a closed 1-in. micrometer is opened $4\frac{1}{2}$ turns, what will the reading be?

**9-10.** Convert the following to metric measure. Remember the use of the space instead of the comma and the comma instead of the decimal.

Area
a. 1 in.² to mm²
b. 2 in.² to mm²

Stress
a. 3000 psi to kPa
b. 30,000 psi to MPa

Energy
a. 300 lbf to J
b. .3 lbf to J

Length, the following to mm, cm, or m— whichever is the most appropriate
a. 8 in.     d. 20 ft
b. .008 in.  e. .5 ft
c. 2 ft

Liquid Measure
a. 10 U.S. gal. to liters
b. 10 quarts to liters

**9-11.** An angle plate is assumed to be 38°52′. You decide to check it with a 10-in. sine bar. What would the height of the gage blocks be for the opposite side?

**9-12.** Five splines have been cut on a shaft with a root diameter of 4.000 in. (Fig. P9-4). The splines are 1 in. wide. What should distance $F$ be?

art p9-4

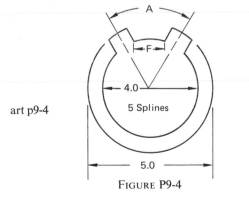

FIGURE P9-4

**9-13.** Sketch a functional gage that would be used to check the location and perpendicularity of four equally spaced $\frac{1}{2}$ in. square slots cut into a 4 in. diameter, $\frac{1}{4}$ in. thick round plate, as shown in Fig. P9-5.

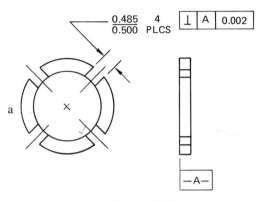

FIGURE P9-5

**9-14.** What instruments would be used to check the part shown in Fig. P9-5 in small quantities, as to:
(1) The depth of the slots to within 0.001 in.
(2) The outside diameter to within 0.001 in.
(3) The 90° angular relationship of the slots, to degrees and minutes.
(4) The minimum diameter to 0.001 in.
(5) The roundness of the part to within 0.0001 in.
(6) The width of the slots to within 0.0001 in. What instruments would be used to check the following dimensions of the part shown in Fig. P9-5 in large quantities?
    1. The outside diameter to 0.001 in.
    2. The minimum diameter to 0.001 in.
    3. The width of the slots to 0.001 in.

**9-15.** In constructing an average-and-range chart the grand average ($\overline{\overline{X}}$) was computed to be 8.502 in. and the range average was 0.008 in. What would the upper and lower control limits be on the $\overline{X}$ chart if the sample size was 3.

**9-16.** What would the difference in size be between a 3.000000-in. steel gage block measured at 68°F (20°C) and at 78°F?

**9-17.** Two plates, each 2.00 in. ±0.006 in. thick, are assembled on a production line. (a) What is the total natural tolerance ($\mu_{na}$) of the assembly? (b) What is the total range or variation of the assembled plates that would be considered satisfactory? (c) Assume that the actual process, when run, showed a total tolerance of ±0.003 in.

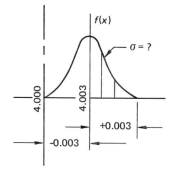

FIGURE P9-6

Make two sketches showing how the normal distribution curves would appear, theoretical and actual. (d) Would the distribution as shown in Fig. P9-6 be satisfactory?

Why or why not? (e) What would the natural tolerance be for the assemblies for the case shown at (d)? (f) What would the standard deviation be for the case shown at (d)?

**9-18.** A designer places the following surface finish specification symbol on a given surface:

$$\frac{0.003-2}{\sqrt[10]{\begin{array}{l}0.030\\ \pm 0.002\end{array}}}$$

(a) Is the roughness width cutoff consistent with the roughness width? (b) What processes would be needed to achieve the surface roughness specified? (c) What would the roughness be in metric measurement? (d) What may be a cause of the waviness?

**9-19.** In an $\bar{X}$ and $R$ control chart, can the $R$ portion serve as a process capability chart?

**9-20.** What is the depth of the scratch on the gage block shown in Fig. 9-40 assuming a helium-type light was used?

# BIBLIOGRAPHY

**Books**

ASTM, *Metric Practice Guide*, ASTM E380-72, Philadelphia, 1972.

ASTM, Handbook of Industrial Metrology, Prentice-Hall, Inc., Englewood Cliffs, New Jersey, 1967.

GALYER, J. F. W., and C. R. SHOTBELT, *Metrology for Engineers*, Cassell & Co. Ltd., London, 1964.

HUME, K. J., *Metrology with Autocollimators*, Hilger & Watts Ltd., London, 1965.

MECHTLY, E. A., *The International System of Units, Physical Constants and Conversion Factors*, NASA, Office of Technology Utilization or Gov. Printing Office, Washington, D.C., Catalog No. C13.10:330.

RAMASWAMY, G. S., and V. V. Rao, *SI Units*, Tata-McGraw-Hill Publishing Company Ltd., Bombay, 1971.

RONNINGEN, H. A., *Weights and Measures, U.S. to Metric—Metric to U.S.*, Ronningen Metric Co., Paradise Valley, Arizona, 1969.

*Seminar of Precision Measurement*, The DoAll Company, Des Plaines, Illinois, 1962.

**Periodicals**

BITTENCE, J. C., "Dimensional Accuracy, What It Is and How To Get It," *Machine Design*, June 14, 1973.

FARAGO, F., "Dimensional Metrology in The Service of Roller Bearing R & D," *Paper No. SP65-49.*, ASTM.

FEINBERG, B., "Product Liability—And A Strategy for Survival," *Manufacturing Engineering & Management*, May 1973.

FREUND, R. A., "Control Charts Eliminate Disturbance Factors," *Chemical Engineering*, Jan. 31, 1966.

GASIECKI, E. A., "Scanning Electron Microscope and Computer: New Tools for Surface Metrology," *Modern Machine Shop*, June 1974.

GELLMAN, N., "How to Set Up a Reliability Control Program," *Machine Design*, March 26, 1964.

KLEIN, S. J., "Product Liability in the 1970's—The Engineer's View," *Mechanical Engineering*, Jan. 1974.

KOEHLER, T. L., "Quality Control: Gap in CPI Armor," *Chemical Engineering*, Jan. 3, 1966.

McADAMS, H. T., Scanning Electron Microscope and the Computer: New Tools for Surface Metrology, *Modern Machine Shop*, June 1974.

NICOLLS, M. O., "Analyzing Surface Finish," *American Machinist*, May 13, 1974.

WELKER, E. L., "The Basic Concepts of Reliability Measurement and Prediction," Paper presented at the 5th Annual West Coast Reliability Symposium, Los Angeles, Feb. 1964.

WISE, C. E., "Products Liability," *Machine Design*, March 28, 1968.

# 10

## *Protective Surface Treatments*

Protective coatings for metals are selected for a number of reasons, such as appearance or customer appeal, cost, corrosion resistance, wear or abrasion resistance, and toughness or resistance to impact. To help ensure lasting qualities of the finish, a thorough preparation, mainly cleaning, of the base metal is necessary.

## METAL CLEANING

During manufacture, many types of contaminants are encountered by the product. These usually include cutting oils, greases, waxes, tars, rust preventatives, dirt, heat-treat scale, and drawing and buffing compounds.

Not all contaminants can be removed by any one easy cleaning method. The problem must be met with a knowledge of each of the contaminants involved. A broad classification of cleaner types is: alkalines, solvents, and acids.

### Alkaline Cleaners

Alkaline cleaners are the most widely used of the three basic types of cleaners because they remove most types of soil and because they can be applied by most any method.

Alkaline cleaners consist mainly of detergents dissolved in water at the time of use. The effectiveness is dependent to a great extent on the detergent or wetting properties of the solution that break the common boundary wherever soil and metal meet and to a lesser extent on the agitation or movement of the bath relative to the parts being cleaned. Common methods of applying alkaline cleaners include electro-cleaning, spray cleaning, dip-tank cleaning, and steam cleaning.

*Electrocleaning.* Before electroplating, the metal is electrocleaned. This is done by making it the anode in a hot alkaline solution. As the current passes through the metal, thousands of tiny oxygen bubbles are formed on the surface. As these bubbles burst, they perform a scrubbing, scouring, agitating motion over the entire surface of the metal. Thus the dirt can be said to be "unplated" from the metal. This process produces the cleanest surface possible from a conventional alkaline solution.

*Spray Cleaning.* Spray-rotary washers are used for small and medium stampings and small machined parts such as nuts and bolts. The alkaline cleaning solution is sprayed on the parts as the drum rotates. After the cleaning spray, a rinsing spray is used. This operation may include some rust inhibitor. The time required by either spray or electrolytic cleaning is usually 0.5 to 1 min at temperatures ranging from 130 to 170°F (54–77°C) for spraying and 160 to 210°F (71–99°C) for the electrolytic process.

*Dip-Tank Cleaning.* Parts with light soil can be cleaned by dipping in an alkaline solution. Parts that are too small to put on racks and that are in quantities too large for handling in baskets are frequently cleaned by the use of rotating barrels. A barrel cleaning line should have at least two cleaning tanks. The first stage removes the gross soil. This allows the subsequent stage to operate more effectively in a comparatively unpolluted solution.

*Steam Cleaning.* Steam-gun cleaning is used mostly on machinery that is too large for soak tanks and spray-washing machines. The alkaline steam solution used on steel or cast iron surfaces may be strong in caustic soda with additions of orthosilicates.

*Ecological Considerations.* An important facet of today's cleaning is disposition of the fluid waste product. Once the soil has been dispersed by the solution, it should reunite away from the work to precipitate out or be skimmed off. The ability of a cleaning solution to perform this function increases its life and produces cleaner work. Most chemical suppliers are now offering biodegradable surfactants (surface-active agents) and phosphate-free cleaners.

### Solvent Cleaning

When heavy oils, grease, and dirt are to be removed from the product, either straight or emulsifiable solvents are used. Straight solvents may be either of the petroleum-base or chlorinated types.

*Petroleum Solvents.* Petroleum solvents are distillates having sufficiently high flash points, 100 to 200°F (38–93°C), to permit use at room temperatures. They are less expensive than the chlorinated hydrocarbons, but they are a fire hazard and must be handled accordingly. The equipment used can be simply a tank for dipping the part, brushes for removing the contaminant, and air drying. Solvents are quickly contaminated and, when this occurs, they do not produce a high degree of surface cleanliness.

***Chlorinated Solvents.*** Chlorinated solvents do not have the fire hazard of petroleum solvents. They are usually used in vapor degreasers with the parts placed in wire baskets suspended in vapors above the boiling solvents. Trichloroethylene boils at 188°F (85°C) and perchloroethylene at 250°F (121°C). A cooling jacket in the upper portion of the tank causes the vapors to condense on the metal surfaces and carry off the grease and soil. This method has the advantage of making the most effective use of the solvent over the entire surface of the part. Good penetration is obtained in even the deepest recesses.

Solvents can become contaminated with water, cyanides, or other reactive materials that have lodged in the recesses of the part as a result of preceding processes. The designer can often prevent this by providing for drainage without functional interference.

***Emulsifiable Solvents.*** The emulsifiable solvents have received a great deal of attention because of the comparative ease of disposal as described in Chapter 1. These solvents are compounded so they can be diluted with water. They are less toxic than the chlorinated solvents. After being agitated in a tank or sprayed with a hot emulsifiable solution, the parts are given an alkaline spray rinse. Wherever process soils are exceptionally heavy, solvents should be considered.

## Acidic Cleaners

Acid cleaners are solutions used for pickling, deoxidizing, and bright dipping. These operations are performed after the oil, grease, and other substances not soluable in acids have been removed from the metal surfaces.

***Pickling.*** Pickling refers to the removal of oxidation, mill scale, flux residues, etc., by solutions of sulfuric or hydrochloric acid. Sulfuric acid is cheaper, and its rate of scale removal can be stepped up when heated to 150°F (66°C). Hydrochloric acid is used at room temperature. Inhibitors have been developed to reduce the harmful action of the acid fumes on plant equipment. The pickling process must be followed by a neutralizing dip and rinse.

Alkaline pickling solutions that are less severe in nature have been developed to remove oxides, scale, oil, and paint. They are also recommended for combined cleaning and deoxidizing of steel before electroplating or other finishes.

***Bright Dipping.*** Bright dipping, as the name implies, is a process used to produce highly reflective surfaces on nonferrous metals. To bright dip copper, for example, the metal is first dipped in sulfuric acid, which acts to oxidize the surface. This is followed by a nitric acid dip to dissolve the newly formed oxides. By this process, the microscopic hills and valleys on the surface are leveled off, increasing the reflectivity of the metal. The bright dip is followed by a thorough rinsing and rapid drying to avoid tarnishing.

Proprietary solutions are obtainable for both copper and aluminum. The solutions used for aluminum consist basically of phosphoric and nitric acids.

### Cleaning Equipment

Most machines used in cleaning metals are combination types. The monorail degreaser shown in Fig. 10-1 utilizes solvent vapors as the initial step in removing grease. The parts then move to the next station, where they are sprayed with warm solvent. The last station consists of a pure vapor rinse to ensure complete removal of all soil.

Other machines are made to handle parts in baskets (Fig. 10–2). The baskets, suspended from crossbars, can be made to rise and fall alternately or to advance

FIGURE 10-1. *A conveyorized degreaser used for both vapor and spray-type cleaning. Courtesy Detrex Chemical Industries, Inc.*

FIGURE 10-2. *Automatically operated cross-bar containers used in both dip and vapor degreasing applications. Courtesy Detrex Chemical Industries, Inc.*

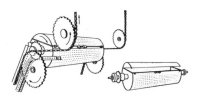

longitudinally in steps from one end of the machine to the other and back again. The baskets may be exposed to vapor rinsing, immersion in solvent, spraying from nozzles, or combinations of these cleaning cycles.

### Ultrasonic Cleaning

It is often necessary to remove insoluble particles from hard to reach cavities, indentations, slots, and small holes. Precision parts such as hypodermic needles, automatic transmissions, carburetors, and hydraulic needle valves must be perfectly clean to function properly. One of the most efficient ways of ensuring their cleanliness is to employ an ultrasonic scrubbing action.

. The principle parts required for ultrasonic cleaning are shown in Fig. 10-3. The generator produces the high-frequency electrical energy. A transducer changes the electrical energy impulses into high-frequency sound waves that vibrate the cleaning agent and parts to be cleaned.

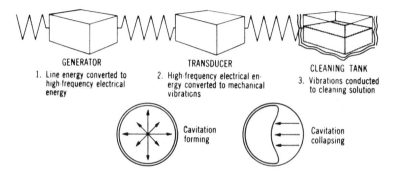

FIGURE 10-3. *Schematic diagram to illustrate the principles of ultrasonic cleaning. Courtesy Turco Products, Inc.*

When sound passes through an elastic medium such as metal, both the sound energy and the medium vibrate at the same rate. When sound passes through a liquid, it, being inelastic, ruptures or *cavitates*, forming small vacuum pockets which almost immediately collapse or *implode*. The rapid action results in scrubbing of a speed and vigor impossible by conventional means.

Ultrasonic cleaning can be very effective in removing scale, rust and tarnish provided the proper chemical is used. The cleaning agents may be acids, solvents or detergents. Acid cleaners are now being used for radioactive scale removal, hard water scale, and oxide removal, welding and brazing oxides, heat treat and forging scale and plasma deposited coatings from gas turbine blades.

Many acid cleaners cannot be used directly in the stainless steel ultrasonic tank since they are too corrosive. In this case, an insert tank, usually polypropylene, is used to contain the acid. Water or an alkaline solution is used in the ultrasonic tank to couple the ultrasonic energy to the insert tank.

Although cavitation is dependent on high-frequency energy, there is a point at which it becomes ineffective. That is, a point may be reached in the frequencies per second in which the liquid is incapable of accepting increases in power since the liquid becomes elastic, thus eliminating further transmission of energy. On the other end of the scale, there is a certain threshold level below which cavitation will not occur. For technological reasons, the low-frequency systems (20 to 25 kHz) are more difficult to engineer and manufacture than the high frequencies. However, the lower frequencies produce better cleaning results because of the stronger cavitation field. If the frequency is within the lower audible range, noise becomes a problem. As the frequency approaches 20 kHz or lower, operating noise not only becomes substantial but may exceed maximum safe limits as specified by OSHA or other regulatory measures. For this reason, ultrasonic equipment which operates at a frequency of 40 kHz is increasingly recommended for industrial use even though it is somewhat less efficient.

### Mechanical Cleaning

Mechanical cleaning of metal surfaces consists of an abrading action produced by abrasive blasting (wet or dry), power brushing, and tumbling.

*Abrasive Blasting.* Dry-abrasive blasting equipment consists of a cleaning cabinet, an abrasive reservoir, and an air nozzle. The abrasive is drawn through the hose to join the high-velocity airstream at the nozzle and impinge on the metal surface. The used abrasive slides down the centrally sloped bottom of the cabinet and returns to the reservoir.

The abrasive used may be shot, steel grit, or coarse, sharp sand. Fine sands do not produce a good cutting action. They also present more of a feeding problem.

Abrasive blasting machines are especially good for cleaning stubborn or hard to remove contaminants such as heavy scale, corrosion, or welding slag. They are also useful in providing an excellent surface with "tooth" for protective coatings.

The wet-abrasive blasting machines are quite similar in appearance to the dry machines. The main difference is that the wet machines use an abrasive slurry that is kept in suspension by an agitator. The abrasive is fed to the blasting gun by a centrifugal pump. Once at the gun, the abrasive is forced out at extremely high velocities with compressed air. The abrasives commonly used are silica, quartz, garnet, aluminum oxide, and novaculite (soft type of silicon dioxide).

The abrasive blasting machine is used both for cleaning and finishing metals. The action obtained depends on the size and type of abrasive. The wet abrasive machine is especially suited for finishing. In fact, the manufacturer of one type refers to it as a liquid honing machine. A variety of surfaces can be produced, as shown by the machine in Fig. 10-4.

*Power Brushing.* Power brushing utilizes both fiber and wire wheels to remove heat-treat scale, weld flux, machining burrs, and other unwanted surface contaminants. Some applications of wire brushing can impart fine microfinishes to metal parts. Several applications of wire brushing are shown in Fig. 10-5.

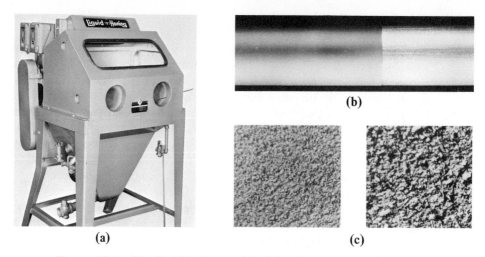

(a)

(b)

(c)

FIGURE 10-4. *The liquid-honing machine (a). A comparison of a conventionally honed (left) and liquid-honed (right) drill rod is shown at (b). Photomicrographs showing the effect of various abrasives on metallic surfaces (c). Courtesy Vapor Blast Manufacturing Co.*

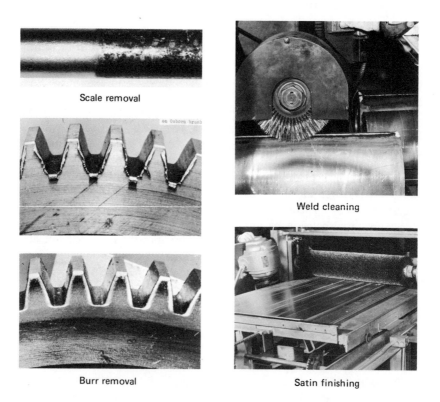

Scale removal

Weld cleaning

Burr removal

Satin finishing

FIGURE 10-5. *Some applications of power brushing. Courtesy Osborn Manufacturing Co.*

FIGURE 10-6. *The main types of power brushes are wheel, cup, end, side action, and wide-faced. Courtesy Osborn Manufacturing Co.*

The principle kinds of power brushes as shown in Fig. 10-6 are wheel brushes, cup brushes, end brushes, side-action brushes, and wide-face brushes. The wheel brushes range in size from 1 to 18 in. (45.72 cm) in diameter with wheel widths from $\frac{1}{8}$ in to 48 in. (3.175 to 121.92 cm). Fiber brushes consist of natural fibers, including Tampico (treated and untreated), horsehair, and various cord materials. Synthetic fibers are also used. Wire wheels are made of steel, stainless steel, and brass.

The larger-sized brushes are usually operated at 4500 to 6000 sfpm. When used with fine abrasive compounds, they can produce fine microinch finishes on metal surfaces.

Shown in Fig. 10-6 are the two main types of wire wheels: crimped and knot-type. The twisted knot type is very versatile. At low surface speeds, it is quite flexible but at high speeds it becomes very hard and can be used for fast cutting of surface contaminants. It may also be used to impart a deep orange-peel finish.

Power brushing is also used to blend sharp contours left by grinding so that uniform plating can result. Polishing marks, draw marks, and scratches can also be blended. Even a few seconds of power brushing can reduce microscopic V-notch cracks that go undetected and greatly increase the fatigue life.

## PROTECTIVE METAL COATINGS

The variety of protective coatings available today covers a wide range in terms of length of life, ease and difficulty of application, effectiveness of protection, and cost. In order to select the optimum coating system, it is necessary to establish the protective requirements. There is generally a direct relationship between the applied cost of a coating and the degree of duration it will provide. From this standpoint, coatings can be divided into three main types: interim, durable, and "permanent."

### Interim Coatings

Interim coatings are those that provide a rust preventative while the parts are in process. Some also provide lubricating qualities to the surface as an aid to further forming operations. In other cases just the opposite is required since adhesives will be used in assembly. The best known and most widely used interim coating process is that of chemical conversion, or simply conversion coatings.

### Conversion Coatings

Conversion coatings get their name from the fact that certain metals are capable of reacting with various chemical agents in such a way that solid chemical compounds are precipitated on the surface or, more accurately, *with* the surface. The metallic and ionic nature of the metal at the surface is altered by conversion into nonmetallic chemical compounds through mechanisms in which the metal atoms become part of the compounds. The process is thus a transformation of the outer atomic or ionic layers of metal into insoluble chemical compounds of the metal, the result of a precipitation reaction. The best known of the conversion coatings are those employing phosphoric acid.

**Phosphate Coatings.** It is said that Roman soldiers stationed in Germany found certain deposits of phosphate-bearing earths which proved effective in removing rust from their armor. Thus, the art of protecting metals by depositing a phosphate coating on them is not new, but the broad scientific base of the process is a more modern achievement.

A phosphating installation requires the removal of all soils. A simple test of cleanliness is to hold a part under cold running water and watch for any appearance of a water break or water separation on the surface. If a water break occurs, the part is not clean and must be reprocessed. Normally, five successive tanks are used for phosphating: cleaning, rinsing, phosphating, rinsing, and a final inhibitive rinsing. Where pickling or other operations are necessary, the number of stages may be as high as nine or more.

Several types of phosphate salts are used: iron phosphate, zinc-iron phosphate, and manganese-iron phosphate.

The iron phosphates provide a light crystal coating for steel. A detergent can be incorporated into the phosphate bath so that cleaning can be carried on simultaneously with the process. The thin phosphate coating under paint is especially good for impact resistance.

The zinc-iron phosphate provides a good base for paint as well as nondrying oils and waxes. It also provides a heavier coating for more extensive outside exposure.

The manganese-iron phosphate provides a heavy black coating that is used on frictional surfaces to prevent galling, scoring, or seizing of parts.

In brief summary, the outstanding characteristics of phosphate coatings are their ability to retain paint and provide lubrication when necessary. Of course the primary purpose is to provide interim corrosion resistance, as shown in Fig. 10-7(*a*). Machines are available that can accomplish vapor degreasing, phosphatizing, and painting [Fig. 10-7(*b*)] in a single pass.

**Chromate Coatings.** Chromate conversion coatings are used on zinc, cadmium, aluminum, copper, silver, magnesium and alloys of these metals. The process consists of a simple chemical dip, spray, or brush treatment. The chromate bath is an acidic solution containing hexavalent chromium compounds known as activators or catalysts.

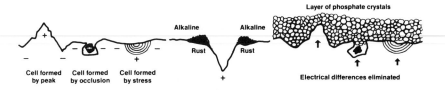

Metal surface without phosphate coat (left) is positively and negatively charged. When moisture is present, electrolytic cells form and rust builds up (center). Phosphate coat neutralizes these electrical differences (right)

(a)

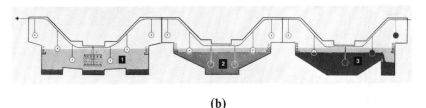

(b)

FIGURE 10-7. *Metal surface without phosphate coat (left) is positively and negatively charged. When moisture is present, electrolytic cells form and rust builds up (center). Phosphate coat neutralizes these electrical differences (right),(a). Vapor degreasing takes place at (1), phosphatizing at (2), and dip painting at (3),(b). If baking-type enamels are used, a fourth but shortened baking step can be added. Courtesy Oakite Products, Inc.*

The principle use of chromate coatings is for corrosion resistance. The chromate furnishes a base for painting but it may also be a decorative final finish. Various colors may be added by subsequent dipping in organic-dye baths.

Most chromate treatments are proprietary processes.

### Anodic Coatings

Anodized films are produced by electrochemical reaction with the base metal and are another form of conversion coating. Anodizing gets its name from the fact that the base metal becomes the anode instead of the cathode as is the usual case in the electroplating process. The process is used on aluminum and magnesium. An electrolyte capable of yielding oxygen on electrolysis is used. As current is passed, oxygen is liberated at the surface, forming an oxide film. Following the anodic treatment, the oxide film may be sealed with boiling water. This closes the pores and destroys the absorptive characteristics of the coating.

Hard-coat anodizing is much the same process, but with a higher voltage to produce a harder surface. Thus items made from light metal can have a hard-wear resistant surface.

The use of colored anodized aluminum has made rapid progress in recent years, particularly in the giftware and home-appliance field. Color imitations of precious and semiprecious metals such as gold, copper, brass, etc., as well as a wide variety of other colors, are used.

Dying the anodized aluminum is done by dipping in a water solution of dyestuff at a temperature of 150°F (66°C) for about 10 min. The work is then sealed and is resistant to further dying or staining.

### Durable Coatings

Durable finishes may also be classified as organic and inorganic finishes. There are several dozen major chemical types of organic coatings that derive from resins such as alkyds, vinyls, and epoxies. The inorganic coatings are porcelain enamels and various plasma-sprayed materials.

*Organic Coatings.* For the initial selection of an organic coating, all environmental requirements should be clearly defined. Consideration must be given to anticipated exposure (whether indoor or outdoor, marine or tropical), corrosive ambients (such as solvents, chemicals, vapors, or salt spray), and mechanical abuse (such as impact, vibration, abrasion, bending, or elongation).

A brief summary of some of the main types of thermosetting polymer coatings and their applications are given in Table 10-1. The main thermoplastic types and their applications are given in Table 10-2.

*Inorganic Coatings.* Porcelain enamels are mixtures of alumina-borosilicate glass to which coloring oxides and other inorganic materials are added. The powder is spread on the steel sheet and heated up to about 400°F (204°C) to fuse it to the steel sheet. The fusion process forms a tenacious enamel-to-steel bond.

Porcelain enamels are supplied in a wide range of finishes, ranging from the smooth nontextured to the special textures. They have excellent resistance to environmental conditions of fading, mechanical damage, abrasion, and corrosion.

Porcelain enamels are extensively used on all major appliances, plumbing ware, domestic and commercial heating equipment, lighting fixtures, and heat exchangers.

Plasma-sprayed materials consist of borides, beryllides, carbides, and nitrides. These coatings, particularly carbides, have proven to be very wear-resistant. In addition to their high hardness they have good chemical stability.

The plasma-spray method is used for applying these coating materials. It furnishes the first controlled temperatures above 20,000°F ever used in industry. This process is discussed in the chapter on materials joining. Carbides and nitrides may be plasma sprayed on Inconel, stainless steel, zirconium, and graphite.

### Priming

The application of the best coating will be of little value unless the proper primer is used first. Primers provide adequate adhesion and compatability of the topcoat with the surface. They are usually of a different chemical type than the topcoat.

Primers are formulated to provide the most stable bond to the substrate, to smooth out the surface for topcoat applications, or to bring anticorrosive or inhibitive

TABLE 10-1. *Properties of Organic Coatings, Thermosetting Types*

| Coating Type | Continuous Service Temp.,°F (°C) | Weather-ability | Adhesion to Metals | Abrasion Resistance | Cost Classification | Typical Applications |
|---|---|---|---|---|---|---|
| Acrylics | 250 (121) | Ex. | Ex. | Ex. | Low | Washing machines, refrigerators, etc. |
| Alkyds Alkyd-amine | 200 (93) 250 (121) | Good to ex. | Ex. | Ex. | Low | Auto primer, cam & drum coatings, appliance coatings, machinery subject to chemical attack |
| Epoxy ester | 300 (149) | Pigmented, ex.; clear, good | Ex. | Good | Intermediate to low | Chem.- & solvent-resistant linings for tanks, pipes, and containers |
| Polyester | 200 (93) | Very good | Good to ex. on rough surfaces | Good | High | Electrical coils for motors, generators, transformers |
| Polyimide | 700 (371) | Good radiation resistance | Good | Good | High | Wire-insulation enamels for high temperatures |
| Silicone | 500 (260) | Ex. | Varies with formulation | Fair to ex. | High | High-temp. coating for furnace pipes, engine blocks, insulative coating for electronics |

TABLE 10-2. *Properties of Organic Coatings, Thermoplastic Types*

| Coating Type | Maximum Continuous Service Temp. °F (°C) | Weather-ability | Adhesion to Metals | Abrasion Resistance | Cost Classification | Typical Applications |
|---|---|---|---|---|---|---|
| Acrylic | 180 (82) | Ex. | Good | Fair | Low | Automobiles, aircraft, ornamental iron |
| Cellulose nitrate | 180 (82) | Good to ex. | Good | Fair | Low | Automobile lacquers, fast-dry finish for machines |
| TFE fluorocarbon | 500 (260) | Ex. | Ex., primers usually required | Ex. | High | Chemical-resistant coatings, low-friction and antistick coatings |
| Chlorinated polyether | 250 (121) | Ex. | Good | Ex. | Low | Coatings for chemical-resistant equipment and parts such as pump housings & impellors, tanks |
| Phenoxy | 180 (82) | Good to ex. | Ex. | Ex. | Low | Specialty coatings for postforming & corrosion-resistant magnetic-tape coating |
| Vinyls | 150 (66) | Fair to good | Ex. | Ex. | Intermediate | Coatings for chemical-resistant equipment, beverage-can lining, postformed metal parts, paper, foil |
| Vinyl fluoride | 300 (149) | Ex. | Ex. | Ex. | Intermediate to high | Circuit boards, irrigation pipe, metal siding |

elements such as zinc chromate into contact with the metal. They also are used to help provide several desired properties simultaneously with one top coat. For example, a clear unfilled resin may adhere well to a surface but a filled version (having marginal adhesion) is necessary to give the surface a smoother profile. The use of the primer in this case can provide the required adhesion and still allow the application of a filled topcoat.

### Application of Organic Coatings

Until recent years, one of the main considerations in applying coating materials to metals was the cost of the operation, since the application often cost more than the coating. Now, however, an important consideration is environmental cleanliness. Today the most popular methods of applying coatings are by several variations of spraying for the liquids and by fluidized bed and electrostatic spray for powders.

#### *Spray Painting*

One of the reasons spray painting is one of the most popular finishing methods is that it offers a great deal of flexibility. The three principal spray-painting processes are: air atomization, airless atomization, and electrostatic atomization. The common denominator of these processes is "atomization," which is the action by which a fluid is divided into small droplets.

*Air Atomization Spraying.* Air atomization is conventional spraying. Over half of all spray painting done in the world today utilizes this method. It is capable of producing the finest finishes available, as witnessed by the automobile industry.

One of the major variables in spray painting is the change in viscosity of the coating material due to temperature fluctuations. This variable can be controlled by the addition of heat. Using suitable heating equipment, an operator can stabilize paint viscosity in a matter of 15 min at the beginning of the day, without fear of undesirable effects due to ambient temperature changes. Of course not all paint-spray materials can be subjected to heat.

Some of the fringe benefits obtained from hot spraying are: (1) the ability to spray a higher-solids material, giving a better film build per pass; (2) fog and overspray can be reduced appreciably through the use of lower atomizing pressure.

*Airless Spraying.* Airless spraying does not use compressed air directly to atomize the paint or other coating material. Instead hydraulic pressure is used to atomize the fluid by pumping the fluid at high pressure (500 to 4500 psi; 34 to 306 kg/mm$^2$) through an orifice in the spray nozzle. Fluid released at these high pressures is separated into small droplets which form a fine atomized spray. Both hot and cold airless systems are shown schematically in Fig. 10-8.

The primary advantage of airless spraying is that it greatly reduces fog or rebound. This makes it possible to use high-production spray equipment where brush painting would normally be required because air-atomized spray would create excessive fog.

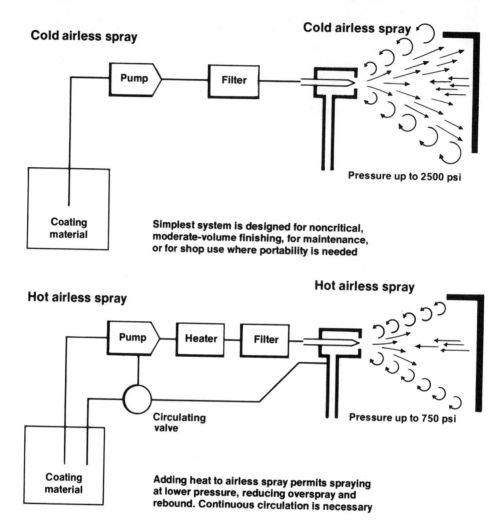

**Cold airless spray**

Pump

Filter

Coating material

Pressure up to 2500 psi

**Cold airless spray**

Simplest system is designed for noncritical, moderate-volume finishing, for maintenance, or for shop use where portability is needed

**Hot airless spray**

Pump

Heater

Filter

Circulating valve

Coating material

Pressure up to 750 psi

**Hot airless spray**

Adding heat to airless spray permits spraying at lower pressure, reducing overspray and rebound. Continuous circulation is necessary

FIGURE 10-8. *Schematics of cold and hot airless paint-spraying systems. Hot airless spraying operates at a lower pressure, giving less overspray and rebound. Courtesy* American Machinist *and Nordson Corp.*

Airless spray can also be used to apply a heavy coating thickness (10–20 mils, wet). Many materials can be sprayed in their normal, unthinned state at very high production rates.

Although fog is reduced, this does not mean that spray booths or air exhaust equipment can be eliminated. Some fog remains and therefore protection from the solvent fumes must be provided for the operator and others in the area.

Airless-spray systems are more costly than air spray and lack some of the versatility of changing spray patterns. An airless gun is either on or off. To change spray patterns, the nozzle must be changed.

*Electrostatic Atomization.* Electrostatic spraying utilizes an electrical differential to cause the paint to separate. A disc or bell rotating at 900 to 1800 rpm is used as the atomization device. The paint flows to the periphery of the disc by centrifugal force and levels out. As this action occurs, 120,000 volts are applied, resulting in the paint particles taking on an electrical surface charge.

The charge particles seek out the nearest grounded surface and attach themselves to it. This surface of course is the article to be painted, which is at a ground potential of zero voltage. The electrical charge of the particles quickly dissipates upon contact with the surface. Thus, other particles tend to seek out unexposed surfaces, providing a uniform buildup.

Although this method is very good for high-production painting, it is not as fine as with more conventional methods and sometimes an orange-peel effect is produced. The chief advantage of the process is in paint savings. Properly applied, paint costs can be reduced by as much as 50 %.

Air electrostatic spray is another version of electrostatic atomization. In this process, the paint is atomized by the conventional compressed-air method or by the airless technique. As the particles leave the gun, they are directed under an electrode that is charged to about 62,000 volts negative (Fig. 10–9). The particles pick up a negative surface charge and proceed toward the workpiece, which is electrically grounded. As the particles approach the work, they rapidly lose velocity and will be almost floating as they reach their destination. The strong negative charge is attracted to the nearest grounded object. Some of the particles may pass the object and the attraction is strong enough to bring them back, depositing paint on the backside of the object. This effect is known as *wraparound* and is one of the biggest advantages of the electrostatic process. Outside corners and edges are usually hard to paint by conventional methods, but are covered very well by electrostatic methods.

### Powder Coating

The use of dry powder coatings has expanded considerably in recent years. Part of this has been because they produce virtually no air pollutants and the other is that a denser film can be formed than with solvent-based materials. Not only can thicker coats be applied in one application, but the coatings have better chemical resistance than paints and they cover edges and blemishes better and do not drip or sag.

Powder coatings are classified as either thermosetting or thermoplastic. Thermosetting powders form coatings that cannot be altered or reworked. Examples are epoxies, polyesters, and acrylics. Thermoplastic coatings soften when heated and harden when cooled. Among those used for powder coatings are polyvinylcholride (PVC), cellulose acetate butyrate, polyethylene, nylon, chlorinated polyether, flourocarbons, and certain types of polyesters and acrylics.

The most widely used powders are the epoxies. They are formulated to be tough, highly durable, and highly decorative. They are available in an assortment of colors, glosses, and textures to meet all requirements.

**Air-electrostatic for water-reducible coatings**

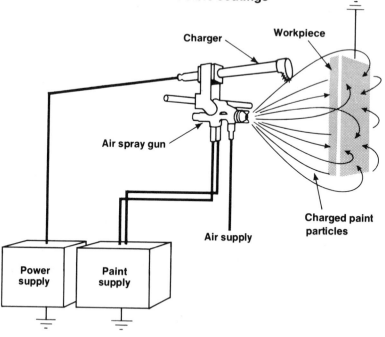

FIGURE 10-9. *Air-electrostatic spray system is designed to handle the increasingly popular water-reducible coatings. Charger creates electrostatic field between the nozzle and the workpiece, charging passing paint particles.* Courtesy American Machinist *and Nordson Corp.*

Polyvinylchloride is the most widely used thermoplastic powder. It is relatively low in cost and provides good chemical, abrasive, and impact resistance. It is also the best coating for electrical insulation.

### Application of Powdered Coatings

Although a number of techniques exist for applying powder coatings, fluidized-bed and electrostatic-spray methods are used to coat most parts.

***Fluidized Bed.*** Preheated parts are immersed in a fluidized bed which consists of a tank filled with dry plastic powder, as shown in Fig. 10-10(*a*). The parts are preheated to a temperature above the melting point of the powder. Particles contacting the surface melt and adhere. Low-pressure air is blown through the diffuser at the bottom of the container. The immersion time is generally from 1 to 30 sec. After dipping, the excess powder should be removed immediately by air blast or tapping.

**Fluidized-bed dipping**

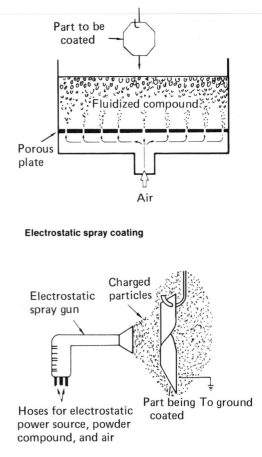

**Electrostatic spray coating**

FIGURE 10-10. *Schematic representations of the two main methods of applying powder coatings to metals. Courtesy* American Machinist *and Nordson Corp.*

Parts coated with thermosetting powders, such as epoxies are postheated to fuse and cure the resin. Parts coated with thermoplastic powders are usually postheated also, to obtain optimum coating flow and fusion.

The fluidized-bed method produces relatively thick coatings, generally from 6 to 60 mils. Uniform thicknesses are difficult to obtain on parts of varying cross sections because the heat retention is not uniform. Thin sections lose their heat faster than thick sections and therefore receive a thinner coat. Coatings have a tendency to build up in holes and recesses.

***Electrostatic Spray.*** In this process, cold parts are sprayed with unheated, electrically charged plastic powder [Fig. 10-10(*b*)]. This process is especially good for large parts and thin coatings. It is also advantageous for parts that need to be coated only on one side. Coatings as thin as 1 mil can be applied by electrostatic spray. After spraying, the coating is fused in an oven.

In general the surface preparation for powder coating is the same as that for painting. Sandblasting and acid etching increase the adhesion of the coatings, especially if no primer is used. Phosphate pretreatment gives maximum adhesion for some resins.

***Advantages and Limitations.*** This process can be used to provide less costly metal protection. As an example, mild steel coated with Nylon 11 has replaced corrosion-resistant nickel-chrome alloy for a spline-gear application. Since only water-borne materials are used, electrostatic spraying is inherently clean and safe. Toxicity and fire hazards are eliminated.

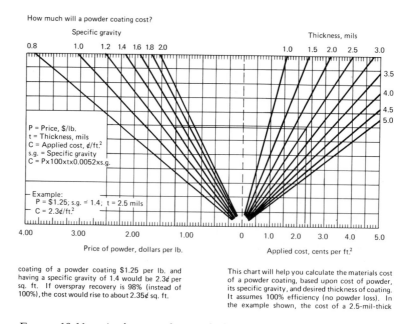

How much will a powder coating cost?

P = Price, $/lb.
t = Thickness, mils
C = Applied cost, ¢/ft.²
s.g. = Specific gravity
C = P x 100 x t x 0.0052 x s.g.

Example:
P = $1.25; s.g. = 1.4; t = 2.5 mils
C = 2.3¢/ft.²

coating of a powder coating $1.25 per lb. and having a specific gravity of 1.4 would be 2.3¢ per sq. ft. If overspray recovery is 98% (instead of 100%), the cost would rise to about 2.35¢ sq. ft.

This chart will help you calculate the materials cost of a powder coating, based upon cost of powder, its specific gravity, and desired thickness of coating. It assumes 100% efficiency (no powder loss). In the example shown, the cost of a 2.5-mil-thick

FIGURE 10-11. *A chart used to calculate the cost of powder-coating applications. Courtesy American Machinist.*

Powder coating is more costly than spray painting, both in material and installation. Curing requires baking, normally at 250 to 400°F (121–204°C). Colors and appearance textures are limited. Shown in Fig. 10-11 is a chart that can be used to determine the cost of powder coatings.

## "Permanent" Metal Coatings

In this discussion, metal or inorganic coatings shall be considered permanent. Zinc, tin, lead, copper, brass, chromium, aluminum, and stainless steel are the principle metal-coating materials and are applied by a variety of processes, namely by hot dipping, plating, metal spraying, vacuum metallizing, and sputtering.

### Hot Dipping

Hot dipping is used to apply a substantial portion of the total amount of all metal coatings. It consists of first thoroughly cleaning the base metal by acids or other methods, followed by proper fluxing, usually with zinc ammonium chloride, after which the metal is immersed in a bath of the molten coating metal. All of the coatings applied by this method are dependent upon an alloying action between the coating metal and the base metal for the bonding action.

The interface alloy is usually more brittle than the base metal or the coating metal. This may interfere in some forming operations during subsequent manufacturing.

The big advantage is that the costs are relatively low. Hot dip may imply a batch process, but high-speed lines are used for coating wire, strip, or continuous sheet coil.

Common applications of the hot-dip process are aluminum, zinc, and tin to steel and lead, and zinc alloys to nonferrous alloys. Zinc coatings on steel are commonly referred to as galvanized coatings.

To ensure an even coating on small parts such as nuts and bolts, after being taken from the bath they are centrifuged until the coating is hard.

### Electroplating

Electroplating is the most widely used method of applying metallic coatings. The controlled thickness of deposit and uniformity of quality are the method's greatest assets. In practice, coatings of 0.0002 to 0.0003 in. (0.005 to 0.008 mm) are not uncommon.

Surfaces to be plated must be buffed smooth to eliminate scratches and unevenness. The quality of the plated surface depends to a great extent on the preparation of the base metal. The work is then cleaned in suitable cleaning solutions to remove all grease, oxides, buffing compounds, etc. After rinsing, the part is ready for plating.

The four essential elements of a plating process are the cathode, anode, electrolyte, and direct current. Current leaves the anode, which is a bar of plating material, and migrates through the electrolyte (salts of the plating metal in solution) to the cathode, or part to be plated. As the ions are deposited on the cathode, they give up their charge and are deposited as metal on the cathode. The current density largely determines the rate at which the metal is deposited. For example, in chromium plating, good deposits may be obtained at cathode current densities ranging between 0.25 amps/sq in. to 6 or more amps/sq in. The current density in turn is determined by the operating temperatures, the size and shape of the workpiece, the condition and nature of the surface, and other factors.

Chromium will provide a good corrosion-resistant film directly over stainless steel, nickel-silver, or Monel. It should never be used directly on brass or zinc die castings due to the dezincifications that will take place under the coating causing it to lift off at some future date.

Hard chromium, as differentiated from decorative chromium, is applied for functional reasons. If a 0.00005 in. (0.0013 mm) deposit is applied over a hardened surface, then it is a hard film because of its backing.  If the same thickness is applied over softer metal. then the chromium will be no harder than the base.  If, however, a deposit of 0.0015 in. (0.04 mm) to 0.002 in. (0.05 mm) is applied over the same unhardened surface, it will be very hard.

Chromium plate has a lower coefficient of friction than any other metal and has good antigalling properties as long as it is not used against itself.

Parts to be plated should be designed with generous fillets and radii instead of sharp corners since current concentrations occur at sharp points, resulting in excessive deposits.  Plating thicknesses should be taken into consideration when designating design tolerances on mating parts.

***Plating of Die Castings.***  Most die castings are finished by electroplating.  Simple shapes can be plated with about 1.3 mils minimum of copper plus nickel, and 0.01 mil minimum of chromium, in approximately 50 min.  Complex shapes require longer times and often special fixturing.  In complicated shapes, more metal is plated unnecessarily on projections and other high-current-density areas.  Shown in Fig. 10-12 are a few sample designs with comments as to the platability of that geometry.  In general, flat surfaces should be crowned, edges should be rounded, and inside angles should be given generous radii to improve the platability.

For good corrosion resistance outdoors, the minimum thickness of copper and nickel should be 0.2 and 1.0 mil, respectively, with a 0.03 mil of chromium.  For indoor parts, the nickel and chromium can be less, about 0.8 and 0.01 mils, respectively.

### Metal Spraying

Although metal spraying is a coating process, it is more commonly associated with welding equipment and is discussed in Chapter 6.

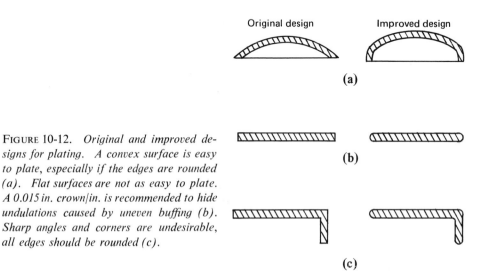

FIGURE 10-12.  *Original and improved designs for plating.  A convex surface is easy to plate, especially if the edges are rounded (a).  Flat surfaces are not as easy to plate. A 0.015 in. crown/in. is recommended to hide undulations caused by uneven buffing (b). Sharp angles and corners are undesirable, all edges should be rounded (c).*

### Vacuum Metallizing

Vacuum metallizing is a process whereby metals and nonmetals are deposited under high vacuum on to prepared surfaces of plastics, metals, glass, paper, textiles, and other materials to produce finishes that are both decorative and functional.

The process consists of placing objects to be metallized in a vacuum chamber with a small amount of selected "plating material." The plating material, depending upon the application, may be aluminum, cadmium, silicon, monoxide, silver, selenium, magnesium fluoride, or a number of other materials. When the required high vacuum has been reached in the chamber, the plating material is heated to its vaporization temperature. The material vaporizes, radiates throughout the chamber, and deposits by condensation on the objects to be metallized. Objects are held in the chamber on fixtures, the design of which enables either entire surfaces or selective areas to be metallized.

After metallizing, the products are dipped in topcoat lacquer. A wide range of colors can be produced by dipping products in dye solutions, which will color the topcoat. Dyes may also be combined with the topcoat. The topcoat serves to protect the very thin metal film from abrasion and wear. It is standard procedure when metallizing plastics to put a base coat of lacquer on the part before metallizing. A metallized finish is then, in effect, an ultrathin film of aluminum, or other metal, sandwiched between two layers of coatings. Depending upon service requirements, topcoats are air-drying or baking types.

The main advantages of vacuum metallizing are: (1) It provides a method of placing a thin film of metal on a nonconductive material. (2) The deposit may be less than five millionths of an inch (0.005 mils), yet it completely conceals the color of the object. (3) It is much less expensive than electroplating. (4) The corrosion resistance of lacquered metal surfaces is high. Salt-spray tests show 500 hours of endurance.

### Sputtering

Sputtering is a process of applying thin films on a substrate. The name of the process derives from the fact that atoms are sputtered or ejected from a cathodic source and are deposited on nearby surfaces. The process is not new. It was used to deposit thin films as early as 1887. However, intensive development did not begin until 1955.

The simplest type of sputtering apparatus is shown schematically in Fig. 10-13. The source of the material to be used in plating is the cathode and the workpiece is the anode. A discharge is initiated between the two electrodes in an argon atmosphere at a pressure of $10^{-2}$ to $10^{-1}$ torr. (One torr may be defined as 1/760 of a standard atmosphere.) The argon ions ($Ar^+$) formed in the discharge are accelerated toward the cathode source by means of a high-voltage power supply (1–5 kV) which creates the necessary plasma. The bombarding ions transfer sufficient momentum to the atoms of the target to cause them to leave the surface with energies many times higher than those of thermally evaporated atoms. The rate of deposition of sputtered films is typically about 100 to 500 Å per minute.

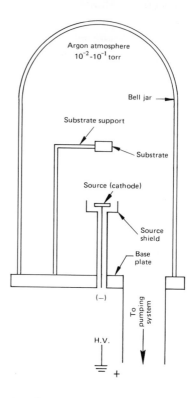

Argon atmosphere
$10^{-2}$-$10^{-1}$ torr

Bell jar

Substrate support

Substrate

Source (cathode)

Source shield

Base plate

(−)

To pumping system

H.V.

+

FIGURE 10-13. *A schematic representation of a cathodic sputtering setup.*

It should be understood that sputtering is not an electrochemical plating process. Deposition of the sputtered films on a substrate does not require that it be used as an anode. A separate anode could be used if more convenient, since the function of the anode is only to establish a gaseous discharge.

Sputtering is a very versatile technique. In addition to the elemental metals, a large number of more complex materials can be deposited with predictable composition. Among the materials deposited are quartz, alumina, carbides, nitrides, borides, and a variety of glasses. Almost any metallic material can be sputtered if it is available in sheet form (not an absolute necessity) for use as a target and if the material is reasonably stable in a vacuum.

The process may also be reversed so that sputter etching results. In this case, the substrate is masked with a photoresist. Then the atoms are removed from the unmasked areas to produce the desired pattern. In photochemical milling, the acid etches under the resist (undercutting). In sputter etching this does not happen. Also, it is universally applicable to all materials, thus eliminating the need to develop and test chemical etchants when different materials are encountered. In some cases, sputter etching is used as a cleaning process prior to sputtering.

Sputtering is useful in producing a dense, solid lubricant in the surface of a metal. Initial work with molybdenum disulfide has shown excellent *stoichiometry* and adherence. Stoichiometry is a measure of the ratio of atoms of various materials in an alloy or compound. A stoichiometric film is one which has the proper ratio. Sputtered $MoS_2$ films deposited on nickel, nichrome, and niobium disks have shown an average friction coefficient of 0.05 and endurance life of over a million cycles. The sputtered film is noncrystalline and is about 200 Å thick.

The widely advertised platinum-edged and chromium-edged razor blades are sputter-coated. Their superior performance is attributed to improved corrosion resistance. To make the process economical, 75,000 blades are strung on a skewer, placed in a vacuum chamber, and sputtered simultaneously. The coating is so thin that the blades do not have to be rehoned after the sputter coating has been applied. In spite of the thinness of the coat, it is pinhole-free and extremely adherent.

## Electroforming

Electroforming may be defined as the production or reproduction of parts by electrodeposition on a mandrel or mold that is subsequently separated from the deposit. Occasionally the mandrel may remain in whole or in part as an integral functional element of the electroform. Electroforming is considered an art and its successful practice is largely dependent on the skill and experience of the operator. Although detailed instructions for producing various electroformed objects are available, they do not guarantee the successful production of a part that has not been previously manufactured by this process.

Some sectional views of electroformed parts and mandrels are shown in Fig. 10-14.

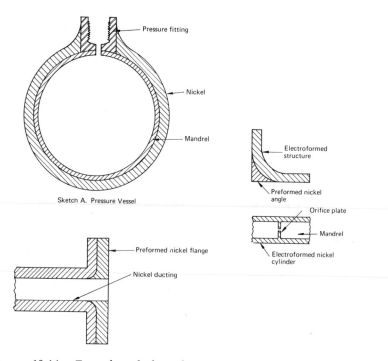

FIGURE 10-14. *Examples of electroformed parts. The electrodeposited nickel may be formed around inserts.*

*Nickel Electroforming.* Most electroformed parts are made of nickel. Studies have been made so that nickel electroforms can now be made with the following property ranges: 130 to 250 DPH, 50 to 200 ksi tensile strength, elongation from less than 1 % to 37 %, and internal stress ranging from compressive to 25 ksi tensile.

One product that is ideally suited to electroforming is that of large, curved metal mirrors. The curved shape, perfect finish, and thin cross section required for precision mirrors can be achieved most economically by the replication of a highly polished glass mandrel. The mandrel may be used either directly in the plating bath (after metallizing the surface) or, more commonly, as a master for casting low-cost, single- or multiple-use replica molds from epoxy resin. The replica is silvered, plated to the desired thickness, and removed to produce a precise metallic reflector.

*Electroformed Composite Structures.* One of the techniques that has been used to fabricate composite electroformed structures is the electrodeposition of metals concurrent with filament winding.

The technique of electroforming composite metal-filament structures is shown in Fig. 10-15. The method depends on the electrodeposition of a matrix metal on

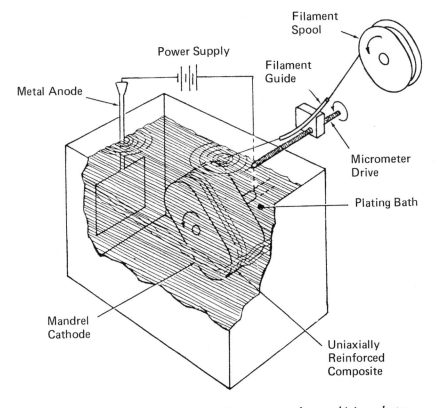

FIGURE 10-15. *Making a composite structure by combining electrodeposition with filament winding.*

which a reinforcing filament composite can be fabricated at room temperature with accurate control over filament spacing and volume-percent loading. Since the electro-deposit grows around the filament, an intimate filament-matrix contact is established at the interface without alloying, resulting in a dense composite structure. Nickel and tungsten wires have been used as the filament. The plating and the filament-winding rate must be coordinated to produce a sound, void-free structure. Shown in Fig. 10-16 is a sectional view of the multilayer formation.

FIGURE 10-16. *A schematic showing a cross section of electroformed composite structure.*

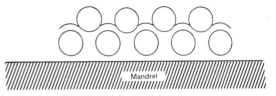

Tungsten filament–nickel composites prepared by this process have shown excellent tensile properties. A structure containing 52 % by volume of tungsten filament showed a tensile strength of about 225 ksi (1551 MPa). Samples made with a tungsten-filament loading higher than 52 % by volume showed excessive void entrapment.

Other research on nickel-boron and nickel–silicon carbide filament systems indicate that electroforming is a better way to make sound fiber-reinforced composites than such metallurgical techniques as liquid-metal infiltration, powder compaction, or extrusion (discussed in Chapter 2).

*Advantages.* A number of basic advantages distinguish electroforming from mechanical forming methods. These advantages enable the manufacturer to: (1) form complex, thin-walled small or large shapes (Shown in Fig. 10-17 is a large, 10 ft long, 12 in. diameter thin-walled (0.004 in. thick) rotary screen being formed on a nickel steel mandrel); (2) laminate similar or dissimilar metals without heat or pressure; (3) reproduce surfaces precisely; (4) incorporate inserts and flanges; (5) form integral cavities; (6) enclose and support nonmetallics; (7) form precise openings (Fig. 10-18).

*Disadvantages.* Parts are limited to materials that can be plated. The mandrel material must be conductive or a nonconductive material such as rubber must be given a very thin conductive coating, usually aluminum or silver, on which the nickel can be electroplated.

### Precoated Metals

Although precoated metals have been available for many years, it has become increasingly attractive in recent years, particularly after the passage of the Clean Air Act of 1970 which created the Environmental Protection Agency. There has been increased monitoring of plant conditions, including paint-storage practices, possible fire and explosion hazards, ventilation conditions, and many other factors incident

FIGURE 10-17. *A large, thin-walled shape, 10 ft long and 12 in. in diameter is being electroformed to 0.004-in. wall thickness on a stainless steel mandrel.* Courtesy Machine and Tool Blue Book.

FIGURE 10-18. *This experimental nozzle was electroformed out of nickel over a pyrolytic graphite liner. It is 3 in. (7.62 cm) long with a wall thickness of $\frac{1}{8}$ to $\frac{3}{16}$ in. (3.17 to 4.76 mm). Note the integral, rectangular, cross-sectional cooling passages.* Courtesy Machine and Tool Blue Book.

to an in-plant paint operation, all of which come under OSHA. Among the 25 most-cited violations of job safety and health standards are those that include failure to meet industry standards for flammable and combustible liquids and their use, spray finishing, and general fire protection. Thus safety regulations have been among the factors that has pushed precoated steel sheet into more than a billion dollar industry.

Advantages of precoated metal often cited are reduced handling and reduced inventory. Substantially lower insurance rates can often be obtained when a finishing line is eliminated.

A wide variety of coatings are available, including tin, zinc, polyesters, vinyls, acrylics, PVC's, epoxies, phenolics, and fluorocarbons. Many of these coatings can be supplied in multicolored printing or embossing of the surface, including woodgrain effects.

### Fabrication of Precoated Metals

Precoated steel sheet can be sheared, punched, stamped, press braked, roll-formed, deep drawn, and spun with the same tools as uncoated steel. Die life is generally longer due to the coating's internal lubricants. The flat metal stock (0.002 in.–0.090 in. thick, 0.5 in.–70 in. wide; 0.05–2.29 mm thick, 12.70–177.8 cm wide) comes in coils.

The use of precoated metal places some demands on the designer. For example, the bend radius must be at least equal to $2t$ and it is better if it can be $3t$ or $4t$. If drawing is required, the radius should be extremely generous.

Exposed edges caused by slitting, notching, or punching presents two problems: appearance and corrosion. Quite often the raw edge can be turned under, as

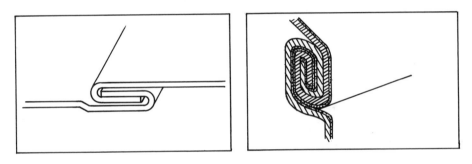

**(a)**

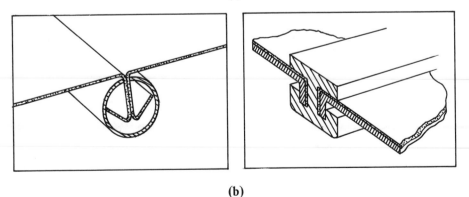

**(b)**

FIGURE 10-19. *Exposed edges of precoated materials may be folded under, as in the seams shown at (a), or covered with snap-on trim, as shown at (b).*

shown by the examples in Fig. 10-19(*a*). Another method is to cover the edges with a molding or snap-on trim (*b*).

Preventing corrosion along the exposed edges is often done by using a galvanized-steel substrate. When the material is cut, the zinc coating tends to "wipe" in the direction of the cutting motion, giving protection to the exposed edge. Where this is not possible, lockseaming provides another alternative. Also, exposed surfaces can be treated if corrosion resistance is essential.

*Welding.* Precoated steels can be welded. In general the temperature should be low (consistent with good welds). The weld cycle should be short. Resistance spot welds and projection welds are the most used (Fig. 10-20).

*Fasteners.* A wide variety of fasteners can be used. The designer can usually select a fastener that will meet the needs of the design and still be hidden from view or used to accent a feature of the product (Fig. 10-21).

*Adhesives.* Both thermoplastic and thermosetting adhesives are used to bond precoated metals. Since there are many adhesives available, the one used should be the

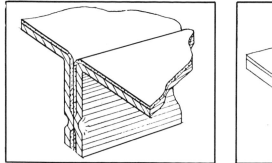
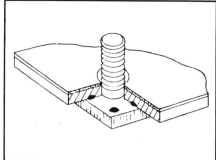

FIGURE 10-20. *Precoated materials can be spot welded, as shown at (a), or projection welded, as shown at (b). In the spot weld the coating is burned off in the weld area.*

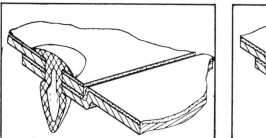

FIGURE 10-21. *A wide variety of fasteners may be used to assemble precoated materials. They may be hidden from view or capped to accent a feature of the product.*

one lowest in cost that meets strength and adhesion requirements, can be easily handled, and is compatible with the coated surface.

The surfaces to be bonded are first coated with adhesive, then brought together, often with heat and/or pressure. The resulting joint seals, fastens, insulates, resists corrosion, dampens vibration, and resists fatigue. Joints can be filled with PVC using hot-air welding for enhanced appearance (Fig. 10-22).

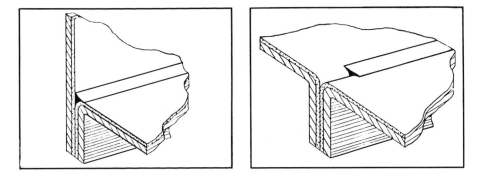

FIGURE 10-22. *Adhesively bonded joints may have a PVC bead placed on top by hot-air welding to enhance the appearance.*

## PROBLEMS

**10-1.** Can ultrasonic cleaning be used to remove oxides or tarnish? Why or why not?

**10-2.** What cleaning steps should be used for parts that are heavily contaminated by oils and grease?

**10-3.** How does bright dipping produce a shiny surface?

**10-4.** Why are ultrasonic cleaners in the 20 kHz range or lower objectionable?

**10-5.** Why are conversion coatings called by that name?

**10-6.** How does a conversion coating prevent corrosion?

**10-7.** What is a big advantage of the electrostatic-spray process?

**10-8.** What is the reason for the increased use of powder coatings in recent years?

**10-9.** What would it cost to powder coat by electrostatic spraying 50 boxes 2 in. high, 3 in. long, and 2 in. wide? The top surface (2 × 3 in.) is open. The cost of the powder per pound is \$1.45. The specific gravity of the powder is 1.6. Assume the overspray recovery was 94 %. The boxes are coated inside and out to a 3 mil thickness.

**10-10.** (a) What three coatings are ordinarily used on die castings and how are they applied? (b) What is the usual thickness of three electroplated coatings on die castings to be used outside?

**10-11.** Can chromium be used directly as a plating surface on any metal?

**10-12.** How thick a coat of chromium is required on a soft material to consider it a hard coat?

**10-13.** What is the big advantage of vacuum metallizing?

**10-14.** Why is the usual application of vacuum-metallized coating able to withstand wear and corrosion?

**10-15.** What is meant by the sputtering process?

**10-16.** (a) What is meant by stoichiometry? (b) How thick a coat (in inches) may a sputtered $MoS_2$ film be? (c) What would this be in millimeters?

**10-17.** What practical application can you think of for sputtering other than that mentioned in the text?

**10-18.** What is meant by electroforming?

**10-19.** A nickel electroformed part has a DPH of 130 to 250. Would you classify this as soft, medium-hard, or hard? Why?

**10-20.** (a) How is an electroformed composite structure fabricated? (b) What is the advantage of this method of fabrication?

**10-21.** Why are chlorinated solvents preferred to petroleum solvents for some applications?

**10-22.** Why is electrocleaning one of the best methods of cleaning?

**10-23.** How may large machine castings be cleaned prior to painting?

**10-24.** Why are emulsifiable solvents used more now than formerly?

**10-25.** What is meant by bright dipping?

**10-26.** In what respect are ultrasonic cleaning and electrocleaning opposite?

**10-27.** How does the vibration frequency effect ultrasonic cleaning?

**10-28.** Is anodizing a conversion coating? Why or why not?

**10-29.** What is the main difference in use between phosphate coatings and chromate coatings?

**10-30.** What type of organic coatings would be the most suitable for the following items: (1) a toaster, (2) a kitchen mixer housing, (3) a garden tractor, (4) a pump in a canning factory, (4) a tractor engine, (6) boiler room equipment in a brewery?

**10-31.** What is one of the big advantages of airless spray painting over conventional spray painting?

# BIBLIOGRAPHY

**Books**

*Metal Finishing*, Metals and Plastics Publications, Inc., Hackensack, New Jersey, 1974.

*The Basics of Sputtering*, Materials Research Corporation, Orangeburg, New York, 1969.

GABE, D. R., *Principles of Metal Surface Treatment and Protection*, Pergamon Press, New York, 1972.

NASA, *Conference on New Technology*, U.S. Government Printing Office, Washington, D.C., 1964.

NASA, *Plating Methods*, SP-5114, U.S. Government Printing Office, Washington, D.C., 1972.

SPIRO, P., *Electroforming*, Robert Draper, Ltd., Teddington, England, 1968.

**Periodicals**

ADAMS, J., "Spray Painting," *Automation*, Sept. 1972.

AZZAM, H. T., "Coatings Without Solvents," *Machine Design*, March 18, 1971.

BLAND, M. R., "What You Should Know About Phosphating," *Automation*, March 1973.

DWYER, J. J., JR., "Modern Coating Methods," *American Machinist*, Nov. 11, 1974.

HARRIS, R., "Precoated Metals," *Appliance*, Nov. 1973.

KUEHNER, M. A., "Selecting Conversion Coatings for Protection and Appearance," *Machine Design*, March 5, 1970.

SQUITERO, A. D., "For Those Complex, High Precision Parts Take Advantage of Modern Electroforming," *Machine and Tool Blue Book*, May 1974.

Staff Report, "Modern Practices for Finishing Aluminum—Anodizing and Sealing," *Metal Process*, Dec. 1966.

WAINDLE, R. F., "Interim Coatings—A Solution to In-Process Corrosion Problems, *The Tool and Manufacturing Engineer*, June 1969.

*Appendix*

TABLE 2A. *Properties of Metals*

| Metal or Alloy | Symbol (Main Elements) | Melting Point °F | Melting Point °C | Tensile Strength ksi | Tensile Strength MPa | Yield Strength ksi | Yield Strength MPa |
|---|---|---|---|---|---|---|---|
| Aluminum 99% | Al | 1220 | 660 | 13 | 89.63 | 5 | 34.47 |
| Aluminum 2011-T3 | Al | | | 55 | 379.2 | 43 | 296.5 |
| Brass, yellow | Cu Zn | 2340 | 900–1100 | 40–110 | 275.8 | 18 | 331.0 |
| Beryllium | Be | | 900–1000 | 70–200 | 70–200 | | |
| Chromium steel (5100) | Cr | 3430 | 1890 | | | | |
| Copper | Cu | 1981 | 1083 | 32 | 220.6 | 10 | 68.95 |
| Cast iron | | | | | | | |
| Grey | Fe C | 2065–2200 | 1130–1204 | 20–60 | 172.4 | | |
| Malleable (pearlitic) | Fe C | | | 60–105 | 413–724 | 40–90 | 275–620 |
| Ductile | Fe C | | | 65–150 | 448–1034 | 30–60 | 206–413 |
| Iron, wrought | Fe | | | 40 | 275.8 | 27 | 186.2 |
| Lead | Pb | 621 | 327 | 3 | 20.68 | 1.9 | 13.09 |
| Magnesium | Mg | 1202 | 650 | 25 | 172.4 | 13 | 89.63 |
| Molybdenum (4150) | Fe Mo C | 4760 | 2625 | 105–124 | 724–855 | 70–100 | 490–690 |
| Silver | Ag | 1760 | 960 | 23 | 158.6 | 8 | 55.16 |
| Steel | Fe | | | | | | |
| Low carbon | Fe C | 2798 | 1510 | 50–75 | 413.7 | 40 | 275.8 |
| Medium carbon | Fe C | 2600–2730 | | 70–120 | 579.2 | 52 | 351.6 |
| High carbon | Fe C | 2600 | | 100–200 | 675.7 | 72 | 496.4 |
| Steel, nickel (2515) | Fe Ni C | | | 90–105 | 724.0 | 69–75 | 579.2 |
| Steel, cast | Fe C | | | 72 | 496.4 | 40 | 275.8 |
| Steel, stainless (304) | Fe Ni Cr | | | 85 | 586.1 | 35 | 241.3 |
| Titanium (Ti-55A) | Ti | 3300 | 1820 | 35 | 241.3 | 25 | 172.4 |
| Tungsten | W | 6170 | 3387 | 500 | 3447 | | |
| Zinc alloy | Zn | 727 | | 41–47 | 282–324 | | |

| Metal or Alloy | Elongation in 2 in. Percent | Modulus of Elasticity $10^6$ PSI | Modulus of Elasticity MPa | Thermal Conductivity BTU/Sq Ft Hr/°F/In. | Thermal Conductivity Cal/Sq Cm/Sec/Deg. C/Cm | Coefficient of Expansion (In/°F) $\times 10^{-6}$ | Coefficient of Expansion (Cm/°C) $\times 10^{-6}$ |
|---|---|---|---|---|---|---|---|
| Aluminum 99% | 35 | 10.2 | 70330 | 1570 | 0.53 | 12 | 25.3 |
| Aluminum 2011-T3 | 15 | 10.2 | 70330 | 1570 | 0.53 | 9.8 | 25 |
| Brass, yellow | | 14.0 | 96530 | 2680 | 0.94 | 11.0–11.6 | 12.4 |
| Beryllium | | | | | | 9.4 | |
| Chromium steel (5100) | | 7.1 | 4891 | | | 9.8 | |
| Copper | | 15.0 | 103400 | 2680 | 0.94 | 9.2 | 16.7 |
| Cast iron | | | | | | | |
| Grey | 0–1 | 13 | 89630 | 310 | 0.106 | 6.0 | 10.2 |
| Malleable (pearlitic) | 10–25 | 12.5 | 86125 | | | 6.6 | 11.88 |
| Ductile | 1–26 | | | | | | |
| Iron, wrought | 3–20 | 29.0 | 199900 | 418 | 0.143 | 6.7 | 12.06 |
| Lead | | 2 | 1379 | 240 | 0.083 | 16.4 | 29.1 |
| Magnesium | 15 | 6.5 | 5128 | 1090 | 0.37 | 14.3 | |
| Molybdenum (4150) | 21–16 | | | | | | |
| Silver | | 10.5 | 72345 | 2900 | 1.00 | 10.6 | 18.8 |
| Steel | | | | | | | |
| Low carbon | 27–38 | 30.0 | 206800 | 460 | 0.18 | 6.7 | 12.06 |
| Medium carbon | 12–32 | 30.0 | 206800 | 460 | 0.18 | 6.7 | 12.06 |
| High carbon | 2–23 | 30.0 | 206800 | 460 | 0.18 | 6.7 | 12.06 |
| Steel, nickel (2515) | 27–24 | 30.0 | 206800 | | | | |
| Steel, cast | | 30.0 | 206800 | 400 | | | |
| Steel, stainless (304) | 50 | 29.0 | 199900 | | 0.137 | 9.3 | 16.74 |
| Titanium (Ti-55A) | 24 | 12.1 | 83369 | | | | |
| Tungsten | | 51 | 351600 | 783–754 | 0.34 | 2.6 | 4.5 |
| Zinc alloy | 10–16 | 18.5 | 127460 | | 0.264 | 15.2 | 26.3 |

TABLE 3A. *Suggested Cutting Speeds for Lathe Work*

| Material | Cutting Speed for High-speed Steel Tools, Surface Feet per Minute |
|---|---|
| Aluminum and alloys | 200–400 |
| Brass and bronze, soft | 100–300 |
| Bronze, high-tensile | 70–90 |
| Cast iron: | |
| Soft | 100–150 |
| Medium | 70–100 |
| Hard | 40–60 |
| Copper | 60–150 |
| Malleable iron | 80–90 |
| Steel: | |
| Low-carbon | 90–150 |
| Medium-carbon | 60–100 |
| High-carbon | 50–60 |
| Tool-and-die | 40–80 |
| Alloy | 50–70 |

TABLE 3B. *Application of Carbides.* *Courtesy* Metalworking

| Material | Operation | Grade | Speed (sfpm) |
|---|---|---|---|
| Cast iron | Roughing | C-1 | 100–250 |
| | Milling | C–1, C-2 | 150–300 |
| | Precision finishing | C-3, C-4 | 200-500 |
| Aluminum | Roughing | C-2 | 400-up |
| | Finishing | C-2, C-3 | 400-up |
| | Grooving | C-3 | 400-up |
| Brass | Roughing | C-2 | 150–300 |
| | Finishing | C-2, C-3 | 250–400 |
| Carbon steel | Roughing | C-5 | 125–200 |
| | Medium cuts | C-6 | 150–325 |
| | Milling | C-5, C-6 | 175–400 |
| | Finishing | C-7, C-8 | 250–350 |
| | Precision finishing | C-7, C-8 | 300–500 |
| Alloy steel | Roughing | C-5 | 100–175 |
| | Finishing | C-6, C-7 | 150–200 |
| | Precision finishing | C-8 | 250–500 |
| Heat-treated steel | Roughing | C-6 | 80–100 |
| | Finishing | C-7 | 100–125 |
| Steel castings | Roughing | C-5, C-6 | 100–200 |
| | Finishing | C-5 | 150–225 |
| | | C-6 | 175–375 |

Drilling speed is 80 % of turning speed.

TABLE 3C. *Materials Machinable by Ceramic. Courtesy* Metalworking

| Material | Tool Life | Best Average Finish (mu-in.) | Usual Speeds (sfpm) |
|---|---|---|---|
| Steel—Soft to $R_c$ 30 | Good | 35 | 1000–2400 |
| —$R_c$ 30 to $R_c$ 45 | Excellent | 25 | 500–1500 |
| —Hard ($R_c$ 45 to $R_c$ 65) | Excellent | 20 | 75– 800 |
| Cast iron | Excellent | 35 | 300–2500 |
| 300 series stainless steel | Poor | 35 | 400–1000 |
| 400 series stainless steel | Good | 35 | 600–1500 |
| Carbon-graphite | Excellent | — | 200–1000 |
| Plastics | Good | — | 300–3000 |
| Tungsten | Excellent | — | 150– 800 |
| Stellite | Excellent | — | 60– 600 |
| Tantung | Excellent | — | 60– 140 |
| Molybdenum | Fair | — | 500– 800 |
| Copper | Excellent | — | 600–2000 |
| Titanium | Poor | — | Not recommended |
| Brass | Good | — | To 3000 |
| Bronze | Good | — | To 3000 |
| Aluminum bronze | Poor | — | Not recommended |
| High-temp alloys (iron base) | Poor | — | Not recommended |
| High-temp alloys (nickel base) | Poor | — | Not recommended |
| High-temp alloys (cobalt base) | Good | — | 60– 600 |

TABLE 5A. *General Properties, Cost, and Fabricating Methods of Plastics. Courtesy Machinery*
**Thermoplastics**

| Name of Plastic Material | Cost | Typical Applications | Maximum Tensile Strength psi | Maximum Service Temperature, °F | Significant Material Characteristics | Fabricating Methods |
|---|---|---|---|---|---|---|
| ABS | medium | pipe, wheels, valve bodies, refrigerator parts, automobile parts, pump impellers, tool handles, football helmets | 4000 to 8000 | 175 to 212 | High impact strength—2 to 10 foot-pounds per inch; tough to −60°F; resists acids, alkalis and salts | Injection molding, extrusion, calendering, vacuum forming |
| Acetal | medium | instrument clusters, carburetor parts, gears, bearings and bushings | 9000 to 10,000 | 185 to 220 | Very tough and rigid with good fatigue properties; resists most solvents | Forming, molding, extrusion, machining |
| Acrylic | medium | aircraft canopies, camera lenses, automobile tail-light lenses, pump parts, sprinkler heads, signs, and lighting diffusers | 6000 to 10,000 | 140 to 200 | Good impact strength—.4 to 2 foot-pounds per inch; optical clarity; weather resistant; tough at low temperatures | Extrusion, casting, injection molding, compression molding, machining |
| Cellulosics (Cellulose Acetate) | medium | rigid packing, recording tape, combs, shoe heels | 8000 | 120 to 170 | Resists moderate heat and weak acids, alkalis, many solvents | Injection, compression or blow molding, vacuum forming, machining |
| (Cellulose Acetate Butyrate) | | packaging, pipe, tubing, tool handles, and steering wheels | 7000 | | Resists heat, weather, and weak acids, alkalis, many solvents | Blow injection and compression molding, extrusion, machining and drawing |
| (Cellulose Acetate Propionate) | | packaging, steering wheels, appliance housings, telephone cases, and pens | 6000 | | Resists heat, weather, and weak acids, alkalis, many solvents | Injection and compression molding, extrusion, machining |
| (Ethyl Cellulose) | | cabinet edge molding, flashlights | 7000 | | Resists heat and weak acids, alkalis, and many solvents | Injection and compression molding, extrusion and drawing, machining |

| Material | Cost | Uses | | | Properties | Machining / Processing |
|---|---|---|---|---|---|---|
| (Cellulose Nitrate) | | billiard balls and fabric coatings | 8000 | | Resists moderate heat and weak acids, alkalis, and many solvents | Machining |
| Ethylene—Vinyl Acetate | medium | syringe bulbs, tubing, gloves, dampening pads, pool liners | 3600 | 150 | Flexible, high impact strength; useful to −150°F | Injection and blow molding, extrusion |
| Fluorocarbons (TFE, PTFE) (FEP, PVF₂) | high | cooking utensil coatings, gaskets, piston rings, lining, low-friction surfaces | 6500 / 7000 | 500 / 400 | Very high chemical resistance; useful to cryogenic temperatures; very low friction | Molding and extrusion; hot or cold formed; machined |
| Ionomer | medium | skin packaging, toys, bottles, wire insulation, trays, covers | | | Transparent, tough at low temperatures, good insulator | Injection or blow molding, extrusion, thermoforming |
| Nylon | high | washers, gears, fasteners, motor parts, zippers, electrical parts and supports, rollers and wheels, bearings | 8500 to 12,500 | 250 to 300 | High impact strength, wear resistant, resilient, useful down to 0°F, resists all common chemicals | Injection, compression and blow molding, cold forming, extrusion, machining |
| Parylene | medium | insulating and protective coatings for paper, cloth, ceramics, metals, micro-circuits | | 200 to 240 | Best insulating plastic known at temperatures near absolute zero; resists all organic solvents | Coating |
| Phenoxy | high | electrical coatings and insulation | 8500 | 170 | Clear, tough, rigid, hard; impact strength—2 foot-pounds per inch | Blow and injection molding, extrusion, coating |
| Polyallomer | medium | automobile panels, pipe fittings, hinges, shoe lasts | 4000 | 200 | Good impact strength—1.5 foot-pounds per inch | Injection molding, extrusion, vacuum forming |

| Name of Plastic Material | Cost | Typical Applications | Maximum Tensile Strength psi | Maximum Service Temperature °F | Significant Material Characteristics | Fabricating Methods |
|---|---|---|---|---|---|---|
| Polycarbonate | medium | armor for small arms fire, structural parts, insulating housings, safety helmets, rocket launcher handles, pump parts, gears, bearing balls | | 250 | Extremely good impact strength—16 foot-pounds per inch; rigid and transparent | All molding techniques, machining |
| Polyethylene | low | kitchen ware, refrigerator parts, pipe, containers, seal rings, structural housings | 1000 to 5000 | 175 to 250 | Tough, flexible or rigid, resists weathering and common chemicals | Molding, extrusion, calendering, coating, casting, and vacuum forming |
| Polyphenylene Oxide | high | electrical insulation, battery cases, medical instruments, hot-water plumbing equipment, nose cones | 11,600 | 225 | High impact strength—1.2 foot-pounds per inch; useful to −275°F; good dielectric; high modulus and low creep | All conventional techniques |
| Polypropylene | low | helmets, pipes and fittings, valves, appliance parts, hinges | 5000 | 230 | Excellent electrical insulator over a wide range of temperatures | Injection and blow molding, extrusion |
| Polystyrene | low | toys, wall tile, food trays, appliance housings, battery cases, instrument panels | 3000 to 8000 | 125 to 165 | Hard, rigid, poor impact resistance; resists most foods, household fluids | Injection and compression molding, extrusion and machining |
| Polysulfone | high | automobile parts, power tool and appliance housings, pipe, computer parts | 10,000 | 300 | Rigid and ductile with good electrical properties over a wide temperature range down to −150°F | Injection and blow molding, extrusion and thermo-forming |
| Vinyl | medium | raincoats, upholstery, hose, wire insulation, gaskets | | 130 | Abrasion and wear resistant; resists most common chemicals | All molding methods, extrusion, casting, and calendering (machining for PVC) |

**Thermosetting Plastics**

| Name of Plastic Material | Cost | Typical Applications | Maximum Tensile Strength psi | Maximum Service Temperature °F | Significant Material Characteristics | Fabricating Methods |
|---|---|---|---|---|---|---|
| Alkyd | low | automobile starter parts, vacuum tube supports, insulators | 8000 | 300 | Resists weak acids and most solvents; good electrical insulator | Compression molding |
| Allylic | medium | lenses, aircraft windows, electronic parts, laminates | 6000 | 350 | High dielectric strength; resists most solvents; optically clear | Injection, compression or transfer molding and extrusion |
| Amino | medium | melamine tableware, distributor caps and counter tops; urea appliance housings | 5000 to 10,000 | 170 to 210 | Colorful, hard, scratch-resistant, resists detergents, cleaners, alcohol, oil | Compression, transfer and plunger molding |
| Casein | | buttons, toys and novelty items | | | Smooth surface, rigid, tough, resists gasoline, organic solvents and chemicals; poor temperature, humidity resistance | Machining |
| Cold molded | low | small gears, jigs, dies, plugs, handles | 12,000 | | Resists alkalis, solvents, water and oils | High pressure molding and post curing |
| Epoxy | | adhesives, tanks, laminated tooling | | 500 | Flexible and resistant to many chemicals | Molding, extrusion, casting, and potting |
| Phenolics | medium | distributor caps, telephones, insulators, appliance parts, grinding wheel bonds | 4000 to 10,000 | 250 to 450 | Hard, good thermal insulator, electrical insulator and good chemical resistance | Injection, compression, transfer or plunger molding, and casting |
| Polyester | low | impregnant for cloth, paper, etc., and for reinforced automobile bodies, luggage | 1000 to 17,000 | 150 to 300 | Tough, resists most solvents, acids, bases and salts | Molding, casting |

**Unclassed Plastics**

| Name of Plastic Material | Cost | Typical Applications | Maximum Tensile Strength psi | Maximum Service Temperature °F | Significant Material Characteristics | Fabricating Methods |
|---|---|---|---|---|---|---|
| Polyimide | high | aerospace parts, valve seats, bearings, seals, retainer rings, hose and tubing, piston rings, compressor vanes | 10,000 to 12,000 | 425 to 450 | High heat resistance, low friction and good wear resistance. Has thermoplastic structure and non-melting character of a thermoset | Machining |
| Polyurethane | medium | foams, coatings, auto parts, tire treads, aircraft structures | over 5000 | | Tough, impact resistant, resists most common chemicals | Molding, extrusion, casting, and calendering |
| Silicone | high | switch parts, coil forms, motor and generator insulation, aircraft radomes and ductwork, electronic component encapsulation | 4000 to 6000 | over 500 | High heat stability; highly resistant to mineral acids and corrosive salts; good dielectric. Either thermoplastic or thermosetting forms are available | Compression and transfer molding, extrusion, calendering, casting, coating and foam |

TABLE 6A.  *Metric Conversion Factors*

| Property | Unit | Symbol | Exact Conversion | | | Approximate Equivalency |
|---|---|---|---|---|---|---|
| | | | From | To | Multiply by | |
| Length | metre<br>centimetre<br>millimetre | m<br>cm<br>mm | inch<br>inch<br>foot | mm<br>cm<br>mm | $2.540 \times 10$<br>2.540<br>$3.048 \times 10^{-4}$ | $25\,mm = 1\,in$<br>$300\,mm = 1\,ft$ |
| Mass | kilogram<br>gram<br>tonne (megagram) | kg<br>g<br>t | ounce<br>pound<br>ton (2000 lb) | g<br>kg<br>kg | $2.835 \times 10$<br>$4.536 \times 10^{-1}$<br>$9.072 \times 10^{2}$ | $2.8\,g = 1\,oz$<br>$kg = 2.2\,lbs\ (35\,oz)$<br>$1\,t = 2200\,lb$ |
| Density | kilogram per<br>cu metre | $kg/m^3$ | pounds per<br>cu ft | $kg/m^3$ | $1.602 \times 10$ | $16\,kg/M^3 = 1\,lb/ft^3$ |
| Temperature | deg Celsius | °C | deg Fahrenheit | °C | $(°F - 32) \times 5/9$ | $0\,°C = 32\,°F$<br>$100\,°C = 212\,°F$ |
| Area | square metre<br>square millimetre | $m^2$<br>$mm^2$ | sq inch<br>sq ft | $mm^2$<br>$m^2$ | $6.452 \times 10^{2}$<br>$9.290 \times 10^{-2}$ | $645\,mm^2 = 1\,in^2$<br>$1\,m^2 = 11\,ft^2$ |
| Volume | cubic metre<br>cubic centimetre<br>cubic millimetre | $m^3$<br>$cm^3$<br>$mm^3$ | cu in<br>cu ft<br>cu yd | $mm^3$<br>$m^3$<br>$m^3$ | $1.639 \times 10^{4}$<br>$2.832 \times 10^{-2}$<br>$7.645 \times 10^{-1}$ | $16400\,mm^3 = 1\,in^3$<br>$1\,m^3 = 35\,ft^3$<br>$1\,m^3 = 1.3\,yd^3$ |
| Force | newton<br>kilonewton<br>meganewton | N<br>kN<br>MN | ounce (Force)<br>pound (Force)<br>Kip | N<br>kN<br>MN | $2,780 \times 10^{-1}$<br>$4.448 \times 10^{-3}$<br>4.448 | $1\,N = 3.6\,oz$<br>$4.4\,N = 1\,lb$<br>$1\,kN = 225\,lb$ |
| Stress | megapascal<br>newtons/sq m | MPa<br>$N/M^2$ | pound/in² (psi)<br>Kip/in² (ksi) | MPa<br>MPa | $6.895 \times 10^{-3}$<br>6.895 | $1\,MPa = 145\,psi$<br>$7\,MPa = 1\,ksi$ |
| Torque | newton-metres | Nm | in-ounce<br>in pound<br>ft pound | Nm<br>Nm<br>Nm | $7.062 \times 10^{3}$<br>$1.130 \times 10^{-1}$<br>1.356 | $1\,Nm = 140\,in\,oz$<br>$1\,Nm = 9\,in\,lb$<br>$1\,Nm = 75\,ft\,lb$<br>$1.4\,Nm = 1\,ft\,lb$ |

Table 6B. *Standard Welding Symbols*

AMERICAN WELDING SOCIETY ◈ STANDARD WELDING SYMBOLS

(COURTESY AMERICAN WELDING SOCIETY)

**TABLE 7A.** *Common g-Words or Preparatory Functions*

| Word (Code) | Explanation |
|---|---|
| g00 | Used with combination point-to-point and contouring systems for expressing a point-to-point operation. |
| g01 g10 g11 | Used for linear interpolation and reserved for contouring Not needed where linear interpolation is used exclusively. g01 is used for blocks of normal dimensions, i.e., the maximum length of machine-tool travel can be 9.9999 in. along one axis. g10 is used for incremental mode where distances are long, up to 99.9990 in. In this case the fourth-place decimal figure must be zero. g11 is for a short dimension block on any one axis up to .9999 in. |
| g02 g20 g21 | Used with *circular interpolation* clockwise direction, normal dimension. g20, the same except used for long dimension. g21, the same except used for short dimension. |
| g03 g30 g31 | Used with *circular interpolation* counterclockwise direction, normal dimension. g30 same except used for long dimension. g31 same only used for short dimension. |
| g04 | Calculated time delay or dwell. It can be regulated by the feed-rate number (f word). Not cyclic or sequential. |
| g05 | *Hold* code, restricts machine's motion until terminated by an operator or by an interlock motion. |
| g06 g07 | Unassigned. May be used at the discretion of the machine-tool system builder. Could become standardized at a future date. |
| g08 | Acceleration code. Increase feed rate to programmed rate starting immediately. |
| g09 | Deceleration code. Smooth deceleration up to 10% of programmed rate starting immediately. |
| g12 | Unassigned. |
| g13 g14 g15 g16 | Axis selection. Used to direct the control system to operate on a particular set of axes. An example would be a machine tool having two heads which are not to operate simultaneously. If head A were to be used, for example, it could be preceded by g14 and the X, Y, and Z words. |
| g17 g18 g19 | Used to identify the plane for such functions as *circular interpolation* or *cutter compensation* which can only operate in two dimensions simultaneously. |

*continued overleaf*

TABLE 7A—*continued*

| Word (Code) | Explanation |
|---|---|
| g22 to g29 | Unassigned. |
| g32 | Unassigned. |
| g33 g34 g35 | Used for machines equipped with thread-cutting facilities, generally lathes. |
| g36 to g39 | Reserved for control use. |
| g40 | Used to discontinue any *cutter compensation*. |
| g41 g42 | Used for *cutter compensation*, g41 when the cutter is on the left side of the work surface, looking in the direction of the cutter, and g42 when the cutter is on the right side of the work surface. |
| g50 to g59 | Unassigned. |
| g60 to g79 | Unassigned but reserved for point-to-point systems. |
| g80 | Cancel fixed cycle. |
| g81 to g89 | Reserved for fixed cycles. Usually used to initiate a series of events such as drilling or counterboring, where a slow feed-*in* is required that may be followed by a dwell and a rapid feed-*out*. |
| g90 to g99 | Unassigned. |

TABLE 7B.  *Some Miscellaneous Functions*

| | |
|---|---|
| m00 | Program stop. |
| m01 | Optional stop.  Acted upon only when the operator has previously signaled for this command by pushing a button. |
| m02 | End of program.  Stops the machine and may include tape rewind. |
| m30 | End of tape rewind.  This code will rewind the tape (assuming the control has this facility). |
| m03 | Start spindle, clockwise rotation. |
| m04 | Start spindle, counterclockwise rotation. |
| m05 | Spindle off. |
| m06 | Tool change, manual or automatic. |
| m07 | Tool compensation, X axis. |
| m08 | Coolant on. |
| m09 | Coolant off. |
| m10 | Automatic clamping of machine slides, workpiece, fixtures, etc. |
| m11 | Automatic unclamping. |
| m12 | Program stop, restart from other reader only. |
| m13 | Start spindle clockwise, coolant on. |
| m14 | Start spindle counterclockwise, coolant on. |
| m15 m16 | Rapid traverse of feed motion in either the +m15 or −m16 direction. |

TABLE 9A. *Metric Equivalents*

| Inches | mm | Inches | mm | Inches | mm |
|--------|--------|--------|--------|--------|--------|
| 0.0001 | 0.0025 | 0.270 | 6.858 | 0.640 | 16.256 |
| 0.0002 | 0.0051 | 0.280 | 7.112 | 0.650 | 16.510 |
| 0.0003 | 0.0076 | 0.290 | 7.366 | 0.660 | 16.764 |
| 0.0004 | 0.0102 | 0.300 | 7.620 | 0.670 | 17.018 |
| 0.0005 | 0.0127 | 0.310 | 7.874 | 0.680 | 17.272 |
| 0.0006 | 0.0152 | 0.320 | 8.128 | 0.690 | 17.526 |
| 0.0007 | 0.0178 | 0.330 | 8.382 | 0.700 | 17.780 |
| 0.0008 | 0.0203 | 0.340 | 8.636 | 0.710 | 18.034 |
| 0.0009 | 0.0229 | 0.350 | 8.890 | 0.720 | 18.288 |
| 0.001 | 0.025 | 0.360 | 9.144 | 0.730 | 18.542 |
| 0.002 | 0.051 | 0.370 | 9.398 | 0.740 | 18.796 |
| 0.003 | 0.076 | 0.380 | 9.652 | 0.750 | 19.050 |
| 0.004 | 0.102 | 0.390 | 9.906 | 0.760 | 19.304 |
| 0.005 | 0.127 | 0.400 | 10.160 | 0.770 | 19.558 |
| 0.006 | 0.152 | 0.410 | 10.414 | 0.780 | 19.812 |
| 0.007 | 0.178 | 0.420 | 10.668 | 0.790 | 20.066 |
| 0.008 | 0.203 | 0.430 | 10.922 | 0.800 | 20.320 |
| 0.009 | 0.229 | 0.440 | 11.176 | 0.810 | 20.574 |
| 0.010 | 0.254 | 0.450 | 11.430 | 0.820 | 20.828 |
| 0.020 | 0.508 | 0.460 | 11.684 | 0.830 | 21.082 |
| 0.030 | 0.762 | 0.470 | 11.938 | 0.840 | 21.336 |
| 0.040 | 1.016 | 0.480 | 12.192 | 0.850 | 21.590 |
| 0.050 | 1.270 | 0.490 | 12.446 | 0.860 | 21.844 |
| 0.060 | 1.524 | 0.500 | 12.700 | 0.870 | 22.098 |
| 0.070 | 1.778 | 0.510 | 12.954 | 0.880 | 22.352 |
| 0.080 | 2.032 | 0.520 | 13.208 | 0.890 | 22.606 |
| 0.090 | 2.286 | 0.530 | 13.462 | 0.900 | 22.860 |
| 0.100 | 2.540 | 0.540 | 13.716 | 0.910 | 23.114 |
| 0.110 | 2.794 | 0.550 | 13.970 | 0.920 | 23.368 |
| 0.120 | 3.048 | 0.560 | 14.224 | 0.930 | 23.622 |
| 0.130 | 3.302 | 0.570 | 14.478 | 0.940 | 23.876 |
| 0.140 | 3.556 | 0.580 | 14.732 | 0.950 | 24.130 |
| 0.150 | 3.810 | 0.590 | 14.986 | 0.960 | 24.384 |
| 0.160 | 4.064 | 0.600 | 15.240 | 0.970 | 24.638 |
| 0.170 | 4.318 | 0.610 | 15.494 | 0.980 | 24.892 |
| 0.180 | 4.572 | 0.620 | 15.748 | 0.990 | 25.146 |
| 0.190 | 4.826 | 0.630 | 16.002 | 1.000 | 25.400 |
| 0.200 | 5.080 | | | | |
| 0.210 | 5.334 | | | | |
| 0.220 | 5.588 | | | | |
| 0.230 | 5.842 | | | | |
| 0.240 | 6.096 | | | | |
| 0.250 | 6.350 | | | | |
| 0.260 | 6.604 | | | | |

**Example: 0.8564 inch in mm**

**0.850    = 21.590**

**0.006    =    .152**

**0.0004 =    .0102**

**0.8564 in. = 21.7522 mm**

# Index